SPRINGER
STUDY
EDITION

Ernst-Detlef Schulze
Martyn M. Caldwell (Eds.)

Ecophysiology of Photosynthesis

With 163 Figures

 Springer

Professor Dr. Ernst-Detlef Schulze
Lehrstuhl für Pflanzenökologie
Universität Bayreuth
Postfach 10 1251
D-95447 Bayreuth
Germany

Professor Dr. Martyn M. Caldwell
Department of Range Science and Ecology Center
Utah State University
Logan, UT 84322-5230
USA

Design of the cover illustration by Roswitha Asche

ISBN 3-540-58571-0 Springer-Verlag Berlin Heidelberg New York

Title of the original edition (hardcover)
Ecological Studies, Volume 100
ISBN 3-540-55952-3 Springer-Verlag Berlin Heidelberg New York

© Springer-Verlag Berlin Heidelberg 1995
Printed in Germany

Production Editor: Herta Böning, Heidelberg
Typesetting: Best-set Ltd., Hong Kong and K+V Fotosatz GmbH, Beerfelden
31/3130-5 4 3 2 1 0 – Printed on acid-free paper

Preface

The increasing concern of global climate change has caused an increasing interest in all aspects of carbon exchange of natural ecosystems. In a world of atmospheric increase of CO_2, we are becoming aware that natural ecosystems, especially the boreal forest, might be major sinks for carbon, and may slow down the effects of anthropogenic CO_2 emissions on climate. Fortunately enough, following 30 years of intensive research in the ecophysiology of photosynthesis, we may be just about ready to actually model global photosynthesis to the necessary level of complexity for predictions of effects of CO_2 change on global vegetation. For many parameters this seems to be actually possible. However, it is also quite clear that important information is still missing, such as the uncertainty about adaptations of plants to high CO_2, and about the unknown balance between respiratory and assimilatory processes at the canopy level. The present volume will review the progress which has been made in understanding photosynthesis in the past decades. It will attempt to develop a global map of photosynthesis, which would not have been possible without the recent history of intensive research on the ecophysiology of photosynthesis, which started in the early 1950s, and was intensified during the International Biological Program.

Research on photosynthesis of plants under field conditions is closely connected to the name of *Otto Ludwig Lange*, who carried out pioneer work on gas exchange under field conditions. O. L. Lange was born in 1927 in Dortmund, Germany. He obtained his doctorate in Göttingen with a thesis on plant heat resistance. Based on this work, he was engaged by Otto Stocker at Darmstadt, the pioneer of measuring photosynthesis of plants in tropical and arid regions with the KOH absorption method of those days. Otto Lange was strongly attracted by this line of research and started an expedition to the Avdat runoff farm of M. Evenari in Israel, using improved technology. It was this expedition which stimulated the study of the ecophysiology of photosynthesis worldwide. It attracted physiologists as well as agriculturists to study plant metabolism in the field. The main concept was that environmental conditions, especially the dryness of the air, have a strong direct impact on plant performance and productivity, which has a human dimension. Otto Lange continued this research with a second

Negev expedition in 1971 and with research in the mediterranean scrub of Portugal. For his international pioneer work and his stimulation of the whole field research, he received numerous international honors for his impaction ecophysiology, such as the Balzan Prize (1988). He is a member of the German Academy of Science (Leopoldina) and of the European Academy of Science (1991), and foreign honorary member of the American Academy of Arts and Sciences (1994). In 1991 he was awarded the Bavarian Order of Science and Art, Bayerischer Maximiliansorden für Wissenschaft und Kunst, which only ten botanists have thus far received, including Hugo von Mohl (1853), Alexander Braun (1874), Julius von Sachs (1888), Wilhelm Pfeffer (1895), Simon Schwendener (1897), and Carl Correns (1932).

This volume is dedicated to Otto Ludwig Lange on the occasion of his retirement. His friends and colleagues, who have joined and followed Otto Lange since his early beginnings, take this opportunity to review this field of research, ranging from molecular biology to global modeling. We think that this research is not only a mirror of the rapidly evolving discipline of ecophysiology, but also a basis for future global climate modeling. In this sense we hope that this review may be used by graduate students, to transfer the knowledge from one generation to the next in a field of research which due to global change has become increasingly important.

Bayreuth and Logan   E.-D. Schulze
                                                          M. M. Caldwell

Contents

Part D: Global Aspects of Photosynthesis

Part E: Perspectives in Ecophysiological Research
of Photosynthesis

Contributors

Amthor, Jeffrey S., Woods Hole Research Center,
13 Church Street, P.O. Box 296, Woods Hole, MA 02543, USA

Ball, Marilyn C., RSBS, ANU, GPO Box 475,
Canberra, ACT, Australia

Beyschlag, Wolfram, Lehrstuhl Botanik II, Universität Würzburg,
Mittlerer Dallenbergweg 64, D-97082 Würzburg, Germany

Bilger, Wolfgang, Lehrstuhl Botanik II, Universität Würzburg,
Mittlerer Dallenbergweg 64, D-97082 Würzburg, Germany

Björkman, Olle, Department of Plant Biology,
290 Panama Street, Stanford, CA 94305-4101, USA

von Caemmerer, Susanne, RSBS, ANU, GPO Box 475,
Canberra, ACT, Australia

Caldwell, Martyn, M., Department of Range Science,
Utah State University, Logan, UT 84322-5230, USA

Cowan, Ian R., Research School of Biological Sciences,
GPO Box 475, Canberra, ACT 2601, Australia

Demmig-Adams, Barbara, Department of Environmental Population,
and Organismic Biology, University of Colorado,
Boulder, CO 80309-0334, USA

Ehleringer, James R., Department of Biology, University of Utah,
Salt Lake City, UT 84112, USA

Fichtner, Klaus, Department of Biological Sciences,
Stanford University, Stanford, CA 94305, USA.
Present address: Immenser Straße 15, D-31303 Burgdorf, Germany

Field, Christopher B., Department of Plant Biology,
Carnegie Institution of Washington, Stanford, CA 94305-1297,
USA

Gamon, John A., Department of Plant Biology,
Carnegie Institution of Washington, Stanford, CA 94305, USA

Green, T. G. Allan, Biological Sciences, Waikato University,
Hamilton, New Zealand

Hattersley, Paul, RSBS, ANU, GPO Box 475,
Canberra, ACT, Australia

Heber, Ulrich, Botanisches Institut I, Universität Würzburg,
Mittlerer Dallenbergweg 64, D-97082 Würzburg, Germany

Henderson, Sally, Research School of Biological Sciences,
The Australian National University, GPO Box 475,
Canberra, ACT 2601, Australia

Kaiser, Werner, Botanisches Institut I, Universität Würzburg,
Mittlerer Dallenbergweg 64, D-97082 Würzburg, Germany

Kappen, Ludger, Botanisches Institut, Universität Kiel,
Olshausenstraße 40, D-24118 Kiel, Germany

Kindermann, Gerald, Botanisches Institut I, Universität Würzburg,
Mittlerer Dallenbergweg 64, D-97082 Würzburg, Germany

Koch, George W., Department of Biological Sciences,
Standford University, Stanford, CA 94305, USA

Körner, Christian, Botanisches Institut, Universität Basel, Schön-
beinstr. 6, CH-4056 Basel, Switzerland

Lange, Otto L., Botanisches Institut II, Universität Würzburg,
Mittlerer Dallenbergweg 64, D-97082 Würzburg, Germany

Larcher, Walter, Botanisches Insitut, Universität Innsbruck,
A-6020 Innsbruck, Austria

Lösch, Rainer, Abteilung Geobotanik, Universität Düsseldorf,
Geb. 26-13, Universitätsstraße 1, D-40225 Düsseldorf, Germany

Luwe, Michael, Botanisches Institut I, Universität Würzburg,
Mittlerer Dallenbergweg 64, D-97082 Würzburg, Germany

Mooney, Harold A., Department of Biological Sciences,
Stanford University, Stanford, CA 94305, USA

Neubauer, Christian, Lehrstuhl Botanik I, Universität Würzburg,
Mittlerer Dallenbergweg 64, D-97082 Würzburg, Germany

Oberbauer, Steven F., Department of Biological Sciences,
Florida International University, Miami, FL 33199, USA

Osmond, C. Barry, RSBS, ANU, GPO Box 475,
Canberra, ACT, Australia

Passioura, John, CSIRO, Division of Plant Industry,
GPO Box 1600, Canberra, ACT 2601, Australia

Pearcy, Robert W., Dept. of Botany, University of California,
Davis, CA 95616, USA

Peñuelas, Josep, Dept. of Biology, Instituto de Recerca
i Tecnologia Agroalimentaries, Centre de Cabrils s/n,
E-08348 Cabrils (Barcelona), Spain

Pereira, João S., Universidade Técnica de Lisboa,
Instituto Superior de Agronomia, Tapada da Ajuda,
P-1399 Lisboa Codex, Portugal

Pfanz, Hardy, Lehrstuhl für Botanik II, Universität Würzburg,
Mittlerer Dallenbergweg 64, D-97082 Würzburg, Germany

Pfitsch, William A., Department of Biology, Hamilton College,
Clinton, NY 13323, USA

Raven, John A., Department of Biological Sciences,
University of Dundee, Dundee, DD14HN, Scotland

Ryel, Ronald J., Department of Range Science,
Utah State University, Logan, UT 84322-5230, USA

Schreiber, Ulrich, Lehrstuhl Botanik I, Universität Würzburg,
Mittlerer Dallenbergweg 54, D-97082 Würzburg, Germany

Schultz, Gudrun, Botanisches Institut, Universität Kiel,
Olshausenstraße 40, D-24118 Kiel, Germany

Schulze, Ernst-Detlef, Lehrstuhl Pflanzenökologie,
Universität Bayreuth, Postfach 101251, D-95440 Bayreuth,
Germany

Schulze, Waltraud, Lehrstuhl Pflanzenökologie,
Universität Bayreuth, Postfach 101251, D-95440 Bayreuth,
Germany

Siegwolf, Rolf T.W., Paul Scherrer Institut,
CH-5232 Villigen (PSI), Switzerland

Slovik, Stefan, Botanisches Institut I, Universität Würzburg,
Mittlerer Dallenbergweg 64, D-97082 Würzburg, Germany

Smith, Tom M., Department of Environmental Sciences,
Clark Hall, University of Virginia, Charlottesville, VA 22903, USA

Tenhunen, John, BITÖK-Pflanzenökologie, Universität Bayreuth,
Dr. Hans-Frisch-Str. 1–3, D-95448 Bayreuth, Germany

Trebst, Achim, Lehrstuhl für Biochemie der Pflanzen,
Abteilung Biologie, Ruhr-Universität Bochum, D-44780 Bochum,
Germany

Vanselow, Renate, Botanisches Institut, Universität Kiel,
Olshausenstraße 40, D-24118 Kiel, Germany

Veljovic-Janovic, Sonja, Botanisches Institut I,
Universität Würzburg, Mittlerer Dallenbergweg 64,
D-97082 Würzburg, Germany

Woodward, Ian F., Department of Animal and Plant Sciences,
The University of Sheffield, Sheffield S102UQ, UK

Yin, Zhua, Botanisches Institut I, Universität Würzburg,
Mittlerer Dallenbergweg 64, D-97082 Würzburg, Germany

Ziegler, Hubert, Lehrstuhl für Botanik,
Technische Universität München, Arcisstraße 21,
D-80333 München, Germany

Part A
Molecular and Physiological Control and Limitations

1 Dynamics in Photosystem II Structure and Function

A. Trebst

1.1 Introduction

In photosynthetic electron flow, light supplies the driving force for NADP reduction, oxygen evolution and ATP formation. Three integral protein complexes in the thylakoid membrane – photosystem I, photosystem II, and the cytochrome b_6/f complex – with several peripheral polypeptides attached, bind the pigments and redox systems that participate in electon flow: chlorophylls in the antenna and the reaction centers of the two photosystems, pheophytin in PS II, plastoquinone bound to PS II and in the quinone pool and menaquinone to PS I, iron in iron sulfur centers (of the Fe_2S_2 and Fe_4S_4 type: of very low potential in ferredoxin and the acceptor site of PS I, and of high potential in the Rieske Fe_2S_2 center in the cytochrome b_6/f complex) and in histidine-bound Fe in PS II, iron in the hemes in cytochromes f and b_6 in the b/f complex and in b_{559} in PS II, copper in plastocyanin, FAD in the Fd-NADP reductase and a manganese cluster in PS II. The proteins bind and orient these components in the membrane, thus allowing transmembrane vectorial electron flow. Each integral membrane complex with an average of about 350 kDa size contains several polypeptide subunits from very low molecular weight (4 kDa) up to 64 kDa. Peripheral hydrophilic proteins are attached to the membrane proteins or are soluble in the matrix (ferredoxin) or lumen (plastocyanin) space. For recent reviews on the functional structure of PS II see Barber (1987, 1992).

1.2 Function of Photosystem II

Photosystem II oxidizes water to evolve oxygen and provides electrons and protons. The energy is conserved in reducing equivalents (plastoquinol) and in a proton motive force. The protons are released in the inner lumen of the thylakoid and contribute to the driving force for ATP synthesis. The electrons are transported across the membrane after the light-dependent charge separation in the (excited) reaction center P_{680} via pheophytin to the first bound plastoquinone Q_A. Then the electron is passed on to the second plastoquinone Q_B. After two excitations the plastoquinol formed has taken up two protons from the matrix space. The essential components of the primary charge separation are a chlorophyll a dimer P_{680}, two monomeric chlorophylls and two pheophytins. These six compounds are oriented symmetrically in the membrane and are bound to two protein subunits of photosystem II, called the D1 and D2 polypeptide. The reaction center P_{680} appears to be connected functionally to only one monomeric chlorophyll and one pheophytin. This is called the active arm of electron flow from the reaction center to Q_A.

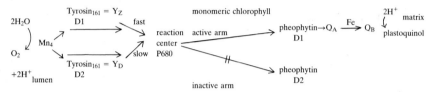

The second symmetrical arm with the other monomeric chlorophyll and the second pheophytin remains inactive (Fig. 1.1).

The tilt of the transmembrane helices of the two subunits (discussed below) places the Q_A side on the D2 protein exactly above the pheophytin

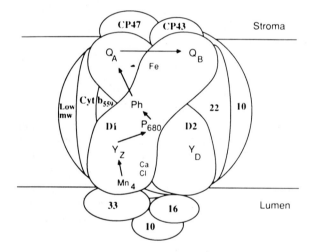

Fig. 1.1. The integral membrane complex of photosystem II oriented in the thylakoid membrane. Several transmembrane polypeptide units carry the prosthetic groups: the reaction center P_{680}, pheophytin (*Ph*) and Fe on the D1 and D2 protein, Q_A on the D2 and Q_B on the D1 protein, antenna chlorophylls on the CP 43 and CP 47 kDa polypeptides. Managanese is attached on the D1/D2 proteins, stabilized by the peripheral polypeptides of 33, 16, und 10 kDa and by Ca and Cl ions. The role of the cytochrome b_{559} is not clear, nor is that of several other polypeptides. Tyrosin 161 in the D1 protein is the electron donor Y_Z between the manganese cluster and P_{680}. A much slower donor tyrosin Y_D is in the D2 protein. After the charge separation of excited P_{680}, the electron travels via pheophytin and a monomeric chlorophyll on the D1 protein to Q_A bound to the D2 protein; this is called the active arm. The symmetrical groups on the D2 protein remain inactive

on the D1 protein with tryptophan 254 of the D2 protein channeling the electrons.

Two excitation cycles and two protons (from the matrix space) are needed for the two electron reduction of Q_B to the hydroquinone. A total of four excitation cycles are needed to oxidize two molecules of water to oxygen and four protons released inside.

$$4H^+_{out} + 2H_2O + 2Q_B + 4h\upsilon \rightarrow O_2 + 4H^+_{in} + 2Q_BH_2.$$

A histidine-bound Fe with no obligatory redox changes participates in the reaction from Q_A to Q_B. Four bound Mn atoms cluster in the oxygen evolution system. The manganese cluster changes valency four times when oxidized by consecutive excitations in the reaction center. Two water molecules are then split to one O_2, possibly via a bound peroxide. The reduced manganese cluster reduces the oxidized $P_{680}+$ via a bound tyrosine radical called Y_Z (tyrosine at position 161 in the sequence of the D1 protein). An equivalent tyrosine in the D2 protein, called Y_D) is also oxidized but much slower. The role of the cytochrome b_{559} in the photosystem II complex remains to be clarified.

The photosystem II complex furthermore contains about 30 core antenna chlorophylls (only chlorophyll a) that channel excitation energy to the reaction center chlorophylls. The additional light harvesting system (LHCP) of about 300 chlorophylls (a and b) and with accessory pigments are functionally and structually attached to the core antenna polypeptides.

The minimum polypeptide composition of a functional oxygen evolving PS II complex consists of seven hydrophobic integral proteins (of 47 and 43 molecular weight, the D1 and D2 proteins, two polypeptides subunits of cytochrome b_{559} and the product of the *psb*I gene) (Barber 1992; Satoh 1992) and three hydrophobic peripheral polypeptides. The core composition of a photosystem II still able to perform the primary charge separation of P_{680} is five polypeptides (D1 and D2 protein, the two subunits of cytochrome b_{559} and the product of the *psb*I gene) (Nanba and Satoh 1987). A preparation of just the D1 and D2 protein has been reported that can still oxidize P_{680} (Satoh 1992). This then shows that just the two polypeptides D1 and D2 carry the reaction center P_{680}, its primary electron acceptor (pheophytin) and the primary donor (the tyrosin radical Y_Z), as well as the two bound plastoquinones Q_A and Q_B and the histidine bound Fe.

The reaction center of photosystem II is homologous in its functional mechanism and cofactor composition to that of purple bacteria and of the green nonsulfur bacteria *Chloroflexus* (Barber 1992; Blankenship 1992). These also reduce quinones (usually ubiquinone in both the Q_A and Q_B site) with bacteriochlorophyll in the reaction center and bacterio-pheophytin as the primary acceptor. They also contain histidine-bound Fe. However, the bacterial systems do not contain a cytochrome b_{559} nor the manganese cluster. The electron donor in the bacterial system is, of course, different from that of photosystem II. As bacteria cannot oxidize water to oxygen, it is a cytochrome c_2 instead of the manganese cluster of the water splitting system. The basic functional structural unit for the reaction center in the bacterial photosystem catalyzes the oxidation of P_{870} and reduction of Q_B via Q_A and pheophytin. It consists of three hydrophobic polypeptides called L, M, and H (for their relative mobility in the SDS gel electrophoresis). Of these, the L and M subunits bind the reaction center bacteriochlorophyll dimer, two monomeric bacterio chlorophylls, two bacterio-pheophytins and the histidine-bound Fe. Q_A is bound to the M and Q_B to the L subunit. In this composition of redox components, the D1 and D2 protein of photosystem II are very homologous (for review see Barber 1992; Table 1.1).

Low molecular weight polypeptides, attached to the photosystem carry the antenna bacteriochlorophylls instead of the higher molecular weight chlorophyll-binding proteins of photosystem II.

The amino acid sequence of all subunits from plant and purple bacteria systems are known and the genetic origin has been established.

1.3 Structure of Photosystem II

The exact topology of the integral proteins of photosystem II as listed in Table 1.1 in the membrane is known only in principle from an analysis of their amino acid sequences. A detailed description of the three-dimensional folding of their amino acid sequences and the orientation and function of amino acid residues in the redox reaction is, however, already possible for the two reaction center polypeptides – the D1 and D2 protein. This knowledge is due to the homology of the D1 and D2 proteins to the two reaction center polypeptides L and M of the purple bacteria.

The photosystem of the purple bacteria *Rps. viridis* and later *Rps. sphaeroidis* has been crystalized (Deisenhofer et al. 1985; Allen et al. 1987). An X-ray structure of about 2.3 Å resolution was possible for the *Rps. viridis* system (Deisenhofer and Michel 1989). This X-ray structure shows how the three hydrophobic proteins L, M, and H span the membrane several times in (11) transmembrane helices and how the redox systems are attached to them. Accordingly, the principal architecture of an integral membrane protein consists of a cluster of hydrophobic amino acids in the sequence of the protein. There are five such clusters of about 20 to 25 hydrophobic amino acids in both the L and M subunit. They form a highly hydrophobic α-helix that has about the length of the hydrophobic part of the lipid bilayer of a membrane of about 35 to 40 Å. There are five helices in each of the L and M subunits, long enough to span the membrane. Hydrophilic segments in the amino acid sequence connect the transmembrane helices on either side of the membrane. If short, the hydrophilic connections will remain in the equivalent phase of the head groups of the

Table 1.1. Comparison of the subunits and prosthetic groups of photosystem II to those of the photosystem of purple bacteria

	Photosystem II	Photosystem of purple bacteria
Polypeptides in the reaction center	D1 subunit D2 subunit	L subunit M subunit
Further subunits	43 kDa (core antenna) 47 kDa (core antenna) Products of the genes psbH-psbN Two cytochrome b_{559} Three peripheral proteins for Mn_4 stabilization	H subunit[a]
Electron carriers	P_{680} chlorophyll dimer	P_{870} bacteriochlorophyll dimer
	Two monomeric chlorophylls	Two monomeric bacteriochlorophylls

	Two pheophytins
	Two plastoquinones
		Two bacteriopheophytins
		Two ubiquinones in *Rb.*
		   sphaeroides
		One ubi- + one menaquinone
		   in *Rps. viridis*

Metal ions	One Fe Four Mn Ca	One Fe
O_2 evolution electron donor	Yes Water	No Cytochrome c_2

[a] In *Rps. viridis*, but not in the other purple bacteria, an additional tetraheme protein is attached to the photosystem; there is also no H-subunit in *Chloroflexus*.

lipids, if longer, they protrude into the hydrophilic space and are accessible to either matrix or lumen components.

The amino acid sequences of L and M subunit of the bacterial system were found to be of a certain homology to the D1 and D2 subunits of photosystem II, respectively. (There is no homologous H subunit in photosystem II). Even more homologous is the equivalence of the hydropathy plots of the amino acid sequence. Such an algorithm is used to indicate hydrophobic helices in an amino acid sequence. They show that there are also five transmembrane helices in the D1 and D2 protein. It was therefore proposed that the D1 and D2 protein form the reaction center of PS II (Trebst 1986; Deisenhofer and Michel 1989). This was subsequently shown experimentally to be correct (Nanba and Satoh 1987). It followed that the D1 protein is the Q_B (as already known in 1985) and the D2 protein the Q_A binding protein of photosystem II.

From the homology in structure and function of the purple bacterial photosystem to that of photosystem II identification of conserved amino acids in the binding of essential redox components – like the two histidines

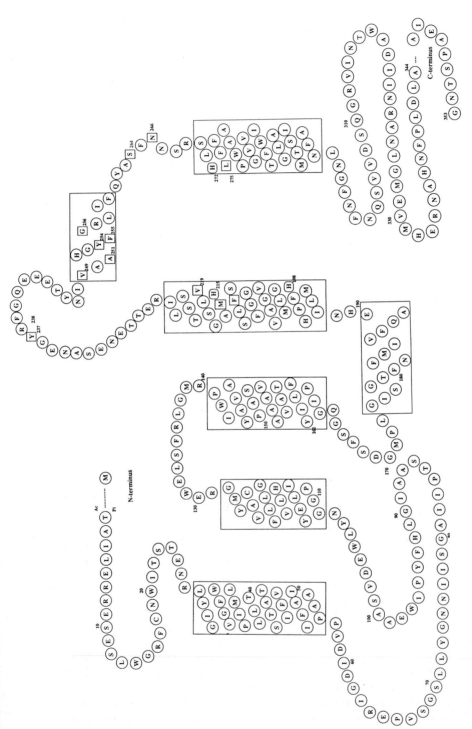

Fig. 1.2. The folding model of the amino acid sequence of the D1 protein, one of the reaction center polypeptides of photosystem II. It is involved in the binding of P_{680} and Fe and is the Q_B and herbicide-binding protein. Five transmembrane and two parallel helices are proposed following the homology of photosystem II to the photosystem of purple bacteria. The polypeptide is processed at both the N- and C-terminus. The threonin at the N-terminus is acetylated and may be phosphorylated

for reaction center chlorophyll binding (his198) and the four histidines for
Fe binding in both the D1 (his215, his272) and D2 (his215, his 270) subunits
– a detailed model of the three-dimensional structure of photosystem II was
proposed (Trebst 1986) (Fig. 1.2). This was particularly successful for the
Q_B-binding niche in PS II.

Instrumental in the support of the model were studies with herbicide-
tolerant mutants of higher plants and algae. Inhibitors, some of them herbicides, were known to displace the plastoquinone from the Q_B site and
occupy the place on the D1 protein instead (for review on herbicides in PS
II see Draber et al. 1991). The molecular biology of herbicide tolerance
identified those amino acids in the D1 protein – by sequencing the mutated
*psb*A gene that encodes it – that participate in herbicide binding (Rochaix
and Erickson 1988). Their location in the amino acid sequence could be well
rationalized by the proposed folding model and in this way described the Q_B
binding niche in great detail (Fig. 1.3). The folding of the D1/D2 protein
model proved to be sufficiently accurate to allow a prediction of the
tyrosine(s) implicated by the ESR signal of the radical to be the electron
donor Y_Z for $P_{680}+$. Sitedirected mutagenesis in Cyanobacteria indeed
identified tyrosin 161 in the D1 protein as the primary electron donor in the
active arm and tyrosin 161 in the D2 protein as the one in the inactive arm
which is $= Y_D$ of electron flow through photosystem II (Debus et al. 1988;
Vermaas et al. 1988; Fig. 1.1).

Further herbicide-tolerant mutants of higher plants, green algae, and
cyanobacteria, isolated after site-selected screening or site-directed mutagenesis, led to the identification of many amino acid substitutions in the
Q_B binding niche (Fig. 1.3). Photoaffinity labeling with radioactive azido
derivatives of herbicides followed by protease digestion and protein sequencing indicated further amino acids in herbicide binding. Antibodies
with site-directed epitops and the trypsin accessibility of arg238 and of the
cleavage site in the "rapid turnover' of the D1 protein and photoinhibition
(see below) are further methodical approaches and provided further insights
into the three-dimensional folding of the D1 protein.

Accordingly, these amino acids of the D1 protein are of significance in
the Q_B and herbicide binding: phe211, met214, his215, leu218, val219,
tyr237, arg238, ile248, val249, tyr254, phe255, gly256, ser264, asn266,
leu271, his272, leu275.

His215 and his272 are ligands to Fe binding and his215 and ser264 form
hydrogen bridges in quinone binding, deduced from the homology to the
purple bacteria system and its X-ray structure (Deisenhofer et al. 1985;
Deisenhofer and Michel 1989). Asn247, his252 and ser264 appear to participate in the proton channel in the protonization of Q_B when it becomes
reduced to quinol. They are at homologous positions to those amino acids in
the proton channel in the bacterial Q_B reduction.

Herbicide-tolerant mutants with a specific point mutation in the amino
acid sequence of the D1 protein can be further analyzed. For example, the

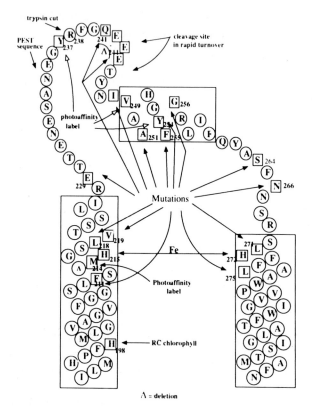

Fig. 1.3. Amino acids in the D1 protein involved in Q_B binding. Only two of the transmembrane helices, a parallel helix and the interconnecting sequence are of immediate significance for Q_B and for herbicide binding. They are identified by amino acid substitutions in herbicide-tolerant plants with point mutation in the *psb*A gene and by photoaffinity labeling. In the extended loop between transmembrane helix 4 and the parallel helix, a specific easily accessible trypsin cut at arg238 and a proposed cleavage site and a protease-recognizing PEST sequence in rapid turnover of the D1 protein are indicated

dependence of the cross-resistance of herbicides on the size of substituents in the chemical structure of the herbicides in various mutants of *Chlamydomonas rh.*, isolated by screening against just one herbicide (Rochaix and Erickson 1988), allowed the refining at the folding model by computer analysis (Draber et al. 1991). Recently, this was possible also for phenol type herbicides (Draber et al. 1993).

Some of the amino acids involved in the Q_B and herbicide-binding niche are in a part of the sequence of the D1 protein that is called the extended loop. This is because about 14 amino acids between transmembrane helix IV and the parallel helix (see Figs. 1.2, 1.3) are not present in the L subunit of purple bacteria, whose Q_B site is otherwise very homologous to that of the

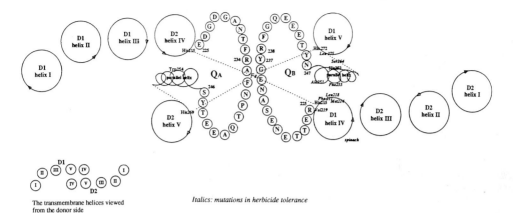

Italics: mutations in herbicide tolerance

Fig. 1.4. A blow up view of the extended loop in the reaction center polypeptides of D1 and D2 photosystem II in the membrane. Viewed from the matrix (acceptor) side. A possible folding of the extended loops that are in contact above the quinone-binding sites is indicated out of scale. Ten transmembrane helices of the two polypeptides are indicated as *circles*. As these helices are tilted in the membrane, their order is somewhat different when viewed either from the matrix or lumen side. (*Italics*: mutations in herbicide tolerance)

D1 protein. As will be discussed below, it is in this extended loop that dynamic properties of possibly regulatory significance in photosynthesis take place. The extended loop in the D1 and also the D2 protein will fold differently from the shorter sequence in the homologous L and M subunits. It has been proposed (Trebst 1991) that they fold back above the Q_A and Q_B sites and provide a contact point where the structural and functional information between the two quinone-binding sites can be exchanged (Fig. 1.4).

It appears that the correlation of structure and function of photosystem II is well based and makes it possible to address many detailed questions on the molecular mechanism in a photosystem at a highly sophisticated biophysical level (Lauterwasser et al. 1992). This chapter will not address this, but instead focus on an aspect of regulation of photosystem II activity.

1.4 Dynamics in the D1 Protein in Rapid Turnover and Stress-Enhanced Photoinhibition

Photosystem II has a suprising property. Although designed to absorb light and split water to oxygen, it is also light-sensitive, and this particularly in the presence of oxygen. Photosystem II function is easily inactivated by excess light. This is particularly true when there is no electron acceptor present. In vivo the terminal electron acceptor of photosynthesis is, of course, CO_2. The bleaching of the photosystems in vivo in the absence of CO_2 is the long-known physiological phenomenon of photoinhibition. (For a recent review see Demmig-Adams and Adams 1992). We are beginning to understand the molecular mechanism of photoinhibition (for review see Barber and Andersson 1992; Prasil et al. 1992). It becomes apparent that the inactivation of photosystem II is due to damage by a modification of an amino acid in the D1 protein. This inactivation of photosystem II can be repaired, but an exchange of an amino acid in a peptide sequence is not possible. For the repair, the damaged D1 protein has to be fully degraded first and then newly synthesized and reassembled back to functionally active photosystem II. This damage and repair cycle is a costly process. It is assumed that photo-inhibition will occur at any light intensity in the field (Greer et al. 1991). It is estimated that it might reduce the efficiency of photosynthesis by up to 15%. Under low light intensities, degradation is masked in vivo by the repair process. A continuous degradation and resynthesis of the D1 protein is called the "rapid turnover" of the D1 protein (Mattoo et al. 1984, 1989). Only when the light-dependent destruction of the D1 protein can no longer be compensated for by the repair process, does the inactivation of photo-synthetic electron flow become apparent as loss of the photosynthesis capacity. This has been intensively studied in *Chlamydomonas rh.* (Schuster et al. 1988, Prasil et al. 1992).

Rapid turnover has been discovered by radioactive pulse chase ex-periments (Mattoo et al. 1984). Of the many protein subunits of PS II discussed above, only the D1 protein is labeled quickly in a light pulse – in a matter of minutes in higher plants – with a suitable radioactive precursor like radioactive sulfate or an amino acid. The label also disappears quickly by a chase in the light, indicating the degradation of the protein.

The mechanism of photodegradation of the D1 protein is not fully understood in all aspects. Probably in the absence of an electron acceptor that draws the electron from the plastohydroquinone pool, the Q_B site plastoquinone remains reduced or leaves the site empty. Then Q_A also becomes fully reduced, possibly even to the double reduced form, and also may leave its site (Barber and Andersson 1992; Vass et al. 1992). Then the electron from the charge separation of the reaction center finds no quinone to reduce. On recombination, the triplet of the P_{680} is formed and reacts with oxygen to form the highly toxic singlet oxygen. It can be quenched to a certain extent by a carotene. The singlet oxygen may modify amino acid residues in the D1 protein, rendering the protein inactive. The modified D1 protein has to be removed completely and replaced by a new copy.

The primary cleavage site in the D1 protein during photoinhibition is a matter of debate at present (Barber and Andersson 1992). In low light rapid turnover, the actual degradation of the D1 protein begins at a cleavage site in the amino acid sequence assumed to be in the extended loop (Mattoo et al. 1984; Jansen et al. 1993) – those amino acids that close up the Q_A and Q_B site towards the matrix space, as discussed above.

The cleavage site of the D1 protein as part of the rapid turnover can be prevented by an inhibitor in the Q_B site, like DCMU (Mattoo et al. 1984; Schuster et al. 1988), other classical herbicides, and also certain phenol-type inhibitors (Jansen et al. 1993). This supports the notion that the Q_B site controls the conformation of the cleavage site and in this way allows or prevents access of a protease in the primary cleavage of the D1 protein in rapid turnover.

After complete degradation of the cleaved D1 protein by so-called house-keeping proteases, the translation machinery of the chloroplast will resynthesize the D1 protein. This is also light-dependent. The new protein will be processed, reinserted into the membrane and reassembled into functional photosystem II units closing the cycle in rapid turnover.

Viewing this process as a regulatory phenomenon, it should be noted that an amino acid sequence in a functional membrane protein contains regulatory sites. In rapid turnover these are those of the extended loop in the D1 protein (Figs. 1.3 and 1.4).

1.5 Photoinhibition and Environmental Stress

As discussed in the rapid turnover of the D1 protein, the cleavage and degradation of the D1 protein already proceeds at moderate light intensity. As degradation is balanced by resynthesis and reassembly of new D1 protein, it does not lead to an impairment of photosynthesis. However, under higher excess light conditions, imbalance in the repair cycle resynthesis does not compensate the degradation. This leads to a rapid disappearance of all photosystem II subunits and a permanent stall of photosynthesis. This impairment of photosynthesis by the light sensitivity of photosystem II is enhanced by stress conditions. Each of the processes – inactivation, the degradation of the D1 protein, resynthesis and reassembly of PS II components – is subject to control and will be influenced by environmental factors. This is the case in the stress-induced increase of photoinhibition. For a recent review see Demmig-Adams and Adams (1992). For example, air pollutants result in light-dependent bleaching and a decrease in photosynthetic activity which is most likely a consequence of the primary light inactivation of PS II and degradation of the D1 protein (Godde and Buchwald 1992). The noxious gases do not react necessaily directly with the primary events in photosystem II, for a review see Lange et al. (1989), but they affect the balance of rapid turnover and the repair cycle.

A stressed spruce needle can no longer compensate for the inactivation and degradation process by its repair capacity, and the system collapses (Godde and Buchwald 1992; Lütz et al. 1992). Overstraining the balance between degradation and repair can come about suddenly just by a further small increase in the level of stress. Photosystem II can tolerate an increase in its inactivation to quite some extent; but it can do so only up to the capacity of its repair system.

1.6   Regulation of Photosystem II by Phosphorylation

A number of proteins of the thylakoid membrane are phosphorylated usually at a threonine at or close to the N-terminus see (Allen 1992).

Phosphorylation is a positioning of a strong charge on the protein that may be essential for its proper folding and insertion into the membrane. The phosphorylation of the 25 kDa light-harvesting protein results in a conformational change that in turn disconnects the complex from the PS II core antenna system. In this way phosphorylation controls the light distribution of excitation energy between the two photosystems, and at the same time also the extent of grana stacking (Allen 1992). The significance of phosphorylation for the other membrane proteins is less clear. Among the core photosystem II proteins, the 9 kDa product of the *psb*H gene appears to be always phosphorylated. The phosphorylation of the 43 kDq antenna subunit and of the D1 and D2 proteins appears to be intermittent. It appears that a phosphorylated D1 protein is functionally not different from the unphosphorylated form, but it may play an important role in photoinhibition by stabilizing the D1 protein (Aro et al. 1992; Elich et al. 1992).

The kinase that is phosphorylating the light-harvesting protein has been purified. It is very interesting that it is a redox-controlled kinase by its attachement to the cytochrome b_6/f complex (Gal et al. 1990), most likely by a quinone-binding site on the kinase. This might be true also for the control of the still unknown protease in the rapid turnover of the D1 protein (Bracht and Trebst, in preparation).

The redox control of the phosphorylating/dephosphorylating system may open a new understanding how the capacity and actual efficiency of the electron flow system in photosynthesis is regulated and adjusted to the need of the assimilatory process coupled to it.

1.7   Conclusions

Site-specific mutations in the *psb*A gene greatly help in modeling the three-dimensional folding of the D1 reaction center polypeptide of photosystem II. The phenomenon of rapid turnover of the D1 protein followed by its cleavage, and then the complete degradation of D1 protein and of photoinhibition of photosystem II is discussed on the molecular and structural level. A primary event in the effect of excess light on the PS II will lead to the modification of the D1 protein followed by its cleavage, and then to complete degradation of the D1 protein. This light-dependent process appears to be enhanced under stress conditions like in forest decline. When the light-dependent repair system that specifically synthesizes the D1 protein is no longer compensating the light degradation, photosystem II is quickly fully inactivated. Phosphorylation of the reaction center polypeptides of PS II may play an important role in regulation of PS II turnover.

Acknowledgment. Work at Bochum was supported by the Deutsche Forschungs-gemeinschaft.

References

Allen JF (1992) Protein phosphorylation in regulation of photosynthesis. Biochim Biophys Acta 1098: 275–335

Allen JP, Feher G, Yeates TO, Komiya H, Rees DC (1987) Structure of the reaction center from *Rhodobacter sphaeroides* R-26: The cofactors. Proc Natl Acad Sci USA 84: 5730–5734

Aro EM, Kettunen R, Tyystjärvi E (1992) ATP and light regulate D1 protein modification and degradation. FEBS Letters 297: 29–33

Barber J (1987) Photosynthetic reaction centres: a common link. Trends Biochem Sci 12: 321–236

Barber J (ed) (1992) The photosystems: structure, function and molecular biology. Topics in photosynthesis, vol 11. Elsevier, Amsterdam

Barber J, Andersson B (1992) Too much of a good thing: light can be bad for photosynthesis. Trends Biochem Sci 17: 61–66

Blankenship RE (1992) Origin and early evolution of photosynthesis. Photosynth Res 33: 91–111

Debus RJ, Barry BA, Babcock GT, McIntosh L (1988) Directed mutagenesis indicates that the donor to P^{+}_{680} in photosystem II is tyrosine-161 of the D1 polypeptide. Biochemistry 27: 9071–9074

Deisenhofer J, Michel H (1989) Das photosynthetische Reaktionszentrum des Pupur-bakteriums *Rhodopseudomonas viridis*. Angew Chem 101: 872–892

Deisenhofer J, Epp O, Miki K, Huber R, Michel H (1985) Structure of the protein subunits in the photosynthetic reaction center of *Rhodopseudomonas viridis* at 3 Å resolution. Nature 318: 618–624

Demmig-Adams B, Adams WW (1992) Photoprotection and other responses of plants to high light stress. Annu Rev Physiol Plant Mol Biol 43: 599–626

Draber W, Kluth JF, Tietjen K, Trebst A (1991) Herbicides in photosynthesis research. Angew Chem Int Ed Engl 30: 1621–1633

Draber W, Hilp U, Likusa K, Trebst A (1993) Inhibition of photosynthesis by 4-nitro-6-alkylphenols: structure-activity studies in wildtype and five mutants of *Chlamydomonas reinhardtii* thylakoids. Z Naturforsch 48c: 308–318

Elich TD, Edelman M, Mattoo AK (1992) Identification, characterization, and resolution of the in vivo phosphorylated form of the D1 photosystem II reaction center protein. J Biol Chem 267: 3523–3529

Gal A, Hauska G, Herrmann R, Ohad I (1990) Interaction between light harvesting chlorophyll-a/b protein (LHC II) kinase and cytochrome b_6/f complex. J Biol Chem 265: 19749–19749

Godde D, Buchwald J (1992) Effect of long term fumigation with ozone on the turnover
   of the D1 reaction center polypeptide of photosystem II in spruce (*Picea abies*). Physiol
   Plant 86: 568–579
Greenberg BM, Gaba V, Mattoo AK, Edelman M (1987) Identification of a primary in
   vivo degradation product of the rapidly turning-over 32 kd protein of photosystem II.
   EMBO J 6: 2865–2869
Greer DH, Ottander C, Öquist G. (1991) Photoinhibition and recovery of photosynthesis
   in intact barley leaves at 5 °C and 20 °C. Physiol Plant 81: 203–210
Jansen M, Depka B, Trebst A, Edelman M (1993) Engangement of specific sites in the
   plastoquinone niche regulates degradation of the D1 protein in photosystem II. J Biol
   Chem 268: 21246–21252
Lange OL, Heber U, Schulze E-D, Ziegler H (1989) Atmospheric pollutants and plant
   metabolism. In: Schulze E-D, Lange OL, Oren R (eds) Forest decline and adr pollution. Springer, Berlin Heidelberg New York, pp 238–273
Lauterwasser, Finkele U, Struck A, Scheer H, Zinth W (1992) Primary electron transfer
   kinetics in bacterial reaction centers with modified bacteriochlorophyll. In: Murata N
   (ed) Research in photosynthesis, Vol I Kluwer, Dordrecht, pp 3.429–3.432
Lütz C, Steiger A, Godde D (1992) Influence of air pollutants and nutrient deficiency
   on D-1 protein content and photosynthesis in young spruce trees. Physiol Plant 85:
   611–617
Mattoo AK, Hoffmann-Falk H, Marder JB, Edelman M (1984) Regulation of protein
   metabolism: coupling of photosynthetic electron transport in vivo degradation of the
   rapidly metabolized 32-kilodalton protein of the chloroplast membranes. Proc Natl
   Acad Sci USA 81: 1380–1384
Mattoo AK, Marder JB, Edelman M (1989) Dynamics of the photosystem II reaction
   center. Cell 56: 241–246
Nanba O, Satoh K (1987) Isolation of a photosystem II reaction center consisting of D-1
   and D-2 polypeptides and cytochrome b-559. Proc Natl Acad Sci USA 84: 109–112
Ottander C, Hundal T, Andersson B, Huner NPA, Öquist G (1993) Photosystem II
   reaction centres stay intact during low temperature photoinhibition. Photosyn Res 35:
   191–200
Prasil O, Adir N, Ohad I (1992) Dynamics of photosystem II: mechanism of photoinhibition and recovery processes. In: Barber J (ed) Topics in photosynthesis. The
   photosystems: structure, function and molecular biology, Elsevier, Amsterdam, pp
   295–348
Rochaix J-D, Erickson J (1988) Function and assembly of photosystem II. Genetic and
   molecular analysis. Trends Biochem Sci 13: 56–59
Satoh K (1992) Structure and function of photosystem II reaction center. In: Murata N
   (ed) Research in photosynthesis vol II. Kluwer, Dordrecht, pp 5.3–5.12
Schuster G, Timberg R, Ohad I (1988) Turnover of thylakoid photosystem II proteins
   during photoinhibition of *Chlamydomonas reinhardtii*. Eur J Biochem 177: 403–410
Trebst A (1986) The topology of the plastoquinone and herbicide-binding peptides of
   photosystem II in the thylakoid membrane. Z Naturforsch 41c: 240–245
Trebst A (1991) A contact site between the two reaction center polypeptides of photosystem II is involved in photoinhibition. Z Naturforsch 46c: 557–562
Vass I, Styring St, Hundal T, Koivuniemi A, Aro E-M, Andersson B (1992) Reversible
   and irreversible intermediates during photoinhibition of photosystem II: Stable reduced
   Q_A species promote chlorophyll triplet formation. Proc Natl Acad Sci USA 89: 1408–
   1412
Vermaas WFJ, Rutherford AW, Hansen O (1988) Site-directed mutagenesis in photosystem II of the Cyanobacterium *Synechocystis sp*. PCC 6803: Donor D is a tyrosine
   redidue in the D2 protein. Proc Natl Acad Sci USA 85: 8477–8481

2 Regulation of Photosynthetic Light Energy Capture, Conversion, and Dissipation in Leaves of Higher Plants

O. Björkman and B. Demmig-Adams

2.1 Introduction

In nature, the intensity of light, or photon flux density (PFD), shows great variation, both temporally and spatially. For example, a leaf in the under-story can experience changes in the incident PFD up to 100-fold within a few seconds (Chazdon and Pearcy 1991). Large changes in PFD are also experienced by exposed leaves when intermittent clouds obscure the sun. In addition, the total daily integrated photon flux varies greatly among habitats as well as within the canopy of a given plant stand. Plants on the floor of a tropical rainforest (Björkman and Ludlow 1972) or redwood forest (Björkman and Powles 1981) may receive as little as 1% of the daily photon flux above the plant canopy.

This great variation in the light environment imposes a great demand on the responsiveness of the photosynthetic system. At PFDs that are limiting to photosynthesis, light must be captured and utilized with the highest possible efficiency. On the other hand, it is imperative that overexcitation of the photosynthetic reaction centers be avoided when the light becomes excessive, as such overexcitation can result in severe damage to these centers (photoinhibitory damage). This problem is further exacerbated in the presence of environmental stresses that lower the capacity for light-saturated photosynthesis as this results in an increased level of *excessive* light energy. Examples of such stresses are unfavorably low temperatures, which directly decrease the rates of photosynthetic electron transport and the turnover of the carbon reduction-oxidation cycles, and drought, which leads to a decrease in the CO_2 available for carbon fixation because of its effect on stomatal conductance.

On these grounds it is clear that plants must possess the ability to regulate the level of excitation energy, in both the long and the short term. In this chapter we will illustrate the various ways in which this may be achieved. Regulation can occur by changes in light interception, by changes in photo-synthetic capacity, and by dissipation of excess excitation energy through various pathways within the chloroplasts. Wherever possible, we will use examples relevant to actual field conditions and for the most part these examples are taken from published and unpublished studies in which the authors have been active participants.

2.2   The Concept of Excess Photon Flux Density

Figure 2.1 illustrates the degree to which light becomes excessive as the photon flux density (PFD) incident on a leaf gradually increases to that of full sunlight ($2000\,\mu mol\,m^{-2}\,s^{-1}$). At low PFDs the photosynthetic rate (curve b) increases linearly with the incident PFD, and in this range there is no excess light as all photons absorbed by the chlorophyll are utilized and thus all the excitation energy is dissipated in photosynthesis. As the PFD is further increased, the rate of photosynthesis starts to deviate from the initial slope (a) of curve b, i.e., some of the light is not used in photosynthesis. The amount of excess PFD is depicted by the horizontal arrows and by curve d. As full sunlight is reached, as much as 75% of the incident PFD is excess PFD.

In the absence of any processes regulating the interception of light energy or the dissipation of excitation energy through pathways other than photosynthesis, the fraction of the photosystem (PS II) centers that are in the reduced state (closed centers) would follow curve c. In full sunlight, only some 20% of the PS II centers would remain in the oxidized state (open centers) and be able to carry out photochemistry. The remaining 80% of the centers would be closed and therefore incapable of photochemistry, a situation thought to predispose these centers to photoinhibitory damage.

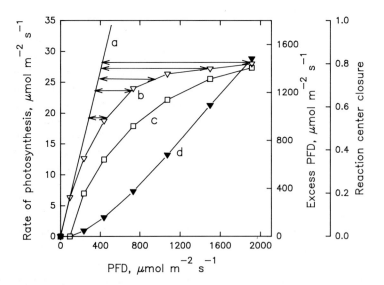

Fig. 2.1. Development of excess light energy in response to an increase in the PFD incident on a leaf. The rate of photosynthesis is depicted by curve *b* and the initial slope of this curve by line *a*. The excess PFD is given by the *horizontal arrows* and by curve *d*. The calculated degree of closure of the photosystem II reaction centers (curve *c*) is based on the (false) assumption that there was no regulation of light interception or dissipation of excitation energy. (Based on data by Schäfer and Björkman 1989)

The example shown in Fig. 2.1 represents a leaf with a moderately high photosynthetic capacity and in the absence of any other stress. The amount of excess PFD, and hence the degree of center closure, obviously would be greater in leaves having a lower photosynthetic capacity or in the presence of environmental stresses that impose a further limitation on the photosynthetic rate. We wish to emphasize that these are predictions based on the assumption that there were no regulation of energy interception or dissipation.

2.3 Regulation of Light Interception

There are at least four different ways in which changes in light interception can be achieved: changes in leaf orientation relative to the solar beam, changes in leaf reflectance, rearrangements of the chloroplasts within the leaf, and changes in the amount of chlorophyll per leaf area. Of these means of regulating light interception, changes in leaf orientation are probably the most powerful.

2.3.1 Changes in Leaf Orientation

Developmental changes in leaf angle in response to the light environment have long been recognized as a phenomenon common to many land plants. In general, leaves developing in the shade have a predominantly horizontal orientation and are arranged in a single layer, whereas those developing in full sun often have steeper angles, especially under conditions of water stress (Ehleringer 1988). A steeply angled leaf obviously intercepts considerably less of the solar radiation during the midday hours when the radiation load is maximal. This is especially important in water-stressed leaves when stomatal closure leads to a severely reduced photosynthetic rate and prevents the dissipation of the absorbed radiation as latent heat. A near-vertical leaf orientation may therefore prevent both overexcitation of the photosynthetic reaction centers and overheating the leaf, which may be lethal. A disadvantage is that such developmental changes in leaf angle are largely irreversible when the stress is relieved.

Passive Leaf Movements. The passive wilting of many species such as sunflower and the leaf rolling in many grasses, such as *Sorghum* spp., are common consequences of water stress in tissues without secondary cell walls. Such wilting can be quite effective in reducing the energy load on the plant, and may decrease water loss by as much as 50 to 70% (Begg 1980). Since leaves without secondary wall thickenings are very sensitive to changes in leaf water deficit, this enables the plant to respond rapidly to periods of peak radiation load and then to recover when the radiation load is low and water relations improve.

Active Leaf Movements. Leaves of many species belonging to different taxonomic groups have the ability to actively change their orientation during the course of the day, thereby changing the angle of the leaf relative to that of the solar beam (Ehleringer and Forseth 1980). The physiology of these movements has recently been reviewed by Koller (1990). Leaves capable of such active movements are equipped with a pulvinus attaching the leaf blade to the petiole or stem. The movements are light-driven and under the control of a blue-light-absorbing pigment system. The identity of this photo-receptor pigment has not been established but the action spectrum for the response indicates that it is a flavoprotein or, possibly, a carotenoid. The leaf or leaflets unfold or fold by simultaneous expansion/contraction in opposite sectors of their pulvinar motor tissue, resulting from massive fluxes of (mainly potassium) ions and water.

Especially rapid and effective active leaf movements are present in shade species of the genus *Oxalis* and a striking example is provided by *Oxalis oregana*, a species native to the floor of redwood forests. In this habitat, the prevailing daytime PFD can be as low as 3μmol photons $m^{-2} s^{-1}$, but occasional sunflecks can suddenly increase the PFD to as much as 1800μmol $m^{-2} s^{-1}$. These leaves have a very low capacity for light-saturated photosynthesis ($3-5 \mu$mol CO_2 $m^{-2} s^{-1}$) and reach 90% of light saturation at a PFD of about 120μmol $m^{-2} s^{-1}$. Each of the three leaflets is attached to the petiole by a pulvinus, which in this species is responsible for both the photoreceptor and the motor functions. In the absence of sunflecks, each leaflet moves such that it continuously faces the brightest area of the canopy above, thereby maximizing light interception. However, when a bright sunfleck strikes the pulvinus and the incident PFD rises from, e.g., 4 to 1600μmol $m^{-2} s^{-1}$, the leaflet rapidly folds down (Fig. 2.2). After a lag of about 10 s the leaf angle changes at a rate of $20° min^{-1}$, and within 6 min the leaflets reach a position essentially parallel to the solar beam. The leaflets remain in this position until the sunfleck disappears, then slowly return to their original orientation.

The fraction of incident radiation intercepted by a leaf is given by the cosine of the leaf angle in relation to the light beam. Hence the light interception changed from over 90% to less than 10% during exposure to a sunfleck in the experiment shown in Fig. 2.2. Subsequent studies under controlled conditions showed that the steady-state position of the leaflets is closely tuned to the PFD over a wide range (Fig. 2.3). Clearly, each pulvinus adjusts the leaflet angle such that light interception is sufficient to saturate photosynthesis while avoiding supersaturation.

As in the Oxalidaceae, active leaf movements are common in species in the families Malvaceae and Fabaceae. Unlike *Oxalis oregana*, many of these

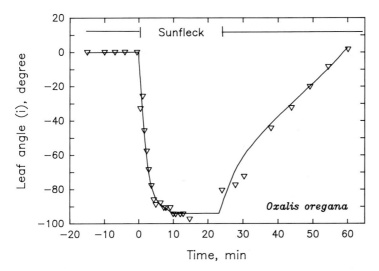

Fig. 2.2. Time course of leaf folding in *Oxalis oregana* in response to the arrival of an intense sunfleck (1590 μmol photons m^{-2}s^{-1}) and of unfolding after departure of the sunfleck. The PFD before the arrival, and after departure of the sunfleck was about 4 μmol photons m^{-2}s^{-1}. (After Björkman and Powles 1981)

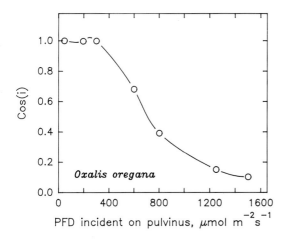

Fig. 2.3. Cosine of the leaf angle in relation to the light beam at steady state as a function of the PFD incident on the pulvinus of a leaflet of *Oxalis oregana*. The light beam was from a collimated xenon arc source. (Original data by O. Björkman)

species occupy predominantly open, sunny habitats. In *Lupinus* spp. (Wainwright 1977), the tropical legumes Siratro (Ludlow and Björkman 1984), and Townsville stylo (Begg and Torssell 1974), the leaves of plants with ample water supply move diaheliotropically throughout the day so that they

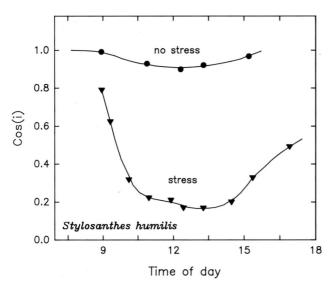

Fig. 2.4. Diurnal variation in the cosine of the leaf angle in relation to the solar beam in leaves of *Stylosanthes humilis*, growing under well-watered conditions (*no stress*), and at the beginning of the dry season (*stress*) during a cloudless day. The minimum water potentials of the unstressed and stressed plants were -0.2 and -2.6 MPa, respectively. (Based on data from Begg and Torssell 1974)

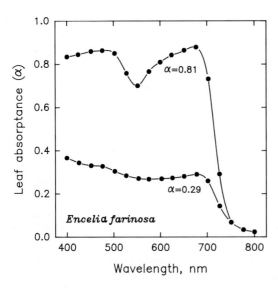

Fig. 2.5. Spectral leaf absorptance (α) of a lightly pubescent, greenish, leaf and of a heavily pubescent, white, leaf of *Encelia farinosa*. (Ehleringer and Björkman 1978)

are kept essentially normal to the solar beam, thereby maximizing light interception. However, when the plants are gradually subjected to water stress, the leaves begin instead to move paraheliotropically. The onset of such movements usually coincides with stomatal closure. During soil drying cycles, paraheliotropic movements first become apparent for short periods just after solar noon but the leaves are diaheliotropic for the rest of the day. The duration of the paraheliotropic periods increases as the drought intensifies until the leaves are paraheliotropic for most of the day. This mode of response is illustrated in Fig. 2.4 in which the diurnal changes in cos(i) of well-watered plants of Townsville stylo (*Stylosanthes humilis*) are compared with those of a plant in the field at the beginning of the dry season when the minimum leaf water potential had fallen to $-2.6\,\text{MPa}$. During midday, the radiation intercepted by the leaves of the stressed plant was only about 20% of that intercepted by those of the unstressed plant.

These light evasion responses can be induced in the absence of water stress by direct exposure of the pulvini to high irradiance (Koller and Shak 1990), indicating that the primary elicitor of the paraheliotropic response is high irradiance, and that water deficit and high temperature act by lowering the threshold for that response.

2.3.2 Changes in Leaf Reflectance

Many species, mostly from arid climates, are capable of regulating light interception in the longer term by changing leaf reflectance. This is achieved by varying the degree of pubescence, i.e., air-filled hairs or trichomes, and in some species through a deposit of salt crystals on the leaf surface. A striking example is *Encelia farinosa*, a native of the low interior California deserts. Leaves that develop during the months with moderate temperatures and relatively favorable water relations are only weakly pubescent and the leaf hairs are alive and water-filled. Leaf reflectance is low and hence the absorptance is high (Fig. 2.5, upper curve). As drought and temperature increase, the hairs increase in both density and length and gradually become filled with air, resulting in a strongly increased reflectance (Ehleringer and Björkman 1978). The development of pubescence is strongly correlated with declining precipitation, increasing heat load and, especially, midday water potential (Ehleringer 1982). Under extreme conditions, such as those found in Death Valley, California, the heavily pubescent leaves appear nearly white and absorb as little as 29% of the photosynthetically active radiation (Fig. 2.5, lower curve). Because the leaf hairs lose their water, the absorptance of near-infrared radiation is also strongly reduced, resulting in a further reduction of the heat load.

Responses similar to those in *E. farinosa* were also found in the South American desert species *E. canescens*. In both western North America and

South America, numerous species along gradients of decreasing precipitation
either increase their leaf pubescence in response to increasing aridity or are
replaced by closely related more pubescent species (Ehleringer et al. 1981).

A disadvantage associated with an increased leaf pubescence is that it
leads to a reduction in photosynthetic rate at times when light is limiting,
and once the highly reflective pubescence has developed, it is irreversible
(Ehleringer 1982). Moreover, it appears that when an active apical meristem
has been stressed, all subsequently produced leaves develop pubescence to a
degree that matches that of the "direst" conditions to which the plant has
been exposed (Ehleringer 1982).

In contrast to the situation in *Encelia*, increases in leaf reflectance are
evidently largely reversible where they are partly caused by a deposit of salt
crystals on the leaf surface, as is the case in many halophytic species of
the genus *Atriplex* or in salt-secreting mangrove species such as *Aegialitis
annulata*. In Death Valley, *Atriplex hymenelytra* leaves produced during the
winter months reflect less than 30% of the photosynthetically active radiation, whereas summer leaves reflect as much as 60%. The reflectance of
these summer leaves decreased to that of winter leaves within a few days
after the leaves were allowed to rehydrate by covering the plants with
reflective plastic bags. The rehydration caused dilution of the salt with no
change in salt content per leaf dry weight (Mooney et al. 1977).

2.3.3 Chloroplast Movements

It is well established that the position and orientation of the chloroplasts
within each cell depends on the PFD incident on the leaf (Haupt and
Scheuerlein 1990). At low PFDs, the chloroplasts are arranged to provide
maximum light interception. Typically, they are assembled perpendicular to
the light direction. As the PFD increases, they gradually line up along the
vertical cell walls, parallel to the light direction, thereby allowing more light
to be transmitted through the leaf. The rapidity of these movements depends
on the extent of change in PFD; it is also strongly temperature dependent.
A typical half-time for the response is 5–10 min (Brugnoli and Björkman
1992). The action spectrum is similar to that for leaf movements and the
photoreceptor is therefore presumably also similar.

The maximum change in absorptance caused by chloroplast movements
among the species studied so far was obtained in *Oxalis oregana* (Fig. 2.6).
The change integrated over the photosynthetic range is 19–20%. Changes in
the order of 10–15% were obtained in shade leaves of several other species
but were considerably smaller in most of the other species (Inoue and
Shibata 1974; Brugnoli and Björkman 1992). Even in *Oxalis* the decrease in
light interception caused by chloroplast movement upon transition from low
to high light is modest in comparison with those caused by the leaf move-

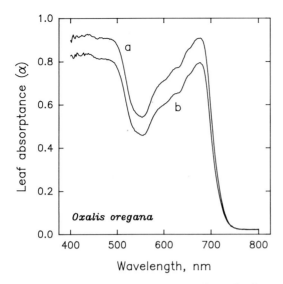

Fig. 2.6. Spectral leaf absorptance of an *Oxalis oregana* leaf having its chloroplasts arranged such that light interception is maximized (curve *a*) and such that it is minimized (curve *b*). The change in chloroplast arrangement was obtained by exposing a predarkened leaflet to a PFD of 940 μmol photons $m^{-2}s^{-1}$ for 30 min. (After Brugnoli and Björkman 1992)

ments described above. However, even where the effect on overall leaf absorptance is relatively small, the chloroplast rearrangement elicited by high PFDs may have other beneficial effects such as providing a more uniform light *distribution* among the individual chloroplasts within the leaf.

2.3.4 Changes in Chlorophyll Content and Photosynthetic Capacity

It is well known that shade leaves tend to have a much greater density of grana stacks per chloroplast volume, larger chloroplasts, and a greater ratio of light-harvesting chlorophyll to stromal enzymes than do leaves that have developed in bright light (see reviews by Björkman 1981; Anderson et al. 1988). This is accompanied by a high capacity for the harvesting of light in relation to the photosynthetic capacity. It is also well established that even fully developed leaves are able to respond to an altered light environment by changing this relationship within a few days. Thus, there is no doubt that, in nature, a high degree of light regulation is achieved by changes in the amount of photosynthetic enzymes and electron carriers relative to chlorophyll. An increase in this ratio results in a higher photosynthetic capacity and therefore decreases the amount of excess excitation energy.

Although each chloroplast in a sun leaf intercepts a relatively smaller fraction of the incident light than each chloroplast in a shade leaf, this is compensated for by a greater number of chloroplasts across the leaf section, and there is no consistent difference in the amount of chlorophyll per unit leaf area between the two types of leaves. Moreover, the principal effect of an increased chlorophyll content per leaf area is a broadening of the absorption bands, resulting in an increased absorptance in the green and the far-red regions, with only a small effect at wavelengths at which the light-harvesting pigments have a high absorption coefficient. Therefore, the increase in α, integrated over the photosynthetically active spectral range, is not linearly dependent on Chl content per leaf area. For example, in *Hedera canariensis*, leaves having Chl contents of 175, 350, and 700 mg m^{-2}, the corresponding α values were 0.70, 0.80, and 0.90 (Björkman and Demmig 1987), i.e., a 59% reduction in chlorophyll content resulted in only a 10–13% reduction in α. We therefore conclude that changes in chlorophyll content per unit leaf area are probably not a major factor in regulating light interception in natural situations.

2.4 Regulation of Energy Dissipation

There are four different ways in which the excitation energy resulting from the absorption of light by chlorophyll can be dissipated. The first is by conversion into chemically bound energy contained in the final products of photosynthesis. The second is by energy consumption in metabolic processes that do not result in energy storage. The third is by reemission of photons as fluorescence (radiative dissipation). The fourth is by conversion of light energy into heat in the pigment bed (thermal or nonradiative dissipation). Strictly speaking, energy dissipation via metabolic processes is nonradiative as well but we will reserve this term for nonmetabolic energy dissipation occurring within the pigment bed itself.

The amount of excitation energy that can be dissipated by fluorescence is quite small, at most 3–4% of the total, and can be neglected for most practical purposes. However, the chlorophyll fluorescence emitted from a leaf can serve as a very powerful intrinsic probe to assess the efficiency of energy conversion or the degree of reaction center closure in photosystem II as well as the extent of nonradiative energy dissipation (Sect. 2.4.2).

2.4.1 Dissipation in Metabolic Processes

Only part of the reduced NADPH and ATP generated in the photoacts of photosynthesis is stored in final stable photosynthetic products. A considerable fraction is also consumed in processes that do not result in carbon fixation. Among these are photorespiration and transport processes involved in the maintenance of ionic balance in the different compartments of the leaf; the latter may be especially significant during salinity stress. Some NADPH and ATP may also be consumed in nitrogen reduction and in "repair" processes. (A small amount of NADPH is also consumed during the operation of the xanthophyll cycle but this can probably be neglected; it is *not* the mechanism by which this cycle regulates energy dissipation; see Sect. 2.4.3).

Another potentially important process of metabolic energy dissipation which may have a regulatory role is the Mehler-ascorbate peroxidase reaction (Neubauer and Schreiber 1989; Neubauer and Yamamoto 1992). This reaction involves the reduction of O_2 by photosystem I (PS I), leading to the formation of H_2O_2 which in turn reacts with ascorbate to form H_2O and O_2 (see also Chap. 1, this Vol). This reaction does not result in any net gas exchange and, to our knowledge, quantitative estimates of the amount of energy dissipated by this mechanism in leaves are lacking at this time. However, in addition to providing any direct energy dissipation, this process may have an important role in supporting the electron transport needed for development of nonradiative dissipation in the pigment bed (Sect. 2.4.3; see also Chap. 1, this Vol.).

Photorespiration is the only major metabolic process for which good estimates of energy dissipation may presently be obtained. It is well established that instead of reacting with CO_2 to form 2 mol phosphoglycerate (a net gain of 1 mol C), a considerable fraction of the ribulosebisphosphate pool reacts with O_2, leading to the formation of phosphoglycolate, part of which is converted to CO_2 (a net loss of 0.5 C). Both the carboxylation and the oxygenation are catalyzed by the enzyme Rubisco and, since O_2 competes with CO_2, the rate of oxygenation depends on the relative concentration of these gases. The rate of photorespiration relative to CO_2 fixation also increases with increasing leaf temperature.

Values for energy dissipation by photosynthetic CO_2 fixation, $P[CO_2]$, and photorespiration, PR, at midday in fully exposed leaves of a cotton crop growing in the San Joaquin Valley, California, are shown in Table 2.1. The plants were grown under normal agricultural irrigation practices as well as under two levels of long-term water stress, obtained by terminating irrigation early in the growing season. In irrigated plants which had high rates of CO_2 fixation, dissipation via this process was estimated to account for 26% of the total excitation energy. Because of the high leaf temperatures, photorespiration was also relatively high and accounted for the dissipation of some 20% of the total excitation energy. Hence the remaining excess would be some 55% of the total. Increased water stress resulted in a progressive decrease in photosynthetic rate and, in severely stressed plants, energy dissipation via $P[CO_2]$ fell to 10% or less of the total energy absorbed. Since water stress caused partial stomatal closure, the leaf temperature rose and the intercellular CO_2 pressure declined. As a result, the ratio of PR to $P[CO_2]$ increased. However, the decline in CO_2 pressure was relatively modest, as the intrinsic (i.e., CO_2-saturated) photosynthetic

Table 2.1. Rate of net CO_2 uptake ($P[CO_2]$), photorespiration (PR), and estimated percent energy dissipation via these processes during midday in cotton leaves exposed to full sunlight in the field in the San Joaquin Valley, California. The midday water potentials were approx. -1.0 MPa (irrigated), -1.7 to -2.1 MPa (moderate stress), and -2.7 to -2.9 MPa (severe stress). Leaf temperatures ranged from approx. $33\,°C$ in irrigated plants to as high as $40\,°C$ in severely stressed plants. Rates of photorespiration were calculated from simultaneous measurements of CO_2 uptake rate, CO_2 ambient pressure, leaf conductance, and leaf temperature, according to von Caemmerer and Farquhar (1981). Percentage dissipation values are based on the rate of $P[CO_2]$ + PR that would have been obtained if this rate had continued to follow the initial slope of the light curve. (Original data by Björkman, Schäfer, and Shih, unpubl.; for an extended abstract of this study, see Björkman and Schäfer 1989)

	Irrigated	Moderate stress	Severe stress
$P[CO_2]$, $\mu mol\ m^{-2}\ s^{-1}$	44.0	27.0	13.0
PR, $\mu mol\ m^{-2}\ s^{-1}$	33.4	24.0	~13.0
$P[CO_2]$ + PR, $\mu mol\ m^{-2}\ s^{-1}$	77.4	51.0	~26.0
$PR/P[CO_2]$	0.76	0.89	~1.0
$P[CO_2]$, % dissipation	26.0	16.4	~9.0
PR, % dissipation	19.6	14.5	~9.0

capacity also declined with increased water stress. This decline in intrinsic capacity was associated with a decline in leaf nitrogen content (data not shown) and presumably a decreased level of Rubisco. Under severe stress, $P[CO_2]$ and PR each contributed about 10% of the total energy dissipation. The remaining excess of excitation energy would thus be 80% of the total.

The rate of photorespiration and hence its contribution to energy dissipation would be expected to be greater in temporarily water-stressed leaves in which the intrinsic photosynthetic capacity remained unchanged and the decline in net CO_2 uptake is mainly due to a decrease in intercellular CO_2 pressure. However, such partial compensation for dissipation via CO_2 uptake is probably simply an inevitable consequence of the properties of the enzyme Rubisco. A truly regulatory process should not result in a reduced photosynthetic efficiency when there is no excess excitation energy. This criterion is not met by photorespiration, as it causes large decreases in carbon gain even when light is limiting to photosynthesis and there is no need for energy dissipation.

2.4.2 Efficiency of Photochemical Energy Conversion and Extent of Nonradiative Energy Dissipation

Estimates of the efficiency of energy conversion in photosystem II (PS II), the fraction of open or closed centers, and the extent of nonradiative energy dissipation presented below are based on chlorophyll fluorescence analysis of intact leaves. A brief outline of the fluorescence measurements used in the present context is given below. For a recent comprehensive review of the foundation for such analyses, see Krause and Weis (1991; see also Chap. 1, this Vol.).

2.4.2.1 Basic Chlorophyll Fluorescence Measurements

At ambient temperatures nearly all of the fluorescence emitted from a leaf emanates from PS II. The basic fluorescence parameters measured are F_o, F_m, and F. F_o is the minimal fluorescence yield, (i.e., the intensity of fluorescence divided by the intensity of the light used to excite fluorescence), which occurs when all reaction centers are in the oxidized, or open, state. F_o is measured under conditions where all the reaction centers are rendered open by removing all light that might be captured by the chlorophyll except for a very weak excitation beam which does not cause an accumulation of reduced electron acceptors in PS II. (In some cases it may be necessary to briefly expose the leaf to far red light to insure that all PS II centers are open; the far red light excites PS I only, thereby removing any electrons from the PS II centers). F_m is the maximum fluorescence yield, measured during exposure to a brief light flash of an intensity and duration just sufficient to temporarily close all centers. F is the yield measured under the actual actinic PFD and the actual degree of center closure. When all centers are open then $F = F_o$; conversely, when all centers are closed then $F = F_m$.

These "control" F_o and F_m values are determined in the absence of any lowering (quenching) of fluorescence caused by nonradiative dissipation (NRD). When the actinic PFD is sufficiently high to cause excess light and NRD to develop, then the fluorescence yield declines. This mainly affects F_m, which in this quenched state is termed F_m', but may to a lesser degree also affect F_o which is then termed F_o'. Accordingly, the variable fluorescence, F_v and F_v', is the difference between F_m and F_o and between F_m' and F_o', respectively.

The actual efficiency of energy conversion in PS II is estimated from the expression $\Phi_{II}' = (F_m' - F)/F_m'$. When all centers are open then $F = F_o'$ and $\Phi_{II}' = (F_m' - F_o')/F_m' = F_v'/F_m'$. Furthermore, the maximum, or intrinsic, efficiency, Φ_{II}, is obtained when all centers are open and NRD = 0; hence $\Phi_{II} = F_v/F_m$. The fraction of centers that remain in the open (oxidized) state, Q_{ox}/Q_t is given by $(F_m' - F)/(F_m' - F_o')$; hence the fraction of centers that are in the closed (reduced) state Q_r/Q_t is equal to $1 - Q_{ox}/Q_t$ and to $(F - F_o')/(F_m' - F_o')$.

Finally, the decline in F_m to F_m' caused by NRD is termed nonphotochemical quenching (NPQ). In this chapter, we have chosen to calculate NPQ according to the Stern-Volmer equation, $NPQ = (F_m - F_m')/F_m' = F_m/F_m' - 1$. On theoretical grounds, if the only cause of the quenching of F_m is an increase in NRD, then NPQ calculated in this way should be proportional to NRD. It should also be proportional to the concentration of the quenching agent.

2.4.2.2 Comparison of Efficiencies and Rates of Energy Conversion and Nonradiative Energy Dissipation Among Species

Figure 2.7 shows NPQ as well as the excess PFD in a cotton leaf over a range of incident PFDs. The excess PFD in both Figs. 2.1 and 2.7 is based on the initial slope of net CO_2 uptake versus PFD, and since the intercellular CO_2 pressure and leaf temperature were kept nearly constant over a wide range of PFD, dissipation via photorespiration is already taken into account in the calculation of excess PFD. The close relationship obtained between the increase in NPQ and in the increase in excess PFD shows that NPQ is essentially linearly related to the excess PFD over a wide range of incident PFDs. Also shown in Fig. 2.7 is the resulting degree of reaction center closure, Q_r/Q_t. Clearly, the increase in NRD causes a much greater fraction of the reaction centers to remain open than would be the case in the absence of such dissipation (cf. Fig. 2.1, curve c).

In the top panel of Fig. 2.8 are shown the PFD dependence of the efficiency of energy conversion in PS II at steady state for four species, as determined by chlorophyll fluorescence measurements. Rates of PS II

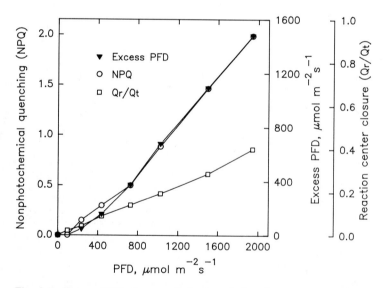

Fig. 2.7. *Excess PFD*, degree of closure of the photosystem II reaction centers, and nonphotochemical fluorescence quenching (*NPQ*) in relation to the PFD incident on a leaf. The curve for excess PFD is the same as in Fig. 2.1. Note the difference between the curves for center closure in these two figures. (Based on data by Schäfer and Björkman 1989)

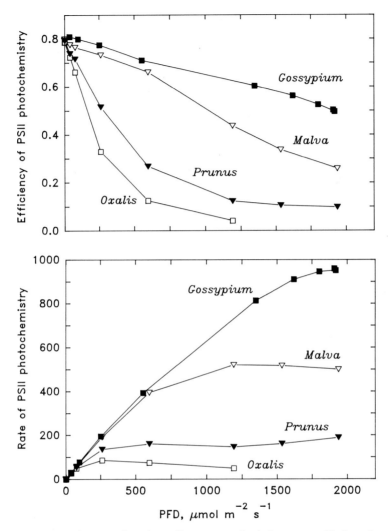

Fig. 2.8. Efficiency (*top panel*) and rate (*bottom panel*) of photosystem II photochemistry at steady state in relation to the PFD incident on the leaf. Responses of *Oxalis oregana*, *Prunus ilicifolia*, and *Malva parviflora* were determined on field-grown plants in March, 1992 at Stanford, California. Leaf temperature was 23–25 °C. (B. Demmig-Adams and O. Björkman, original data). Responses of *Gossypium hirsutum* (cotton) leaves were determined on irrigated plants in a cotton field in the San Joaquin Valley, California in the summer of 1987. Conditions were as given in Table 2.1. (Based on unpublished data of Björkman and Schäfer)

photochemistry (Fig. 2.8, lower panel) were obtained by multiplying the efficiency by the PFD of the actinic light. The plants are listed in the order of increasing photosynthetic capacity which ranged from 4 to 40 μmol CO_2 $m^{-2} s^{-1}$. *Oxalis oregana* was growing in deep shade; the evergreen shrub

Prunus ilicifolia in partial shade in the winter; the weed *Malva parviflora* in the open in the winter; and *Gossypium hirsutum* (cotton) under intense summer sunlight in the San Joaquin Valley. All plants had an adequate supply of water. During the experimental treatments the leaves were restrained to prevent leaf movements and were held normal to the light beam. Wavelengths <500 nm were excluded from the actinic beam in experiments with leaves in which chloroplast movements can induce significant changes in absorptance.

No differences were detected among the four species in the efficiency of energy conversion, determined in the absence of any excess light. This is in accordance with previous studies involving a large number of ecologically diverse species which showed that this efficiency is remarkably constant as long as the plants have not been subjected to severe stress (see Björkman and Demmig 1987). However, the responses to an increased PFD were strikingly different among the different species. In *Oxalis*, the efficiency fell precipitously as the PFD was increased, even at quite modest PFDs, and the rate of PS II photochemistry reached a maximum at PFDs of around $250 \mu mol$ photons $m^{-2} s^{-1}$; moreover, further increases in PFD led to a decline in this already very low maximum rate. In *Prunus*, the efficiency showed a strong decline already at very moderate PFDs and the rate reached

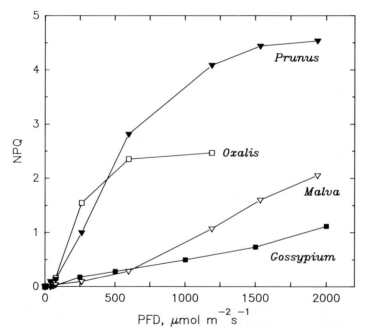

Fig. 2.9. Nonphotochemical fluorescence quenching (*NPQ*) at steady state in relation to the PFD incident on the leaf in different species. Sources and conditions as in Fig. 2.8

a plateau above $500\,\mu$mol photons m^{-2} s^{-1}. In *Malva* the decline in efficiency was much less pronounced and the rate did not reach its maximum up to PFDs of about $1000\,\mu$mol photons m^{-2} s^{-1}. Lastly, in field-grown cotton the decline in efficiency was even more gradual and the rate did not reach saturation until the PFD approached full sunlight. It is noteworthy that the light responses of the rate of PS II photochemistry and the relative light-saturated rates among the different species show strong resemblances to the corresponding light responses and light-saturated rates of CO_2 uptake (not shown). These findings give further support to the notion that simple determinations of $(F_m' - F)/F_m'$ can provide a rapid and convenient method for nondestructive estimates of photosynthetic performance in the field.

The responses of NPQ to increasing PFD in these four species are shown in Fig. 2.9. In *Oxalis*, a high NPQ was present even at quite low PFDs but

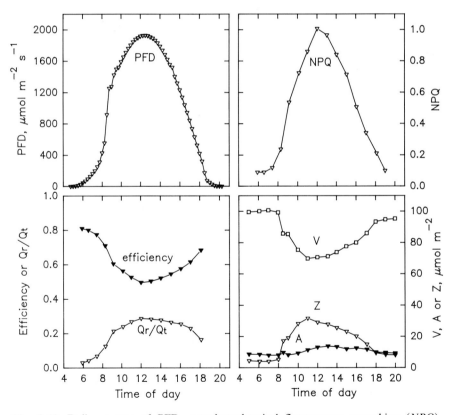

Fig. 2.10. Daily courses of PFD, nonphotochemical fluorescence quenching (*NPQ*), degree of closure of the photosystem II centers (Q_r/Q_t) and contents of the xanthophyll cycle pigments (*V*, *A*, and *Z*) in leaves of cotton growing in the field at Stanford, California in July, 1988. (Based on original data by B. Demmig-Adams, O. Björkman, W. Bilger, W.W. Adams III, S.S. Thayer and C. Shih)

NPQ saturated already at moderate PFDs, indicating that the maximum capacity is reached. Therefore, these shade leaves evidently were unable to further counteract the accumulating excess excitation energy when the incident PFD increased beyond 500 µmol photons $m^{-2}s^{-1}$. In *Prunus*, NPQ did not reach saturation until about 1500 µmol photons $m^{-2}s^{-1}$ and the maximum level of NPQ was about twice as high as in *Oxalis*. In contrast to *Oxalis* and *Prunus*, the *Malva* and cotton leaves showed only small increases in NPQ up to 500 µmol photons $m^{-2}s^{-1}$ and, beyond this point, NPQ continued to rise more or less linearly with increasing PFD. Evidently, the maximum capacity for NRD was not reached even in full sunlight in these leaves, and there remained a "reserve" capacity for even higher NRD under conditions where the photosynthetic capacity would be reduced by environmental stress.

In Fig. 2.10 are shown the diurnal courses of PFD, the efficiency of energy conversion in PS II, the degree of reaction center closure, and NPQ in fully exposed leaves of a cotton crop, growing in the field at Stanford, California. (Also shown are the corresponding changes in the components of the xanthophyll cycle; these will be referred to in Sect. 2.4.3.2) As expected, NPQ closely followed the development of excessive light: it was very low in the early morning, increased as the PFD increased to reach a peak around solar noon, then gradually declined as the PFD decreased in the afternoon. Conversely, the efficiency of energy conversion was highest in the morning reached a minimum around noon and then rose again in the afternoon. Although the fraction of closed centers, Q_r/Q_t, increased as the PFD increased, and then fell as the PFD decreased, it remained much lower than one would expect in the absence of NRD, and at noon only 30% of the centers were in the closed state.

Much larger diurnal changes in the photochemical efficiency of PS II and in NPQ than in this cotton crop can be seen in plants such as the cactus *Nopalea cochenillifera* growing in the field (Fig. 2.11). The actual degree of PS II center closure in a southeast facing cladode was maintained very low even at peak PFD, presumably due to the very high NPQ level. The hypothetical degree of reaction center closure that would have been reached if there had been no NRD was 68%, compared with an actual closure of only 20%.

2.4.3 Nonradiative Energy Dissipation and the Xanthophyll Cycle

There is much evidence that one prerequisite for the development of NPQ is an acidification of the thylakoid lumen resulting from a buildup of a proton gradient across the thylakoid membrane (Krause et al. 1982; Gilmore and Yamamoto 1991, 1992). The extent of this gradient is determined by the balance between the rate of electron transport driving the proton pumping

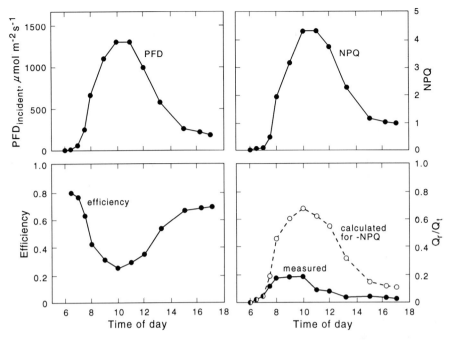

Fig. 2.11. Incident PFD, efficiency of photosystem II photochemistry, nonphotochemical fluorescence quenching (*NPQ*) and degree of closure of the photosystem II centers (Q_r/Q_t) for a south-facing cladode of the cactus *Nopalea cochenillifera* during the course of a clear day in northern Venezuela in March, 1988. The calculated level of Q_r/Q_t that would have been reached in the absence of nonradiative dissipation (calculated for -NPQ) is also shown. (Based on data by Adams et al. 1989)

and the rate of dissipation of the gradient by ATP consumption in the carbon reduction/oxidation cycles and other ATP-consuming processes. Therefore, whenever the potential for ATP generation exceeds the rate of ATP consumption, the proton gradient would supposedly increase, thus promoting the development of NPQ. The Mehler ascorbate peroxidase reaction, mentioned in Section 2.4.1, may assist in maintaining a substantial and persistent proton gradient since it consumes NADPH but no ATP; it may also allow linear electron flow to continue under conditions of limited electron consumption through the carbon reduction cycle as, for example, would be the case when the availability of CO_2 is restricted (Neubauer and Yamamoto 1992). The buildup of a proton gradient, or rather the resulting acidification of the lumen, is thought to induce a conformational change in the thylakoid membrane. The conformational change may alter the association of the various pigments in the chlorophyll-carotenoid-protein complexes such that thermal deactivation of the excited chlorophyll molecules is promoted (see below).

2.4.3.1   The Xanthophyll Cycle

In addition to chlorophyll, carotenoids constitute a major component of the pigment-protein complexes of the thylakoid membranes and some of these carotenoids, especially β-carotene, have long been considered to pro- tect the membrane against destructive events caused by overexcitation of the chlorophyll (Krinsky 1979; Siefermann-Harms 1987; Young 1991). Of particular interest with respect to NRD are the carotenoids violaxanthin (diepoxy-zeaxanthin), antheraxanthin (monoepoxy-zeaxanthin) and zeaxan- thin (dihydroxy-β-carotene), which are the carotenoid components of the xanthophyll cycle. When there is an excess of excitation energy, violaxan- thin (V) is de-epoxidized to zeaxanthin (Z) via antheraxanthin (A). This de- epoxidation occurs when light-driven proton pumping causes a lowering of the lumen pH because the enzyme catalyzing the V to Z de-epoxidation (violaxanthin de-epoxidase) has its optimum activity around pH 5 (Yamamoto 1979). Conversely, when light becomes limiting to photosyn- thesis, the lumen pH rises and Z is reepoxidized to V by the action of the enzyme zeaxanthin epoxidase (Hager 1980). A recent study confirms that V, A, and Z are all localized within the light-harvesting complexes of PS II as well as in PS I and indicates that deepoxidation of V to Z takes place in these complexes (Thayer and Björkman 1992).

2.4.3.2   Dynamics of Changes in the Epoxidation State and Pool Size of the Xanthophyll Cycle Pigments

Changes in the Epoxidation State. It is now well established that both the time course and the extent of NPQ development in leaves are strongly correlated with Z formation under a wide range of conditions, irrespective of whether excess light was imposed by exposure to high PFDs or to water stress, unfavorable leaf temperature, or subatmospheric CO_2 pressures (see reviews by Demmig-Adams 1990; Demmig-Adams and Adams 1992a). An example of this close relationship is illustrated in Fig. 2.12, which compares the steady state levels of NPQ and Z at different PFDs in a sun leaf of *Hedera canariensis* and in Fig. 2.13 which shows that the temperature dependence of the maximum rate of NPQ in a *Malva* leaf closely follows that of Z formation. (Also see the relationship between the rate of devel- opment of NPQ and Z formation in Fig. 2.18, below).

The comparison between the diurnal changes in NPQ and in V, A, and Z in leaves of cotton (Fig. 2.10) provides an example of the finding that a close relationship between the Z content and NPQ also holds during the daily course of natural variation in PFD in nature. That the epoxidation state of the xanthophyll cycle pigments changes in response to diurnal changes in PFD has been shown for many other C3 species (Demmig-Adams et al. 1989a; Thayer and Björkman 1990; Adams and Demmig-Adams 1992;

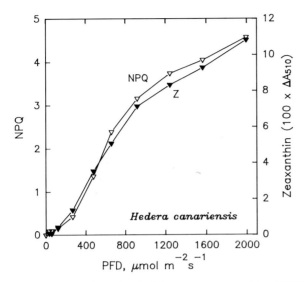

Fig. 2.12. Nonphotochemical fluorescence quenching (*NPQ*) and relative zeaxanthin content (Z) at steady state in relation to the PFD incident on a *Hedera canariensis* leaf. Z formation was monitored by increases in leaf absorbance at 510 nm (see Fig. 2.17). Leaf temperature ranged from 25 to 29 °C. (Based on data from Bilger and Björkman 1990)

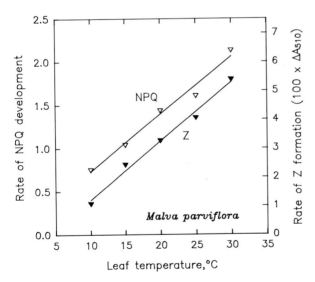

Fig. 2.13. Temperature dependence of the maximum *rates* of NPQ development and zeaxanthin formation (Z) upon exposure of predarkened *Malva parviflora* leaves to a PFD of 2000 µmol photons $m^{-2}s^{-1}$. These rates are the increases in NPQ and in the absorbance at 510 nm obtained at 1.5 min after the onset to the light exposure. (After Bilger and Björkman 1991)

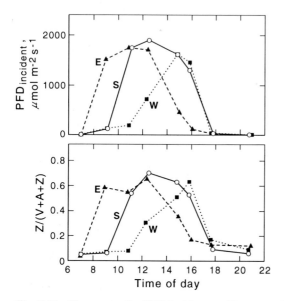

Fig. 2.14. Changes in the PFD incident on *Euonymus kiautschovicus* leaves having three different orientations and the ratios of zeaxanthin (Z) to the total xanthophyll cycle pool ($V + A + Z$) during the course of a clear, warm day in October, 1990 in Boulder, Colorado. *E* East-facing; *S* south-facing, *W* west-facing. (Based on data from Adams et al. 1992)

Adams et al. 1992; Demmig-Adams and Adams 1992b). A recent study on seven species of C4 plants (representing three subgroups of the C4 pathway) confirmed that PFD-dependent diurnal changes in the epoxidation state of the xanthophyll pigments take place also in C4 plants. It is noteworthy that changes in epoxidation state occurred in both mesophyll and bundle-sheath cells and that the extent of these changes was similar in the two cell types (Yamamoto and Björkman, unpubl. data).

Under full midday sunlight in the field, leaves with low rates of photosynthesis exhibited higher levels of zeaxanthin than did leaves with high rates of photosynthesis (Thayer and Björkman 1990; Adams and Demmig-Adams 1992; Demmig-Adams and Adams 1992b). This is also evident when the maximum $Z/(V + A + Z)$ ratios in *Euonymus* leaves shown in Fig. 2.14 (relatively low photosynthetic rate) are compared with those of the cotton leaves shown in Fig. 2.10 (high photosynthetic rate). Figure 2.14 also shows that each leaf, whether east-, south-, or west-facing, exhibited a maximal $Z/(V + A + Z)$ ratio precisely at the time when it received maximal PFD.

The responses of plants to a combination of high light and other environmental stresses such as drought, or chilling temperatures can involve a *sustained* high level of NRD throughout the day and night cycle (Adams et al. 1987; Björkman et al. 1988; Demmig et al. 1988). Such sustained NRD is accompanied by a retention of a considerable amount of Z in the leaves even at dawn (Demmig et al. 1988; Adams and Demmig-Adams, unpubl. data). This is in contrast to the daily courses seen under favorable conditions, where the Z formed during exposure to high daytime PFDs is gradually reepoxidized to V as the PFD decreases in the late afternoon and the amount of Z present in the leaves in the early morning is usually quite low.

Changes in the V + A + Z Pool Size. On the assumption that substitution of Z and A for V (directly or indirectly) determines the upper limit of NRD that can be reached in a leaf under excess light, this limit would ultimately depend on the total pool size of V + A + Z. Hence, one would expect the response to a prolonged exposure to excess light to be an increase in the size of this pool. There is now much evidence that this is indeed the case. All studies to date consistently show that the V + A + Z pool changes in response to the growth light regime, both under controlled conditions (Demmig-Adams et al. 1989c) and in natural situations (Thayer and Björkman 1990; Demmig-Adams and Adams 1992b) and also in response to drought (Demmig-Adams et al. 1988) and unfavorable temperatures (Bilger and Björkman 1991). Sun leaves consistently have a larger pool size than shade leaves (up to four times). The very small V + A + Z pool size present in extreme shade leaves, such as those of *Oxalis oregana*, could thus be the cause of the low maximum capacity for NRD (Fig. 2.9).

In leaves of plants normally found in open environments, the V + A + Z pool increases with increasing daily photon receipt during leaf development (see Adams et al. 1992). An example of this response is given in Fig. 2.15, which shows the carotenoid contents of cotton cotyledons growing in the field over a wide range of different total daily photon receipts. As was found in other species studied thus far, the V + A + Z pool size shows a very large and continuous increase with increased photon receipt up to full daily solar radiation while the changes in other carotenoids are much smaller. Figure 2.16 shows that also in leaves or cotyledons that have developed either under a low, or a high photon receipt, the V + A + Z pool size can respond dynamically to changes in light regime. When cotton cotyledons that had developed under the lowest light regime used in these field experiments were exposed to full daylight (Fig. 2.16, left panel), the pool size began to increase within several hours and within 6 days approached that of the cotyledons that had developed under full daily solar radiation. Perhaps more surprisingly, upon shading of cotyledons that had developed under the highest light regime (Fig. 2.16, right panel), the pool size exhibited a decline with a similar time course and extent as that of the rise shown in Fig. 2.16, left panel. Quantitatively similar results as shown for the cotyledons were obtained with cotton *leaves* when exposed to the same changes in light regime.

It is also noteworthy that changes in the V + A + Z pool size appear not to respond to changes in PFD as such but rather to changes in excess PFD.

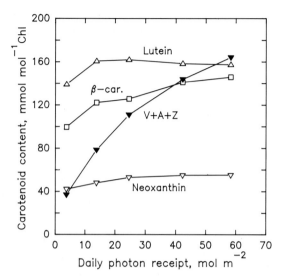

Fig. 2.15. Influence of the daily photon receipt during the development of cotton cotyledons on the pool size of the xanthophyll cycle pigments ($V + A + Z$) and the contents of other major carotenoids, expressed on a chlorophyll a + b basis. (Chlorophyll content per unit leaf area was little affected over this range of photon receipts.) Experiments were conducted in the field at Stanford, California in June-July, 1989. Attenuation of the photon receipt was obtained by shading the seedlings with neutral-density filters of varying transmittance. (Original data by O. Björkman and C. Shih)

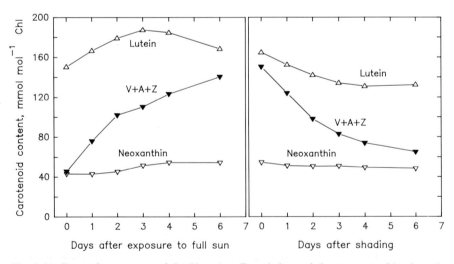

Fig. 2.16. Dynamic responses of the $V + A + Z$ pool size and the contents of lutein and neoxanthin to sudden changes in light regime. In the *left panel* the daily photon receipt was increased from 3.6 to 58 mol photons $m^{-2} d^{-1}$ and in the *right panel* it was decreased from 58 to 3.6 mol photons $m^{-2} d^{-1}$. Other conditions were as in Fig. 2.15. (Original data by O. Björkman and C. Shih)

Thus, one may speculate that the events leading to an increased carotenoid biosynthesis are triggered by a factor associated with excessive light energy such as a low epoxidation state of the xanthophyll cycle. The finding that, among the bulk pigments of the thylakoid membrane, the xanthophyll cycle components are the only ones that exhibit large dynamic changes in response to excessive light both in the short term (reversible deepoxidation) and in the longer term (pool size adjustment) provides indirect evidence that this cycle plays a role in the protection against excess light.

2.4.4 Mechanism of Nonradiative Dissipation

In Fig. 2.12 and Fig. 2.13 the de-epoxidation of V to Z was monitored in intact leaves by measurements of changes in absorbance. These measurements are based on the fact that V, A, and Z differ in their absorption spectra. The spectral changes caused by light-induced Z formation in a *Prunus* leaf are shown in Fig. 2.17, curve a. The maximum change caused by the deepoxidation occurs around 510 nm. Another absorbance change which peaks around 540 nm (Fig. 2.17, curve b) is thought to reflect the conformational change of the thylakoid membrane considered to be involved in

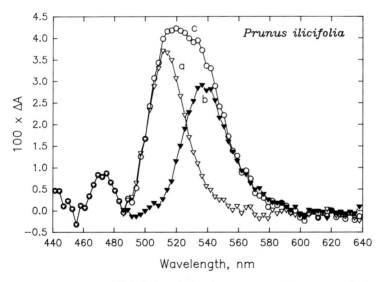

Fig. 2.17. Spectra of light-induced absorbance changes in a *Prunus ilicifolia* leaf caused by de-epoxidation of violaxanthin to zeaxanthin (curve *a*) and by "light-scattering" changes thought to reflect a conformational change in the thylakoid membranes (curve *b*). The total spectral change caused by exposure of a predarkened *Prunus* leaf to a PFD of 650 μmol photons $m^{-2} s^{-1}$ is shown by curve *c*. Curve *b* is the spectral change observed during the relaxation upon return to darkness. Measurements were made with the instrumentation described by Brugnoli and Björkman (1992). (Original data by O. Björkman and C. Shih)

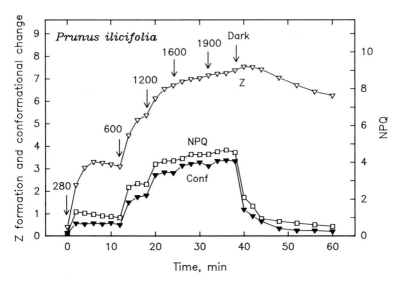

Fig. 2.18. Time course of changes in NPQ and of changes in absorbance caused by Z formation (*Z*) and putative conformational changes (*Conf*) in response to stepwise increases in the PFD incident on a *Prunus ilicifolia* leaf. The PFD is indicated by the *number above each arrow*. The values for the aborbance changes are those obtained at 510 nm (Z formation) and at 541 nm (conformational change). The latter values were adjusted for the direct contribution of Z formation to the total absorbance change at 541 nm (see Fig. 2.17). (Original data by B. Demmig-Adams and O. Björkman)

nonradiative dissipation. It is therefore possible to simultaneously and non-destructively monitor changes in the epoxidation state of the xanthophyll cycle, putative membrane conformational changes, and NPQ in the same leaf. An example of the results of such measurements is given in Fig. 2.18, which shows the responses obtained when a *Prunus* leaf was exposed to stepwise increases in PFD. Clearly, there is a close relationship between the kinetics of zeaxanthin formation, the development of the putative confor-mational change, and the development of NPQ. However, it is also evident from the relaxation kinetics seen after darkening of the leaf that both the putative confirmational change and NPQ exhibit a relatively rapid relaxation even though there was little change in Z. Similar results have been obtained in studies with leaves of other species (e.g., Bilger et al. 1989; Demmig-Adams et al. 1989b; Bilger and Björkman 1991). Several studies with isolated chloroplasts have also shown that the presence of a high Z level alone is insuffucent to cause a high NPQ level (Gilmore and Yamamoto 1991, 1992). One might therefore ask whether the close relationship obtained between Z formation and NPQ is simply a consequence of the fact that both have similar dependences on lumen pH. For a number of reasons, some of which are discussed below, the answer is most probably no.

Perhaps the strongest evidence for a causal involvement of the xanthophyll cycle in mediating energy dissipation in the pigment bed (NRD) comes from studies in which V to Z de-epxidation was blocked by application of dithiothreitol (DTT). DTT is a powerful inhibitor of violaxanthin deepoxidase (Yamamoto and Kamite 1972). Application of a DTT solution to leaves through the cut petioles effectively prevented Z formation and caused a strong suppression of NPQ and an increased degree of reaction center closure (Bilger et al. 1989; Bilger and Björkman 1990, 1991; Demmig-Adams et al. 1990). Such blocking of Z formation also resulted in an apparently increased susceptibility to photoinhibition during longer-term exposure to high PFD (Winter and Königer 1989; Adams et al. 1990; Bilger and Björkman 1990). However, DTT treatments had no detectable short-term effects on photosynthetic O_2 or CO_2 exchange, nor on the efficiency of energy conversion in PS II. Moreover, subsequent studies have shown that application of DTT to cotton leaves in which high levels of zeaxanthin are already present does not suppress the development of NPQ nor of the putative conformational change as determined by absorbance changes (Bilger and Björkman, unpubl. data). Similar results were obtained with isolated lettuce chloroplasts by Gilmore and Yamamoto (1992). No significant NPQ development was obtained upon exposure to conditions that caused the proton gradient to rise if Z formation was prevented by DTT and the preparations initially did not contain any Z. However, full NPQ development occurred when DTT was applied to preparations in which high levels of Z had been preformed. The above studies indicate that the main effect of DTT on the development of NPQ was through its prevention of Z formation alone both in leaves and isolated chloroplasts. Moreover, according to Gilmore and Yamamoto (pers. comm.), at any given lumen proton concentration NPQ was found to be linearly dependent on the Z concentration or perhaps even more closely on the Z + A concentration, in both lettuce and pea chloroplasts. Conversely, at any given Z or Z + A concentration, NPQ was linearly dependent on the lumen proton concentration. Hence, NPQ was directly proportional to the product of $[H^+]$ and [Z] or [Z + A].

As was discussed above, in the presence of excess light the lumen proton concentration rises, causing a high activity of violaxanthin deepoxidase and subsequent depoxidation of V to A and Z. A unifying hypothesis is that the resulting substitution of Z (and perhaps also A) for V in thylakoid membranes, together with the increased proton concentration, causes a conformational change which alters the association among the pigment molecules. As a result of this change, NRD is favored over fluorescence and energy transfer to the reaction centers, thus causing an increase in NPQ. Horton et al. (1991) proposed that protonation of the lumen surface causes an aggregation of the major light harvesting pigment-protein complex of PS II and that this aggregation is amplified by Z. While it still needs to be established whether Z is involved directly or indirectly in NRD, this is consistent with the finding that, upon depoxidation, Z and A do indeed replace V in the PS II light-harvesting complexes (Thayer and Björkman 1992). The finding that deepoxidation of V to Z also takes place in PS I may indicate that a simultaneous increase in NRD occurs in this photosystem as well.

2.5 Conclusions

In this chapter we have documented that plants have evolved a remarkable array of mechanisms that enable them to regulate both the interception and the dissipation of light energy, both in the short term (second to minutes) and in the longer term (hours to days). Some of these mechanisms, such as rapid leaf movements, are present in relatively few species, while other processes, such as nonradiative dissipation, are most probably ubiquitous in green plants. At present, our understanding of the molecular mechanism of nonradiative energy dissipation in the pigment bed is only rudimentary; current hypotheses are likely to be subject to change and yet other mechanisms of light regulation may also be discovered. It is certain, however, that if one has seen one mechanism of light regulation, one has not seen them all.

The fact that such a wide array of photoprotective processes have evolved is fully consistent with the assumption that overexcitation of the photosynthetic system is very harmful and must be avoided. Moreover, laboratory experiments have provided strong evidence that such overexcitation can cause injury to the photosystems, especially to the PS II reaction center complex. However, little is known about the actual occurrence of photoinhibitory damage in nature and even less about the consequences with regard to primary productivity. Our present guess is that photoinhibitory damage is uncommon in natural plant stands. Responses that in the past were thought to be indicative of photoinhibitory damage (reduced photon yield of CO_2 uptake and O_2 evolution and quenching of the maximum fluorescence yield) in many cases now appear to be reflections of the operation of photo-protective processes.

References

Adams WW III, Demmig-Adams B (1992) Operation of the xanthophyll cycle in higher plants in response to diurnal changes in incident sunlight. Planta 186: 390–398
Adams WW III, Smith SD, Osmond CB (1987) Photoinhibition of the CAM succulent *Opuntia basilaris* growing in Death Valley: evidence from 77K fluorescence and quantum yield. Oecologia 71: 221–228

Adams WW III, Díaz M, Winter K (1989) Diurnal changes in photochemical efficiency, the reduction state of Q, radiationless energy dissipation, and non-photochemical fluorescence quenching in cacti exposed to natural sunlight in northern Venezuela. Oecologia 80: 553–561

Adams WW III, Demmig-Adams B, Winter K (1990) Relative contributions of zeaxanthin-related and zeaxanthin-unrelated types of "high-energy-state" quenching of chlorophyll fluorescence in spinach leaves exposed to various environmental conditions. Plant Physiol 92: 302–309

Adams WW III, Volk M, Hoehn A, Demmig-Adams B (1992) Leaf orientation and the response of the xanthophyll cycle to incident light. Oecologia 90: 404–410

Anderson JM, Chow WS, Goodchild DJ (1988) Thylakoid membrane organisation in sun/shade acclimation. Aust J Plant Physiol 15: 11–26

Begg JE (1980) Morphological adaptations of leaves to water stress. In: Turner NC, Kramer PJ (eds) Adaptation of plants to water and high temperature stress. Wiley-Interscience, New York, pp 33–42

Begg JE, Torssell BWR (1974) Diaphotonastic and parahelionastic leaf movements in *Stylosanthes humilis* HBK (Townsville Stylo). R Soc N Z Bull 12: 277–283

Bilger W, Björkman O (1990) Role of the xanthophyll cycle in photoprotection elucidated by measurements of light-induced absorbance changes, fluorescence and photosynthesis in leaves of *Hedera canariensis*. Photosynth Res 25: 173–185

Bilger W, Björkman O (1991) Temperature dependence of violaxanthin de-epoxidation and non-photochemical fluorescence quenching in intact leaves of *Gossypium hirsutum* L. and *Malva parviflora* L. Planta 184: 226–234

Bilger W, Björkman O, Thayer SS (1989) Light-induced spectral absorbance changes in relation to photosynthesis and the epoxidation state of xanthophyll cycle components in cotton leaves. Plant Physiol 91: 542–551

Björkman O (1981) Responses to different quantum flux densities. In: Lange OL, Nobel PS, Osmond CB, Ziegler H (eds) Physiological plant ecology I. Encyclopedia of plant physiology, NS, vol 12A. Springer, Berlin Heidelberg New York, pp 57–107

Björkman O, Demmig B (1987) Photon yield of O_2 evolution and chlorophyll fluorescence characteristics at 77 K among vascular plants of diverse origins. Planta 170: 489–504

Björkman O, Ludlow MM (1972) Characterization of the light climate on the floor of a Queensland rainforest. Carnegie Inst Wash Yearb 71: 85–94

Björkman O, Powles SB (1981) Leaf movement in the shade species *Oxalis oregana*. I. Response to light level and light quality. Carnegie Inst Wash Yearb 80: 59–62

Björkman O, Schäfer C (1989) A gas exchange-fluorescence analysis of photosynthetic performance of a cotton crop under high-irradiance stress [Extended abstract.] Philos Trans R Soc Lond B 323: 309–311

Björkman O, Demmig B, Andrews TJ (1988) Mangrove photosynthesis: response to high-irradiance stress. Aust J Plant Physiol 15: 43–61

Brugnoli E, Björkman O (1992) Chloroplast movement in leaves: influence on chlorophyll fluorescence and measurements of light-induced absorbance changes related to ΔpH and zeaxanthin formation. Photosynth Res 32: 23–35

Chazdon RL, Pearcy RW (1991) The importance of sunflecks for forest understory plants. BioScience 41: 760–766

Demmig-Adams B (1990) Carotenoids and photoprotection in plants. A role for the xanthophyll zeaxanthin. Biochim Biophys Acta 1020: 1–24

Demmig-Adams B, Adams WW III (1992a) Photoprotection and other responses of plants to high light stress. Annu Rev Plant Physiol Plant Mol Biol 43: 599–626

Demmig-Adams B, Adams WW III (1992b) Carotenoid composition in sun and shade leaves of plants with different life forms. Plant Cell Environ 15: 411–419

Demmig B, Winter K, Krüger A, Czygan F-C (1988) Zeaxanthin and the heat dissipation of excess light energy in *Nerium oleander* exposed to a combination of high light and water stress. Plant Physiol 87: 17–24

Demmig-Adams B, Adams WW III, Winter K, Meyer A, Schreiber U, Pereira JS, Krüger A, Czygan F-C, Lange OL (1989a) Photochemical efficiency of photosystem II,

photon yield of O_2 evolution, photosynthetic capacity, and carotenoid composition
during the "midday depression" of net CO_2 uptake in *Arbutus unedo* growing in
Portugal. Planta 177: 377–387

Demmig-Adams B, Winter K, Krüger A, Czygan F-C (1989b) Zeaxanthin and the
induction and relaxation kinetics of the dissipation of excess excitation energy in leaves
in 2% O_2, 0% CO_2. Plant Physiol 90: 887–893

Demmig-Adams B, Winter K, Winkelmann K, Krüger A, Czygan F-C (1989c) Photosynthetic characteristics and the ratios of chlorophyll, β-carotene, and the components
of the xanthophyll cycle upon a sudden increase in growth light regime in several plant
species. Bot Acta 102: 319–325

Demmig-Adams B, Adams WW III, Heber U, Neimanis S, Winter K, Krüger A, Czygan
F-C, Bilger W, Björkman O (1990) Inhibition of zeaxanthin formation and of rapid
changes in radiationless energy dissipation by dithiothreitol in spinach leaves and
chloroplasts. Plant Physiol 92: 293–301

Ehleringer JR (1982) The influence of water stress and temperature on leaf pubescence
development in *Encelia farinosa*. Am J Bot 69: 670–675

Ehleringer JR (1988) Changes in leaf characteristics of species along elevational gradients
in the Wasatch Front, Utah. Am J Bot 75: 680–689

Ehleringer JR, Björkman O (1978) Pubescence and leaf spectral characteristics in a desert
shrub, *Encelia farinosa*. Oecologia 36: 151–162

Ehleringer JR, Forseth I (1980) Solar tracking by plants. Science 210: 1093–1098

Ehleringer JR, Mooney HA, Gulmon SL, Rundel PW (1981) Parallel evolution of leaf
pubescence in *Encelia* in coastal deserts of North and South America. Oecologia 49:
38–41

Forseth I, Ehleringer JR (1980) Solar tracking response to drought in a desert annual.
Oecologia 44: 159–163

Gilmore AM, Yamamoto HY (1991) Zeaxanthin formation and energy-dependent fluorescence quenching in pea chloroplasts under artificially-mediated linear and cyclic
electron transport. Plant Physiol 96: 635–643

Gilmore AM, Yamamoto HY (1992) Dark induction of zeaxanthin-dependent nonphotochemical fluorescence quenching mediated by ATP. Proc Natl Acad Sci USA 89:
1899–1903

Hager A (1980) The reversible, light-induced conversions of xanthophylls in the chloroplast. In: Czygan F-C (ed) Pigments in plants. Fischer, Stuttgart, pp 57–79

Haupt W, Scheuerlein R (1990) Chloroplast movement. Plant Cell Environ 13: 595–614

Horton P, Ruban AV, Rees D, Pascal AA, Noctor G, Young AJ (1991) Control of
the light-harvesting function of chloroplast membranes by aggregation of the LHCII
chlorophyll-protein complex. FEBS Lett 292: 1–4

Inoue Y, Shibata K (1974) Comparative examination of terrestrial plants leaves in terms
of light-induced absorption changes due to chloroplast rearrangements. Plant Cell
Physiol 15: 717–721

Koller D (1990) Light-driven leaf movements. Plant Cell Environ 13: 615–632

Koller D, Shak T (1990) Light-driven movements in the solar-tracking leaf of *Lupinus
palaestinus* Boiss. Photochem Photobiol 52: 187–196

Krause GH, Weis E (1991) Chlorophyll fluorescence and photosynthesis: the basics.
Annu Rev Plant Physiol Plant Mol Biol 42: 313–349

Krause GH, Vernotte C, Briantais J-M (1982) Photoinduced quenching of chlorophyll
fluorescence in intact chloroplasts and algae. Resolution into two components. Biochim
Biophys Acta 679: 116–124

Krinsky NI (1979) Carotenoid protection against oxidation. Pure Appl Chem 51: 649–660

Ludlow MM, Björkman O (1984) Paraheliotropic leaf movement in Siratro as a protective mechanism against drought-induced damage to primary photosynthetic reactions:
damage by excessive light and heat. Planta 161: 505–518

Mooney HA, Ehleringer JR, Björkman O (1977) The energy balance of leaves of the
evergreen desert shrub *Atriplex hymenelytra*. Oecologia 29: 301–310

Neubauer C, Schreiber U (1989) Photochemical and non-photochemical quenching of
chlorophyll fluorescence induced by hydrogen peroxide. Z Naturforsch 44c: 262–270

Neubauer C, Yamamoto HY (1992) Mehler-peroxidase reaction mediates zeaxanthin formation and zeaxanthin-related fluorescence quenching in intact chloroplasts. Plant Physiol 99: 1354–1361

Powles SB, Björkman O (1981) Leaf movement in the shade species *Oxalis oregana*. II. Role in protection against injury by intense light. Carnegic Inst Wash Yearb 80: 63–66

Schäfer C, Björkman O (1989) Relationship between photosynthetic energy conversion efficiency and chlorophyll fluorescence quenching in upland cotton (*Gosspyium hirsutum* L.). Planta 17: 367–376

Siefermann-Harms D (1987) The light harvesting and protective functions of carotenoids in photosynthetic membranes. Physiol Plant 69: 561–568

Thayer SS, Björkman O (1990) Leaf xanthophyll content and composition in sun and shade determined by HPLC. Photosynth Res 23: 331–343

Thayer SS, Björkman O (1992) Carotenoid distribution and deepoxidation in thylakoid pigment-protein complexes from cotton leaves and bundle-sheath cells of maize. Photosynth Res 33: 213–225

von Caemmerer S, Farquhar GD (1981) Some relationships between the biochemistry of photosynthesis and the gas exchange of leaves. Planta 153: 376–387

Wainwright CM (1977) Suntracking and related leaf movements in a desert lupin (*Lupinus arizonicus*). Am J Bot 64: 1032–1041

Winter K, Königer M (1989) Dithiothreitol, an inhibitor of violaxanthin de-epoxidation, increases the susceptibility of leaves of *Nerium oleander* L. to photoinhibition of photosynthesis. Planta 180: 24–31

Yamamoto HY (1979) Biochemistry of the violaxanthin cycle in higher plants. Pure Appl Chem 5: 639–648

Yamamoto HY, Kamite L (1972) The effects of dithiothreitol on violaxanthin deepoxidation and absorbance changes in the 500-nm region. Biochim Biophys Acta 267: 538–543

Young AJ (1991) The photoprotective role of carotenoids in higher plants. Physiol Plant 83: 702–708

3 Chlorophyll Fluorescence as a Nonintrusive Indicator for Rapid Assessment of In Vivo Photosynthesis

U. Schreiber, W. Bilger, and C. Neubauer

3.1 Introduction

In the past, ecophysiologically oriented photosynthesis research has been governed by gas-exchange measurements, mainly involving sophisticated (and costly) systems for simultaneous detection of CO_2 uptake and H_2O evaporation (see, e.g., Field et al. 1989). With the help of these methods, fundamental knowledge on in situ photosynthesis has been gained. Only recently, progress has been made in the development of alternative practical methods for nonintrusive assessment of in vivo photosynthesis which have the potential of not only evaluating overall quantum yield and capacity, but also allowing insights into the biochemical partial reactions and the partitioning of excitation energy (see, e.g., Snel and van Kooten 1990). As a consequence, photosynthesis research at the level of regulatory processes has been greatly stimulated, leading to important new concepts (see reviews by Foyer et al. 1990; Demmig-Adams 1990; Melis 1991; Allen 1992). In particular, chlorophyll fluorescence has evolved as a very useful and informative indicator for photosynthetic electron transport in intact leaves, algae, and isolated chloroplasts (reviews by Briantais et al. 1986; Renger and Schreiber 1986; Schreiber and Bilger 1987, 1992; Krause and Weis 1991; Karukstis 1991).

Chlorophyll fluorescence provides a large signal, which can be measured at some distance from the sample investigated. Furthermore, highly selective modulation fluorometers are now available, with which fluorescence yield can be measured in full sunlight, and which are small and low in power consumption, such that they are well suited for field measurements. As a result of intensive research in a number of laboratories, methods are now available by which the fluorescence information can be quantitatively analyzed and evaluated such that quantum yields and relative electron transport rates can be obtained. Computerized systems are available which relieve the user from tedious measuring protocols and calculations of the relevant parameters derived from fluorescence data. Hence, ecophysiologists are offered a convenient alternative to gas exchange measurements.

The present chapter is not another general review on chlorophyll fluorescence and the large amount of literature published recently on this topic (for this purpose, see, e.g., Krause and Weis 1991; Schreiber and Bilger 1992). Also, this chapter is not written for specialists in the field of fluorescence research. Rather, it is intended to give an introduction to nonspecialists on major aspects of the chlorophyll fluorescence indicator function, on the methods developed for quenching analysis and on the physiologically relevant information obtained by fluorescence measurements.

3.2 Indicator Function of Chlorophyll Fluorescence

Contrary to chlorophyll in solution, chlorophyll in vivo displays large changes in fluorescence yield upon illumination (Kautsky and Hirsch 1931). In vivo, chlorophyll exists in the form of pigment protein complexes which are embedded in the thylakoid membrane and funnel their excitation energy into the reaction centers (P_{680} and P_{700}), where energy conversion by charge separation takes place (see scheme in Fig. 3.1). Two major factors cause changes in fluorescence yield: the rate of photochemical energy conversion and the rate of nonradiative energy dissipation. For reasons so far unknown, at room temperature the variable fluorescence originates almost exclusively from PS II. Hence, fluorescence changes reflect primarily the state of PS II.

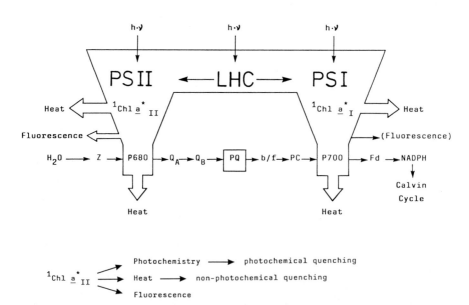

Fig. 3.1. Schematic illustration of primary energy conversion in photosynthesis which governs in vivo chlorophyll fluorescence yield. Variable fluorescence originates almost exclusively from PS II. Maximal fluorescence yield is lowered by photochemical charge separation and heat dissipation. See text for further explanations

In addition, variations in energy distribution between the two photosystems will also influence fluorescence yield. As PS I is essentially nonfluorescent, any increase in energy transfer from PS II to PS I may be considered equivalent to nonradiative dissipation. Fluorescence emission competes with photochemistry and heat dissipation. Therefore, two basic types of fluorescence quenching, photochemical and nonphotochemical, can be distinguished (see Sect. 3.5).

3.3 Rapid Fluorescence Induction Kinetics

The existence of two fundamentally different types of fluorescence quenching has until recently prevented the practical use of chlorophyll fluorescence in steady-state investigations. However, provided that changes in heat dissipation are relatively slow, the *rapid* fluorescence induction transients occurring upon a dark-light transition can be interpreted in terms of changes in photochemical quenching only, which is determined by the concentration of open reaction centers (Kautsky et al. 1960; Duysens and Sweers 1963). When all reaction centers are open (Q_A fully oxidized), the minimal fluorescence yield, F_o, is observed, whereas the maximal fluorescence yield, F_m, is found when all centers are closed (Q_A fully reduced). The difference between F_o and F_m is called variable fluorescence, F_v. For a wide variety of dark-adapted plants the F_m/F_o ratio amounts to values of 5–6, corresponding to F_v/F_m values of 0.8–0.833 (Björkman and Demmig 1987).

When a dark-adapted sample is illuminated with actinic light, Q_A reduction rate initially is higher than the rate of reoxidation by plastoquinone and by PS I activity. The resulting fluorescence rise reflects the exhaustion of the PS II acceptor pool (Fig. 3.2). A measure for the acceptor pool size is given by the area between the induction curve and the F_m-line (Murata et al. 1966). In the presence of DCMU, which blocks electron transfer from Q_A to Q_B, a steep fluorescence rise with a complementary area corresponding to one electron is observed. In a control sample, the area amounts to approximately 20 electron equivalents.

The relationship between Q_A reduction and variable fluorescence yield is nonlinear (Joliot and Joliot 1964), presumably due to cooperativity between PS II units at the antenna level. The increase in fluorescence yield is smaller than predicted from the extent of Q_A reduction, as excitons absorbed in closed units may be transferred to neighboring open units (see Fig. 3.3). The degree of cooperativity between PS II units in vivo is under debate (Keuper and Sauer 1989). The issue is complicated by the existence of PS II heterogeneity, with PS IIα units displaying a high extent of cooperativity and PS II β operating as separate units (Melis and Schreiber 1979; see review by Melis 1991).

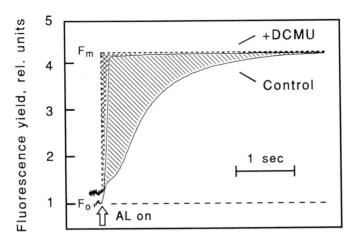

Fig. 3.2. Rapid induction kinetics upon onset of continuous actinic illumination. Spinach chloroplasts with and without 10^{-5}M DCMU present in the suspension medium. Modulated fluorescence yield measured with a PAM Fluorometer (Walz). Already in the dark, DCMU addition causes some increase in the minimal fluorescence yield, F_o. Maximal yield, F_m, is induced upon illumination showing largely different kinetics with the two samples. The *hatched areas*, bound by the *broken line* and the *two curves*, correspond to the size of the acceptor pools available to PS II in control and DCMU-inhibited samples. *AL* 100 μE m^{-2} s^{-1} 650 nm actinic light

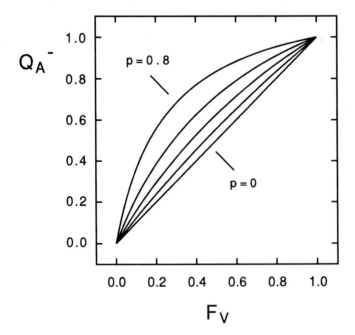

Fig. 3.3. Theoretical relationship between reduction of primary PS II acceptor, Q_A, and variable fluorescence yield, F_v, based on the function $F_v = (1 - p)Q_A^-/(1 - Q_A^- \cdot p)$, where the parameter p corresponds to different probabilities of energy transfer between PS II units (Joliot and Joliot 1964). Note: in reality, the relationship is more complex due to PS II heterogeneity. See text

The fluorescence rise upon onset of actinic illumination is complex. At moderate light intensities, a rapid initial rise is separated by an intermediate level (or dip) from a major slow rise phase. At high intensity, the rise becomes polyphasic (Delosme 1967; Neubauer and Schreiber 1987; Schreiber and Neubauer 1987). It has been proposed that the fluorescence rise to the intermediate level reflects closure of PS IIβ centers (Melis 1985, 1991; Govindjee 1990). However, an initial fluorescence rise in the presence of an oxidized plastoquinone pool can also be predicted from known properties of charge accumulation at the two-electron gate Q_B (Velthuys and Amesz 1974). In this case, illumination initially causes the semiquinone anion Q_B^- to accumulate, in equilibrium with some Q_A^-, before $Q_B^=$ is formed, which can then be reoxidized by the plastoquinone pool.

At saturating light intensities, when the charge separation rate exceeds the rate of the Q_A–Q_B electron transfer, the I level is raised considerably, but not to the F_m level (Neubauer and Schreiber 1987). The fluorescence increase in strong light consists of so-called "photochemical" and "thermal" phases, with the latter comprising 30–50% of the total rise (see Fig. 3.4). Whereas the photochemical phase follows the intensity-times-time law (Delosme 1967), the rate of the thermal phases, which is limited by electron donation from the PS II donor side (Schreiber and Neubauer 1987), saturates at approximately $1500\,\mu E\,m^{-2}\,s^{-1}$. A minimal time of about 200 ms is required to induce F_m by saturating light. This aspect is of practical relevance for the separation of photochemical and nonphotochemical quenching components by the saturation pulse method (see Sect. 3.5).

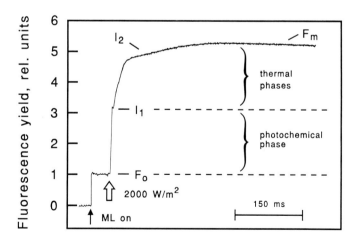

Fig. 3.4. Multiphasic fluorescence rise in saturating white light. Intact spinach leaf; measurement with PAM Fluorometer. The rate of the fluorescence rise from I_1 to F_m (thermal phases) cannot be further increased by higher light intensities. It is limited by electron donation from the H_2O-splitting site. ML Weak modulated measuring light

3.4 Slow Fluorescence Induction Kinetics and Fluorescence Quenching Under Steady-State Conditions

With dark-adapted samples, after onset of illumination the initial fluorescence increase to a peak level is followed by a slower decline with complex kinetics (see Fig. 3.5). This decline reflects the activation of photosynthetic electron flow, involving a variety of regulatory processes. It has been known for a long time that the secondary fluorescence decline requires the presence of dioxygen (Kautsky and Franck 1943; Munday and Govindjee 1969; Schreiber et al. 1971). Recent work has revealed that ascorbate peroxidase activity (see review by Asada and Takahashi 1987) plays an essential role in the activation of electron flow (Neubauer and Schreiber 1989a,b; Schreiber et al. 1991). Calvin cycle activity requires proton gradient formation and stroma alkalization for enzyme activation and ATP formation. It is O_2-dependent electron flow, i.e., the combination of O_2 reduction and consequent H_2O_2 reduction, which appears to be largely responsible for ΔpH formation during the induction period (see also Sect. 3.7).

Once a ΔpH is formed, this not only leads to ATP formation and induction of assimilatory electron flow, but also to conformational changes at the levels of pigment protein complexes and reaction centers (see review by Krause and Weis 1991), as well as to the de-epoxidation of violaxanthin to form zeaxanthin (see review by Demmig-Adams 1990), all of which are closely linked with an increase of nonradiative energy dissipation and the development of so-called nonphotochemical quenching (see also Sect. 3.7). Hence, the secondary fluorescence decline and the low steady-state fluorescence yield eventually reached are determined by two overlapping, fundamentally different quenching components, photochemical (due to

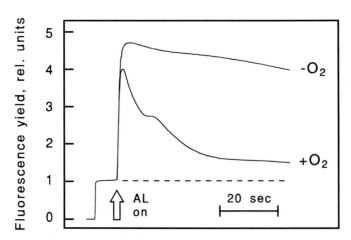

Fig. 3.5. Slow fluorescence induction kinetics in presence and absence of O_2. Dark-adapted spinach leaf: actinic light intensity $250\,\mu E\,m^{-2}s^{-1}$

assimilatory and nonassimilatory electron flow) and nonphotochemical (due to increased heat formation), which have to be separated in order to interpret the fluorescence information. In practice, separation and quantification of the two types of quenching is achieved by the saturation pulse method (Schreiber et al. 1986).

3.5 The Saturation Pulse Method

The rationale of the saturation pulse method is simple: upon application of a sufficiently strong light pulse, Q_A is fully reduced, and hence photochemical fluorescence quenching becomes suppressed; the remaining quenching is nonphotochemical (Bradbury and Baker 1981). In practice, this method requires a particular measuring technique, with exceptional selectivity and sensitivity. On the one hand, it is desirable that the measuring light has no actinic effect, so that the minimal yield, F_o, can be monitored. On the other hand, the saturation pulse has to be very strong to ensure complete Q_A reduction even under conditions of maximal electron transport rates, as encountered in the field. For this purpose, a fluorometer based on a new modulation principle (Pulse Amplitude Modulation, PAM) was developed

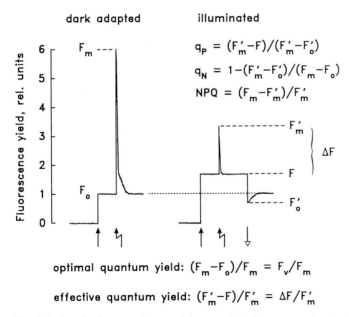

$$q_P = (F'_m - F)/(F'_m - F'_o)$$

$$q_N = 1 - (F'_m - F'_o)/(F_m - F_o)$$

$$NPQ = (F_m - F'_m)/F'_m$$

optimal quantum yield: $(F_m - F_o)/F_m = F_v/F_m$

effective quantum yield: $(F'_m - F)/F'_m = \Delta F/F'_m$

Fig. 3.6. Standard nomenclature of characteristic fluorescence levels, definition of quenching coefficients, and useful expressions derived from fluorescence parameters. Illustration of fluorescence information obtained with the help of the saturation pulse method. For the sake of clarity, in this presentation F_o quenching is exaggerated

which tolerates a ratio of $1:10^6$ between modulated fluorescence and nonmodulated background signal (Schreiber 1986; Schreiber et al. 1986; Schreiber and Bilger 1987).

In Fig. 3.6, schematic traces of a measurement of modulated chlorophyll fluorescence are shown, with the characteristic fluorescence levels and quenching coefficients being defined in agreement with a proposal for standard nomenclature at a recent fluorescence workshop (van Kooten and Snel 1990).

While the sample is still in the dark-adapted state, the minimal and maximal yields, F_o and F_m, are determined. The ratio $(F_m - F_o)/F_m = F_v/F_m$ is a convenient measure of the potential maximal PS II quantum yield of a given sample (see reviews by Butler 1978; Björkman 1987). During illumination, the fluorescence yield, F, undergoes complex changes. With the help of saturation pulses the changed levels of maximal yield, F_m', are determined, with $F_m - F_m'$ representing nonphotochemically quenched fluorescence and $F_m' - F$ representing photochemically quenched fluorescence.

So-called quenching coefficients q_P and q_N have been defined (originally called q_Q and q_E) for quantification of photochemical and nonphotochemical quenching, respectively (Dietz et al. 1985; Schreiber et al. 1986). In the original definitions it was assumed that only variable fluorescence is quenched nonphotochemically. However, it has been shown that F_o is also lowered by ΔpH-dependent nonphotochemical quenching (Bilger and Schreiber 1986). Therefore, correct calculation of q_P and q_N requires previous determination of F_o'.

F_o' can be determined upon sample darkening and application of weak far-red background light for PS I-driven Q_A oxidation. Knowledge of F_o' is indispensable in order to obtain information on the extent of PS II "openness" via q_P-calculation. On the other hand, nonradiative dissipation can be also assessed by the following expression which does not involve F_o' (Bilger and Björkman 1990):

$$NPQ = (F_m - F_m')/F_m'. \qquad (2)$$

It should be realized that this expression of nonphotochemical quenching is based on the matrix model of antenna organization (see reviews by Lavorel and Etienne 1977; Butler 1978), assuming the existence of nonphotochemically quenching traps, e.g., zeaxanthin (Demmig-Adams 1990; see also Björkman and Demmig, Chap. 2, this Vol.). The actual mechanism of nonphotochemical quenching is still controversial (see Sect. 3.6).

3.6 Quantum Yield and Rate Determination by Fluorescence Measurements

Weis and Berry (1987) derived an empirical equation, containing only fluorescence parameters, which gave an excellent linear correlation between electron transport rate calculated from fluorescence and CO_2 fixation rate:

$$J = I \cdot q_P(m - b \cdot q_N), \tag{1}$$

where J is the rate of electron transport, I is the incident photon flux density, and m and b are empirically derived, plant-specific constants. Hence it follows for the quantum yield of PS II:

$$\phi_{II} = J/I = q_P (m - b \cdot q_N). \tag{2}$$

Then $m - b \cdot q_N$ corresponds to the quantum yield of open PS II centers, ϕ_P:

$$\phi_P = \phi_{II}/q_P = m - b \cdot q_N. \tag{3}$$

From this expression, it is apparent that the constant m corresponds to the maximal quantum yield of a dark-adapted sample (for $q_N = 0$) and that $m - b$ represents the minimal quantum yield (at $q_N = 1$). Actually, more recent work has suggested that $m - b$ may amount to zero (Weis and Lechtenberg 1989; Snel et al. 1990; Krieger 1992). In this case, the expression for ϕ_P is further simplified:

$$\phi_P = \phi_{Po} \cdot (1 - q_N), \tag{4}$$

where ϕ_{Po}, which is identical to the constant m, represents the maximal quantum yield with all PS II centers open, in practice showing values around 0.3–0.35 (Weis and Berry 1987; Sharkey et al. 1988). It should be noted that these values relate to the *overall* quantum yield considering *incident* quantum flux density. Hence, a value of 0.35 corresponds to a *PS II* quantum yield of approximately $0.35 \times 2 \times 0.84 = 0.833$, where the factor 2 accounts for the fact that two photoreactions are involved and the factor 0.84 takes into consideration that only 84% of the incident light is absorbed (see also Sect. 3.9). The resulting relationship is not only of great practical importance, but also has theoretical implications concerning the mechanism of nonphotochemical quenching. According to the definition of q_N (see Fig. 3.6),

$$1 - q_N = (F_m' - F_o')/(F_m - F_o) = F_v'/F_v, \tag{5}$$

and upon substitution in Eq. (4)

$$\phi_P = \phi_{Po} \cdot F_v'/F_v. \tag{6}$$

Hence, a decline in quantum yield parallels the relative loss in variable fluorescence. This suggests that complete PS II centers are inactivated by a nonradiative dissipation process. Weis and co-workers (Weis et al. 1990; Krause and Weis 1991) proposed that upon internal acidification (below pH 5.5) of the thylakoids, PS II centers are rendered inactive and nonfluorescent by a strong nonradiative dissipation process. Recent data suggest that release of Ca^{2+} from the water-splitting site of the reaction center complex may be involved (Krieger 1992). Charge separation at Ca^{2+}-depleted centers is assumed to be followed by recombination between Q_A^- and P_{680+} leading to quantitative transformation of excitation energy into heat.

Charge recombination had been previously suggested to be responsible for nonphotochemical quenching by Schreiber and Neubauer (1987, 1989, 1990) on the basis of the finding that a number of factors, which cause a weakening of the PS II donor side and a stimulation of recombination luminescence, also induce a lowering of F_m and a corresponding increase in q_N.

Genty et al. (1989) demonstrated that PS II quantum yield can also be very satisfactorily determined using the following simple expression:

$$\phi_{II} = q_P \cdot F_v'/F_m' = (F_m' - F)/F_m' = \Delta F/F_m' \tag{7}$$

or

$$\phi_P = \phi_{II}/q_P = F_v'/F_m'. \tag{8}$$

From a practical point of view, Eq. (7) has the great advantage of not requiring knowledge of F_o', determination of which may be problematic, particularly under field conditions. As to the theoretical background, the quantum yield is considered to be determined by the product of reaction center "openness" (represented by q_P) and the excitation capture efficiency of open centers (represented by F_v'/F_m'). The latter aspect implies a mechanism of nonradiative energy dissipation in the antenna system (Butler 1978), which is in contrast to the concept of reaction center quenching deduced by Weis and co-workers (see above).

Genty et al. (1989) observed a linear relationship between the quantum yield of CO_2 fixation and $\Delta F/F_m'$ at saturating CO_2 and 1% O_2. As was shown by Harbinson et al. 1990 and Krall et al. 1991, addition of 20% O_2 induces considerable nonlinearity, suggesting that not only CO_2 fixation but also photorespiration and other forms of O_2-dependent electron flow are indicated by $\Delta F/F_m'$. Seaton and Walker (1990) reported a curvilinear relationship between the quantum yield of *O_2 evolution* and $\Delta F/F_m'$, which was practically identical for a large variety of different leaves. As these authors used 5% CO_2 in their experiments, it may be assumed that the oxygenase function of the Rubisco was suppressed and that the nonlinearity was caused by O_2 reduction in the Mehler reaction and the consequent reduction of the formed H_2O_2 (see following section).

3.7 Fluorescence as an Indicator of Nonassimilatory Electron Flow

Whereas gas exchange measurements require net changes in external CO_2- or O_2 concentration, fluorescence also responds to internally balanced carboxylation/decarboxylation and O_2 uptake/O_2 release reactions, the importance of which increases under conditions of closed stomata, when CO_2 becomes limiting. An important protective function of such nonassimilatory electron flow against photoinhibition has been postulated (Heber et al. 1990;

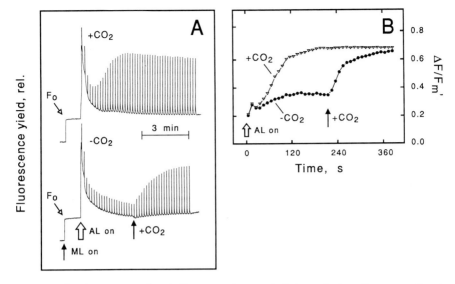

Fig. 3.7A,B. Assessment of assimilatory and nonassimilatory electron flow upon illumination of a dark-adapted spinach leaf in the presence and absence of external CO_2. **A** Original traces. **B** Calculated values of the effective quantum yield, $\Delta F/F_m'$. The leaf sample was flushed by air with and without CO_2. See text for further explanations

Wu et al. 1991). Two essential features of this protective flow may be distinguished by fluorescence, the relief of electron pressure at the PS II acceptor side (reflected in increased photochemical quenching) and thylakoid membrane energization caused by ΔpH-formation (reflected by increased nonphotochemical quenching).

In Fig. 3.7, two induction curves are compared in the presence and absence of external CO_2 (see also Sect. 3.3). The original fluorescence traces are displayed in Fig. 3.7A, while Fig. 3.7B shows the kinetics of $\Delta F/F_m'$ calculated from the F_m' values obtained with repetitive application of saturation pulses. It is apparent that in the presence of CO_2 an initial decrease in F_m' is followed by relaxation of F_m', which is accompanied by an increase in $\Delta F/F_m'$ (see Fig. 3.7B). The initial decrease of F_m' depends on the presence of O_2 (not shown, but see Fig. 3.5) and reflects ΔpH formation. The relaxation of F_m' parallels the onset of Calvin cycle activity and O_2 evolution (not shown). In the absence of CO_2, when Calvin cycle activity is prevented, a quasi-steady state is reached which is characterized by high nonphotochemical quenching and an appreciable electron flow rate, as reflected by $\Delta F/F_m'$. As expected, CO_2 addition relaxes nonphotochemical quenching and stimulates flux.

The example of Fig. 3.7 may serve to illustrate how under special experimental conditions information on photosynthetic electron flow can be obtained by fluorescence quenching analysis, where conventional gas exchange methods like O_2 polarography and infrared gas analysis encounter principal difficulties. In particular, this is true for the combination of Mehler reaction and ascorbate peroxidase reaction, which results in zero net O_2 exchange as outlined by the following reactions:

(1) $2H_2O \xrightarrow{\text{PS II}} 4e^- + 4H^+ + O_2 \uparrow$ ⎫

                                               ⎬ Mehler reaction causing q_p

(2) $4O_2 \downarrow + 4e^- \xrightarrow{\text{PS I}} 4\dot{O}_2^-$ ⎭

(3) $4\dot{O}_2^- + 4H^+ \xrightarrow{\text{SOD}} 2H_2O_2 + 2O_2 \uparrow$

(4) $2H_2O_2 + 4e^- + 4H^+ \xrightarrow[\text{MDR}]{\text{APO}} 4H_2O$ ⎫

                                      ⎬ Ascorbate peroxidase

(5) $2H_2O \xrightarrow{\text{PS II}} 4e^- + 4H^+ + O_2 \uparrow$ ⎭ reaction causing q_P

Balance: $4O_2 \uparrow$ and $4O_2 \downarrow$

Reactions (1)–(2) represent the classical Mehler reaction, which causes photochemical quenching, as it involves flux through PS II. Reactions (3)–(4) are dark enzymic processes catalyzed by superoxide dismutase (SOD), ascorbate peroxidase (APO), and monodehydroascorbate reductase (MDR). It is the monodehydroascorbate which serves as acceptor of electrons produced in reaction (5), which again results in photochemical fluorescence quenching. Hence, the depicted sequence of combined Mehler peroxidase reaction, in which eight electrons are transported through the two photo-systems, when four O_2 are evolved and four O_2 are taken up, can be monitored by fluorescence whereas net O_2-exchange is zero.

This tool has been applied in a series of studies to characterize O_2-dependent electron flow in intact chloroplasts (Neubauer and Schreiber 1988, 1989a,b; Schreiber and Neubauer 1990; Schreiber et al. 1991), which were recently extended to intact leaves by parallel measurements of photoacoustic signals (Reising and Schreiber 1992). This work has revealed that the role of O_2-dependent electron flow has been underestimated in the past (see also Walker 1992). For a long time, the role of "pseudocyclic electron flow" had been primarily seen in a contribution to ATP production to satisfy the stoichiometric needs of the Calvin cycle. In addition, the Mehler reaction had been considered synonymous with pseudocyclic flow, and the produced H_2O_2 was supposed to be detrimental, for which reason generally catalase was added in studies with isolated chloroplasts. However, recent fluorescence work has confirmed previous conclusions based on mass spectroscopy (Asada and Badger 1984), that in vivo the Mehler reaction is tightly coupled with the ascorbate-peroxidase reaction (Schreiber and Neubauer 1990). Hence, the H_2O_2 formed by the Mehler reaction may be viewed as a "natural Hill reagent". Furthermore, it now appears that light-dependent reduction of monodehydroascorbate is decisive for membrane energization, with all its implications for regulation and photoprotection (Schreiber et al. 1991).

In Fig. 3.8, a scheme is presented which summarizes the concept of the regulation of photosynthesis by O_2-dependent electron flow which has evolved from fluorescence studies. Whenever CO_2 reduction is limited (e.g., before Calvin cycle activation, at excess light intensities, after stomata closure, following stress treatment, etc.), O_2- and H_2O_2 reduction serve as valve reactions to release electron pressure. Not only do O_2 and H_2O_2 (via monodehydroascorbate) act as electron acceptors, but they also cause the buildup of a ΔpH, which may be considered the key for regulated dissipation of excess excitation energy. Of course, assimilatory and photorespiratory electron flow are coupled to H^+-translocation as well. However, both reactions require more ATP than they produce such that additional electron flow is required for ΔpH buildup and maintenance. It is this ΔpH which causes down-regulation of PS II (see Sect. 3.4 and 3.6) which is reflected in nonphotochemical ("energy-dependent") quenching.

Besides O_2-dependent linear electron flow, cyclic electron flow around PS I may also contribute to membrane energization. This appears particularly

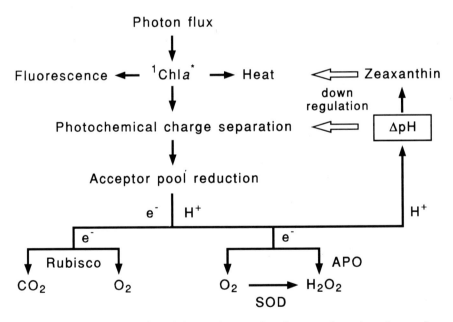

Fig. 3.8. Schematic illustration of the regulatory roles of oxygen-dependent electron flow and of the transthylakoidal ΔpH. When the photosyntheic apparatus is exposed to a photon flux density which exceeds the capacity of the Calvin cycle, strong acceptor pool reduction would lead to photoinhibitory damage, unless alternate electron acceptors become available and PS II is down-regulated. While photorespiration provides an alternate electron sink, O_2- and H_2O_2 reduction in the Mehler-ascorbate peroxidase reaction sequence are mainly responsible for ΔpH- and consequent zeaxanthin formation, which cooperate in the down-regulation of PS II

feasible in systems which display high ratios of PS I vs. PS II activity, as, e.g., in cyanobacteria (Melis 1991; Mi et al. 1992) and bundle-sheath chloroplasts. In systems with balanced PS I vs. PS II, any backcycling of electrons from the PS I acceptor side into the plastoquinone pool will inevitably lead to "overreduction" of the intersystem chain (Asada et al. 1990). Hence, although there can be no doubt about the existence of cyclic PS I electron flow per se (Arnon and Chain 1975; Moss and Bendall 1984), its contribution to membrane energization under conditions of excess radiation in vivo is still uncertain. In any case, even cyclic flow requires "poising" by O_2 (Arnon and Chain 1977) and the O_2-dependent flux involved in the poising reaction may not be insignificant when light is applied which drives both photoreactions.

Although important questions, like the role of cyclic PS I in membrane energization and the actual molecular mechanism of nonphotochemical quenching, still await final clarification, the practicability of the fluorescence quenching analysis, as outlined in Sections 3.5 and 3.6, has been well established. Hence, the point has been reached for the fluorescence method to be applied in ecophysiology to study photosynthesis in situ. An example of in situ measurements is given in the following section.

3.8   In Situ Measurements of $\Delta F/F_m'$ and of Relative Electron Transport Rate

Since it was realized that reliable information on the quantum yield of photosynthesis can be obtained by the saturation pulse method, particularly in conjunction with the approach of Genty et al. (1989), great efforts have been made to develop portable systems for in situ on-line determinations of $\Delta F/F_m'$ and of the relative electron transport rate $PFD \times \Delta F/F_m'$. Currently at least one suitable system is commercially available (PAM-2000, Walz), which is an extremely miniaturized computer-controlled version of the PAM Fluorometer (Schreiber et al. 1986). With this instrument, all relevant fluorescence parameters (F_o, F_o', F_m, F_m', F) are measured and the resulting values of F_v/F_m, $\Delta F/F_m'$, NPQ, q_P, and q_N are calculated. The incident light intensity (PFD) can be continuously monitored in the leaf plane at the same site where fluorescence is assessed. The resulting value of $PFD \times \Delta F/F_m'$ closely reflects relative electron transport rate. In this way, assessment of in situ photosynthesis has become rather simple and efficient.

Figure 3.9 shows an example of in situ measurements of light saturation curves of the relative electron transport rate ($PFD \times \Delta F/F_m'$) obtained with two leaves of *Phaseolus coccineus*, one of which had been previously exposed for 5 min to 43 °C. Heat pretreatment was in the dark (2 h after sunset) and the actual measurements were at noon of the following day. Natural daylight

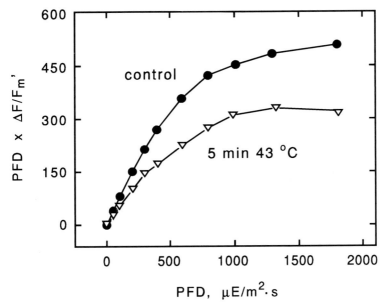

Fig. 3.9. Light-response curves of control and previously heat-treated leaves of *Phaseolus coccineus* measured in situ with a portable modulation fluorometer (PAM-2000, Walz). Photosynthetically active photon flux density (PFD) was measured by a microquantum sensor (sensitive from 380 to 720 mm) incorporated in the leaf clip holder at the same site where fluorescence was measured. Before start of the measurements, the leaves were dark-adapted for 10 min. Then light intensity was increased stepwise from 50 to 1800 μE m^{-2}s^{-1}, with 5 min illumination at each intensity before sampling $\Delta F/F_m'$. See text for further explanations

was supplemented by appropriate intensities of light from a halogen lamp to yield PFD values between 50 and 1800 μEm^{-2}s^{-1}. The resulting light response curves demonstrate that the relatively moderate heat pretreatment has caused a substantial decrease in electron transport rates. The maximal capacity, as well as the quantum yield at low intensities, are affected.

In Table 3.1, some relevant fluorescence parameters of control and previously hat-treated leaves are compared for dark and light states. It is apparent that the heated sample displays an increase in F_o and a decrease in F_m while still in the dark. The lowering of F_m is considerably enhanced by daylight. Even weak actinic light (50 μEm^{-2}s^{-1}) induces further nonphotochemical quenching in the heat-treated sample, whereas it is almost ineffective in the control sample. It is known from previous studies (Bilger et al. 1987) that heat pretreatment stimulates the nonradiative energy dissipation process, a phenomenon also reflected in the stimulation of light scattering, which is another indicator of thylakoid membrane energization (Heber 1969; Bilger et al. 1988). At 1000 μEm^{-2}s^{-1} the relative electron transport rate in the heat-treated leaf is suppressed by 32% with respect to the control leaf.

Table 3.1. Characteristic fluorescence parameters and derived expressions measured with a control and a previously heat-treated leaf of *Phaseolus coccineus*

	Control		5 min 43 °C	
	Night[a]	Noon[b]	Night[a]	Noon[b]
F_o	0.33	0.33	0.39	0.38
F_m	1.94	1.57	1.44	0.97
F_v/F_m	0.83	0.79	0.73	0.61
F_m' 50 $\mu E\,m^{-2}\,s^{-1}$	–	1.52	–	0.69
$\Delta F/F_m'$ 1000 $\mu E\,m^{-2}\,s^{-1}$	–	0.45	–	0.31
qp 1000 $\mu E\,m^{-2}\,s^{-1}$	–	0.80	–	0.65
F_v'/F_m' 1000 $\mu E\,m^{-2}\,s^{-1}$	–	0.56	–	0.48
L (1000)	–	0.46	–	0.63

[a] "Night" measurements were started 2 h after sunset and in the case of the heated leaf, 30 min after heat treatment.
[b] "Noon" measurements were started at noon of the day following heat treatment. For assessment of dark fluorescence parameters, the sample was covered by a dark cloth and the measurements were started after 10-min dark adaptation.

As can be concluded from the q_P and F_v'/F_m' values, this overall suppression is due to decreases of PS II "openness" (q_P) as well as of the energy capture efficiency of open centers (F_v'/F_m'). The significance of the L(1000) values, which relate to the relative limitation of quantum yield, will be discussed in the following section.

3.9 Yield Limitation and Excessive Photon Flux Density

The optimal quantum yield F_v/F_m is close to a mean value of 0.83 among unstressed leaves of many different species and ecotypes (Björkman and Demmig 1987). This reflects a fundamental agreement between different species in the basic organization of the light-converting primary processes (see also Björkman 1987). Hence, an optimal quantum yield as well as an optimal electron transport rate at a given photon flux density of photosynthetically active radiation (PFD) can be predicted for an unstressed "model leaf":

Maximal quantum yield = 0.83.   (9)

Maximal relative rate = PFD × 0.83.   (10)

To predict the actual quantum yield and rate of electron flow, it has to be considered that two photoreactions are involved and approximately 84% of the incident light is absorbed (see also Sect. 3.6).

Therefore:

$$\text{Estimated maximal rate} = \text{PFD} \times 0.83 \times 0.84 \times 0.5$$
$$= \text{PFD} \times 0.35. \qquad (11)$$

The value of 0.35 closely corresponds to experimentally derived values for the parameter $m = \phi_{Po}$ in the empirical equation of Weis and Berry (1987) (see also Sharkey et al. 1988).

Once it is established that a certain species does not constitute an exception from the general rule, the actually observed relative electron transport rate, $\text{PFD} \times \Delta F/F_m'$, may be compared with the response predicted for a "model leaf" and conclusions on the apparent limitations can be drawn. This is done in Fig. 3.10 for a control and previously heat-treated sample, data of which were already presented in Fig. 3.9 and Table 3.1. The deviation from the optimal quantum yield may be quantified in the previously defined sense of "limitation" (Schreiber and Bilger 1987), i.e., in the example of Fig. 3.10 for the control sample by the line segment ratio AB/AB'. Corresponding to this ratio of rate values is an identical ratio of PFD values, BC/B'C', which represents the "excessive PFD" (Demmig and Winter 1988; see also Björkman and Demmig, this Vol.). The following general expression for limitation at a given value of PFD characterizes the relative

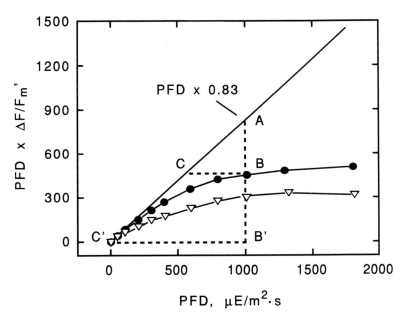

Fig. 3.10. Experimental light-response curves in comparison with the optimal quantum yield line of a "model leaf" (PFD $\times$ 0.83). Control, *upper curve* .5 min 43 °C sample, *lower curve* (see also Fig. 3.9). Definition of "limitation" (line segment ratio AB/AB') and of "excessive PFD" (equivalent line segment ratio BC/B'C') for the control sample. See text for further explanations

restriction in photosynthetic performance of a sample in comparison with a "model leaf":

$$L_{(PFD)} = 1 - \Delta F/(F_m' \times 0.83) \tag{12}$$

or

$$L_{(PFD)} = 1 - (q_P \times F_v'/F_m')/0.83. \tag{13}$$

Equation (13) makes clear that limitation with respect to the optimal quantum yield (0.83) is caused by the combination of a lowering of q_P (from optimal value of 1) and of F_v'/F_m' (from optimal value of 0.83). Hence $L_{(PFD)}$ becomes zero only, if $q_P = 1$ *and* $F_v'/F_m' = 0.83$. In the example of Figs. 3.9 and 3.10, L(1000) amounts to 0.46 in the case of the control and to 0.63 with the heat-treated sample (see Table 3.1).

Any deviation from the optimal quantum yield line, whether it is expressed in terms of "limitation", "excessive PFD", or simply "quantum yield lowering", may be considered an integrative measure of the stress to which a sample is exposed in a given situation. The immediate response revealed by fluorescence is that of "light stress" in a broad sense. Actinic light is used as a diagnostic tool to assess the efficiency of the photosynthetic apparatus "under light pressure". In this way, certain deficiencies, e.g., at the sink or Calvin cycle levels, are expressed which would remain unnoticed at moderate light intensity. The extent to which a given light intensity is excessive depends on a large number of factors, including light and temperature acclimation, water status and stomatal opening, the developmental state, and previous stress treatments. Hence, in practice, in order to draw reliable conclusions on the photosynthetic performance of a plant from fluorescence measurements, knowledge and control of environmental factors is essential.

3.10 Conclusions

In conclusion, it may be stated that chlorophyll fluorescence analysis, which in the past has primarily served as a pioneering tool in basic photosynthesis research, has now reached a point of sophistication and reliability to be applied for the assessment of in vivo photosynthesis under field conditions. Instrumentation and methods are available with which it is possible to judge various aspects of photosynthetic performance of plants in their natural environment. Relevant information which can be obtained within seconds includes the maximal quantum yield of PS II (F_v/F_m), the effective quantum yield ($\Delta F/F_m'$), the relative electron transport rate (PFD $\times \Delta F/F_m'$), the excitation capture efficiency (F_v'/F_m'), photochemical and nonphotochemical quenching coefficients (q_P and q_N), as measures of PS II openness and downregulation of PS II, and eventually $L_{(PFD)} = 1 - \Delta F/(F_m' \times 0.83)$, representative for the limitation that a sample is experiencing at a given light intensity. It is clear that to obtain all this information, a sample does not even have to be touched, i.e., the methods employed are nondestructive and the same sample can be followed in its physiological development and changes over extended periods of time. On the basis of such information, essential conclusions on the performance of a plant in a given natural or artificial environment can be drawn. In particular, in the future, these methods will be useful for rapid screening of genetically modified plant cultivars and selecting for optimal performance in environments with various forms of stress. It appears that the first response of a plant to environmental stress is an increase of non-radiative energy dissipation, which is readily reflected in chlorophyll fluorescence. In this manner, the plants attempt to avoid photoinhibitory damage, repair of which is rather costly (see contribution by Björkman and Demmig, Chap. 2, this Vol.). Exactly how heat dissipation is stimulated remains to be resolved. However, there is little doubt that O_2-dependent electron flow, ΔpH formation, violaxanthin de-epoxidation, and PS II reaction center inactivation are involved.

Acknowledgments. We wish to thank Murray Badger for reading the manuscript. Support by the Deutsche Forschungsgemeinschaft (SFB 176 and 251) is gratefully acknowledged.

References

Allen JF (1992) Protein phosphorylation in regulation of photosynthesis. Biochim Biophys Acta 1098: 275–335

Arnon DI, Chain RK (1975) Regulation of ferredoxin-catalyzed photosynthetic phosphorylations. Proc Natl Acad Sci USA 72: 4961–4965

Arnon DI, Chain RK (1977) Role of oxygen in ferredoxin-catalyzed cyclic phosphorylations. FEBS Lett 82: 297–302

Asada K, Badger M (1984) Photoreduction of $^{18}O_2$ and $^{18}H_2O_2$ with a concomitant evolution of $^{16}O_2$ in intact spinach chloroplasts. Evidence for scavenging of hydrogen peroxide by peroxidase. Plant Cell Physiol 25: 1169–1179

Asada H, Takahashi M (1987) Production and scavenging of active oxygen in photosynthesis. In: Kyle DJ, Osmond CB, Arntzen CJ (eds) Photoinhibiton. Elsevier, Amsterdam, pp 227–287

Asada K, Neubauer C, Heber U, Schreiber U (1990) Methyl viologen-dependent cyclic electron transport in spinach chloroplasts in the absence of oxygen. Plant Cell Physiol 31: 557–564

Bilger W, Björkman O (1990) Role of the xanthophyll cycle in photoprotection elucidated by measurements of light-induced absorbance changes, fluorescence and photosynthesis in leaves of *Hedera canariensis*. Photosynth Res 25: 173–186

Bilger W, Schreiber U (1986) Energy-dependent quenching of dark-level chlorophyll fluorescence in intact leaves. Photosynth Res 10:303–308

Bilger W, Schreiber U, Lange OL (1987) Chlorophyll fluorescence as an indicator of heat-induced limitation of photosynthesis in *Arbutus unedo*. In: Tenhunen J, Catarino FM, Lange OL, Oechel WC (eds) Plant response to stress. Springer, Berlin Heidelberg New York, pp 391–399

Bilger W, Heber U, Schreiber U (1988) Kinetic relationship between energy-dependent fluorescence quenching, light scattering, chlorophyll luminescence and proton pumping in intact leaves. Z Naturforsch 43cc: 377–887

Björkman O (1987) Low-temperature chlorophyll fluorescence in leaves and its relationship to photon yield of photosynthesis in photoinhibition. In: Kyle DJ, Osmond CB, Arntzen CJ (eds) Photoinhibiton. Elsevier, Amsterdam, pp 123–144

Björkman O, Demmig B (1987) Photon yield of O_2-evolution and chlorophyll fluorescence characteristics at 77 K among vascular plants of diverse origins. Planta 170: 489–504

Bradbury M, Baker NR (1981) Analysis of the slow phases of the in vivo chlorophyll fluorescence induction curve. Changes in the redox state of photosystem II electron acceptors and fluorescence emission from photosystem I and II. Biochim Biophys Acta 63: 542–551

Briantais JM, Vernotte C, Krause GH, Weis E (1986) Chlorophyll a fluorescence of higher plants: chloroplasts and leaves. In: Govindjee, Amesz J, Fork CD (eds) Light emission by plants and bacteria. Academic Press, New York, pp 539–583

Butler WL (1978) Energy distribution in the photochemical apparatus of photosynthesis. Annu Rev Plant Physiol 29: 345–378

Delosme R (1967) Etude de l' induction de fluorescence des algues vertes et des chloroplasts at début d'une illumination intense. Biochim Biophys Acta 143: 108–128

Demmig B, Winter K (1988) Light response of CO_2-assimilation, reduction state of Q and radiationless energy dissipation in intact leaves. Aust J Plant Physiol 15: 151–162

Demmig-Adams B (1990) Carotenoids and photoprotection in plants: a role for the xanthophyll zeaxanthin. Biochim Biophys Acta 1020: 1–24

Dietz KJ, Schreiber U, Heber U (1985) The relationship between the redox state of Q_A and photosynthesis in leaves at various carbon dioxide, oxygen and light regimes. Planta 166: 219–226

Duysens LNM, Sweers HE (1963) Mechanism of the two photochemical reactions in algae as studied by means of fluorescence. In: Studies on microalgae and photosynthetic bacteria. University of Tokyo Press, Tokyo, pp 353–372

Field CB, Ball JT, Berry JA (1989) Photosynthesis: principles and field techniques. In: Pearcy RW, Ehleringer J, Mooney HA, Rundel PW (eds) Plant physiological ecology. Chapman and Hall, London, pp 209–253

Foyer C, Furbank R, Harbinson J, Horton P (1990) The mechanisms contributing to photosynthetic control of electron transport by carbon assimilation in leaves. Photosynth Res 25: 83–100

Genty B, Briantais JM, Baker N (1989) The relationship between the quantum yield of photosynthetic electron transport and quenching of chlorophyll fluorescence. Biochim Biophys Acta 990: 87–92

Govindjee (1990) Photosystem II heterogeneity: the acceptor side. Photosynth Res 25: 151–160

Harbinson J, Genty B, Baker NR (1990) The relationship between CO_2 assimilation and electron transport in leaves. Photosynth Res 25: 213–224

Heber U (1969) Conformational changes of chloroplasts induced by illumination of leaves in vivo. Biochim Biophys Acta 180: 302–319

Heber U, Neimanis S, Setlikova E, Schreiber U (1990) Why is photorespiration a necessity for leaf survival under water stress? In: Sinha SK, Sane PV, Agrawal PK, Bhargave SC (eds) Proc of the Int Congress of Plant Physiol. Soc Plant Physiol Biochem, New Delhi, pp 581–592

Joliot P, Joliot A (1964) Etude cinétique de la reaction photochimique libérant l' oxygène au cours de la photosynthèse. CR Acad Sci Paris 258: 4622–4625

Karukstis KK (1991) Chlorophyll fluorescence as a physiological probe of the photosynthetic apparatus. In: Scheer H (ed) Chlorophylls. CRC Press, Boca Raton, pp 769–795

Kautsky H, Franck U (1943) Chlorophyllfluoreszenz und Kohlensäureassimilation. Biochem Z 315: 139–232

Kautsky H, Hirsch A (1931) Neue Versuche zur Kohlenstoffassimilation. Naturwissenschaften 19: 964

Kautsky H, Appel W, Amann H (1960) Die Fluoreszenzkurve und die Photochemie der Pflanze. Biochem Z 332: 277–292

Keuper HJK, Sauer K (1989) Effect of photosystem II reaction center closure on nanosecond fluorescence relaxation kinetics. Photosynth Res 20: 85–103

Krause GH, Weis E (1991) Chlorophyll fluorescence and photosynthesis: the basics. Annu Rev Plant Physiol 42: 313–349

Krall JP, Edwards GE, Ku MSB (1991) Quantum yield of photosystem II and efficiency of CO₂-fixation in *Flaveria* (Asteraceae) species at varying light and CO₂. Aust J Plant Physiol 18: 369–383

Krieger A (1992) pH-abhängige Regulation von Photosystem II: Einfluß von Calcium. Thesis, Heinrich-Heine Universität, Düsseldorf

Lavorel J, Etienne AL (1977) In vivo chlorophyll fluorescence. In: Barber J (ed) Primary processes of photosynthesis. Elsevier, Amsterdam, pp 203–268

Melis T (1985) Functional properties of photosystem IIβ in spinach chloroplasts. Biochim Biophys Acta 808: 334–342

Melis T (1991) Dynamics of photosynthetic membrane composition and function. Biochim Biophys Acta 1058: 87–106

Melis T, Schreiber U (1979) The kinetic relationship between the C550 absorbance change, the reduction of Q (A₃₂₀) and the variable fluorescence yield change in chloroplasts at room temperature. Biochim Biophys Acta 547: 47–57

Mi H, Endo T, Schreiber U, Ogawa T, Asada K (1992) Electron donation from cyclic and respiratory flows to photosynthetic intersystem chain is mediated by pyridine nucleotide dehydrogenase in the cyanobacterium *Synechocystis* PCC 6803. Plant Cell Physiol 33: 1233–1237

Moss DA, Bendall DS (1984) Cyclic electron transport in chloroplasts. The Q-cycle and the site of action of antimycin. Biochim Biophys Acta 767: 389–395

Munday JC, Govindjee (1969) Light-induced changes in the fluorescence yield of chlorophyll a in vivo point III. The dip and the peak in the fluorescence transient of *Chlorella pyrenoidosa*. Biophys J 9: 1–21

Murata N, Nishimura M, Takamiya A (1966) Fluorescence of chlorophyll in photosynthetic systems. II. Induction of fluorescence in isolated spinach chloroplasts. Biochim Biophys Acta 120: 23–33

Neubauer C, Schreiber U (1987) The polyphasic rise of chlorophyll fluorescence upon onset of strong continuous illumination. I. Saturation characteristics and partial control by the photosystem II acceptor side. Z Naturforsch 42: 1246–1254

Neubauer C, Schreiber U (1988) Induction of photochemical and nonphotochemical quenching of chlorophyll fluorescence by low concentrations of m-dinitrobenzene. Photosynth Res 15: 233–246

Neubauer C, Schreiber U (1989a) Photochemical and non-photochemical quenching of chlorophyll fluorescence induced by hydrogen peroxide. Z Naturforsch 44c: 262–270

Neubauer C, Schreiber U (1989b) Dithionite-induced fluorescence quenching does not reflect reductive activation in spinach chloroplasts. Bot Acta 102: 314–318

Reising H, Schreiber U (1992) Pulse-modulated photoacoustic measurements reveal strong gas-uptake component at high CO₂-concentrations. Photosynth Res 31: 227–238

Renger G, Schreiber U (1986) Practical applications of fluorometric methods to algae and higher plant research. In: Govindjee, Amesz J, Fork CD (eds) Light emission by plants and bacteria. Academic Press, New York, pp 587–619

Schreiber U (1986) Detection of rapid induction kinetics with a new type of high-frequency modulated chlorophyll fluorometer. Photosynth Res 9: 261–272

Schreiber U, Bilger W (1987) Rapid assessment of stress effects on plant leaves by chlorophyll fluorescence measurements. In: Tenhunen JD, Catarino FM, Lange OL, Oechel WD (eds) Plant response to stress. Springer, Berlin Heidelberg New York, pp 27–53

Schreiber U, Bilger W (1992) Progress in chlorophyll fluorescence research: major developments during the last years in retrospect. Prog Bot 54: 151–173

Schreiber U, Neubauer C (1987) The polyphasic rise of chlorophyll fluorescence upon onset of strong continuous illumination. II. Partial control by the photosystem II donor side and possible ways of interpretation. Z Naturforsch 42: 1255–1264

Schreiber U, Neubauer C (1989) Correlation between donor-side-dependent quenching and stimulation of charge recombination at PS II. FEBS Lett 258: 339–342

Schreiber U, Neubauer C (1990) O_2-dependent electron flow, membrane energization and the mechanism of non-photochemical quenching of chlorophyll fluorescence. Photosynth Res 25: 279–293

Schreiber U, Bauer R, Franck UF (1971) Chlorophyll fluorescence induction in green plants at oxyten deficiency. In: Forti G, Avron M, Melandri A (eds) Proc 2nd Int Congr Photosynth. Junk, The Hague, pp 169–179

Schreiber U, Schliwa U, Bilger W (1986) Continuous recording of photochemical and non-photochemical chlorophyll fluorescence quenching with a new type of modulation fluorometer. Photosynth Res 10: 51–62

Schreiber U, Reising H, Neubauer C (1991) Contrasting pH-optima of light-driven O_2- and H_2O_2-reduction in spinach chloroplasts as measured via chlorophyll fluorescence. Z Naturforsch 46c: 635–643

Seaton GGR, Walker DA (1990) Chlorophyll fluorescence as a measure of photosynthetic carbon assimilation. Proc R Soc Lond B 242: 29–35

Sharkey TD, Berry JA, Sage RF (1988) Regulation of photosynthetic electron transport in *Phaseolus vulgaris* L., as determined by room-temperature chlorophyll a fluorescence. Planta 176: 415–424

Snel JFH, van Kooten O (eds) (1990) The use of chlorophyll fluorescence and other non-invasive spectroscopic techniques in plant stress physiology. Photosynth Res (Spec Iss) 25: 146–332

Snel JFH, van Ieperen W, Vredenberg WJ (1990) Complete suppression of oxygen evolution in open PS 2 centers by non-photochemical fluorescence quenching? In: Baltscheffsky M (ed) Current research in photosynthesis, vol II. Kluwer, Dordrecht, pp 911–914

van Kooten O, Snel JFH (1990) The use of chlorophyll fluorescence nomenclature in plant stress physiology. Photosynth Res 25: 147–150

Velthuys BR, Amesz J (1974) Charge accumulation at the reducing side of system 2 of photosynthesis. Biochim Biophys Acta 333: 85–94

Walker DA (1992) Excited leaves. New Phytol 121: 325–345

Weis E, Berry JA (1987) Quantum efficiency of photosystem II in relation to "energy"-dependent quenching of chlorophyll fluorescence. Biochim Biophys Acta 894: 198–208

Weis E, Lechtenberg D (1989) Fluorescence analysis during steady state photosynthesis. Philos Trans R Soc Lond Ser B 323: 253–268

Weis E, Lechtenberg D, Krieger A (1990) Physiological control of primary photochemical energy conversion in higher plants. In: Baltscheffsky M (ed) Current research in photosynthesis, vol IV. Kluwer, Dordrecht, pp 307–312

Wu J, Neimanis S, Heber U (1991) Photorespiration is more effective than the Mehler reaction in protecting the photosynthetic apparatus against photoinhibition. Bot Acta 104: 283–291

4 Higher Plant Respiration and Its Relationships to Photosynthesis

J.S. Amthor

The heat produced by a respiring cell is an inescapable component of cellular metabolism, the cost which Nature has to pay for creating biological order out of physical chaos in the environment of plants and animals.
J.L. Monteith (1972)

4.1 Introduction

Respiration is the complement of photosynthesis in higher plants.[1] The primary function of photosynthesis is to assimilate CO_2 and radiant energy in the formation of carbohydrates. A significant portion of those carbohydrates become the main substrates of respiration (James 1953; Krotkov 1960; ap Rees 1980), but often after some period of storage or distance of transport. The function of respiration is to convert photoassimilate into substances usable by growth, maintenance, transport, and nutrient assimilation processes (Beevers 1961). Respiration does this by breaking down sugars into smaller molecules (carbon skeleton intermediates), phosphorylating ADP and other nucleosides, and reducing nucleotides – respiration does not only generate ATP. Some of the carbon skeleton intermediates become the precursors of growth and are diverted away from respiratory metabolism and used in biosynthetic reactions, whereas the ATP and NAD(P)H formed during respiration are used in all heterotrophic energy-requiring processes (Fig. 4.1).

During respiration and growth, CO_2 is released as a byproduct. Indeed, it is commonly surmised (Kira 1975; Amthor 1989; Ryan 1991) that up to half, or even more, of the carbon assimilated in photosynthesis (less photorespiratory decarboxylations) is eventually released during plant respiration, albeit "accurate and relevant estimates of [the ratio of respiration to photosynthesis] are rare because plant physiologists have seldom tried to measure the respiration rate of whole plants" (Monteith 1972). Heat is another important byproduct of respiration and growth, with perhaps half of the energy contained in photosynthate released as heat during plant heterotrophic metabolism. The fraction of carbon and energy in photoassimilate that is "lost" during subsequent metabolism depends on the pathways of respiration and mitochondrial ADP:O, *relative rates* of growth and main-

---

[1] The focus herein is on terrestrial higher plants. Moreover, except where specifically noted, the discussion is limited to C_3 plants, which comprise 95% of known plant species. Gardeström and Edwards (1985) and Dry et al. (1987) have discussed various aspects of mitochondria and respiration in C_4 and CAM plants. Geider (1992) has reviewed respiration in phytoplankton.

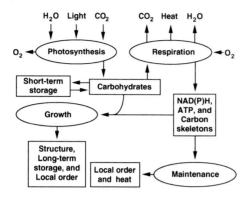

Fig. 4.1. Simplified scheme of the roles of photosynthesis and respiration in the carbon and energy economy of a plant. Here, *photosynthesis* means the balance of photosynthetic carboxylations and photorespiratory decarboxylations (carboxylations will exceed decarboxylations in the light under normal conditions). *Short-term storage* refers to starch in chloroplasts, sucrose in vacuoles, and the like. *Carbohydrates* include compounds such as cytosolic hexoses that are immediately available to metabolic processes such as respiration. *Growth* includes the processes of nutrient uptake and assimilation, transport of photosynthate from sources to sinks, and biosynthesis of new structure (cell wall, plasmalemma, and protoplasm) and long-term storage such as starch and protein in seeds and tubers. *Maintenance* refers to active processes associated with the turnover (breakdown and replacement) of existing structure and intracellular metabolite transport to counteract leakage through membranes. Processes other than respiration and photorespiration can release CO_2, but their quantitative contribution to the carbon balance of a plant is usually small and is not shown

tenance processes, and *efficiencies* of energy use in maintenance and of carbon and energy retention in growth (see Appendix). The latter is, in part, a function of tissue composition (Penning de Vries et al. 1974; McDermitt and Loomis 1981; Williams et al. 1987; Lafitte and Loomis 1988).

Fig. 4.2. Carbon metabolism phase of higher plant respiration. All the reductant and some of the ATP formed during respiration are associated with the carbon metabolism phase. Glucose and fructose may arise from compounds other than sucrose and starch. Glycolysis and the oxidative pentose phosphate network occur in the cytosol, and at least in part in plastids, and are linked by the common metabolites G-6-P, F-6-P, and Gly-3-P. Pyruvate is probably the main carbon substrate of the TCA cycle, but malate can also serve as a substrate, for example, via the *malate shunt*. Much of glycolysis may be catalyzed by a multienzyme complex (Srere 1987) associated with the outer mitochondrial membrane. Similarly, the TCA cycle may be largely confined to a multienzyme complex associated with the matrix side of the inner mitochondrial membrane, perhaps at complex I sites. Under physiological conditions, pyruvate and malate cross the inner mitochondrial membrane via carriers. According to Douce (1985), a plant cell is likely to contain hundreds to thousands of mitochondria, and these can occupy about 7% of the cytoplasmic volume. Abbreviations: *CoA* coenzyme A; *DiHOAcP* dihydroxyacetone-P; *1,3-DiPGA* 1,3-diphosphoglycerate; *E-4-P* erythrose 4-P; *F-1,6-P₂* fructose 1,6-P₂; *F-6-P* fructose 6-P; *GL-6-P* glucono-δ-lactone 6-P; *G-1-P* glucose 1-P; *G-6-P* glucose 6-P; *Gly-3-P* glyceraldehyde 3-P; *α-KG* α-ketoglutarate; *OAA* oxaloacetate; *PEP* phosphoenolpyruvate; *6-PG* 6 phosphogluconate; *2-PGA* 2-phosphoglycerate; *3-PGA* 3-phosphoglycerate; *Pᵢ* orthophosphate (inorganic); *PPᵢ* pyrophosphate (inorganic); *R-5-P* ribose 5-P; *Ru-5-P* ribulose 5-P; *Su-7-P* sedoheptulose 7-P; *UDP-G* UDP-glucose; *Xu-5-P* xylulose 5-P

4.2 Pathways and Controls of Respiration

Respiration is composed of glycolysis, the oxidative pentose phosphate network, the TCA cycle, mitochondrial electron transport, oxidative phosphorylation, and related reactions. (Photorespiration is distinguished from respiration, although the two may interact in photosynthesizing cells.) In general, the pathways (Figs. 4.2 and 4.3) and control of respiration are similar among higher plants and other organisms, indicating the conservation of the

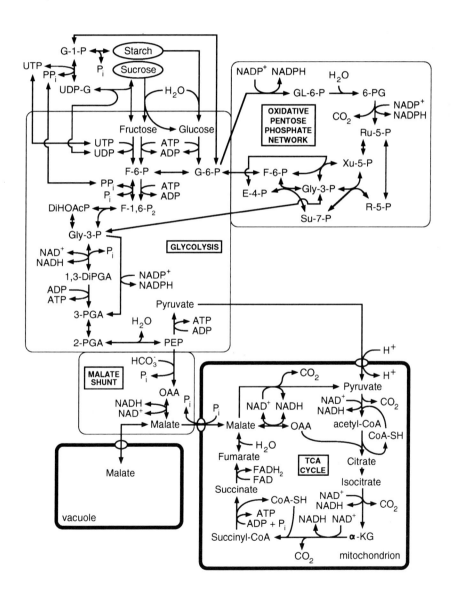

prevalent features of respiration during evolution (Chapman and Ragan 1980). The reactions and structures of respiration evolved under a wide range of environmental and biological conditions, often building upon metabolism and structures already in place. Glycolysis is "the most universal of metabolic pathways" (Prosser 1986) and developed early in the evolution of life. Parts of the TCA cycle (anaerobic, reductive, perhaps lacking α-ketoglutarate dehydrogenase) and the oxidative pentose phosphate network arose later. The photosynthetic carbon reduction cycle probably followed this, but due to the extensive oxidation of iron and other metals, it was some time before O_2 released by photosynthetic light reactions accumulated in the atmosphere (Cloud 1976). Under the aerobic conditions that eventually existed, α-ketoglutarate dehydrogenase may have appeared, perhaps from pyruvate dehydrogenase, and the TCA cycle was complete and functioning oxidatively (Gest 1987). Later, electron transport to O_2 and oxidative phosphorylation greatly increased the yield of usable energy from a unit of

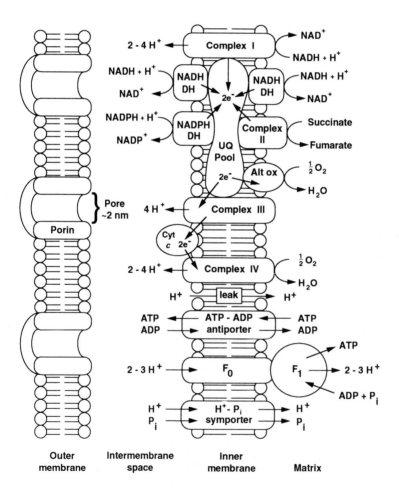

substrate such as glucose. It is possible, even probable, that mitochondria (and chloroplasts) originated from free-living aerobic eubacteria that invaded a primitive eukaryote, leading to a permanent symbiotic relationship (Lehninger 1965; Margulis 1970; Douce 1985; Sitte and Eschbach 1992).

4.2.1 Unique Properties of Plant Respiration and Mitochondrial Metabolism

While respiratory biochemistry is similar among eukaryotes, higher plant respiration differs from that of many other organisms in notable ways (see, e.g., Ikuma 1972; Palmer 1979; ap Rees 1985; Douce and Neuburger 1989). For example: (A) in plants, sucrose is a major product of photosynthesis and a primary form of carbon translocated between organs, and it is often appropriate to take sucrose (rather than, say, glucose) as the starting point of respiration. (B) In addition to the nonreversible conversion of fructose 6-P to fructose 1,6-P_2 by phosphofructokinase, many plants contain PP_i-requiring fructose 6-P 1-phosphotransferase that catalyzes the same reaction, but reversibly and using PP_i (forming P_i) rather than ATP (forming ADP). (C) Plant succinate-CoA ligase phosphorylates ADP rather than GDP (but see Weitzman 1987). (D) The rate of mitochondrial O_2 uptake per unit protein can be faster in plants than in animals. (E) Rotenone-resistant electron transport (bypass of complex I) is possible in many plants. (F) Rapid cyanide-resistant O_2 uptake is possible in plants. (G) Plant mitochondria can oxidize cytosolic NADH and NADPH (a separate dehydrogenase exists for

Fig. 4.3. Higher plant mitochondrial respiratory chain (drawn sizes and shapes are arbitrary). *Complexes I, II, III,* and *IV* are not known to differ greatly in plants and other organisms. Cytochrome c (*Cyt c*) is a peripheral protein linking complexes III and IV. The matrix-facing NADH dehydrogenase (*NADH DH*) other than complex I is insensitive to rotenone and is not known to exist in animals. That rotenone-resistant NADH dehydrogenase may actually be a second UQ binding site on complex I (within the membrane) that is not coupled to proton pumping (Soole et al. 1990, 1992). The *alternative pathway* refers to electron transport from ubiquinol (reduced ubiquinone) to O_2 via the alternative oxidase (*Alt ox*). Succinate and fumarate are intermediates of the TCA cycle. The NADH and NADPH dehydrogenases facing the intermembrane space are found in plants and fungi but not mammals although mammals can oxidize cytosolic reducing equivalents indirectly via metabolite shuttles. Compared to the dehydrogenases facing the intermembrane space, succinate (*Complex II*) may have preferential access to the alternative oxidase, perhaps due to some spatial "association," or more simply, a shorter diffusion-path length between the two (Day et al. 1991). Oxidative phosphorylation is presumably driven by the movement of protons from the intermembrane space to the mitochondrial matrix via $F_1 \cdot F_0$ ATPases passing through the inner membrane (Nicholls and Ferguson 1992). The outer membrane pores are large enough to be freely permeable to ATP and NAD(P)H. The plant mitochondrial outer membrane is discussed by Mannella (1985) and the inner membrane by Douce (1985). Proposed "respiratory" electron transport in chloroplasts in the dark apparently does not pump protons (e.g., Singh et al. 1992) – a major function of mitochondrial electron transport – so it is not similar to this figure

each nucleotide). (H) Rapid transport of oxaloacetate (OAA) across the inner membrane of plant mitochondria can occur under physiological conditions. (I) NAD^+ can be actively accumulated from an external medium by plant mitochondria, apparently via a specific NAD^+ transporter. (J) Plant mitochondria readily oxidize malate in the presence of OAA and absence of pyruvate. (K) Plant mitochondria contain some unique cytochromes and the composition of common cytochromes can differ among plants and other organisms. (L) In leaf mitochondria, glycine decarboxylase may account for as much as half of the matrix protein and rapid glycine decarboxylation, a component of photorespiration, is possible. (M) Fatty acid oxidation is generally slow in plant mitochondria. (N) Plants contain the largest, most complex mitochondrial genome. Many of these "unique" properties confer a large degree of metabolic flexibility to plants.

4.2.2   Control of Respiration Rate

Respiration rate can be regulated by the amount of respiratory machinery (enzymes and transporters), the amount of respiratory substrate (carbohydrates and O_2), the rate of ATP and NAD(P)H use [ADP and $NAD(P)^+$ regeneration], or the rate of respiratory intermediate use. There has also been considerable interest in the control of respiration by calcium in plants (Wellburn and Owen 1991) and mammals (Brown 1992). In rapidly growing plant cells, the amounts of respiratory machinery or carbohydrate may limit respiration because the demands for respiratory products (carbon skeletons, ATP, and reductant) are high and those products are used as rapidly as they are produced. Moreover, the amount of respiratory machinery in very young cells may be small, being in a state of construction itself, and the supply of carbohydrate may be limited by incompletely developed phloem near growing cells. In mature or slowly growing plant cells, short-term (seconds to hours) control of respiration rate is probably brought about by the rate of use of respiratory products, and in particular ATP,[2] rather than a lack of metabolic machinery or carbohydrates (French and Beevers 1953; Beevers 1961, 1970; Copeland and Turner 1987; Dry et al. 1987; Farrar and

---

[2] Control of respiration rate by ADP availability (use of ATP) can exist with respect to glycolysis, the TCA cycle, and oxidative phosphorylation. Generally, the term *respiratory control* is used to refer to a feedback inhibition of respiratory chain (Fig. 4.3) activity due to a large proton motive force (Δp) arising when ADP levels are low and oxidative phosphorylation is slow. Respiratory control has been studied with the use of uncoupling agents, which not only disengage the link between respiratory chain activity and Δp and therefore oxidative phosphorylation, but also limit pyruvate and P_i uptake by mitochondria because that transport is driven by Δp (see Figs. 4.2 and 4.3). This may underlie the observation that uncouplers stimulate glycolysis to a greater extent than they do the TCA cycle (Wiskich and Dry 1985) and indicates that the capacity of respiration is underestimated under the influence of uncouplers.

JHH Williams 1991). That is, in all but the youngest cells, respiration may be pulled along at a rate appropriate to the rate of growth, transport, nutrient uptake and assimilation, and maintenance processes, and this is generally below the rate limitation set by the amount of carbohydrates and respiratory enzymes. Because respiratory products are used by nearly all heterotrophic processes, and because the respiratory network is flexible and contains many branch points (see Figs. 4.2 and 4.3), "it seems inevitable that control will be found to be extremely complex and to be shared among a variety of reactions and a range of regulators" (ap Rees 1988).

Even when respiration is coupled to other processes via the use of its products, it does not follow that those other processes obligatorily use respiratory products with maximum efficiency. Although it is difficult to imagine extensive inefficiencies in higher plants following their long evolution, there is no compelling evidence that plant metabolism is optimal, nor need there be any, for "contrary to a widely held belief, the anatomical and physiological features of . . . extant life forms are not necessarily optimal solutions from the *synchronic* point of view, which takes into account only the present situation. . . . [But instead,] it appears that evolution used, not what was theoretically optimal, but whatever happened to be available to it and could be appropriated to serve a needed function" (Delbrück 1986) (see also Maynard Smith 1978). Moreover, "consumption of ATP by simple hydrolysis in which there is no outcome useful to the plant probably occurs in all cells at a finite rate and would yield a minimal background of idling respiration" (Beevers 1970). Also, in most plant communities, no more than a few percent of the energy in solar radiation absorbed by leaves is contained in the plant mass resulting from growth (Larcher 1983). Nevertheless, the efficiency of higher plant heterotrophic metabolism is, by many measures, impressive.

4.2.3 Energy Conservation During Plant Respiration

As many as 30 to 36 ADP can be phosphorylated per hexose oxidized completely during plant respiration (see Figs. 4.2 and 4.3; Nicholls and Ferguson 1992; Amthor 1993a). When ATP use is rapid but carbon skeleton use is slow, the mitochondrial ADP:O can be expected to be high (perhaps near 3) and the ratio of carbon diverted away from the respiratory network to that released as CO_2 will be small. That is to say, carbohydrate metabolism in respiration may yield mostly CO_2, heat, and ATP in mature cells rather than carbon skeletons. Under different circumstances, e.g., in rapidly growing tissue, NADH produced during carbon skeleton use in growth may exceed that required for ADP phosphorylation (for ATP use) via the respiratory chain and oxidative phosphorylation (Penning de Vries et al. 1974). In such a case, the circumvention of respiratory control by ADP availability may be required. The rotenone-resistant complex I bypass (item

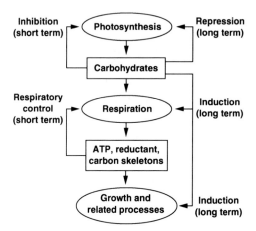

Fig. 4.4. Proposed place of carbohydrates and respiratory products in regulating photosynthesis (net of photorespiratory decarboxylations), respiration, and growth. In the short term, carbohydrate accumulation in leaves can lead to feedback inhibition of photosynthesis. In the long term, sugars may limit photosynthetic capacity via the repression of photosynthetic genes. Conversely, carbohydrates may elicit an increase in both respiratory capacity and the rate of growth and related heterotrophic metabolism. The products of respiration must be used in order for respiration to continue so that, e.g., growth, can pull respiration along at a rate appropriate to the need for respiratory products. The control of respiration is unlikely to be "absolute" in vivo and some "unneeded" respiration is likely to occur. Notably, stress may uncouple photosynthesis, respiration, growth, and carbohydrate level (see Amthor 1993b)

E above; see Fig. 4.3) and cyanide-resistant bypass of complexes III and IV (item F above; see Fig. 4.3) facilitate such a bypass of respiratory control (Bryce et al. 1990). The complex I bypass is engaged when matrix NADH levels are high (Soole et al. 1990). The cyanide-resistant, alternative pathway is common among higher plants, but is also found in some animals, fungi, bacteria, and algae (Henry and Nyns 1975). It is engaged when the ubiquinone (UQ) pool is highly reduced (Dry et al. 1989). Also, as Δp increases, the conductance of the inner membrane to passive proton transport or "leaks" increases. Thus, a large Δp aids the oxidation of NAD(P)H under ADP limited conditions via passive dissipation of the proton gradient due to a large driving force (Δp) and a large conductance. Moreover, the $NADP^+$- linked 3-phosphoglyceraldehyde dehydrogenase reaction in glycolysis (Kelly and Gibbs 1973; Duff et al. 1989) bypasses ADP phosphorylation. When "excess" ATP is formed, futile cycling or an adenylate kinase system coupled to fatty acid processing (Fricaud et al. 1992) can regenerate ADP.

On the whole, it remains worthwhile to consider respiration as being coupled to the rate of use of its products even though there are opportunities for uncoupling and these may well come into play during the normal course of metabolism. In support of this generalization, it is commonly observed that the rate of plant respiration displays a stoichiometric relationship with the rates of, e.g., growth, translocation, nutrient uptake, and nitrogen assimilation (Farrar 1985; Amthor 1993b).

4.2.4 Respiration Rate and Carbohydrate Level

A strong positive correlation often exists between plant or tissue carbohydrate content and respiration rate, suggesting that carbohydrate level and respiration rate are related (Williams and Farrar 1990; Farrar and JHH Williams 1991; Amthor 1993b). Respiration can be under short-term (minutes to hours) control by ADP availability, however, even though it is correlated with carbohydrate level over the longer term (Journet et al. 1986; Brouquisse et al. 1991; Douce et al. 1991). Notably, the addition of sugars to a tissue increases respiration in the short term when that tissue has been starved or excised from a supply of endogenous sugars, but not usually otherwise (ap Rees 1988; Rebeille 1988; Williams and Farrar 1990). Moreover, the relationship between carbohydrate level and respiration rate is often strongest in growing tissues, and may be absent in mature organs, i.e., growth respiration appears to be controlled by carbohydrate level but maintenance respiration does not (Amthor 1989).

It has been inferred that carbohydrates induce processes consuming respiratory products so that over the long term (hours to days) respiration is positively related to carbohydrate level, but through respiratory control mechanisms (Bingham and Farrar 1988; Farrar and JHH Williams 1991; Farrar and ML Williams 1991). For example, an increase in (specific) sugars may stimulate root nitrate uptake and assimilation (Hänisch ten Cate and Breteler 1981; Aslam and Huffaker 1984) and perhaps enhance cell division and differentiation (Williams and Farrar 1990; Farrar and JHH Williams 1991). Each of these increase the demand for carbon skeletons and energy. Protein levels may be regulated by carbohydrate concentration (Baysdorfer and van der Woude 1988; Wenzler et al. 1989) with a decrease in cellular carbohydrate level leading to a loss of respiratory machinery (Journet et al. 1986) and an increase in sugars resulting in increased respiratory capacity (Avelange et al. 1990; Farrar and JHH Williams 1991). Conversely, sugars can lead to short-term inhibition of photosynthesis (Foyer 1988) and *repress* photosynthetic genes (Sheen 1990; see also Krapp et al. 1991; Schäfer et al. 1992). The motif is that carbohydrates – even specific sugars – can act as "messages" as well as substrates of growth and respiration (Williams and Farrar 1990). The messages, which are products of photosynthesis and common forms of carbon translocated from sources to sinks, would coordinate carbon and energy fluxes through photosynthesis and on to growth and respiration (Fig. 4.4).

4.3   Respiration in Photosynthesizing Leaves

Respiration, photosynthesis, and photorespiration occur simultaneously in photosynthesizing cells. The most immediate spatial and temporal inter- actions between respiration and photosynthesis therefore occur in photosyn- thesizing cells where respiration and photorespiration are concurrently releasing CO_2 in mitochondria; chloroplast metabolism as well as respiration can be contributing ATP and reductant to the cytosol; and carbon skeletons are produced in respiration, photosynthesis, and photorespiration.

In leaves subject to moderate to high light and normal temperature and CO_2 concentration, photosynthetic CO_2 assimilation may proceed 10 to 20 times faster than CO_2 release in respiration in the dark at the same temperature. Nonetheless, the extent of respiration in leaves during the day has been of considerable interest from the earliest studies of plant respiration and photosynthesis, and opinions concerning the effects of light or photo- synthesis on respiration have varied extensively, even up to the present. Because a principal function of respiration is to supply reductant and usable energy for active processes, often via the cytosol, and because photosyn- thesis can supply the cytosol with both reductant and ATP (Fig. 4.5), some respiration might be superfluous in the light. Thus, the needs for respiration can differ in light and dark, and respiration may slow in the light compared

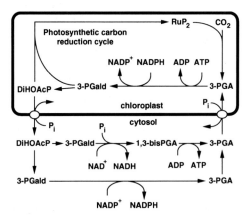

Fig. 4.5. Transport of photosynthetically generated ATP and reductant from a chloroplast to the cytosol via the DiHOAcP/3-PGA shuttle. The ATP and NADPH in the chloroplast are formed from ADP and NADP$^+$ during the light reactions of photosynthesis. Not all the intermediates of the photosynthetic carbon reduction cycle are shown. The cytosolic conversions of 3-PGald to 3-PGA are also (alternative) components of glycolysis, and are the "site" of a direct interaction between photosynthesis and respiration in the light. The NADP$^+$-linked 3-PGald dehydrogenase reaction, which can be induced by low P_i (Duff et al. 1989), might affect activity of the oxidative pentose phosphate net- work via the NADP$^+$:NADPH (Copeland and Turner 1987). Abbreviations: *DiHOAcP* dihydroxyacetone-P; *3-PGA* 3-phosphoglycerate; *1,3-bisPGA* 1,3-bisphosphoglycerate; *3-PGald* 3-phosphoglyceraldehyde; *RuP$_2$* ribulose 1,5-bisphosphate

to the rate in the dark at the same temperature, although total mitochondrial CO_2 release could increase due to photorespiration. Although enzymes of glycolysis and the oxidative pentose phosphate network exist in chloroplasts (ap Rees 1985), it is reasonable to assume that those pathways are inhibited or reversed during photosynthesis.

A high cytosolic ATP:ADP can inhibit oxidative phosphorylation (Dry and Wiskich 1982), and although the ratio may increase in the light under physiological conditions (Gardeström and Wigge 1988), it may not be high enough to be inhibitory. The availability of free (unbound) ADP to respiratory reactions in the light and dark has not, apparently, been well characterized and is probably important, and dynamic, in vivo. Photosynthesis may produce mostly NADPH in the cytosol, compared to ATP and NADH; Krömer and Heldt (1991b) cite evidence that NADPH may exceed NADH by 250 times in the cytosol of photosynthesizing *Spinacia oleracea* cells. This suggests an inhibition of the cytosolic oxidative pentose phosphate network, but not necessarily glycolysis or the TCA cycle unless the NADPH is oxidized by the respiratory chain (see Fig. 4.3).

Because CO_2 is assimilated in photosynthesis while being released in respiration, and O_2 is generated by photosynthesis while being consumed in respiration, simple gas exchange techniques cannot be used to estimate rates of respiration in leaves in the light. The use of labeled CO_2 or O_2 will not fully overcome this difficulty for several reasons. The photosynthetic assimilation of respired CO_2 – mitochondria and chloroplasts are generally in close association in leaves – is of particular importance. A similar situation has existed for the direct measurement of photorespiration, resulting in a range of "unreliable" estimates of its rate (Sharkey 1988). Fortunately for the study of photorespiration, the kinetics of a single enzyme (Rubisco) determines the ratio of RuP_2 carboxylation to RuP_2 oxygenation, and it is possible to calculate the ratio of photosynthesis to photorespiration if chloroplast CO_2 concentration is known (Sharkey 1988). Diversion of photorespiratory glycine to processes other than serine synthesis, which results in photorespiratory decarboxylation (Fig. 4.6), will alter the ratio of photosynthetic carboxylation to photorespiratory decarboxylation for a given chloroplast CO_2 concentration, but nonetheless, accurate estimates of the concurrent CO_2 fluxes associated with photosynthesis and photorespiration are possible because of the central role of Rubisco in those processes. When the chloroplast CO_2 concentration is equal to the CO_2 compensation point in the absence of respiration (Γ_*; Farquhar et al. 1980), respiration rate in photosynthesizing leaves can be estimated from gas exchange rate (see, e.g., Kirschbaum and Farquhar 1987). The value of Γ_* is a function of the CO_2/O_2 specificity of Rubisco (see Laing et al. 1974; Farquhar et al. 1980; Jordan and Ogren 1984). If respiration rate is affected by CO_2 concentration (e.g., Amthor et al. 1992), effects of CO_2 on respiration must be accounted for to estimate respiration rate at CO_2 levels different from Γ_*.

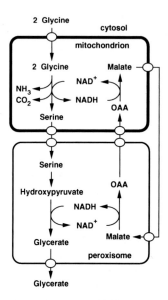

Fig. 4.6. Photorespiratory pathway from glycine to glycerate and a malate/OAA shuttle between a mitochondrion and a peroxisome. Conversion of glycine to serine is a two-step process catalyzed by glycine decarboxylase and serine hydroxymethyltransferase. The mitochondrial NAD^+/NADH cycle and conversion of OAA to malate are direct interactions between photorespiration and respiration occurring in the light and during the early phase of a dark period. Most reactions found in mitochondria and peroxisomes have been omitted for simplicity

The Kok effect, an apparent abrupt decrease in the quantum yield of photosynthesis as radiation increases in the vicinity of the light compensation point (Kok 1948), is evidence of an inhibition of respiration by light. In careful experiments, Sharp et al. (1984) and Kirschbaum and Farquhar (1987) observed Kok effects, whereas Björkman and Demming (1987) did not. Björkman and Demming (1987), however, used high CO_2 concentrations which may have inhibited respiration (Amthor et al. 1992) and overridden effects of light on respiration. Based on measurements of leaf CO_2 exchange and a mechanistic model of RuP_2 carboxylation/oxygenation, Brooks and Farquhar (1985) concluded that respiration was slowed by light to a considerable degree. The bulk of the effect of light on respiration occurred as light increased from darkness to 10 to 50 µmol photons $m^{-2} s^{-1}$.

After reviewing the literature, Graham (1980) concluded that cytosolic glycolysis can, apparently, operate in the light, and that "the provision of essential carbon skeletons required in synthetic reactions is the raison d'être for the continued operation of 'dark' respiration in the light". Graham and Chapman (1979) stated that in the light the TCA cycle "can operate at a rate comparable with that in the dark," and that owing to its function as a source of carbon skeletons for growth, the TCA cycle might be little affected by light in growing tissue, but perhaps slowed by light in mature tissue. McCashin et al. (1988) presented evidence that TCA cycle activity was slowed only 20% (but see Gemel and Randall 1992) by light in growing *Triticum aestivum* leaves, but they used nonphysiological, i.e., nonphotorespiratory, conditions. Evidence that the TCA cycle (but not necessarily mitochondrial electron transport and oxidative phosphorylation) is slowed in the light comes from the partial inactivation of mitochondrial pyruvate dehydrogenase complex (mtPDC) by light. The light inactivation is apparently transduced through some product of photosynthesis or photorespiration resulting in the phosphorylation of the mtPDC (Gemel and Randall 1992).

Graham (1980) surmised that "inhibition of oxidative phosphorylation in the light seems probable." Conversely, the results of Krömer et al. (1988) and Krömer and Heldt (1991a) indicate that oxidative phosphorylation is actually required for rapid photosynthesis (measured as O_2 evolution). They suggested that mitochondria contribute needed ATP to the cytosol in photosynthesizing cells and can oxidize excess photosynthetic redox equivalents (see also Ebbighausen et al. 1985). Oxidative phosphorylation may be most important to photosynthesis in leaves not capable of rapidly producing starch – respiratory ATP is apparently required to support rapid cytosolic sucrose-P synthesis (Hanson 1992) and the consequent release of P_i for further use in photosynthetic carbon metabolism (Sharkey 1985), whereas rapid starch synthesis (and P_i release) can occur without concurrent respiration. Active mitochondrial electron transport might limit photoinhibition of photosynthesis through the lessening of over-reduction of the photosynthetic apparatus (Saradadevi and Raghavendra 1992). Also, respiratory chain activity (which will be slowed if oxidative phosphorylation is inhibited as in the experiments of Krömer and his colleagues) may be needed for continued operation of photorespiration via the turnover of mitochondrial NADH required by glycine decarboxylase.

Clearly, photosynthetic metabolism can alter the demands for respiratory products. Indeed, mutual interactions between photosynthesis and respiration in the light are likely. Quantitative effects of light on respiration remain largely unknown, however. A diversity of estimates of effects of photosynthesis on respiration are not necessarily contradictions, for they might result from differing needs for respiration in the light in different studies; photosynthesis will not inhibit respiration by some constant fraction irrespective of environmental, physiological, and ontogenetic factors. Regarding the significance of effects of photosynthesis on respiration, it was suggested (Amthor 1989) that respiration by mature leaves is not overly important to the whole-plant carbon balance, especially in crops during seed growth (see, e.g., Gaastra 1963). In forests, however, where net primary production may be a smaller fraction of gross primary production (Kira 1975), leaf respiration may account for a majority of whole-plant respiration (Allen and Lemon 1976; Hagihara and Hozumi 1991), and effects of light on respiration are more important to productivity. To the extent that photosynthetic reductant and ATP are generated in excess of that used in CO_2 assimilation, and that they serve to "replace" the need for some respiratory metabolism, the resulting reduction in respiration is a benefit to the carbon and energy balance of a plant. But, a reduction in respiration without replacement sources of ATP, reductant, and carbon skeletons is likely to be a detriment to the plant as growth and maintenance processes will be deprived of substrates.

4.4   Photorespiration and Mitochondrial Metabolism

Mitochondrial glycine decarboxylation and the linked NAD^+ reduction is central to photorespiratory carbon metabolism (Tolbert 1980) with CO_2 release by photorespiration probably exceeding that of respiration under moderate to high light and ambient CO_2. Reduction of mitochondrial NAD^+ (forming $NADH + H^+$) by glycine decarboxylase could deprive NAD-linked TCA cycle dehydrogenases of that cosubstrate and limit TCA cycle activity. Moreover, a supply (regeneration) of NAD^+ is required for photo-respiration as well as TCA cycle activity. If carbon flux through the photo-respiratory cycle does not match and rate of RuP_2 oxygenation, which is a function of chloroplast CO_2 and O_2 partial pressures and the temperature-dependent substrate specificity of Rubisco (Jordan and Ogren 1984), photo-synthesis is slowed (Dry et al. 1987). The fate of mitochondrial NADH during daytime is therefore important to rates of respiration, photorespi-ration, and photosynthesis.

4.4.1   Oxidation of Photorespiratory NADH by the Respiratory Chain

Matrix NADH formed by glycine decarboxylase (Fig. 4.6) can be oxidized by the mitochondrial respiratory chain and coupled to as many as three sites of proton translation (Douce 1985; Fig. 4.3), but that oxidation might compete with TCA cycle-generated NADH and succinate for access to the UQ pool. Photorespiratory NAD^+ reduction might elicit rotenone-resistant complex I bypass activity because of high matrix NADH levels and also enhance alternative pathway engagement due to increased reduction of the UQ pool. Dry et al. (1987), however, consider the latter to be relatively unimportant to normal daytime activity of the respiratory chain.

   Glycine may have preferential access to some mitochondrial NAD^+, or the resulting NADH may have preferential access to the respiratory chain, compared to NAD^+-linked TCA cycle enzymes and the NADH formed by them. Glycine decarboxylation will not saturate the respiratory chain how-ever, for when a second substrate such as malate is added to mitochondria supplied with glycine, O_2 uptake can increase markedly (Dry et al. 1987;

Wiskich et al. 1990). Thus, the TCA cycle can continue to operate during glycine decarboxylation, although perhaps at a reduced rate.

Cytosolic ATP:ADP declines if photorespiration is slowed, indicating that photorespiration contributes to ATP production (Gardeström 1987; Gardeström and Wigge 1988). This implicates the mitochondrial respiratory chain as one mechanism of photorespiratory NADH oxidation. Photorespiratory NADH production may exceed the capacity of the respiratory chain to oxidize it, however, in which case other means of oxidizing NADH, such as substrate shuttles, are also required (Dry et al. 1987).

4.4.2   Oxidation of Photorespiratory NADH via Substrate Shuttles

If carbon is conserved in photorespiration – i.e., 3/4 of it – so that glycerate is formed from hydroxypyruvate in peroxisomes and then transported to chloroplasts, an amount of NADH equal to the amount formed in mitochondria by glycine decarboxylase is required in peroxisomes. Carbon conservation need not, however, be complete (Grodzinski 1992) as serine and related compounds may be exported from leaves and this can increase with low CO_2 (high photorespiration). Nonetheless, NADH is required in peroxisomes when glycerate is being formed and it might then be beneficial to transfer redox equivalents from mitochondria to peroxisomes. The oxidation of NADH by malate dehydrogenase (MDH; OAA $\rightarrow$ malate) coupled to a malate/OAA shuttle is one mechanism for this transfer (Fig. 4.6). Dry et al. (1987) concluded that such a shuttle is "likely to operate in vivo as an adjunct to the NADH reoxidation capacity of the mitochondrial respiratory chain." NADH oxidation by MDH requires a supply of OAA whereas OAA is also required for activity of the TCA cycle and is formed in the TCA cycle *from* malate. Plant mitochondria can import OAA via a translocator with a high affinity for cytosolic OAA (Ebbighausen et al. 1985) and are able to export malate (Douce 1985).

Oxidation of NADH by MDH and activity of a malate/OAA shuttle might seem impossible if the TCA cycle is engaged because MDH would be catalyzing OAA $\rightarrow$ malate *and* malate $\rightarrow$ OAA reactions simultaneously. One population of mitochondria supporting photorespiration and another the TCA cycle would allow this, but experimental data do not support this notion (Wiskich et al. 1990). Another possibility, and one supported by experimental evidence (Dry and Wiskich 1985; Wiskich et al. 1990), is that within individual mitochondria, glycine decarboxylase and malate and OAA transporters are spatially separated from TCA cycle enzymes. This separation might be facilitated by enzyme or enzyme complex "attachment" to specific locations on the matrix side of the inner mitochondrial membrane.

To the extent that photorespiratory NADH is oxidized by the respiratory chain and glycerate is formed from glycine, NADH required in peroxisomes must come from nonmitochondrial sources. These sources can include the chloroplast dihydroxyacetone-P/3-phosphoglycerate shuttle (Fig. 4.5), gly-colysis, or a chloroplast malate/OAA shuttle. An α-ketoglutarate/glutamate – malate/aspartate shuttle between mitochondria and peroxisomes is apparently not important in leaves (Krömer and Heldt 1991b).

4.5   Daytime Photosynthesis and Nighttime Respiration

Respiration in a mature leaf at night can be positively related to the previous daytime net photosynthesis of that leaf (Ludwig et al. 1975; Azcón-Bieto and Osmond 1983). Similarly, fruit (Satterlee and Koller 1984) or shoot and whole-plant (Amthor 1933b) nighttime respiration may be positively related to the amount of photosynthesis during the previous daytime. High sugar levels resulting from rapid photosynthesis may increase energy use for compartmentation. Short-term storage of sucrose occurs in vacuoles (ap Rees 1988), and although Kaiser and Heber (1984) reported that sucrose transport across the tonoplast in *Hordeum vulgare* mesophyll protoplasts was not energy-dependent, Getz (1991) observed ATP-dependent sucrose transport into tonoplast vesicles from *Beta vulgaris* root tissue. High sugar levels might also accelerate futile cycling between triose-Ps and hexose-Ps (by mass action or allosteric mechanisms) which is thought to enhance the sensitivity of metabolism to demands for sugars and carbon skeletons (Dancer et al. 1990; Hatzfeld and Stitt 1990; Hatzfeld et al. 1990). Fast respiration after rapid photosynthesis is probably in part related to metabolic costs of translocation (phloem loading, unloading, and related processes; Amthor 1993a) and is conceivably linked to maintenance of enzymes and membranes used during the day for carbon and nitrogen assimilation and processing. In the long term, ample carbohydrates resulting from rapid photosynthesis might induce growth in immature organs as outlined above (see Fig. 4.4), leading to increased respiration to support processes such as biosynthesis and nitrogen assimilation.

According to Azcón-Bieto et al. (1983), respiration rate and engagement of the alternative pathway (Fig. 4.3) are enhanced when sugar levels are high. The implication is that the alternative pathway allows rapid respiration (bypassing respiratory control by ADP) and functions to oxidize "excess" sugars (e.g., Steingröver 1981). Ap Rees (1988), however, has argued that insufficient evidence exists to show that the alternative pathway aids in the disposition of excess carbohydrates. Moreover, if sugars stimulate rapid growth, "overproduction" of NAD(P)H may result (Penning de Vries et al. 1974). Engagement of the alternative pathway, and the rotenone-resistant bypass of complex I, would then facilitate the regeneration of $NAD(P)^+$ in the face of ADP limitations and expedite biosynthetic processes rather than waste carbohydrates.

Work to date relating respiration to previous photosynthesis has been with herbs and with seedlings and small individuals of woody plants, whereas the bulk of global higher plant photosynthesis and respiration occurs in large trees. Temporal relationships between whole-tree photosynthesis and respiration may be loose because of the greater average distance of transport from sources to sinks and slower turnover time of some carbohydrate storage pools. Knowledge of those relationships, however, is scant. Also, differences between on/off and sinusoidal (natural) light treatments have not been well studied with respect to the links between photosynthesis and respiration occurring over a few hours, with most quantitative research employing on/off light treatments. Temporal patterns of leaf metabolite levels differ with sinusoidal and on/off light regimes (Fondy et al. 1989; Servaites et al. 1989a,b) and respiration could respond to those differences.

4.5.1   Light Level

One consequence of long-term high light is fast leaf respiration (Björkman 1981). This is independent of photosynthetic *capacity* (Sims and Pearcy 1991) and may reflect a response to daytime net photosynthesis. Actual CO_2 assimilation (use of machinery) rather than capacity for assimilation (amount of machinery) could be the link to maintenance processes and therefore leaf maintenance respiration. Respiratory acclimation to a change in prevailing light can occur within a few days and disrupts the relationship between mature leaf respiration rate and nitrogen content (Sims and Pearcy 1991), showing that a leaf maintenance coefficient based on nitrogen or protein level ($m_{(N)}$; see Appendix) is a function of prevailing light as well. Also, the true growth yield ($Y_{G(C)}$; see Appendix) of leaves can be inversely related to light level during growth (Williams et al. 1989).

In addition to the relationship between daytime light levels and subsequent respiration caused by different photosynthesis rates, Heichel (1970) found that leaf respiration of the C_4 species *Zea mays* was positively related to previous light level independent of net CO_2 assimilation. Measurements were made in CO_2-free air so CO_2 was assimilated only at the rate at which it was released in respiration and other decarboxylation processes. Heichel (1970) concluded that "light . . . was required to produce a substrate which was subsequently used in . . . respiration," but it is also plausible that non-CO_2-assimilating light-driven reactions related to, e.g., nitrogen metabolism, were supported in part by respiration and that such metabolism (including related maintenance processes) continued into the dark period. Notably, blue light can stimulate respiration by unknown mechanisms in the absence of photosynthesis and this can persist into a following dark period (Kowallik 1982). Both past and present light levels affect respiration rate.

4.5.2 CO_2 Concentration

Ludwig et al. (1975) observed that the relationship between *Lycopersicon esculentum* leaf nighttime respiration and previous daytime net photosynthesis was stronger when photosynthesis was varied by changing light level (CO_2 held constant) compared to changing CO_2 level (light held constant) with on/off light treatments. In other instances too, elevated daytime CO_2 level leads to a decline in the ratio of nighttime respiration to daytime photosynthesis (e.g., Gifford et al. 1985; Du Cloux et al. 1987; Dutton et al. 1988; Gaudillère and Mousseau 1989), but the ratio may eventually return to that of plants under ambient CO_2 (Grodzinski 1992; Morin et al. 1992). Charles-Edwards and Ludwig (1975) attributed this to high levels of "glycollate pathway products" (versus starch) with low CO_2 and that the glycolate pathway intermediates, e.g., glycine, were important substrates for CO_2 releasing reactions during the night. Photorespiratory intermediates do appear to contribute to CO_2 efflux for the first ca. 20–30 min of a dark period following a period of constant light in *Triticum aestivum* leaves (Azcón-Bieto and Osmond 1983), but the differential response of respiration to previous photosynthesis across CO_2 levels is more far-reaching.

It is not surprising that growth and respiration can respond markedly to a change in light. Diurnal and seasonal cycles of light include large amplitudes, and spatial variation in light levels within plant communities is often high. Plants are adapted to such spatial and temporal variations in light, and several photoreceptors are known to exist. Conversely, natural temporal and spatial variations in CO_2 are small. Indeed, no CO_2 sensor is known to exist in higher plants, although carbamate formation on Rubisco might serve such a function (see Lorimer 1983).

Elevated CO_2 increases the ratio of starch to sucrose, with surcose production being linked to growth and other processes using respiratory products, but with starch accumulating in a less coordinated way (Morin et al. 1992). (Wullschleger et al. (1992a) observed decreased leaf sucrose in spite of enhanced photosynthesis due to elevated CO_2 in two field-grown tree species.) Some of the increase in carbon accumulation and decline in the ratio of nighttime respiration to daytime net photosynthesis (increase in growth efficiency, $Y_{(C)}$; see Appendix) with elevated CO_2 is due to increased short-term storage pool size (Rowland-Bamford et al. 1990) rather than only increased efficiency of growth or respiration. A corollary is that 24-h net CO_2 exchange is not a reliable measure of growth because the size of the short-term storage pools can vary considerably from day to day depending on environmental and developmental circumstances (Moldau and Karolin 1977; McCree 1986).

Long-term elevated daytime CO_2 often results in a decrease in specific (dry mass basis) respiration rate even though photosynthesis increases (Amthor 1991; Drake and Leadley 1991). On a nitrogen or protein basis, however, effects of CO_2 history on respiration are less striking or nonexistent (Amthor 1991; Baker et al. 1992; Wullschleger et al. 1992b). Photo-respiration is inversely related to CO_2, so mitochondrial NADH oxidation during the day may also be inversely related to CO_2. Decreased daytime demand for NADH oxidation with elevated CO_2 might lead to decreased mitochondrial oxidative capacity (diminution of respiratory chain components), resulting in the observed decrease in nighttime respiration rate. But, if a decrease in mitochondrial oxidative *capacity* due to slowed mitochondrial photorespiratory NADH oxidation does lead to a decrease in oxidative *activity*, respiration is not normally under tight control by product use, unless the decreased capacity is also related to a decline in, say, ATP use for respiratory chain maintenance.

In addition to negative effects of long-term daytime elevated CO_2 on specific respiration rate, nighttime shoot and leaf CO_2 efflux is negatively and instantaneously related to *nighttime* CO_2 level (Decker and Wien 1958; Gale 1982; Bunce 1990; Amthor et al. 1992; but see Ryle et al. 1992). The apparent inhibition of respiration by CO_2 in the dark is readily reversible (Amthor et al. 1992). Elevated nighttime CO_2 can slow mobilization of leaf starch and this, too, may be related to slowed respiration (Wullschleger et al. 1992a). Both past and present CO_2 levels affect respiration rate.

4.6 Photosynthesis and Root Respiration

Root respiration supports growth and maintenance in roots and active uptake of nutrients from the soil solution. It also supports nutrient assimilation, although a considerable fraction of nitrate assimilation may occur in shoots (Pate and Layzell 1990). Roots are frequently chosen for respiratory studies to avoid concurrent photosynthesis and photorespiration, but root respiration is difficult to study in soil because access to roots is limited and soil microbes also respire. (Soil microbes oxidize primarily dead roots, shoot litter, and root exudates so that soil respiration is linked to long-term previous photosynthesis and plant growth.) Thus, root respiration and carbon budgets are often studied in nutrient solution culture (e.g., Farrar and JHH Williams 1991) whereas little is known of in situ root respiration (but see, e.g., Mogensen 1977; Holthausen and Caldwell 1980; Wagner and Buyanovsky 1989). In solution culture, a significant portion of nutrient uptake may function to replace nutrients lost during efflux from roots whereas in soil this component of uptake may be small (Macduff and Jackson 1992). Theoretical estimates of minimal respiratory requirements for ion uptake, maintenance, and growth have been outlined frequently (e.g., Amthor 1993a).

The linkages between root and shoot growth and metabolism (Brouwer 1983) and stoichiometry between whole-plant photosynthesis and growth (Farrar 1985) dictate a linkage between photosynthesis and root respiration

to support nutrient acquisition for the whole plant and growth and maintenance of the root. In herbs, a change in photosynthesis caused by higher
or lower light results in a respective increased or decrease in root respiration
within a few hours (Hansen 1977; Massimino et al. 1981). Carbohydrate reserves allow some temporal delays in the linkage. At constant temperature,
root respiration may be most rapid during the day (Huck et al. 1962). The
relationship between photosynthesis and root respiration is paralleled by
changes in nutrient uptake (Huck et al. 1962; Massimino et al. 1981; Gastal
and Saugier 1989) and perhaps nitrate reduction (Hansen 1980) and root
growth. Daily root respiration may be linearly related to translocation of
sugar to roots (Hatrick and Bowling 1973), which is likely to depend to daytime photosynthesis. Respiration rate of nodulated *Trifolium repens* roots is
sensitive to changes in shoot photosynthesis in as short a time as 10 min,
which may reflect the time taken to translocate photosynthate from leaves
to roots and indicates that "current photosynthate [rather than reserve
carbohydrate] is the primary source of energy for N_2 fixation" in such plants
(Ryle et al. 1985). On the whole, herb root and nodule respiration responds
positively to shoot photosynthesis; this is not surprising. In keeping with
the previous theme, translocation of sugars to roots will tend to stimulate
processes using sugars and respiratory products, and root activity itself is
likely to facilitate import of photosynthate. In nature, respiration in individual roots or groups of roots is likely to vary from the average rate of the
whole root system due to local differences in soil conditions and root activity
and ontogeny (Holthausen and Caldwell 1980).

4.7 Conclusions

Photosynthesis, photorespiration, and respiration most likely interact in
photosynthetic cells during the day because the three processes share intermediates. Both photorespiration and respiration release CO_2 in mitochondria
and probably compete for access to the respiratory chain. Photosynthesis
and respiration can each supply ATP, reductant, and carbon skeletons to
the cytosol. Moreover, products of the individual processes are substrates
and cofactors of the other processes, e.g., CO_2 released by respiration
and photorespiration is a photosynthetic substrate and sugars produced by
photosynthesis are respiratory substrates. Respiration and photosynthesis
can interact via source-sink relationships, too; an increase in sink activity
can stimulate photosynthesis whereas a decline in sink activity may slow
photosynthesis (Herold 1980) with respiration often coupled to sink activity.
Respiration and photosynthesis are also related because respiration contributes to the construction of photosynthetic machinery.

There is a more or less fixed ratio of daily whole-plant carbon gain in
photosynthesis (less photorespiration), carbon use in growth, and carbon

loss in respiration within a species or genotype (Farrar 1985; Farrar and JHH Williams 1991) although the precise chain of events underlying that relationship is incompletely understood. Indeed, the rate of whole-plant respiration depends in large part on previous photosynthesis and the partitioning of photosynthate to growth or storage. Thus, the physical environment is expected to influence respiration in part by controlling photosynthesis and carbon partitioning – conditions favorable for photosynthesis tend to enhance respiration. The environment can also influence respiration independent of an effect on photosynthesis (Amthor 1993b). Many environmental stresses inhibit growth to a greater extent than they reduce photosynthesis, resulting in an accumulation of nonstructural carbohydrates (Munns 1988). The reduction in growth lessens the demand for growth and maintenance respiration and the ratio between respiration and photosynthesis becomes smaller. This gives rise to an *apparent* increase in the efficiency of photoassimilate use ($Y_{(C)}$, see Appendix; McCree 1986). In healthy plants, the balance between capacities for carbohydrate production in photosynthesis and use in heterotrophic metabolism may be coordinated by induction and repression triggered by carbohydrate status (Fig. 4.4). In sum, photosynthesis, respiration, and growth interact across wide temporal and spatial scales. The success of a plant is dependent on a coordination of those processes across those scales. The significance of photosynthesis to plant success can scarcely be appreciated without a consideration of those interactions.

Appendix

It is traditional for plant ecophysiologists to consider respiration as composed of two or more functional components. The separation of components is based on different processes supported by respiratory products rather than different respiratory pathways. The basic relationships underlying the simple and common two-component view (e.g., Wohl and James 1942; Pirt 1965; Thornley 1970, 1971) begin with the tenet that the rate of use of substrate carbon is the sum of carbon use rate for *growth* ($\dot{C}_G$, mol $C\,s^{-1}$) and for *maintenance* ($\dot{C}_M$, mol $C\,s^{-1}$)

$$\dot{C} = \dot{C}_G + \dot{C}_M,$$

where $\dot{C}$ is the rate of substrate carbon use (mol $C\,s^{-1}$) and is derived from, e.g., "carbohydrates" of Fig. 4.1. All the carbon used for maintenance is respired and released as CO_2. Part of the substrate used for growth, however, is added to plant structure and long-term storage so $\dot{C}_G$ is itself the sum of two components

$$\dot{C}_G = \dot{C}_T + \dot{C}_R,$$

where $\dot{C}_T$ (mol $C\,s^{-1}$) is the rate of growth (i.e., addition of carbon to structure and long-term storage) and $\dot{C}_R$ is the rate of respiration (mol

Cs^{-1}) required to support that rate of growth. From a two-component viewpoint, growth includes translocation and nutrient uptake and assimilation in addition to cellular biosynthesis per se.

Respiration rate (R, mol $CO_2 s^{-1}$) is then given by

$$R = \dot{C}_M + \dot{C}_R,$$

where $\dot{C}_M$ is called maintenance respiration rate and $\dot{C}_R$ is called growth respiration rate. Specific respiration rate (r, mol $CO_2 g^{-1}s^{-1}$) is equal to R/W, where W is plant dry mass (g).

The specific maintenance respiration rate or *maintenance coefficient* (m, mol $CO_2 g^{-1}s^{-1}$) is equal to $\dot{C}_M/W$, but maintenance respiration rate may be better related to plant protein content than to dry mass (Ryan 1991; but see Byrd et al. 1992 concerning leaves) in which case a protein-based maintenance coefficient ($m_{(N)}$, mol $CO_2 g^{-1}$ protein s^{-1}) can be defined by

$$m_{(N)} = \dot{C}_M/N,$$

where N is the protein content (g) of existing phytomass. The maintenance coefficient can be estimated experimentally or theoretically, but all available methods are problematic (Amthor 1989). In any case, given a value for $m_{(N)}$, maintenance respiration rate is equal to $m_{(N)} N$. Because maintenance processes occur continuously in all living cells, R will always be greater than zero. The above relationships can be extended by dividing metabolism into a greater number of classes of process (Amthor 1993a).

The *apparent growth yield* is the amount of new plant structure and long-term storage formed per unit of substrate consumed. It is also called the *growth efficiency* (Tanaka and Yamaguchi 1968; Yamaguchi 1978; Amthor 1989). The apparent growth yield in terms of carbon ($Y_{(C)}$, mol C added to new structure and long-term storage per mol C in substrate used for growth and respiration, or mol C mol^{-1}C) is

$$Y_{(C)} = \dot{C}_T/\dot{C} = \dot{C}_T/(R + \dot{C}_T).$$

The apparent growth yield with respect to energy ($Y_{(E)}$, J J^{-1}) is given by

$$Y_{(E)} = \dot{E}_T/\dot{E},$$

where $\dot{E}_T$ (J s^{-1}) is $\dot{C}_T$ times the energy content (J mol^{-1}C) of the products of growth and $\dot{E}$ (J s^{-1}) is $\dot{C}$ times the energy content (J mol^{-1}C) of the substrates of growth and respiration, e.g., carbohydrates.

The yield of the growth processes per se, or *true growth yield* (Pirt 1965), in terms of carbon ($Y_{G(C)}$, mol C added to new structure and long-term storage per mol C used in growth processes, or mol C mol^{-1}C) is

$$Y_{G(C)} = \dot{C}_T/\dot{C}_G,$$

which is related to the production value (*PV*) of Penning de Vries et al. (1974). Again, a comparable relationship defines the true growth yield in terms of energy ($Y_{G(E)}$, JJ^{-1})

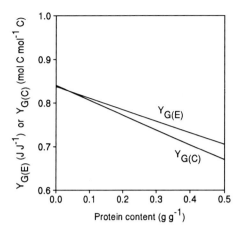

Fig. 4.A1. Calculated true growth yields of energy $[Y_{G(E)}]$ and carbon $[Y_{G(C)}]$ as a function of tissue protein content (the underlying principles of calculation are outlined by Penning de Vries et al. 1989). Here, growth includes biosynthesis of new structure, mineral uptake, translocation, and nitrogen assimilation. The source of nitrogen is nitrate with 25% of nitrogen assimilation taking place heterotrophically (i.e., 75% is supported directly by photosynthetic metabolism). The dry composition (mass/mass) of this hypothetical tissue is 5% fats, 7% lignins, 5% organic acids, and 8% minerals. The protein content is shown on the figure (*abscissa*) and the carbohydrate content is given by the remainder (i.e., 25% to 75%, *from right to left*)

$$Y_{G(E)} = \dot{E}_T / \dot{E}_G,$$

where $\dot{E}_G$ (J s^{-1}) is $\dot{C}_G$ times its initial energy content, i.e., as carbohydrate. An upper limit (theoretical maximum value) of $Y_{G(C)}$, and $Y_{G(E)}$, can be calculated based on stoichiometries of biosynthetic and respiratory pathways. If $Y_{G(C)}$ is determined from such a pathway analysis, a minimum rate of growth respiration can be estimated from

$$\dot{C}_R = (1 - Y_{G(C)}) \, \dot{C}_T / Y_{G(C)}.$$

For many plant tissues, $Y_{G(C)}$ and $Y_{G(E)}$ will take on values between about 0.75 and 0.80 (Fig. 4.A1). That is, no more than 0.75–0.80 of the carbon and energy in photosynthate can be retained in new plant structure because of growth costs. Values of $Y_{(C)}$ and $Y_{(E)}$ are always less than the values of $Y_{G(C)}$ and $Y_{G(E)}$, respectively.

The ratio of heat released to CO_2 released in heterotrophic metabolism (β, kJ mol^{-1} CO_2) will be negatively related to relative growth rate if the composition of growing tissue does not change with growth rate (Fig. 4.A2). Moreover, r should increase as relative growth rate increases (Fig. 4.A2), as is commonly observed (see Amthor 1993b). Specific respiration rate is expected to be positively related to growth efficiency, whereas the ratio of energy release to CO_2 release should be negatively related to growth efficiency (Fig. 4.A3).

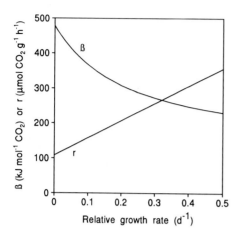

Fig. 4.A2. Calculated specific respiration rate (r) and ratio of heat released to CO_2 released (β) as a function of relative growth rate (RGR) for plant structure that is 55% carbohydrates, 20% proteins, 5% fats, 5% lignins, 5% organic acids, and 10% minerals. The maintenance respiration coefficient is set to $0.15\,\mu$mol $CO_2\,g^{-1}$ protein s^{-1} in this simulation and the growth costs are based on Penning de Vries et al. (1989), Pate and Layzell (1990), and Amthor (1993a). Nitrogen source and assimilation are as in (Fig. 4.A1). Growth respiration is divided among biosynthesis of new structure (59%), translocation (18%), nitrogen assimilation (10%), nitrogen uptake (9%), and non-nitrogen mineral uptake and transport (4%). An RGR of 0 denotes mature tissue, i.e., a state of maintenance. The specific rate of heat production is the product of β and r, i.e., $51.6\,J\,g^{-1}\,h^{-1}$ for RGR = 0 and $82.2\,J\,g^{-1}\,h^{-1}$ for RGR = $0.5\,d^{-1}$ for this hypothetical plant

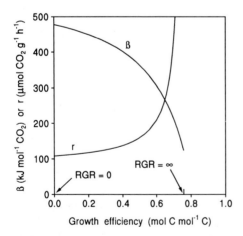

Fig. 4.A3. Calculated specific respiration rate (r) and ratio of heat released to CO_2 released (β) for tissue described in (Fig. 4.A2) as a function of instantaneous growth efficiency or apparent growth yield ($Y_{(C)}$). With RGR = ∞, growth efficiency is about $0.76\,$mol C mol^{-1} C, β is about $123\,kJ\,mol^{-1}\,CO_2$, and r is infinite. In this simulation, which assumes that tissue composition, $Y_{G(E)}$, $Y_{G(C)}$, and $m_{(N)}$ remain constant across growth rates, it is an increase in RGR that increases growth efficiency and r while decreasing β

References

Allen LH Jr, Lemon ER (1976) Carbon dioxide exchange and turbulence in a Costa Rican tropical rain forest. In: Monteith JL (ed) Vegetation and the atmosphere, vol 2. Academic Press, London, pp 265–308

Amthor JS (1989) Respiration and crop productivity. Springer, Berlin Heidelberg New York

Amthor JS (1991) Respiration in a future, higher-CO_2 world. Plant Cell Environ 14: 13–20

Amthor JS (1993a) Respiration and carbon assimilate use. In: Boote KJ (ed) Physiology and determination of crop yield. American Society of Agronomy, Madison (in press)

Amthor JS (1993b) Plant respiratory responses to the environment and their effects on the carbon balance. In: Wilkinson RE (ed) Plant-environment interactions. Marcel Dekker, New York (in press)

Amthor JS, Koch GW, Bloom AJ (1992) CO_2 inhibits respiration in leaves of *Rumex crispus* L. Plant Physiol 98: 757–760

ap Rees T (1980) Assessment of the contributions of metabolic pathways to plant respiration. In: Davies DD (ed) Metabolism and respiration. Biochemistry of plants, vol 2. Academic Press, New York, pp 1–29

ap Rees T (1985) The organization of glycolysis and the oxidative pentose phosphate pathway in plants. In: Douce R, Day DA (eds) Higher plant cell respiration. Encyclopedia of plant physiology, NS, vol 18. Springer, Berlin Heidelberg New York, pp 391–417

ap Rees T (1988) Hexose phosphate metabolism by nonphotosynthetic tissues of higher plants. In: Preiss J (ed) Carbohydrates. Biochemistry of plants, vol 14. Academic Press, San Diego, pp 1–33

Aslam M, Huffaker RC (1984) Dependency of nitrate reduction on soluble carbohydrates in primary leaves of barley under aerobic conditions. Plant Physiol 75: 623–628

Avelange M-H, Sarrey F, Rébillé F (1990) Effects of glucose feeding on respiration and photosynthesis in photoautotrophic *Dianthus caryophyllus* cells. Plant Physiol 94: 1157–1162

Azcón-Bieto J, Osmond CB (1983) Relationship between photosynthesis and respiration. The effect of carbohydrate status on the rate of CO_2 production by respiration in darkened and illuminated wheat leaves. Plant Physiol 71: 574–581

Azcón-Bieto J, Lambers H, Day DA (1983) Effect of photosynthesis and carbohydrate status on respiratory rates and the involvement of the alternative pathway in leaf respiration. Plant Physiol 72: 598–603

Baker JT, Laugel F, Boote KJ, Allen LH Jr (1992) Effects of daytime carbon dioxide concentration on dark respiration in rice. Plant Cell Environ 15: 231–239

Baysdorfer C, van der Woude WJ (1988) Carbohydrate responsive proteins in the roots of *Pennisetum americanum*. Plant Physiol 87: 566–570

Beevers H (1961) Respiratory metabolism in plants. Row, Peterson and Co, Evanston, Illinois

Beevers H (1970) Respiration in plants and its regulation. In: Setlik I (ed) Prediction and measurement of photosynthetic productivity. PUDOC, Wageningen, pp 209–214

Bingham IJ, Farrar JF (1988) Regulation of respiration in roots of barley. Physiol Plant 73: 278–285

Björkman O (1981) Responses to different quantum flux densities. In: Lange OL, Nobel PS, Osmond CB, Ziegler H (eds) Physiological plant ecology I. Responses to the physical environment. Encyclopedia of plant physiology, NS, vol 12A. Springer, Berlin Heidelberg New York, pp 57–107

Björkman O, Demming B (1987) Photon yield of O_2 evolution and chlorophyll fluorescence characteristics at 77 K among vascular plants of diverse origins. Planta 170: 489–504

Brooks A, Farquhar GD (1985) Effect of temperature on the CO_2/O_2 specificity of ribulose-1,5-bisphosphate carboxylase/oxygenase and the rate of respiration in the light. Planta 165: 397–406

Brouquisse R, James F, Raymond P, Pradet A (1991) Study of glucose starvation in excised maize root tips. Plant Physiol 96: 619–626

Brouwer R (1983) Functional equilibrium: sense or nonsense? Neth J Agric Sci 31: 335–348

Brown GC (1992) Control of respiration and ATP synthesis in mammalian mitochondria and cells. Biochem J 284: 1–13

Bryce JH, Azcón-Bieto J, Wiskich JT, Day DA (1990) Adenylate control of respiration in plants: the contribution of rotenone-insensitive electron transport to ADP-limited oxygen consumption by soybeam mitochondria. Physiol Plant 78: 105–111

Bunce JA (1990) Short- and long-term inhibition of respiratory carbon dioxide efflux by elevated carbon dioxide. Ann Bot 65: 637–642

Byrd GT, Sage RF, Brown RH (1992) A comparison of dark respiration between C_3 and C_4 plants. Plant Physiol 100: 191–198

Chapman DJ, Ragan MA (1980) Evolution of biochemical pathways: evidence from comparative biochemistry. Annu Rev Plant Physiol 31: 639–678

Charles-Edwards DA, Ludwig LJ (1975) A model of leaf carbon metabolism. Ann Bot 39: 819–829

Cloud P (1976) Beginning of biospheric evolution and their biogeochemical consequences. Paleobiology 2: 351–387

Copeland L, Turner JF (1987) The regulation of glycolysis and the pentose phosphate pathway. In: Davies DD (ed) Biochemistry of metabolism. Biochemistry of plants, vol 11. Academic Press, San Diego, pp 107–128

Dancer J, Hatzfeld W-D, Stitt M (1990) Cytosolic cycles regulate the turnover of sucrose in heterotrophic cell-suspension cultures of *Chenopodium rubrum* L. Planta 182: 223–231

Day DA, Dry IB, Soole KL, Wiskich JT, Moore AL (1991) Regulation of alternative pathway activity in plant mitochondria. Deviations from Q-pool behavior during oxidation of NADH and quinols. Plant Physiol 95: 948–953

Decker JP, Wien JD (1958) Carbon dioxide surges in green leaves. J Sol Energy Sci Eng 2: 39–41

Delbrück M (1986) Mind from matter? An essay on evolutionary epistemology. Blackwell, Palo Alto

Douce R (1985) Mitochondria in higher plants: structure, function, and biogenesis. Academic Press, Orlando

Douce R, Neuburger M (1989) The uniqueness of plant mitochondria. Annu Rev Plant Physiol Plant Mol Biol 40: 371–414

Douce R, Bligny R, Brown D, Dorne A-J, Genix P, Roby C (1991) Autophagy triggered by sucrose deprivation in sycamore (*Acer pseudoplatanus*) cells. In: Emes MJ (ed) Compartmentation of plant metabolism in non-photosynthetic tissues. Soc Exp Biol Sem Ser, vol 42. Cambridge University Press, Cambridge, pp 127–145

Drake BG, Leadley PW (1991) Canopy photosynthesis of crops and native plant communities exposed to long-term elevated CO_2. Plant Cell Environ 14: 853–860

Dry IB, Wiskich JT (1982) Role of the external adenosine triphosphate/adenosine diphosphate ratio in the control of plant mitochondrial respiration. Arch Biochem Biophys 217: 72–79

Dry IB, Wiskich JT (1985) Characterisation of glycine and malate oxidation by pea leaf mitochondria: evidence of differential access to NAD and respiratory chains. Aust J Plant Physiol 12: 329–339

Dry IB, Bryce JH, Wiskich JT (1987) Regulation of mitochondrial respiration. In: Davies DD (ed) Biochemistry of metabolism. Biochemistry of plants, vol 11. Academic Press, San Diego, pp 213–252

Dry IB, Moore AL, Day DA, Wiskich JT (1989) Regulation of alternative pathway activity in plant mitochondria: nonlinear relationship between electron flux and the redox poise of the quinone pool. Arch Biochem Biophys 273: 148–157

Du Cloux HC, André M, Daguenet A, Massimino J (1987) Wheat response to CO_2 enrichment: growth and CO_2 exchanges at two plant densities. J Exp Bot 38: 1421–1431

Duff SMG, Moorhead GBG, Lefebvre DD, Plaxton WC (1989) Phosphate starvation inducible "bypasses" of adenylate and phosphate dependent glycolytic enzymes in *Brassica nigra* suspension cells. Plant Physiol 90: 1275–1278

Dutton RG, Jiao J, Tsujita MJ, Grodzinski B (1988) Whole-plant CO_2 exchange measurements for nondestructive estimation of growth. Plant Physiol 86: 355–358

Ebbighausen H, Jia C, Heldt HW (1985) Oxaloacetate translocator in plant mitochondria. Biochim Biophys Acta 810: 184–199

Farquhar GD, von Caemmerer S, Berry JA (1980) A biochemical model of photosynthetic CO_2 assimilation in leaves of C_3 species. Planta 149: 78–90

Farrar JF (1985) The respiratory source of CO_2. Plant Cell Environ 8: 427–438

Farrar JF, Williams JHH (1991) Control of the rate of respiration in roots: compartmentation, demand and the supply of substrate. In: Emes MJ (ed) Compartmentation of plant metabolism in non-photosynthetic tissues. Soc Exp Biol Sem Ser, vol 42. Cambridge University Press, Cambridge, pp 167–188

Farrar JF, Williams ML (1991) The effects of increased atmospheric carbon dioxide and temperature on carbon partitioning, source-sink relations and respiration. Plant Cell Environ 14: 819–830

Fondy BR, Geiger DR, Servaites JC (1989) Photosynthesis, carbohydrate metabolism, and export in *Beta vulgaris* L. and *Phaseolus vulgaris* L. during square and sinusoidal light regimes. Plant Physiol 89: 396–402

Foyer CH (1988) Feedback inhibition of photosynthesis through source-sink regulation in leaves. Plant Physiol Biochem 26: 483–492

French RC, Beevers H (1953) Respiratory and growth responses induced by growth regulators and allied compounds. Am J Bot 40: 660–666

Fricaud A-C, Walters AJ, Whitehouse DG, Moore AL (1992) The role(s) of adenylate kinase and the adenylate carrier in the regulation of plant mitochondrial respiratory activity. Biochim Biophys Acta 1099: 253–261

Gaastra P (1963) Climatic control of photosynthesis and respiration. In: Evans LT (ed) Environmental control of plant growth. Academic Press, New York, pp 113–138

Gale J (1982) Evidence for essential maintenance respiration of leaves of *Xanthium strumarium* at high temperature. J Exp Bot 33: 471–476

Gardeström P (1987) Adenylate ratios in the cytosol, chloroplasts and mitochondria of barley leaf protoplasts during photosynthesis at different carbon dioxide concentrations. FEBS Lett 212: 114–118

Gardeström P, Edwards GE (1985) Leaf mitochondria (C_3 + C_4 + CAM). In: Douce R, Day DA (eds) Higher plant cell respiration. Encyclopedia of plant physiology, NS, vol 18. Springer, Berlin Heidelberg New York, pp 314–346

Gardeström P, Wigge B (1988) Influence of photorespiration on ATP/ADP ratios in the chloroplasts, mitochondria, and cytosol, studied by rapid fractionation of barley (*Hordeum vulgare*) protoplasts. Plant Physiol 88: 69–76

Gastal F, Saugier B (1989) Relationships between nitrogen uptake and carbon assimilation in whole plants of tall fescue. Plant Cell Environ 12: 407–416

Gaudillère J-P, Mousseau M (1989) Short term effect of CO_2 enrichment on leaf development and gas exchange of young poplars (*Populus euramericana* cv. I 214) Acta Oecologica 10: 95–105

Geider RJ (1992) Respiration: taxation without representation? In: Falkowski PG, Woodhead AD (eds) Primary productivity and biogeochemical cycles in the sea. Environmental science research, vol 43. Plenum, New York, pp 333–360

Gemel J, Randall DD (1992) Light regulation of leaf mitochondrial pyruvate dehydrogenase complex. Role of photorespiratory carbon metabolism. Plant Physiol 100: 908–914

Gest H (1987) Evolutionary roots of the citric acid cycle in prokaryotes. In: Kay J, Weitzman PDJ (eds) Krebs' citric acid cycle – half a century and still turning. Biochemical society symposium 54. Biochemical Soc, London, pp 3–16

Getz HP (1991) Sucrose transport in tonoplast vesicles of red beet roots is linked to ATP hydrolysis. Planta 185: 261–268

Gifford RM, Lambers H, Morison JIL (1985) Respiration of crop species under CO_2 enrichment. Physiol Plant 63: 351–356

Graham D (1980) Effects of light on "dark" respiration. In: Davies DD (ed) Metabolism and respiration. Biochemistry of plants, vol 2. Academic Press, New York, pp 525–579

Graham D, Chapman EA (1979) Interactions between photosynthesis and respiration in higher plants. In: Gibbs M, Latzko E (eds) Photosynthesis II. Photosynthetic carbon metabolism and related processes. Encyclopedia of plant physiology, NS, vol 6. Springer, Berlin Heidelberg New York, pp 150–162

Grodzinski B (1992) Plant nutrition and growth regulation by CO_2 enrichment. BioScience 42: 517–525

Hagihara A, Hozumi K (1991) Respiration. In: Raghavendra AS (ed) Physiology of trees. Wiley, New York, pp 87–110

Hänisch ten Cate CH, Breteler H (1981) Role of sugars in nitrate utilization by roots of dwarf bean. Physiol Plant 52: 129–135

Hansen GK (1977) Adaption to photosynthesis and diurnal oscillation of root respiration rates for *Lolium multiflorum*. Physiol Plant 39: 275–279

Hansen GK (1980) Diurnal variation of root respiration rates and nitrate uptake as influenced by nitrogen supply. Physiol Plant 48: 421–427

Hanson KR (1992) Evidence for mitochondrial regulation of photosynthesis by a starchless mutant of *Nicotiana sylvestris*. Plant Physiol 99: 276–283

Hatrick AA, Bowling DJF (1973) A study of the relationship between root and shoot metabolism. J Exp Bot 24: 607–613

Hatzfeld W-D, Stitt M (1990) A study of the rate of recycling of triose phosphates in heterotrophic *Chenopodium rubrum* cells, potato tubers, and maize endosperm. Planta 180: 198–204

Hatzfeld W-D, Dancer J, Stitt M (1990) Fructose-2,6-bisphosphate, metabolites and "coarse" control of pyrophosphate: fructose-6-phosphate phosphotransferase during triose-phosphate cycling in heterotrophic cell-suspension cultures of *Chenopodium rubrum*. Planta 180: 205–211

Heichel GH (1970) Prior illumination and the respiration of maize leaves in the dark. Plant Physiol 46: 359–362

Henry M-F, Nyns E-J (1975) Cyanide-insensitive respiration. An alternative mitochondrial pathway. Sub-Cell Biochem 4: 1–65

Herold A (1980) Regulation of photosynthesis by sink activity – the missing link. New Phytol 86: 131–144

Holthausen RS, Caldwell MM (1980) Seasonal dynamics of root system respiration in *Atriplex confertifolia*. Plant Soil 55: 307–317

Huck MG, Hageman RH, Hanson JB (1962) Diurnal variation in root respiration. Plant Physiol 37: 371–375

Ikuma H (1972) Electron transport in plant respiration. Annu Rev Plant Physiol 23: 419–136

James WO (1953) Plant respiration. Oxford University Press, London

Jordan DB, Ogren WL (1984) The CO_2/O_2 specificity of ribulose 1,5-bisphosphate carboxylase/oxygenase. Planta 161: 308–313

Journet E-P, Bligny R, Douce R (1986) Biochemical changes during sucrose deprivation in higher plant cells. J Biol Chem 261: 3193–3199

Kaiser G, Heber U (1984) Sucrose transport into vacuoles isolated from barley mesophyll protoplasts. Planta 161: 562–568

Kelly GJ, Gibbs M (1973) Nonreversible D-glyceraldehyde 3-phosphate dehydrogenase of plant tissues. Plant Physiol 52: 111–118

Kira T (1975) Primary production of forests. In: Cooper JP (ed) Photosynthesis and productivity in different environments. Cambridge University Press, Cambridge, pp 5–40

Kirschbaum MUF, Farquhar GD (1987) Investigation of the CO_2 dependence of quantum yield and respiration in *Eucalyptus pauciflora*. Plant Physiol 83: 1032–1036

Kok B (1948) A critical consideration of the quantum yield of Chlorella photosynthesis. Enzymologia 13: 1–56

Kowallik W (1982) Blue light effects on respiration. Annu Rev Plant Physiol 33: 51–72

Krapp A, Quick WP, Stitt M (1991) Ribulose-1,5-bisphosphate carboxylase-oxygenase, other Calvin-cycle enzymes, and chlorophyll decrease when glucose is supplied to mature spinach leaves via the transpiration stream. Planta 186: 58–69

Krömer S, Heldt HW (1991a) On the role of mitochondrial oxidative phosphorylation in photosynthesis metabolism as studied by the effect of oligomycin on photosynthesis in protoplasts and leaves of barley (*Hordeum vulgare*). Plant Physiol 95: 1270–1276

Krömer S, Heldt HW (1991b) Respiration of pea leaf mitochondria and redox transfer between the mitochondrial and extramitochondrial compartment. Biochim Biophys Acta 1057: 42–50

Krömer S, Stitt M, Heldt HW (1988) Mitochondrial oxidative phosphorylation participating in photosynthetic metabolism of a leaf cell. FEBS Lett 226: 352–356

Krotkov G (1960) The organic materials of respiration. In: Ruhland W (ed) Plant respiration inclusive fermentations and acid metabolism, part 1. Encyclopedia of plant physiology, vol XII. Springer, Berlin Heidelberg New York, pp 47–65

Lafitte HR, Loomis RS (1988) Calculation of growth yield, growth respiration and heat content of grain sorghum from elemental and proximal analysis. Ann Bot 62: 353–361

Laing WA, Ogren WL, Hageman RH (1974) Regulation of soybean net photosynthetic CO_2 fixation by the interaction of CO_2, O_2, and ribulose 1,5-diphosphate carboxylase. Plant Physiol 54: 678–685

Larcher W (1983) Physiological plant ecology, 2nd edn. Springer, Berlin Heidelberg New York

Lehninger AL (1965) The mitochondrion: molecular basis of structure and function. Benjamin, New York

Lorimer GH (1983) Carbon dioxide and carbamate formation: the makings of a biochemical control system. Trends Biochem Sci 8: 65–68

Ludwig LJ, Charles-Edwards DA, Withers AC (1975) Tomato leaf photosynthesis and respiration in various light and carbon dioxide environments. In: Marcelle R (ed) Environmental and biological control of photosynthesis. Junk, The Hague, pp 29–36

Macduff JH, Jackson SB (1992) Influx and efflux of nitrate and ammonium in Italian ryegrass and white clover roots: comparisons between effects of darkness and defoliation. J Exp Bot 43: 525–535

Mannella CA (1985) The outer membrane of plant mitochondria. In: Douce R, Day DA (eds) Higher plant cell respiration. Encyclopedia of plant physiology, NS, vol 18. Springer, Berlin Heidelberg New York, pp 106–133

Margulis L (1970) Origin of eukaryotic cells. Yale University Press, New Haven

Massimino D, André M, Richaud C, Daguenet A, Massimino J, Vivoli J (1981) The effect of a day at low irradiance of a maize crop. I. Root respiration and uptake of N, P and K. Physiol Plant 51: 150–155

Maynard Smith J (1978) Optimization theory in evolution. Annu Rev Ecol Syst 9: 31–56

McCashin BG, Cossins EA, Canvin DT (1988) Dark respiration during photosynthesis in wheat leaf slices. Plant Physiol 87: 155–161

McCree KJ (1986) Whole-plant carbon balance during osmotic adjustment to drought and salinity stress. Aust J Plant Physiol 13: 33–43

McDermitt DK, Loomis RS (1981) Elemental composition of biomass and its relation to energy content, growth efficiency, and growth yield. Ann Bot 48: 275–290

Mogensen VO (1977) Field measurements of dark respiration rates of roots and aerial parts in Italian ryegrass and barley. J Appl Ecol 14: 243–252

Moldau H, Karolin A (1977) Effect of the reserve pool on the relationship between respiration and photosynthesis. Photosynthetica 11: 38–47

Monteith JL (1972) Solar radiation and productivity in tropical ecosystems. J Appl Ecol 9: 747–766

Morin F, André M, Betsche T (1992) Growth kinetics, carbohydrate, and leaf phosphate content of clover (*Trifolium subterraneum* L.) after transfer to a high CO_2 atmosphere or to high light and ambient air. Plant Physiol 99: 89–95

Munns R (1988) Why measure osmotic adjustment? Aust J Plant Physiol 15: 717–726

Nicholls DG, Ferguson SJ (1992) Bioenergetics 2. Academic Press, London

Palmer JM (1979) The "uniqueness" of plant mitochondria. Biochem Soc Trans 7: 246–252

Pate JS, Layzell DB (1990) Energetics and biological costs of nitrogen assimilation. Biochem Plants 16: 1–42

Penning de Vries FWT, Brunsting AHM, van Laar HH (1974) Products, requirements and efficiency of biosynthesis: a quantitative approach. J Theor Biol 45: 339–377

Penning de Vries FWT, Jansen DM, ten Berge HFM, Bakema A (1989) Simulation of ecophysiological processes of growth in several annual crops. PUDOC, Wageningen, The Netherlands

Pirt SJ (1965) The maintenance energy of bacteria in growing cultures. Proc R Soc B 163: 224–231

Prosser CL (1986) Adaptational biology: molecules to organisms. Wiley, New York

Rebeille F (1988) Photosynthesis and respiration in air-grown and CO_2-grown photoautotrophic cell suspension cultures of carnation. Plant Sci 54: 11–21

Rowland-Bamford AJ, Allen LH Jr, Baker JT, Boote KJ (1990) Carbon dioxide effects on carbohydrate status and partitioning in rice. J Exp Bot 41: 1601–1608

Ryan MG (1991) Effects of climate change on plant respiration. Ecol Appl 1: 157–167

Ryle GJA, Powell CE, Gordon AJ (1985) Short-term changes in CO_2 evolution associated with nitrogenase activity in white clover in response to defoliation and photosynthesis. J Exp Bot 36: 634–643

Ryle GJA, Powell CE, Tewson V (1992) Effect of elevated CO_2 on the photosynthesis, respiration and growth of perennial ryegrass. J Exp Bot 43: 811–818

Saradadevi K, Raghavendra AS (1992) Dark respiration protects photosynthesis against photoinhibition in mesophyll protoplasts of pea (Pisum sativum). Plant Physiol 99: 1232–1237

Satterlee LD, Koller HR (1984) Response of soybean fruit respiration to changes in whole plant light and CO_2 environment. Crop Sci 24: 1007–1010

Schäfer C, Simper H, Hofmann B (1992) Glucose feeding results in coordinated changes of chlorophyll content, ribulose-1,5-bisphoshate carboxylase-oxygenase activity and photosynthetic potential in photoautrophic suspension cultured cells of Chenopodium rubrum. Plant Cell Environ 15: 343–350

Servaites JC, Geiger DR, Tucci MA, Fondy BR (1989a) Leaf carbon metabolism and metabolite levels during a period of sinusoidal light. Plant Physiol 89: 403–408

Servaites JC, Fondy BR, Li B, Geriger DR (1989b) Sources of carbon for export from spinach leaves throughout the day. Plant Physiol 90: 1168–1174

Sharkey TD (1985) Photosynthesis in intact leaves of C_3 plants: physics, physiology and rate limitations. Bot Rev 51: 53–105

Sharkey TD (1988) Estimating the rate of photorespiration in leaves. Physiol Plant 73: 147–152

Sharp RE, Matthews MA, Boyer JS (1984) Kok effect and the quantum yield of photosynthesis. Light partially inhibits dark respiration. Plant Physiol 75: 95–101

Sheen J (1990) Metabolic repression of transcription in higher plants. Plant Cell 2: 1027–1038

Sims DA, Pearcy RW (1991) Photosynthesis and respiration in Alocasia macrorrhiza following transfers to high and low light. Oecologia 86: 447–453

Singh KK, Chen C, Gibbs M (1992) Characterization of an electron transport pathway associated with glucose and fructose respiration in the intact chloroplasts of Chlamydomonas reinhardtii and spinach. Plant Physiol 100: 327–333

Sitte P, Eschbach S (1992) Cytosymbiosis and its significance in cell evolution. Prog Bot 53: 29–43

Soole KL, Dry IB, James AT, Wiskich JT (1990) The kinetics of NADH oxidation by complex I of isolated plant mitochondria. Physiol Plant 80: 75–82

Soole KL, Dry IB, Wiskich JT (1992) Partial purification and characterization of complex I, NADH:ubiquinone reductase, from the inner membrane of beetroot mitochondria. Plant Physiol 98: 588–594

Srere PA (1987) Complexes of sequential metabolic enzymes. Annu Rev Biochem 56: 89–124

Steingröver E (1981) The relationship between cyanide-resistant root respiration and the storage of sugars in the taproot in *Daucus carota* L. J Exp Bot 32: 911–919

Tanaka A, Yamaguchi J (1968) The growth efficiency in relation to the growth of the rice plant. Soil Sci Plant Nutr 14: 110–116

Thornley JHM (1970) Respiration, growth and maintenance in plants. Nature 227: 304–305

Thornley JHM (1971) Energy, respiration, and growth in plants. Ann Bot 35: 721–728

Tolbert NE (1980) Photorespiration. In: Davies DD (ed) Metabolism and respiration. Biochemistry of plants, vol 2. Academic Press, New York, pp 487–523

Wagner GH, Buyanovsky GA (1989) Soybean root respiration assessed from short-term ^{14}C-activity changes. Plant Soil 117: 301–303

Weitzman PDJ (1987) Patterns of diversity of citric acid cycle enzymes. In: Kay J, Weitzman PDJ (eds) Krebs' citric acid cycle – half a century and still turning. Biochemical society symposium 54. Biochemical Soc, London, pp 33–43

Wellburn AR, Owen JH (1991) Control of the rate of respiration in shoots: light, calcium and plant growth regulators. In: Emes MJ (ed) Compartmentation of plant metabolism in non-photosynthetic tissues. Soc Exp Biol Sem Ser, vol 42. Cambridge University Press, Cambridge, pp 189–198

Wenzler HC, Mignery GA, Fisher LM, Park WD (1989) Analysis of a chimeric class-I patatin-GUS gene in transgenic potato plants: high-level expression in tubers and sucrose-inducible expression in cultured leaf and stem explants. Plant Mol Biol 12: 41–50

Williams JHH, Farrar JF (1990) Control of barley root respiration. Physiol Plant 79: 259–266

Williams K, Percival F, Merino J, Mooney HA (1987) Estimation of tissue construction cost from heat of combustion and organic nitrogen content. Plant Cell Environ 10: 725–734

Williams K, Field CB, Mooney HA (1989) Relationships among leaf construction cost, leaf longevity, and light environment in rain-forest plants of the genus *Piper*. Am Nat 133: 198–211

Wiskich JT, Dry IB (1985) The tricarboxylic acid cycle in plant mitochondria: its operation and regulation. In: Douce R, Day DA (eds) Higher plant cell respiration. Encyclopedia of plant physiology, NS, vol 18. Springer, Berlin Heidelberg New York, pp 281–313

Wiskich JT, Bryce JH, Day DA, Dry IB (1990) Evidence for metabolic domains within the matrix compartment of pea leaf mitochondria. Implications for photorespiratory metabolism. Plant Physiol 93: 611–616

Wohl K, James WO (1942) The energy changes associated with plant respiration. New Phytol 41: 230–256

Wullschleger SD, Norby RJ, Hendrix DL (1992a) Carbon exchange rates, chlorophyll content, and carbohydrate status of two forest tree species exposed to carbon dioxide enrichment. Tree Physiol 10: 21–31

Wullschleger SD, Norby RJ, Gunderson CA (1992b) Growth and maintenance respiration in leaves of *Liriodendron tulipifera* L. exposed to long-term carbon dioxide enrichment in the field. New Phytol 121: 515–523

Yamaguchi J (1978) Respiration and the growth efficiency in relation to crop productivity. J Fac Agric Hokkaido Univ 59: 59–129

Added in proof: Gly-3-P in Fig. 4.2 is the same as 3-PGald in Fig. 4.5 and 1,3-DiPGA in Fig. 4.2 is the same as 1,3-bisPGA in Fig. 4.5

5   Apoplastic and Symplastic Proton Concentrations and Their Significance for Metabolism

H. Pfanz

5.1   Introduction

Proton concentration is a major factor in restricting life within definite boundaries. In general, life occurs in a neutral or slightly acidic environment, but fungi, bacteria, animals, and higher plants have nevertheless managed to conquer terrestrial and aqueous niches at extreme pH values. The pH extremes for cellular growth are around pH 1 and pH 11 (Souza et al. 1974; Langworthy 1978). Life therefore exists within the enormous H^+ concentration range of 10^{10}. During evolution, higher plants have adapted to the different soil conditions with which their root systems have been confronted. Acidophilic, neutrophilic, and acidophobic plant types have evolved. The pH of the soil water determines the availability of nutrients or heavy metals and thus also determines soil toxicity (Larcher 1980). Some important cultural plants tolerate only a very narrow soil pH range (e.g., *Medicago sativa*), whereas others are very tolerant (e.g., *Secale cereale*). The pH sensitivity of these plants seems to be either due to direct H^+ effects on the roots, to mycorrhizal or rhizobial symbionts (Schubert 1987), or to indirect effects like the release of toxic heavy metals or aluminium from the ion-exchanging compounds in the soil solution (Ulrich 1981; Rehfuess 1981).

Also within plant organs and cells, the hydrogen ion concentration is probably the most fundamental factor for a functional metabolism. Nearly every reaction in a living cell is affected – or even regulated – by pH. The hydrogen ion concentration determines not only the ionic state and consequently the availability of several organic and inorganic metabolites, but also the stability and function of biological macromolecules and membranes. Especially proteins are very sensitive to modifications of hydrogen ion concentration. Many key enzymes – namely those of the chloroplast stroma – are directly regulated by pH (Woodrow et al. 1984). The reason for the strong pH sensitivity of enzymes is either the different ionic state of the substrate or of the active site of the enzyme itself, or, in some (irreversible) cases, the pH stability of the protein.

Naturally, there are many possible disturbances leading to an increase or decrease in cellular pH. The potential alkalization of the cytoplasm due to nitrate reduction and the possible acidification of cells during anaerobiosis are two examples. Nevertheless, as long as the perturbation does not exceed a certain threshold, cellular pH-stat mechanisms are able to cope with excess protons or hydroxyl ions. To maintain cellular pHs within a distinct range, several biochemical and biophysical pathways are possible (see Raven 1986, 1988; Davies 1973, 1986). In the repertoire of the cellular pH-stat mechanisms there is the metabolic destruction of organic acids, the reduction of nitrate, or the pumping of protons from one compartment to another, strategies necessary to reduce a possible proton burden.

In this chapter the principles of pH, the methods of pH determination, and the influence of pH on cellular reactions will be described. The proton concentrations of extracellular and several intracellular leaf cell compartments and the possible changes of pH during development and aging, or due to the influence of air pollutants, are discussed. The pH dependency on lignifying and IAA-oxidizing peroxidases will be shown as an example of apoplastic reactions. Being the most important symplastic reaction, the effects of pH on photosynthesis are examined.

5.2 Definitions

5.2.1 The pH Concept

PH value (potentia Hydrogenii) is a measure of the actual concentration (activity) of protons (or hydronium ions H_3O^+) in a solution. It is defined as

$$pH = -\log [H^+], \tag{1}$$

and thus as the negative decadic logarithm of the H^+ concentration (activity). Per definitionem, a solution is called neutral when its pH value is aproximately 7. Solutions above pH 7 (7–14) or below pH 7 (7–0) are called alkaline (basic) or acidic, respectively (theoretically, pH values below zero or above 14 are possible). In other words, if the proton concentration increases, the pH decreases, and vice versa. A change in the proton concentration by a factor of 10 is equal to a pH change of one unit.

5.2.2 The Buffer Concept

A buffer may be defined as a solution which resists pH changes (within a certain limit) despite the addition of acids or bases. This "resistance" to changes in pH is due to the presence of compounds in the solution able to neutralize the surplus H^+ or OH^-. The principle of buffering can be seen in Eqs. (2) and (3).

$$DH + NaOH \longrightarrow DNa + H_2O \tag{2}$$

Table 5.1. pK values of organic and inorganic acids normally occurring in plants. (After Weast et al. 1986)

Acid	pK_1	pK_2	pK_3
Acetic acid	4.75^a	–	–
Lactic acid (100 °C)	3.08	–	–
Malic acid	3.40^a	5.11^a	–
Citric acid	3.14^b	4.77^b	6.39^b
Oxalic acid	1.23^a	4.19^a	–
Phosphoric acid	2.12^a	7.21^a	12.67^c
Carbonic acid (CO_2)	6.37^a	10.25^a	–
Sulfuric acid	<0	1.92^a	–
Sulfurous acid (SO_2)	1.81^c	6.91^c	–
H_2S	7.04^c	11.96^c	–
HF	3.45^a	–	–

[a] 25 °C; [b] 20 °C; [c] 18 °C.

and

$$DNa + HCl \longrightarrow DH + NaCl, \tag{3}$$

with D being an inorganic or an organic anion. For pH values between pH 4 and 10 (the pH range where most biological reactions take place; cf. Table 5.4) the most effective buffers are mixtures of weak acids and their conjugate bases (and vice versa; cf. Table 5.1).

Buffer capacity (β) is the inverse slope of the titration curve, or in other words, the effectivity of keeping the pH constant despite the addition of increments of H^+ or OH^-. It can be written as

$$\beta = \frac{\Delta_{a,b}}{\Delta pH}, \tag{4}$$

with $\Delta_{a,b}$ being the increment of a monoprotic strong acid (or base). According to what was said above, and independent of the specific function in cellular metabolism, proteins inevitably contribute to the overall buffer capacity of cells. The reason for this buffering is the high content of weakly alkaline (amino and guanidino groups) and acidic (carboxyl groups) residues. In addition to proteins, various other compounds with different pK values and buffering abilities are present in cells and organelles in varying concentrations (see Table 5.1). It is therefore no wonder that titration curves of leaf homogenates do not allow the determination of distinct equivalence points or pK values (see Pfanz et al. 1987).

The pK value is defined as the negative decadic logarithm of the dissociation constant. Depending on the number of dissociation steps possible (e.g., in the case of mono- or polyprotic acids), more than one pK value exists (see Table 5.1). From Eq. (5) it is evident that, at a given pH, the pK value determines the amount of dissociated and undissociated acid or base.

5.2.3   Techniques to Determine Intra- and Intercellular pH

There are several methods available for the determination of inter- and intracellular pH values in animal and plant cells. As there is a vast amount of literature on these techniques (for recent reviews, see Kurkdjian and Guern 1989a; Pfanz and Heber 1989), only the basic principles will be summarized here.

To estimate the extracellular proton concentration of root cells, micro-electrodes localized only a few microns from root surfaces have been used (Newman et al. 1987). Generally, the classical, but still valuable method of using pH-indicating dyes in agar (Marschner et al. 1982) is used for estimating extracellular pH in roots of intact plants, whereas for leaves, flat pH electrodes are pressed to partially abraded epidermal tissue (Lürsen 1978). The magnitude and the changes of fluxes of weak acids across the plasmalemma are also used for extracellular pH estimations (Grignon and Sentenac 1991). Fluorescent dyes infiltrated into leaves allow qualitative and quantitative estimations of extracellular pH (Pfanz and Dietz 1987; Canny 1988; Edwards et al. 1988). Other attempts use fluids extracted via centrifugation (Madore and Webb 1981)· or apply pressure to extract an extracellular sap (Hartung et al. 1988). A recent approach using fluorescence was described by Hoffmann et al. in 1992.

The spectrum of methods for measuring symplastic pH values is even broader than that described for the cell wall phase (see Caldwell 1956; Boron and Roos 1976; Kurkdjian and Guern 1989a; Pfanz and Heber 1989). A crude, but quick and easy method to use for all plant materials is pH measurement in homogenized extracts of plant organs, of isolated cells or protoplasts, or even of isolated organelles (Clevenger 1919; Pfanz 1987; Wagner 1990; Pfanz and Beyschlag 1992). The main limitation is that with this crude method mainly vacuolar pH is measured.

H^+-selective microelectrodes have extensively been used to study cytoplasmic and vacuolar pH values (Felle and Bertl 1986; Steigner et al. 1988). Chemical probes also found broad application. This technique is based on the different permeation characteristics of undissociated and dissociated probes. Generally, fluorescent or radiolabeled compounds have been used as probes in such procedures (for details see Kurkdjian and Guern 1989a; Pfanz and Heber 1989 and references therein). One of the most widely used compounds in this context is ^{14}C-DMO (5.5-dimethyl-oxazolidine-2.4-dione; Waddell and Butler 1959; Pfanz et al. 1987; Kurkdjian and Guern 1989b). Recently, pyranine and carboxyfluoresceine have been used to follow the kinetics of intracellular pH changes in C_3 plants during light/dark transitions and during exposure to potentially acidic air pollutants (Yin et al. 1990, 1991).

Furthermore, gas exchange with very fast IRGA systems has been applied to detect chloroplastic pH values (Oja et al. 1986). A very intriguing, noninvasive method for pH estimations in (nearly) intact leaf tissues is ^{31}P

NMR spectrometry. This technique allows simultaneous pH measurements of at least the vacuole and the chloroplast compartment (for review see Roberts 1984). Through NMR, a vast amount of literature has accumulated on symplastic pH values (Guern et al. 1986; Kurkdjian and Guern 1989a,b).

5.3 Cellular pH

5.3.1 The Apoplastic pH

5.3.1.1 The Apoplast – What It Is

One can consider a whole plant cormus as consisting of two main "compartments"; one being the continuum of the protoplasts of all organs of a plant body linked with each other via the plasmodesmata, called the *symplast*, the other, the counterpart, in form of a continuous "water-containing cellulose skeleton" being the *apoplast*. The apoplast consists of two main fractions. These fractions are: (1) the gaseous intercellular air space and (2) the aqueous cell wall phase. The intercellular air space (IAS) borders the external air space at the substomatal cavity. The labyrinth-like IAS penetrates through leaves, petioles, stems, and roots, forming an infinite air-filled continuum. The concentration of carbon dioxide in the intercellular air space is of vital importance for a working photosynthesis and for the calculation of photosynthetic rates (Beyschlag et al., this Vol.). In aquatic plants, a special aerenchyma has evolved to guide oxygen to the roots or rhizomes, thus avoiding anaerobiosis (Schröder et al. 1986).

The cell wall phase again consists of two fractions, one being the aqueous phase and the other being a fibrillar cellulose skeleton (the proper cell wall; see below). The cell wall is thus a "continuous, more or less tough integument . . . , formed mainly by a fibrillar framework within which the interstices are filled with an amorphous material" (Colvin 1981) and an aqueous medium.

The cell wall was long considered to be a dead compartment of minor importance and its only significance was seen in the stabilization of the plant cells. Yet it has been recognized and it is now generally accepted that the plant cell wall is a "living" reaction room with vital functions (Varshney and Varshney 1985; Pfanz and Oppmann 1991; Grignon and Sentenac 1991; Pfanz et al. 1992). The different functions are: (1) The cell wall is a rigid cytoskeleton that allows plant growth up to heights of 100 m and more. All the "naked" protoplasts are embedded in, and surrounded by the cell wall, and there is tight contact between the plasmalemma and the apoplastic solution. The cell wall protects the protoplasts from mechanical injuries like abrasion, desiccation, ultraviolet radiation or excessive light, etc. (Colvin 1981). (2) The apoplast of roots and leaves is the very first target for

Table 5.2. Intercellular air space (IAS) and cell wall volume (CWV) of leaves from different plant species. IAS. was determined by infiltrating water into fully water-saturated, pre-weighed leaves and final re-weighing. Cell wall volume was calculated through microscopical analyses. Data are expressed per fresh weight (FM). (After Pfanz 1987)

Species	CWV $\mu l\, g^{-1}$ FM	IAS $\mu l\, g^{-1}$ FM
Spinacia oleracea	77	680
Citrus limon	175	550
Hordeum vulgare	81	300
Taxus baccata	100	345

microbial attack (Esquerre-Tugaye and Lamport 1979), atmospheric air pollutants (Pfanz et al. 1990; Pfanz and Oppmann 1991) or even heavy metals (Verkleij and Schat 1989), and therefore is an ideal site for stress recognition and for detoxification. (3) If not transported through the symplast (via plasmodesmata), compounds excreted or taken up into a cell inevitably have to pass through the aqueous phase of the cell wall. This includes nutrients as well as toxic compounds. (4) The extracellular proton concentration modifies the ionic state of dissociable acids or bases and thus determines permeability and availability of solutes. Photosynthetic gas exchange or uptake and release of gaseous air pollutants also proceed via this water phase.

5.3.1.2 Cytomorphological Parameters

Table 5.2 gives an impression of the dimensions of the gaseous and the aqueous volumes occupied by the apoplast of leaves. It seems as if the cell wall occupies a higher volume in scleromorphous leaves (dicot: *Citrus*, and polycot: *Taxus*) than in more mesomorphic leaves (dicot: *Spinacia*, and monocot: *Hordeum*). The intercellular air space ranges between 300 to nearly $700\,\mu l\, g^{-1}$ fresh mass of the leaf.

Naturally, these ratios cannot be regarded as being constant throughout the life of a leaf. During the development, maturation, and aging of leaves, changes occur in the proportions of intercellular air space or cell wall volume. In fact, the portion of the cell wall mass per unit leaf mass increases in the annual course in coniferous and deciduous leaves (Table 5.3).

5.3.1.3 Apoplastic pH and Extracellular Enzymes

Coinciding with the various functions of the apoplast, a great number of different enzymes are localized in the cell wall. Among those proteins

Table 5.3. Portion of the leaves that is cell wall material in common forest trees. The cell wall fragments were isolated quantitatively from deciduous and coniferous trees. The data are given in mg dry cell wall per g dry leaf matter. (After Pfanz and Weinerth, unpubl.)

Species	May	July	September
Fagus sylvatica	420	450	450
Acer pseudoplatanus	290	350	350
Picea abies	390	n.d.	410
Pinus sylvestris	400	460	550

sequestered by the protoplasts, covalently bound, ionically bound, and freely permeable ones were described (Mäder et al. 1975; Lamport 1980). Extracellular phosphatases and glucosidases (Keegstra and Albersheim 1970), glucanases (Goldberg 1977), invertases (Baker 1978), cellulases, chitinases, pectinases (Jermyn and Yeow 1975), methyl esterases, and peroxidases (Mäder et al. 1975; Rohringer et al. 1983; Pfanz et al. 1990; Pfanz and Oppmann 1991) were found. Recently an IAA-oxidizing extracellular peroxidase has been partially characterized (Pfanz 1992).

Cell wall peroxidases are thought to be responsible for the lignification process. In vitro they have been shown to catalyze the oxidative polymerization of phenylpropanoids via a radical chain reaction (Elstner and Heupel 1976; Gross et al. 1977) and additionally, it has been shown that these enzymes are able to generate hydrogen peroxide, which is a prerequisite for lignification (Elstner and Heupel 1976). Using fluids isolated from the apoplast of intact leaves of barley, beech, and spruce, we were able to broaden our knowledge on the spectrum of extracellular peroxidases (Pfanz et al. 1990; Pfanz and Oppmann 1991; Oppmann and Pfanz 1992; Pfanz 1992).

Figure 5.1 shows the pH dependence of such apoplastic peroxidases from different leaf sources. In the case of IAA oxidation, optimum pH was around 4 in all the cell wall extracts tested. As the isozymes of peroxidases have different catalytic activities, pH optima also differ (cf. Pedreno et al. 1989; Pfanz and Oppmann 1991). There is still a great lack of knowledge about the physicochemical properties of the apoplast, especially with regards to local pH differences (Pfanz et al. 1988; Pedreno et al. 1989). Furthermore, nearly nothing is known about regulation or modulation of apoplastic enzymes.

The compartmentation of plant growth factors can be regulated by manipulating pH and thereby also changing the activity of IAA-oxidizing peroxidases within the system. A decrease in cytoplasmic pH or an increase in apoplastic pH leads to an efflux of IAA or ABA from the cell interior into the external periplasmic space. In parallel, extracellular IAA catabolism is decreased near neutrality and above (see Fig. 5.1), leading to a decrease

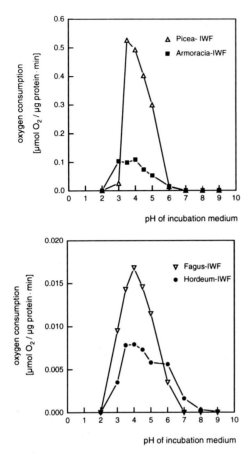

Fig. 5.1. pH dependence of the IAA oxidation of apoplastic peroxidases from different leaf sources. *Upper panel* ($\triangle$) = *Picea abies*; ($\blacksquare$) = *Armoracia lapathifolia*. *Lower panel* (∇) = *Fagus sylvatica*; ($\bullet$) = *Hordeum vulgare*

in the ratio of IAA_{sympl} to IAA_{apopl}. (Kaiser and Hartung 1981; Cowan et al. 1982). To our knowledge, no evidence has been brought forward that ABA is also sequestered in the extracellular space.

5.3.1.4 Apoplastic pH and Growth

One theory for cell expansion is based on extracellular acidification with a concomitant cell wall loosening by cation/proton exchange from the carboxyl groups of the uronic acids. IAA has been postulated to catalyze the apoplastic acidification (Taiz 1984) and is thus directly involved in cellular growth. Nevertheless, it is still a matter of dispute whether or not proton extrusion and the concomitant loosening of the Ca^{2+} from the carboxyl groups indeed weaken the cell wall (Ilan 1973; Cleland 1981). During cell extension and cell wall synthesis, physical as well as biochemical parameters change. Growing and expanding cells need a flexible skeleton to allow osmotic swelling and concomitant volume change, whereas differentiated and mature cells have to be surrounded by a rigid and stiffened wall. For this reason, growing leaves – or growing parts of a leaf – have a more acidic pH than nongrowing, mature parts of leaves.

5.3.1.5   Apoplastic pH and Hormone Action

According to the acid growth theory, cell enlargement through water uptake and osmotic swelling is preceded by IAA-stimulated cell wall loosening (see above). The hormone is thought to promote acidification of the apoplastic space via proton extrusion from the protoplasts, leading to a displacement of Ca (and Mg) ions from the carboxyl groups of the uronides of the middle lamella (Taiz 1984; Cleland 1971). Acidification of the symplast, on the other hand, reduces IAA catabolism (cf. Fig. 5.1). Provided that IAA (or ABA) are moving between the apoplast and the symplast purely by diffusion (Kaiser and Hartung 1981; Cowan et al. 1982; Hartung 1983; Hartung and Slovik 1991), an external acidification would lead to an increase in undissociated IAA and therefore to an influx of the hormone into the cells. As most of the phytohormones have pK values around 4.7 (GA, IAA, and ABA are weak acids), accumulation will take place in alkaline compartments where these hormones are trapped. Cowan et al. (1982) were the first to calculate transport and distribution of ABA on this basis. This excellent study of Cowan's was the basis of a more detailed computer model established by Hartung and Slovik (1991) some 10 years later. It can be shown with those models that metabolic or stress-induced pH changes of the symplast or of the apoplast therefore can lead to a redistribution of slightly acidic plant hormones, with various consequences on metabolism (Kaiser and Hartung 1981; Kremer et al. 1987).

5.3.2   The Symplastic pH

5.3.2.1   Photosynthesis and pH

The photosynthetic reactions of higher plants and algae are very sensitive to changes in the pH of the chloroplast stroma (Heldt et al. 1973; Werdan et al. 1975; Heber and Heldt 1981). Most of the steps involved in the assimilation of CO_2 are catalyzed by enzymes and are therefore strictly pH-dependent (Woodrow et al. 1984). The tight regulation of the proton concentration in the chloroplast therefore plays a dominant role in controlling photosynthetic carbon reduction.

Table 5.4. Proton concentrations of various cellular compartments as determined in intact tissues and isolated cell organelles

Cellular compartment	Reference[a]	pH
Apoplast	1–4	5.25–6.40
Vacuole	5–14	0.67–6.5
Chloroplast stroma	15–18	Light: 8; dark 7.2–7.4
Thylakoid	18, 19	5.0–6.5
Cytosol	10, 20, 21	7.2–7.6
Mitochondria	22, 23	

[a] 1, Pfanz and Dietz (1987); 2, Grignon and Sentenac (1991) and references therein; 3, Hartung et al. (1988); 4, Hoffmann et al. (1992); 5, Meeuse (1956); 6, McClintock et al. (1982); 7, Kurkdjian and Guern (1989a); 8, Kurkdjian and Guern (1989b); 9, Smith and Raven (1979); 10, Martin et al. (1982); 11, Strack et al. (1987); 12, Steigner et al. (1988); 13, Kaiser and Hartung (1981); 14, Matile (1978); 15, Heber and Heldt (1981); 16, Heldt et al. (1973); 17, Oja et al. (1986); 18, Werdan et al. (1975); 19, Falkner et al. (1976); 20, Mathieu et al. (1986); 21, Guern et al. (1986); 22, Addanki et al. (1968); 23, Roos and Boron (1981).

It is known that the pH of the illuminated chloroplast stroma in vivo is approximately pH 8 (Heldt et al. 1973; Enser and Heber 1980; Espie and Colman 1981; Oja et al. 1986). In darkened leaves, stromal pH is around pH 7.4 and thus near the proton concentration of the cytosol embedding the chloroplasts (see Table 5.4). The acidification of the chloroplast stroma during the light/dark transition greatly influences the activity of several Calvin cycle enzymes. The pH optimum of the Calvin cycle enzyme fructose-bisphosphatase (FBPase) is near 8. Below pH 7.5, activity is hardly measurable (Leegood et al. 1982; Woodrow et al. 1984). Alkalization from pH 5.5 to 8.5 shifts the sedoheptulose-bisphosphatase dimer (which is inactive) to the monomeric "active" form of the enzyme (Buchanan et al. 1976).

5.3.2.2 Extracellular pH Changes and Photosynthesis

Within a normal life cycle, a protoplast is very often confronted with situations of varying pH. PH changes occur in the apoplast during cell extension and leaf growth, (Taiz 1984); but also during uptake of potentially acidic or alkaline air pollutants transient pH perturbations cannot be excluded (Mansfield and Freer-Smith 1981; Pfanz and Heber 1989; Yin et al. 1991). The pH of an external solution optimal for protoplast or cell photosynthesis is around pH 7. Figure 5.2 gives an impression of the photosynthetic behaviour of leaves and leaf cells measured under various "external" proton conditions. The experiments were performed with isolated protoplasts of mono- and dicotyledonous plant species (*Hordeum* and *Ficaria*), free-living green algae (*Euglena*), and with leaf tissues of conifers (*Tsuga*), deciduous

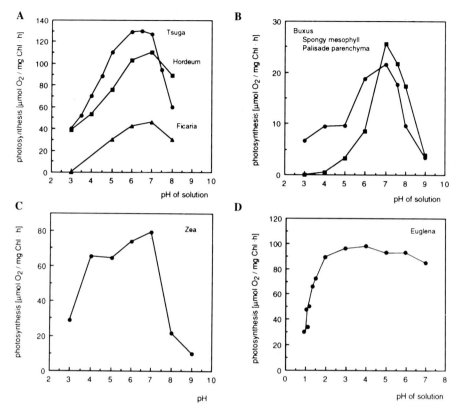

Fig. 5.2 A–D. Photosynthetic oxygen evolution and external pH. The tissues, cells, or protoplasts were kept in buffered solutions at the pH specified. Light was provided by a 150 W lamp ($1000 \, \mu E \, m^{-2} s^{-1}$). Oxygen evolution was measured at $20 \, ^\circ C$ in a Clark-type electrode. **A** Protoplasts of *Hordeum vulgare* (■), *Ficaria verna* (▲), and needles of *Tsuga canadense* (●). **B** Palisade parenchyma (●) and spongy mesophyll (■) of *Buxus sempervirens*. **C** Leaf tissue of *Zea mays*. **D** Unicellular green alga *Euglena viridis*. The data of the leaf tissues were obtained after the epidermis of the leaves was peeled and the leaves were infiltrated with the test solutions. (**A** and **D** after Pfanz 1987; **B** and **C** after Pfanz, Bruch, and Lesch, unpubl.)

trees (*Buxus*), and C_4 plants (*Zea*). The epidermis was peeled from the leaves and the tissue was infiltrated prior to the experiments. The incubation media in which the experiments were performed therefore reflect an "artificial" apoplast. Optimum photosynthetic activity was obtained when pH values around neutrality were applied in the experiments. The values for optimum photosynthesis were similar, irrespective of whether isolated protoplasts or leaf discs (e.g., Fig. 5.2B) were examined. Furthermore, the data were highly similar, independent of the origin of the tissue (e.g., conifer needles or C_4 leaves; Fig. 5.2A and C) under the exposure conditions (short

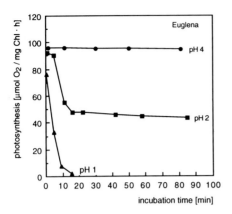

Fig. 5.3. Kinetics of the inhibition of photosynthesis of *Euglena* cells at different acidic conditions in the incubation medium

time: 10–15 min) applied. Compared to maize or barley, *Buxus* revealed a relative sharp pH optimum. The photosynthetic response to different pH conditions in the apoplast was also similar, when palisade parenchyma cells and cells from the spongy mesophyll of single leaves were studied (*Buxus* in Fig. 5.2B). Within a particular leaf, photosynthesis curves of these different tissues paralleled each other at pH values above neutrality, but below pH 6 a difference was to be seen, indicating a higher proton sensitivity of the spongy mesophyll.

Quite in contrast to what was said for box tree (*Buxus*), the unicellular alga *Euglena* showed a very broad pH range of working photosynthesis. Between pH 2 and 8 nearly no difference in CO_2 reduction rates was measured (Fig. 5.2D). Only when pH values were further reduced did photosynthesis decrease. The optimal stromal pH for a working photosynthesis lies in the very narrow pH range of 7.6–7.8 (Heldt et al. 1973; Woodrow et al. 1984). It is generally believed that stromal pH values are similar in all plant species, as the crucial enzymes are thought to have similar pH optima in all plants. A difference in the proton sensitivity of plant tissues is therefore thought to simply reflect different magnitudes in the proton permeability of the biomembranes and in pH-stat mechanisms able to cope with the deviations in pH. That the H^+ permeability of the plasmalemma is a crucial factor in determining acid effects of plant tissue is seen in Fig. 5.3, where the kinetics of the inhibition of *Euglena* photosynthesis is given as a function of external pH. Whereas the alga was able to photosynthetize several hours at pH 4, its photosynthetic rates decreased to 50% within 15 min at pH 2, or within 5 min at pH 1, to be no longer measurable after 15 min at pH 1. It is interesting to note that photosynthesis in the extremely acidophilic alga *Dunaliella acidophila* is optimal at pH 1 and becomes inhibited at pH values higher than 2 (Gimmler et al. 1990).

5.3.2.3 Intracellular pH Changes and Photosynthesis

Potentially acidic molecules can penetrate membranes easily when they are in their neutral (undissociated) form [HA in Eq. (5)]. Whether they are protonated or not is determined by the dissociation constant (K_a) and by the pH of the solution. A useful expression relating K_a of a weak acid to the concentration of the various forms of the acid, and to the pH, is the Henderson-Hasselbalch equation:

$$pH = pK_a + \log\frac{[A^-]}{[HA]},\qquad(5)$$

where A^- is the dissociated and HA the protonated form of the weak acid, and pK_a the negative decadic logarithm of the dissociation constant. The pK itself is influenced by the ionic strength (law of Debye-Hückel; see Pfanz and Heber 1989) and the temperature of the medium (Van't Hoff's law see Morris 1974, Aducci et al. 1982).

The dissociation equation for a bi-protic weak acid is

$$H_2A \leftrightarrow HA^- + H^+ \leftrightarrow A^{2-} + 2H^+.\qquad(6)$$

For carbon dioxide or sulfur dioxide the equations are given below. The pK values for these reactions are listed in Table 5.1.

$$H_2SO_3 \leftrightarrow HSO_3^- + H^+ \leftrightarrow SO_3^{2-} + 2H^+\qquad(7)$$

$$H_2CO_3 \leftrightarrow HCO_3^- + H^+ \leftrightarrow CO_3^{2-} + 2H^+.\qquad(8)$$

During a short-term exposure of the cells to acid stress (10–20 min), the plasmalemma seemed to be relative impermeable to the externally applied "charged" protons (cf. Figs. 5.2 and 5.3). However, the membrane barrier is negligible when undissociated weak acids or bases (acting as proton or hydroxyl ion carriers) easily penetrate the plasmalemma (Pfanz et al. 1987). According to Eq. (6), they liberate protons (or hydroxyl ions) inside the cytoplasm (or the vacuole; for details see Pfanz and Heber 1989). According to the flux equation

$$\Phi = \frac{\Delta C}{R},\qquad(9)$$

permeation of H^+-consuming or H^+-producing compounds is a matter of prevailing pH gradients (under conditions where only the neutral form is freely permeable and the dissociated forms are not transported via special carriers). With pH values between pH 4.2 and 6.4 in the cell wall of leaves (Pfanz and Dietz 1987; Grignon and Sentenac 1991, Hoffmann et al. 1992), and cytoplasmic pH values around neutrality (cf. Table 5.1), potentially acidic molecules are trapped inside the cell (principle of the ion trap). As a consequence of dissociation and proton liberation, pH perturbations are likely to occur. Figure 5.4 gives an example of the inhibition of the pho-

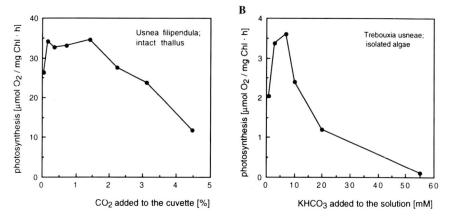

Fig. 5.4 A,B. Photosynthesis of an intact thallus (**A**) and of isolated algae *Trebouxia usneae* (**B**) of the lichen *Usnea filipendula* at various CO_2 or $KHCO_3$ concentrations in the medium. Measurements were performed at pH 6.5 in a Clark-type oxygen electrode either in the gas phase (**A**) or in solution (**B**). The algae were freshly isolated from homogenized material. (After Pfanz and Protte, unpubl.)

tosynthetic carbon reduction of a lichen in the presence of an excess of carbon dioxide. It is seen that, irrespective of whether the CO_2 concentration in air was raised over 1.5% (v/v) in the experiment with an intact *Usnea* thallus, or the bicarbonate concentration was increased in the incubation medium of freshly isolated *Trebouxia* algae to values around 5 mM

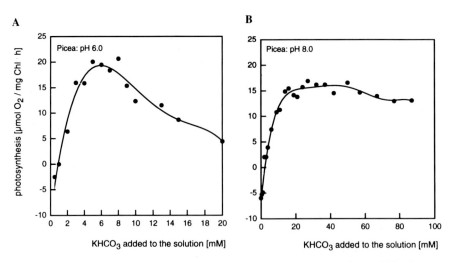

Fig. 5.5 A,B. Photosynthesis of spruce needles at various concentrations of bicarbonate and two different proton concentrations (pH 8 and 6) of the medium. Measurements were performed in an oxygen electrode with 1-year-old needles of *Picea abies* that were cut longitudinally. (After Pfanz and Rumpel, unpubl.)

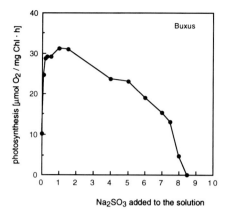

Fig. 5.6. Inhibition of photosynthesis of box tree leaves (*Buxus sempervirens*) after the addition of sulfite to the incubation medium. Measurements were performed with palisade parenchyma cells in an oxygen electrode. (After Pfanz, Lesch, and Bruch, unpubl.)

(at pH 6.4), net photosynthesis decreased in both cases. Inhibition was reversible (not shown) because the inhibitory effect was due only to pure acidification and the original situation could be restored by the cellular pH stat mechanisms after cessation of the stress (Bown 1985, cf. also Wagner 1990). Figure 5.5 clearly demonstrates that the extent of the decrease in photosynthesis is also determined by external pH. The lower the pH of the external solution, the lower the bicarbonate concentration needed to decrease the rate of photosynthesis in the mesophyll cells of spruce needles [cf. Eqs. (6), (8) and Table 5.1].

Not only carbon dioxide can lead to intracellular acidification when applied in excess; other potentially acidic gases can also reduce stromal pH and consequently inhibit photosynthesis (Thomas et al. 1944; Simon and Beevers 1952; Hager and Moser 1985; Pfanz and Heber 1986; Pfanz et al. 1987; Kronberger 1988; Pfanz and Heber 1989; Yin et al. 1990, 1991). The pH-induced inhibition of photosynthesis of a *Buxus* leaf treated with sulfur dioxide is given in Fig. 5.6. Yet, in contrast to the acidifying properties of CO_2, the damaging effects brought about during SO_2 stress are not directly comparable, as sulfur dioxide does not act only as an acid, but also has highly deleterious effects on metabolism due to the toxicity of its anions (cf. Ziegler 1975).

5.4   Conclusions

Although nutrition (Raven 1986, 1988), water stress and desiccation (Hartung et al. 1988), anaerobiosis (Hager and Moser 1985), and pollutant stress

(Pfanz et al. 1987; Kronberger 1988; Yin et al. 1990, 1991) can create proton/hydroxyl imbalances in cells and organelles, plants are equipped with pH-stat mechanisms to cope with the excess H^+ or OH^- (Davies 1973, 1986; Smith and Raven 1979; Pfanz and Heber 1986; Raven 1988). Longer-lasting deviations from the normal "physiological" pH values are rarely recorded, and pH is kept within very narrow limits (Pfanz and Beyschlag 1992; Pfanz and Vollrath 1992; Heber et al., this Vol.). Cellular pH values are tightly regulated and well protected by biochemical and biophysical buffer systems to allow a functioning metabolism, as relatively small pH changes (e.g., less than 0.3 pH units) may lead to severe changes in metabolism.

Acknowledgments. Most of the experiments presented above were performed within the Sonderforschungsbereich 251 of the Universität Würzburg and within the project No. 6495-1053-21632 of the Staatsministerium für Landesentwicklung und Umweltfragen (StMLU) coordinated by the Projektgruppe Bayern zur Erforschung der Wirkung von Umweltschadstoffen (PBWU) and the EUROSILVA/Eureka 447 project No. PT BEO 51/Og 1201 5. The proofreading of Martha Virginia White is gratefully acknowledged.

References

Addanki S, Cahill FD, Sotos JF (1968) Determination of intramitochondrial pH and intramitochondrial-extramitochondrial pH gradient of isolated heart mitochondria by the use of 5.5-dimethyl-2.4-oxazolidinedione. J Biol Chem 243: 2337–2348
Aducchi P, Federico R, Carpinelli G, Podo F (1982) The temperature dependence of intracellular pH in higher plant cells. Planta 156: 579–582
Baker DA (1978) Proton co-transport of organic solutes by plant cells. New Phytol 81: 485–497
Boron WF, Roos A (1976) Comparison of microelectrode, DMO, and methylamine methods for measuring intracellular pH. Am J Physiol 231: 799–809
Bown AW (1985) CO_2 and intracellular pH. Plant Cell Environ 8: 459–465
Buchanan BB, Schürmann P, Wolosiuk RA (1976) Appearance of sedoheptulose-1.7-diphosphatase activity on conversion of chloroplast fructose-1.6-diphosphatase from dimer to monomer form. Biochem Biophys Res Commun 69: 970–978
Caldwell PD (1956) Intracellular pH. Int Rev Cytol 5: 229–277
Canny MJ (1988) Bundle sheath tissue of legume leaves as the site of recovery of solutes from the transpiration stream. Physiol Plant 73: 457–464
Cleland RE (1981) Wall extensibility: hormones and wall extension. In: Tanner W, Loewus FA (eds) Plant carbohydrates II. Encycl Plant Physiol NS, vol 13B. Springer, Berlin Heidelberg New York, pp 255–273
Clevenger CB (1919) Hydrogen-ion concentration of plant juices: II Factors affecting the hydrogen-ion concentration of plant juices. Soil Sci 8: 277–242
Colvin JR (1981) Ultrastructure of the plant cell wall: Biophysical viewpoint. In: Tanner W, Loewus FA (eds) Plant carbohydrates II. Encyc Plant Physiol NS 13B. Springer, Berlin Heidelberg New York, pp 9–23
Cowan IR, Raven JA, Hartung W, Farquhar GD (1982) A possible role for abscisic acid in coupling stomatal conductance and photosynthetic carbon metabolism in leaves. Aust J Plant Physiol 9: 489–498
Davies DD (1973) Control of and by pH. Symp Soc Exp Biol 27: 513–529
Davies DD (1986) The fine control of cytosolic pH. Physiol Plant 67: 702–706

Edwards MC, Smith GN, Bowling DJF (1988) Guard cells extrude protons prior to stomatal opening – a study using fluorescence microscopy and pH microelectrodes. J Exp Bot 39: 1541–1547

Elstner EF, Heupel A (1976) Formation of hydrogen peroxide by isolated cell walls from horse-radish (*Armoracia lapathifolia* Gilib.). Planta 130: 175–180

Enser U, Heber U (1980) Metabolic regulation by pH gradients. Inhibition of photosynthesis by indirect proton transfer across the chloroplast envelope. Biochim Biophys Acta 592: 577–591

Espie GS, Colman B (1981) The intracellular pH of isolated, photosynthetically active *Asparagus* mesophyll cells. Planta 153: 210–216

Esquerre-Tugaye MT, Lamport DTA (1979) Cell surfaces in plant microorganism interactions. I. A structural investigation of cell wall hydroxyproline rich glycoproteins which accumulate in fungus-infected plants. Plant Physiol 64: 314–319

Falkner G, Horner F, Werdan K, Heldt HW (1976) pH changes in the cytoplasm of the blue-green alga *Anacystis nidulans* caused by light-dependent proton flux into the thylakoid space. Plant Physiol 58: 717–718

Felle B, Bertl A (1986) The fabrication of H^+-selective liquid membrane microelectrodes for use in plant cells. J Exp Bot 37: 1416–1428

Gimmler H, Bental M, Degani H, Avron M, Pick U (1990) The H^+ export capacity of *Dunaliella acidophila* and the permeability of the plasma membrane for H^+ and weak acids. In: Baltscheffsky M (ed) Current research in photosynthesis, vol IV. Kluwer, Dordrecht, pp 773–776

Goldberg R (1977) On possible connections between auxin-induced growth and cell wall glucanase activities. Physiol Plant 50: 261–264

Grignon C, Sentenac H (1991) pH and ionic conditions in the apoplast. Annu Rev Plant Physiol Plant Mol Biol 42: 103–128

Gross GG, Janse C, Elstner EF (1977) Involvement of malate, monophenols, and the superoxide radical in hydrogen peroxide formation by isolated cell walls from horse-radish (*Armoracia lapathifolia* Gilib.). Planta 136: 271–276

Guern J, Mathieu Y, Pean M, Pasquier C, Beloeil J-C, Lallemand J-Y (1986) Cytoplasmic pH regulation in *Acer pseudoplatanus* cell. I. ^{31}P-NMR description of acid-load effects. Plant Physiol 82: 840–845

Hager A, Moser J (1985) Acetic acid esters and permeable weak acids induce active proton extrusion and extension growth of coleoptile segments by lowering cytoplasmic pH. Planta 163: 391–400

Hartung W (1983) Die intrazelluläre Verteilung von Phytohormonen in Pflanzenzellen. Hohenheimer Arb 129: 64–80

Hartung W, Slovik S (1991) Physicochemical properties of plant growth regulators and plant tissue determine their distribution and redistribution: stomatal regulation by abscisic acid in leaves. New Phytol 119: 361–382

Hartung W, Radin JW, Hendrix DL (1988) Abscisic acid movement into the apoplastic solution of water-stressed cotton leaves. Role of apoplastic pH. Plant Physiol 86: 908–913

Heber U, Heldt H-W (1981) The chloroplast envelope: structure and function, and role in leaf metabolism. Annu Rev Plant Physiol 32: 139–168

Heldt H-W, Werdan K, Milovancev M, Geller G (1973) Alkalization of the chloroplast stroma caused by light-dependent proton flux into the thylakoid space. Biochim Biophys Acta 314: 224–241

Hoffmann B, Plänker R, Mengel K (1992) Measurement of pH in the apoplast of sunflower leaves by means of fluorescence. Physiol Plant 84: 146–153

Ilan J (1973) On auxin pH drop and on the improbability of its involvement in the primary mechanism of auxin-induced growth promotion. Physiol Plant 28: 146–148

Jermyn MA, Yeow YM (1975) A class of lectins present in the tissues of seed plants. Aust J Plant Physiol 2: 501–531

Kaiser WM, Hartung W (1981) Uptake and release of abscisic acid by isolated photoautotrophic mesophyll cells, depending upon pH gradients. Plant Physiol 68: 202–206

Keegstra K, Albersheim P (1970) The involvement of glycosidases in the cell wall metabolism of suspension-cultured *Acer pseudoplatanus* cells. Plant Physiol 45: 675–678

Kremer H, Pfanz H, Hartung W (1987) Die Wirkung saurer Luftschadstoffe auf Verteilung
und Transport pflanzlicher Wachstumsregulatoren in Laub- und Nadelblättern.
Konsequenzen für streß- und entwicklungsphysiologische Prozesse. Allg Forstztg
27/28/29: 741–744

Kronberger WC (1988) Kinetics of non-ionic diffusion of hydrogen fluoride in plants. II
Model estimations on uptake, distribution, and translocation of F in higher plants.
Phyton (Austria) 28: 27–49

Kurkdjian A, Guern J (1989a) Intracellular pH: Measurement and importance in cell
activity. Annu Rev Plant Physiol Plant Mol Biol 40: 271–303

Kurkdjian A, Guern J (1989b) Intracellular pH in higher plants. I. Improvements in the
use of the 5.5-dimethyloxazolidine $2(^{14}C),4$-dione distribution technique. Plant Sci Lett
11: 337–344

Lamport DTA (1980) Structure and function of plant glycoproteins. In: Stumpf PK, Conn
EE (eds) The biochemistry of plants, vol 3. Academic Press, New York, pp 501–541

Langworthy TA (1978) Microbial life in extreme pH values. In: Kushar DJ (ed) Aerobial
life in extreme environments. Academic Press, New York, pp 279–315

Larcher W (1980) Physiological plant ecology, 2nd edn. Springer, Berlin Heidelberg New
York

Leegood RC, Kobayashi Y, Neimanis S, Walker DA, Heber U (1982) Cooperative
activation of fructose-1.6-bisphosphatase by reductant, pH, and substrate. Biochim
Biophys Acta 682: 168–178

Lürsen K (1978) The surface pH of *Avena* coleoptiles and its correlation with growth.
Plant Sci Lett 13: 309–313

Mäder M, Meyer Y, Bopp M (1975) Lokalisation der Peroxidase-Isoenzyme in Protoplasten
und Zellwänden von *Nicotiana tabacum* L. Planta 122: 259–268

Madore M, Webb JA (1981) Leaf free space analysis and vein loading in *Cucurbita pepo*.
Can J Bot 59: 2550–2587

Mansfield TA, Freer-Smith PH (1981) Effects of urban air pollution on plant growth. Biol
Rev 56: 343–368

Marschner H, Römheld V, Ossenberg-Neuhaus H (1982) Rapid method for measuring
changes in pH and reduction processes along roots of intact plants. Z Pflanzenphysiol
105: 407–416

Martin JB, Bligny R, Rebeille F, Douce R, Leguay J-J, Mathieu Y, Guern J (1982) A ^{31}P
nuclear magnetic resonance study of intracellular pH of plant cells cultivated in liquid
medium. Plant Physiol 70: 1156–1161

Mathieu Y, Guern J, Pean M, Pasquier C, Beloeil J-C, Lallemand J-Y (1986) Cytoplasmic
pH regulation in *Acer pseudoplatanus* cells. II. Possible mechanisms involved in pH
regulation during acid-load. Plant Physiol 82: 846–852

Matile P (1978) Biochemistry and function of vacuoles. Annu Rev Plant Physiol 29:
193–213

McClintock M, Higinbotham N, Uribe EG, Cleland RE (1982) Active, irreversible
accumulation of extreme levels of H_2SO_4 in the brown alga *Desmarestia*. Plant Physiol
70: 771–774

Meeuse BJD (1956) Free sulfuric acid in the brown alga, *Desmarestia*. Biochim Biophys
Acta 19: 372–374

Morris JG (1974) A biologist's physical chemistry, 2nd edn. Edward Arnold, London

Newman JA, Kochian LV, Grusak MA, Lucas WJ (1987) Fluxes of H^+ and K^+ in corn
roots. Characterization and stoichiometry using ion-selective microelectrodes. Plant
Physiol 84: 1177–1184

Oja V, Laisk A, Heber U (1986) Light induced alkalization of the chloroplast stroma in
vivo as estimated from the CO_2 capacity of intact sunflower leaves. Biochim Biophys
Acta 849: 355–365

Oppmann B, Pfanz H (1992) The influence of sulfur dioxide on lignifying processes in the
cell wall of spruce needles. IUFRO Centennial, 15th Int Meet on Air pollution effects,
9–11 Sept 1992, Tharandt, p 51

Pedreno MA, Ros Barcelo A, Sabater F, Munoz R (1989) Control by pH of cell wall peroxidase activity involved in lignification. Plant Cell Physiol 30(2): 237–241

Pfanz H (1987) Aufnahme und Verteilung von Schwefeldioxid in pflanzlichen Zellen und Organellen. Auswirkungen auf den Stoffwechsel. Dissertation, Universität Würzburg

Pfanz H (1992) Oxidation of IAA by extracellular enzymes. Plant Physiol 99: S 835

Pfanz H, Beyschlag W (1993) Photosynthetic performance and nutrient status of Norway spruce (*Picea abies* L. Karst.) at forest sites in the Ore Mountains (Erzgebirge). Trees 7: 115–122

Pfanz H, Dietz K-J (1987) A fluorescence method for the determination of the apoplastic proton concentration in intact leaf tissues. J Plant Physiol 129: 41–48

Pfanz H, Heber U (1986) Buffer capacities of leaves, leaf cells, and leaf cell organelles in relation to fluxes of potentially acidic air pollutants. Plant Physiol 81: 597–602

Pfanz H, Heber U (1989) Determination of extra- and intracellular pH values in relation to the action of acidic gases on cells. In: Linskens HF, Jackson JF (eds) Gases in plant and microbial cells. Mod Meth Plant Anal NS, vol 9. Springer, Berlin Heidelberg New York, pp 322–343

Pfanz H, Oppmann B (1991) The possible role of apoplastic peroxidases in detoxifying the air pollutant sulfur dioxide. In: Lobarzewski J, Greppin H, Penel C, Gaspar T (eds) Biochemical, molecular, and physiological aspects of plant peroxidases. University M Curie-Sklodowska and University of Geneva, Lublin and Geneva, pp 401–417

Pfanz H, Vollrath B (1993) Photosynthese und Nährstoffgehalte von Buchen unterschiedlich stark SO_2-belasteter Standorte. In: Rehfuess KE, Ziegler H (eds) Zustand und Gefährdung der Laubwälder. Rundgespräche Kommission Ökologie, vol 5. Dr Pfeil Verlag, München, pp 129–142

Pfanz H, Martinoia E, Lange OL, Heber U (1987) Flux of SO_2 into leaf cells and cellular acidification by SO_2. Plant Physiol 85: 928–933

Pfanz H, Gsell W, Martinoia E, Dietz K-J (1988) The buffer capacity of the leaf apoplast and its significance in regard to pH changes by acidic air pollutants. Plant Physiol 86: S 899

Pfanz H, Dietz K-J, Weinerth I, Oppmann B (1990) Detoxification of sulfur dioxide by apoplastic peroxidases. In: Rennenberg H, Brunold Ch, De Kok LJ, Stulen I (eds) Sulfur nutrition and assimilation in higher plants; fundamental, environmental and agricultural aspects. SBP Acad Publ, the Hague, NL, pp 229–233

Pfanz H, Würth G, Oppmann B, Schultz G (1992) Sulfite oxidation in, and sulfate uptake from the cell wall of leaves. In vivo studies. Phyton (A) 32: 95–98

Raven JA (1986) Biochemical disposal of excess H^+ in growing plants? New Phytol 104: 175–206

Raven JA (1988) Acquisition of nitrogen by the shoots of land plants: its occurrence and implications for acid-base regulation. New Phytol 109: 1–20

Rehfuess KE (1981) Über die Wirkungen der sauren Niederschläge in Waldökosystemen. Forstwiss Cbl 100: 363–381

Roberts JKM (1984) Study of plant metabolism in vivo using NMR spectroscopy. Annu Rev Plant Physiol 33: 375–386

Rohringer R, Ebrahim-Nesbat F, Wolf G (1983) Proteins in intercellular washing fluids from leaves of barley (*Hordeum vulgare* L.). J Exp Bot 34: 1589–1605

Roos A, Boron WF (1981) Intracellular pH. Physiol Rev 61(2): 296–434

Schröder P, Grosse W, Woermann D (1986) Localization of thermo-osmotically active partitions in young leaves of *Nuphar lutea*. J Exp Bot 37(138): 1450–1461

Schubert E (1987) Der Einfluß des pH-Wertes und der H^+-Pufferung des Bodens auf das Wachstum und die N_2-Fixierung der Ackerbohne (*Vicia faba*). Diss, FB 19 Ernährungswiss, Justus Liebig Univ Gießen

Simon EW, Beevers H (1952) The effect of pH on the biological activities of weak acids and bases. I. The most usual relationship between pH and activity. New Phytol 51: 163–190

Smith FA, Raven JA (1979) Intracellular pH and its regulation. Annu Rev Plant Physiol 30: 289–311

Souza KA, Deal PH, Mack HM, Turnbill EE (1974) Growth and reproduction of microorganisms under extremely alkaline conditions. Appl Microbiol 28: 1066–1068

Steigner W, Köhler K, Simonis W, Urbach W (1988) Transient cytoplasmic pH changes in correlation with opening of potassium channels in *Eremosphaera*. J Exp Bot 198: 23–36

Strack D, Sharma V, Felle H (1987) Vacuolar pH in radish cotyledonal mesophyll cells. Planta 172: 563–565

Taiz L (1984) Plant cell extension: regulation of cell wall mechanical properties. Annu Rev Plant Physiol 35: 585–657

Thomas MD, Hendricks RH, Bryner R, Hill GR (1944) Some chemical reactions of sulphur dioxide after absorption by alfalfa and sugar beets. Plant Physiol 19: 212–226

Ulrich B (1981) Eine ökosystemare Hypothese über die Ursachen des Tannensterbens (*Abies alba* Mill.). Forstwiss Cbl 100: 228–236

Varshney SRK, Varshney CK (1985) Response of peroxidase to low levels of SO_2. Environ Exp Bot 25: 107–114

Verkleij JAC, Schat H (1989) Mechanisms of metal tolerance in higher plants. In: Shaw AJ (ed) Heavy metal tolerance in plants: evolutionary aspects. CRC Press, Boca Raton, pp 179–193

Waddell WJ, Butler TC (1959) Calculation of intracellular pH from the distribution of DMO. Application to skeleton muscle of the dog. J Clin Invest 38: 720–729

Wagner U (1990) Kinetik und Mechanismus der pH-Stabilisierung in grünen Blättern höherer Pflanzen. Diss, Universitität Würzburg

Weast RC, Astle MJ, Beyer WJ (1986) Handbook of chemistry and physics. CRC Press, Boca Raton, Florida

Werdan K, Heldt HW, Milovancev M (1975) The role of pH in the regulation of carbon fixation in the chloroplast stroma. Studies on CO_2 fixation in the light and dark. Biochim Biophys Acta 396: 276–292

Woodrow IE, Murphy DJ, Latzko E (1984) Regulation of stromal sedoheptulose-1.7-bisphosphatase activity by pH and Mg^{2+} concentration. J Biol Chem 259: 3791–3795

Yin Z-H, Neimanis S, Wagner U, Heber U (1990) Light-dependent pH changes in leaves of C_3 Plants. I. Recording pH changes in various cellular compartments by fluorescent probes. Planta 182: 244–252

Yin Z-H, Schmidt W, Heber U (1991) Influence of the air pollutants (SO_2,NO_2,O_3) on cellular pH changes in leaves. In: PBWU (eds) Expertentagung Waldschadensforschung im östlichen Mitteleuropa und in Bayern. GSF Bericht 24/91, pp 574–578

Ziegler I (1975) The effect of SO_2 pollution on plant metabolism. Residue Rev 56: 79–105

6 The Significance of Assimilatory Starch for Growth in *Arabidopsis thaliana* Wild-Type and Starchless Mutants

W. Schulze and E.-D. Schulze

6.1 Introduction

Vegetative growth is generally assumed to reach maximum rates when plants invest as much carbohydrate as possible in the growth of new leaves (Monsi and Murata 1970; Schulze 1983). Only with new leaves are additional production organs established which by compound interest contribute to further growth (Harper 1989; Diemer et al. 1992). Growth rates are eventually limited by supply of resources, such as water and nutrients, or light in the case of self-shading (Schulze and Chapin 1987). Storage of carbohydrates could be expected to compete with growth of leaves and roots and thus reduce the maximum growth rate (Chapin et al. 1990). Nevertheless, carbohydrate storage is a very common phenomenon in the plant kingdom and production of assimilatory starch occurs in chloroplasts of all leaf tissues. Therefore, in this chapter we investigate the functional role of assimilatory starch and its interaction with growth using a starchless mutant compared with a wild type of *Arabidopsis thaliana* (Schulze et al. 1991).

It is not known if assimilatory starch competes with growth processes because methods which perturb the formation of starch by altering the environment generally also have direct effects on growth, and therefore the data are only correlative in nature. For example, reduced light will reduce assimilatory starch and growth, but one cannot distinguish whether or not plants with or without starch formation have increased growth at high light. This problem can only be investigated by using isogenic plants, which have a single mutation in the enzymatic pathway of assimilatory starch formation. Only then can one compare plants with and without the capacity of starch formation under identical conditions of light or nutrition. Here we describe experiments using mutants of *Arabidopsis* with no, or very diminished, starch production in order to study the effect of assimilatory starch on plant performance, and to test to what extent respiration and growth are affected by the lack of starch formation (Schulze et al. 1991; Stitt and Schulze 1993). In this sense, genetics and molecular biology enable us to ask new questions in ecophysiology about regulation of plant growth and its response to environment.

6.2 The Metabolic Pathway of Assimilatory Starch Formation and the Use of Mutants to Circumvent Chloroplast Starch Formation

Carbon dioxide is fixed in the Calvin cycle to build triose phosphates (Triose-P) which may pass through the chloroplast membrane into the cytosol and serve respiration or form sucrose for phloem transport (Fig. 6.1). Otherwise, Triose-P may remain within the chloroplast to supply the Calvin cycle, or to form assimilatory starch in a metabolic pathway which has great similarity to the transformations of Triose-P in the cytosol. The regulation of this branched pathway is rather complicated and involves feedback and feedforward regulation and external effectors of mainly the enzymes fructose-1, 6-bisphosphatase (Fru-1, 6-Pase) and sucrose-phosphate-synthase (SPS). Buildup of Triose-P in the cytosol may feed back on the Fru-1, 6-Pase in the chloroplast and stimulate starch production in the chloroplast (Stitt 1993).

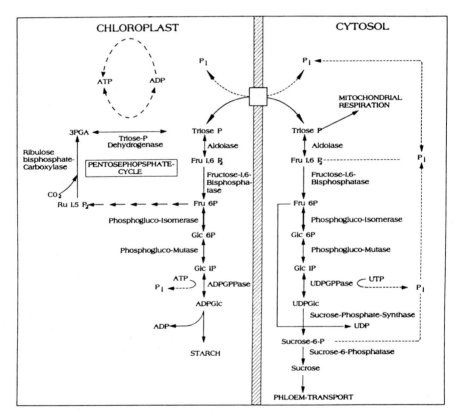

Fig. 6.1. Biochemical pathways in the chloroplast and cytosol involved in starch formation and breakdown

In the present study, three types of *Arabidopsis* mutants were used to have three levels of assimilatory starch under various conditions of light and nutrition. The performance of a wild type (WT) was compared with the growth response of (1) a mutant deficient in phosphogluco-mutase in the chloroplast (PGM mutant) (Caspar et al. 1986), (2) a mutant showing only 7% of the wild-type activity of ADP glucose phosphorylase (7% ADPGPPase mutant) in which starch production is reduced by 70%, and (3) a progeny of a cross between wild-type and 7% ADPGPPase mutant with a 55% ADP glucose phosphorylase activity and a 20% reduction in starch formation (55% ADPGPPase mutant). Using these mutants, we investigated whether the formation of assimilatory starch decreases or increases relative growth rates.

6.3   The Diurnal Starch Turnover

During the course of a day, the experimental plants showed large variations in assimilatory starch formation and turnover (Fig. 6.2). In the wild type, starch concentration increased at high nitrogen supply (6 mM NH_4NO_3) from about 70 μmol g^{-1} fw in the morning to about 200 μmol g^{-1} fw in the evening. At low N supply (0.1 mM NH_4NO_3), the daily increase of starch was similar, but it occurred from a higher base line starch level, i.e., starch concentrations in the morning were 200 μmol g^{-1} fw as compared to 70 μmol g^{-1} fw at high N supply (Fig. 6.2A). The daily increase in starch concentration decreased from 130 μmol g^{-1} fw at high light (photon flux density, PFD = 600 μmol m^{-2} s^{-1}) to about 50 μmol g^{-1} fw in low light (PFD = 80 μmol m^{-2} s^{-1}) (Fig. 6.2D).

In contrast to the wild type, the starch turnover in the 55% ADPGPPase mutant at low light was lower than in the wild type (Fig. 6.2E), and the 7% ADPGPPase mutant also had a reduced starch turnover at high light (Fig. 6.2B). The PGM mutant lacked starch formation, but showed an increase of soluble carbohydrates during the course of a day, which increased with nutrition (Fig. 6.2C and F).

Starch accumulation is highly dependent on leaf age (Fig. 6.3). If leaf weight is taken as indicator for leaf age in the rosette of vegetative *Arabidopsis*, we observe that starch concentrations were larger in leaves of low weight (young leaves) than in leaves of high weight (old leaves). The variation of evening and morning concentrations is large, because different leaves were collected from different plants for each data point. Therefore, it is not clear if the daily increase in starch is smaller or larger in young than in old leaves. However, the high starch level in young leaves may be an indication that sink leaves accumulate starch not only from their own photosynthesis, but also from carbohydrates which are imported from source leaves via the phloem.

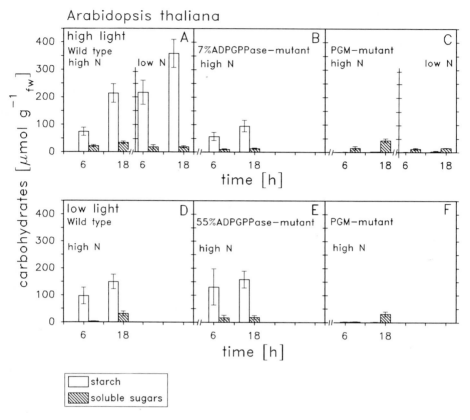

Fig. 6.2 A–F. Contents of starch and soluble sugars in the morning (6 h) and evening (18 h) to show the turnover at high (**A–C**) and low light (**D–F**) as well as at high and low N supply

6.4 Significance of Leaf Starch for Growth

6.4.1 Effects of Leaf Starch on Biomass Formation

The starting point of our experiments was the observation of Caspar et al. (1986) that starchless mutants grew as rapidly as wild type in continuous light, but died if the dark period exceeded 12 to 14 h. Therefore, the plants in this experiment were grown in a 14-h light/310-h dark regime. We investigated whether growth was quantitatively related to the amount of starch moving through the starch pool, and to what extent the significance of leaf starch for growth was influenced by light and nitrogen supply.

At high light and N supply, there was a positive linear, and near stoichiometric, correlation between the daily amount of carbon passing through the leaf starch pool of the different mutants and the daily increment

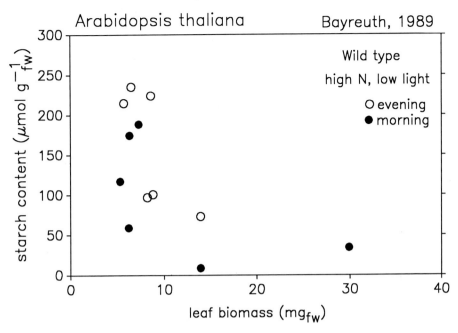

Fig. 6.3. Starch content in leaves as a function of leaf weight used as an indicator of leaf age

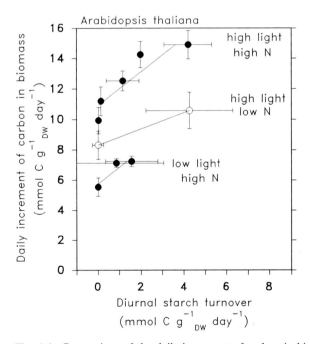

Fig. 6.4. Comparison of the daily increment of carbon in biomass and the daily turnover of starch

Table 6.1. Effect of decreased starch turnover in mutant plants on leaf N concentration and the shoot/root ratio at varying N and light supply. Results are given as mean ± SE

Light [$\mu mol\, m^{-2} s^{-1}$]	NH$_4$NO$_3$ (mM)	Genotype	Biomass (mg$_{dw}$)		Leaf N conc (mmol g$^{-1}_{dw}$)	Shoot/root ratio
			Leaf	Root		
600	6	WT	38.9 ± 4.1	5.1 ± 0.6	2.74 ± 0.2	7.6 ± 1.2
		7% ADPGPPase	20.4 ± 6.8	2.7 ± 0.5	3.47 ± 0.5	7.6 ± 2.9
		PGM	15.8 ± 1.6	4.1 ± 0.2	3.77 ± 0.2	3.9 ± 0.4
600	0.1	WT	6.2 ± 1.7	3.8 ± 1.0	0.44 ± 0.2	1.6 ± 0.3
		PGM	3.4 ± 2.0	4.8 ± 1.0	2.61 ± 0.1	0.7 ± 0.1
80	6	WT	8.5 ± 0.4	0.6 ± 0.1	4.86 ± 0.4	14.6 ± 2.9
		55% ADPGPPase	7.6 ± 0.5	0.6 ± 0.1		15.2 ± 2.9
		PGM	2.9 ± 0.2	0.3 ± 0.1	5.64 ± 0.1	9.6 ± 1.8

of carbon in biomass (Fig. 6.4). The PGM mutant had no diurnal starch turnover, while highest starch turnover was observed in the wild type.

At high light and N supply, the wild type reached 55% higher biomass than the PGM mutant after 20 days of growth (Fig. 6.4, Table 6.1). At low light and high N, the daily starch turnover decreased, but total biomass was 65% higher in the WT than in the PGM mutant. At low N supply, the effect of daily starch turnover decreased, biomass was only 19% higher in the WT than in the PGM mutant. Obviously, growth is promoted when carbon is retained temporarily in leaves and exported at night, and the fraction of starch which was turned over was not as efficiently used for growth at low N as at high N supply. Indeed, the accumulation of starch under conditions of low N supply (Fig. 6.2) indicates that plants do not respond "optimally" to low N. They overinvest in photosythetic machinery, using N which might have been invested more productively elsewhere, e.g., in roots for acquisition of more N. Therefore, the effect of starch formation on partitioning into root and shoot needs further investigations.

6.4.2   Effects of Leaf Starch on Regulation of Shoot/Root Ratios

A decreased availability of N caused the starch concentration to increase in the wild type while shoot weight decreased by 82% and root weight by 16% (Table 6.1). However, for the same change in nutrition, the PGM mutant lacking starch increased root weight by 17%. When starchless mutants were grown on low N, they actually produced larger roots and their leaves contained more N than the WT. This indicates that the reallocation of biomass to root growth allows a large increase in N uptake and decrease in the C/N balance in the plant. From these results, we feel that starch may not play as significant a role as N in these responses. The shoot/root ratio decreased more in starchless mutants than in the starch-accumulating wild type. The absence of a tight modulation of allocation by carbohydrates might explain why plants growing in low N still tend to overinvest in shoot growth.

6.5   The Carbon Balance

In order to demonstrate the overall effect of assimilatory starch formation on growth, we calculated an average balance of carbon partitioning, which is necessarily based on a number of assumptions: 23% of the carbon gain is used for respiration (9% in roots, 14% in shoots; Penning de Vries 1975) and sink leaves have a sixfold higher respiration rate than source leaves (Thornley 1976). In the light of observations of Rawson and Woodward (1976), we expect that the net assimilation rate of source and sink leaves is not vastly different on a dry weight basis. Rates of carbon fixation may even be lower in older than in younger leaves because of the large mass of old leaves (Thornley 1976; Turgeon 1989).

A daily carbon gain of 100% (or 100 mmol C per plant and day) will be partitioned into different compartments (Fig. 6.5). We may assume that

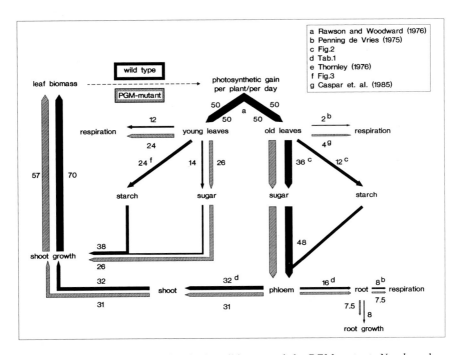

Fig. 6.5. Model of the carbon flux in the wild-type and the PGM mutant. *Numbers* show the amount of C that is transported to the designated regions. *Black arrows* represent the fluxes in the wild type, *gray arrows* show the corresponding fluxes in the PGM mutant

50% of this carbon gain had been assimilated by old and 50% by young leaves (Rawson and Woodward 1976). In the wild type, the carbon gain is either transported to the phloem or stored as starch at different rates, depending on leaf age. Fourteen percent (Penning de Vries 1975) is used for respiration in a ratio of 1:6 for old versus young leaves (Thornley 1976). Old leaves deliver assimilates into the root and support root growth and the necessary fraction of root respiration, and they supply assimilates to sink leaves. These assimilates, together with the assimilatory carbon gain of sink leaves, suport growth. In the case of the wild type, the amount of carbon supporting growth is 70% of the daily carbon gain.

Caspar et al. (1986) showed that dark respiration in leaves of the starchless PGM mutant 69.19 nmol CO_2 $g^{-1} h^{-1}$ compared with 11.5 nmol CO_2 g^{-1} h^{-1} in the wild type, while root respiration was the same. This docs not greatly decrease the supply of assimilates from old to young leaves. However, if starch formation is lacking, young leaves are expected to have much greater respiration rates, which will consume a major fraction of the carbon gain of the young leaves. In total, the carbon supply of young leaves is decreased to 57% in PGM mutants, mainly because of the consumption of carbon by respiration.

We think that the increased respiration which was observed by Caspar et al. (1975) is due to an activation of the alternative pathway of cyanide-resistant respiration, which regulates the level of Triose-P in the cytosol. This may be important for releasing Pi in the case of the starchless mutant, which otherwise may be bound in sugar phosphates, and affect metabolism. We are aware that this is one possible interpretation, because Pi is also regulated by other mechanisms, such as the activity of SPS. This mechanism, however, may not be effective if the sink leaf is not able to store carbohydrates as starch.

6.6 Conclusions

Assimilatory starch promotes growth (1) because it allows a carbohydrate supply over day and night for growth processes, which otherwise may run out of carbohydrates at night, and (2) "protects" carbohydrates from respiration, by keeping the cytosolic level of Triose-P low. The data show that, independent of the daily starch turnover, accumulation of starch occurred at N limitation, and N accumulation occurred at carbohydrate limitaion by low light. This indicates (3) a limited ability of plants to adjust their partitioning pattern for resource acquisition and makes "emergency" reactions, such as increased respiration, necessary to reduce soluble carbohydrate levels in the case of starchless mutants.

References

Caspar T, Huber SC, Somerville C (1986) Alterations in growth, photosynthesis and respiration in a starchless mutant of *Arabidopsis thaliana* deficient in chloroplast phosphoglucomutase activity. Plant Physiol 76: 1–7

Chapin FS III, Schulze ED, Mooney HA (1990) The ecology and economics of storage in plants. Annu Rev Ecol Syst 21: 423–447

Diemer M, Körner C, Prock S (1992) Leaf life span in wild perennial herbaceous plants: a survey and attempts at a functional interpretation. Oecologia 89: 10–16

Geiger DR (1980) Processes affecting carbon allocation and partitioning among sinks. In: Cronshaw J, Lucas WT, Giaquinta RT (eds) Phloem Transport. Liss, New York, pp 375–388

Harper JL (1989) The value of a leaf. Oecologia 80: 53–58

Lambers H (1982) Cyanide-resistant respiration: non-phosphorylating electron transport pathway acting as an energy overflow. Physiol Plant 55: 478–485

Monsi M, Murata Y (1970) Development of photosynthetic system as influenced by distribution of matter. In: Predictions and measurement of photosynthetic productivity. PUDOC, Wageningen, pp 115–139

Penning de Vries FWT (1975) Use of assimilates in higher plants. In: Cooper JP (ed) Photosynthesis and productivity in different environments. Cambridge University Press, Cambridge, pp 459–480

Rawson HM, Woodward RG (1976) Photosynthesis and transpiration on dicotyledonous plants. I. Expanding leaves of tobacco and sunflower. Aust J Plant Physiol 3: 247–256

Rufty TW, Huber SC, Volk RJ (1984) Alterations in leaf carbohydrate metabolism in response to nitrogen stress. Plant Physiol 88: 725–730

Schulze E-D (1983) Plant life forms and their carbon, water and nitrogen relations. Encyclopedia Plant Physiology. 12B: 615–676

Schulze E-D, Chapin FS III (1987) Plant specialization to environments of different resource availability. Ecological Studies 61: 120–148

Schulze W, Stitt M, Schulze E-D, Neuhaus HE, Fichtner K (1991) A Quantification of the significance of assimilatory starch for growth of *Arabidopsis thaliana* L. Heynh. Plant Phyasiol 95: 890–895

Stitt M (1993) Flux control at the level of the pathway, illustrated in studies with mutants and transgenic plants having a decreased activity of enzymes involved in photosynthesis partitioning. In: Schulze ED (ed) Flux control in biological systems. Acodemic Pressa, SanDiago (in press)

Stitt M, Schulze E-D (1993) Plant growth, storage and resource allocation – from flux control in a metabolic chain to the whole plant level. In: Schulze ED (ed) Flux control in biological systems. Academic Press, SanDiago (in press)

Thornley JHM (1976) Mathematical models in plant physiology. Academic Press, New York

Turgeon R (1989) Biosynthesis and degradation of starch in higher plants. Annu Rev Plant Physiol 40: 119–138

7 Photosynthesis, Storage, and Allocation

K. Fichtner, G.W. Koch, and H.A. Mooney

7.1 Introduction

Here we review patterns and consequences of allocation and storage of resources by plants in relation to their photosynthetic capacity. We relate these phenomena to resource availability, plant growth rate, and plant growth form. We concentrate our chapter on the organ and whole plant dimensions of allocation. Various aspects of this problem have been reviewed recently, such as storage relationships to growth rate (Chapin et al. 1990), carbon allocation and tissue costs (Chiariello et al. 1989), allocation and stress (Mooney and Winner 1991), allocation modeling (Bastow-Wilson 1988), and controls on carbon partitioning (Wardlaw 1990), so here we cover these topics lightly or not at all.

In order to evaluate the relationships among resources, photosynthesis, and storage, we concentrate our discussion on results of experiments utilizing cultivars or treatments differing in critical parameters related to either resource availability, resource acquisition, or resources storage. First, we examine tobacco plants that have been engineered to have differing photosynthetic capacities. Then we examine controls on root/shoot allocation in wild radish grown under conditions producing a wide range of shoot activity (photosynthesis) and root activity (nitrate uptake). Finally, we compare resource storage in a wild type and cultivar of lima bean that differ in a number of basic resources processing parameters.

7.2 The Impact of Photosynthesis on Growth, Storage, and Biomass Allocation in Transgenic Tobacco

To understand the ecophysiological relevance of photosynthesis for growth, storage, and allocation, it is desirable to alter the rate of photosynthesis independently of other plant parameters. Changing resource availability in order to change photosynthesis not only alters the rate of photosynthesis but may have secondary effects on whole plant physiology. This is shown, for example, for light, which affects photosynthesis directly as well as having photomorphogenetic effects (Mohr 1972), and nitrogen (N), which also has a strong effect on biomass partitioning (Bastow-Wilson 1988). Recently, genetic engineering offered a new approach for the production of genetically altered well-defined near-isogenic plants. These transgenic plants can be used to address specific questions such as about the role of phytochrome (Sharkey et al. 1991) or carbon metabolism (Stitt et al. 1991) in plants. In our analysis here we utilize findings on transgenic tobacco plants (*Nicotiana tabacum* L.) which were transformed with antisense DNA-sequences to *rbc*S, the small subunit for ribulose-1,5-bisphosphate carboxylase-oxygenase (Rubisco, the enzyme responsible for the primary CO_2 fixation in C_3 plants) (Rodermel et al. 1988). Individuals of these plants have decreased amounts of Rubisco and therefore also decreased rates of photosynthesis (Quick et al. 1991a,b, 1992). By changing Rubisco genetically, photosynthesis is altered as directly as possible with minimal secondary or indirect effects. Thus, these transgenic tobacco plants offer the unique opportunity to investigate plants grown under uniform environmental conditions with the individuals having a wide range of different photosynthetic rates. In the data presented here interactions of photosynthesis and N availability are also explored.

7.2.1 Photosynthesis and Growth

When the rate of photosynthesis was altered genetically, the relative growth rate (RGR) increased linearly with photosynthesis over its entire range ($r^2 = 0.83$) when the plants were grown under high N supply (HN; 5.0 mM NH_4NO_3) (Fig. 7.1A). RGR was much lower and independent of the rate of photosynthesis ($r^2 = 0.23$) when the plants were grown under low N supply (LN; 0.1 mM NH_4NO_3). At an intermediate N supply (MN; 0.7 mM NH_4NO_3) RGR increased linearly with photosynthesis at low rates (but with a smaller slope than at HN) and was constant at higher rates of photosynthesis. The linear increase of RGR with photosynthesis at HN resulted in an exponential increase in biomass in these plants, because the biomass B at a given time is given by $B = B_{initial} \cdot e^{RGR \cdot \Delta T}$. This exponential increase demonstrates the multiplying effect of increased carbon availability for growth if nutrient supply is ample. At LN, N deficiency inhibited further growth when photosynthesis increased (compare also the low N content in these plants in Fig. 7.2D). In this case a high portion of photosynthetic products was stored as starch (Fig. 7.2A) instead of being used for formation of new leaves and roots. Similar interactions between carbon and N availability have been seen qualitatively when plants were grown under varying light (sun and shade plants, Corre 1983; birch, Ingestad and McDonald 1989) or CO_2 (soybean, Cure et al. 1988; tobacco, K. Fichtner, unpubl. results) conditions. Thus, enhanced carbon availability promotes growth only if nutrients are sufficiently supplied. The use of transgenic plants with their broad range of photosynthetic rates allows us to describe this dependency in a more quantitative way.

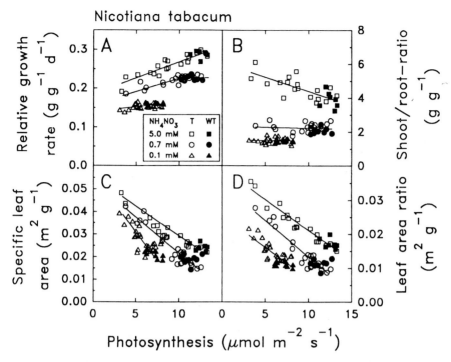

Fig. 7.1A–D. The relationship between the ambient photosynthetic rate of mature source leaves and **A** relative growth rate, **B** shoot/root ratio, **C** specific leaf area, and **D** leaf area ratio in transgenic (*T, open symbols*) and wild-type (*WT, full symbols*) tobacco (*Nicotiana tabacum* L.) grown at $300 \, \mu mol \, m^{-2} s^{-1}$ irradiance, $20/15\,°C$ air temperature at a 12/12 h day/night cycle (see Quick et al. 1992). The plants were cultivated in sand and watered once a day with an otherwise complete nutrient solution containing 0.1 ($\triangle$, $\blacktriangle$), 0.7 ($\bigcirc$, $\bullet$), and 5.0 mM NH_4NO_3 ($\square$, $\blacksquare$). Each *symbol* refers to a separate plant. (Data from Fichtner et al. 1993)

The above results show the relation between photosynthesis and growth using intraspecific variation of photosynthesis, i.e., within a single species. In an interspecific comparison, i.e., within a range of different species under the same environmental conditions, no relation of photosynthesis and growth may be found (e.g., Poorter et al. 1990). In this case different patterns of biomass allocation probably have a greater influence on growth (see Potter and Jones 1977; Poorter and Remkes 1990; Fichtner and Schulze 1992) than the rate of photosynthesis itself.

7.2.2 Photosynthesis and Biomass Allocation

In the wild-type (WT) plants, N nutrition enhanced photosynthesis and thus RGR (Fig. 7.1A); but this was not the only effect of N. Even at the same rate of photosynthesis, plants at HN had a higher RGR than at LN or MN.

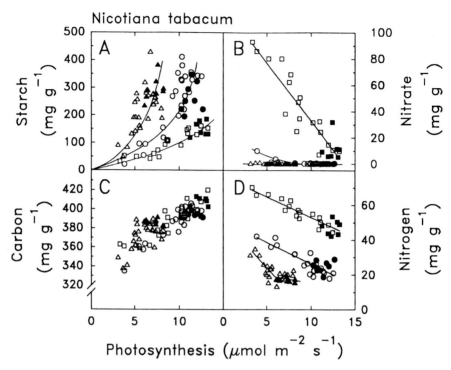

Fig. 7.2A–D. The relationship between the ambient photosynthetic rate and **A** starch, **B** nitrate, **C** carbon, and **D** total nitrogen content per dry weight of the entire leaf material harvested in the middle of the photoperiod. Each *symbol* (see Fig. 7.1) refers to a separate plant. (Data from Fichtner et al. 1993)

Enhanced N supply also increased the allocation of biomass to the above-ground plant parts (Fig. 7.1B), mainly to the leaves, due to less demand for roots for N uptake (Chapin 1980). Thus, even at comparable rates of photosynthesis, more photosynthates were available for the formation of new leaves at HN than at LN, promoting further growth. If the rate of photosynthesis was decreased genetically, at MN and LN the shoot/root ratio was unaffected. At HN the shoot/root allocation pattern changed only slightly in favor of the shoot (one third more than WT), i.e., only when there was a severe carbon limitation due to the combination of high N supply and low photosynthesis. In general, light (*Impatiens*, Evans and Hughes 1961; birch, Margolis and Vézina 1988) and CO_2 (soybean, Cure et al. 1988; tobacco, K. Fichtner, unpubl. results) have a relatively small effect on shoot/root allocation. In the experiment on tobacco shown here, N nutrition altered the shoot/root ratio to a much greater extent (three- to fourfold at the same rate of photosynthesis when HN and LN are compared) than the rate of photosynthesis (Fig. 7.1B). Thus, it can be concluded that N limitation does not influence the allocation between shoot and root simply by limiting carbon availability, but also through other mechanisms (Bastow-Wilson 1988).

The leaf area maintained per leaf dry weight (specific leaf area, SLA, the reciprocal of the specific leaf mass) increased when photosynthesis was inhibited genetically (Fig. 7.1C), i.e., the investment of biomass per leaf area produced declined when photosynthates were limiting. This resembles changes in leaf investment patterns when light (*Impatiens*, Evans and Hughes 1961; sun and shade plants, Corre 1983; birch, Margolis and Vézina 1988) or CO_2 (soybean, Allen et al. 1988; Cure et al. 1988; five annuals, Garbutt et al. 1990) was altered. Lower carbon gains due to lower rates of photosynthesis are in part compensated for by an increased leaf area per leaf biomass invested. In tobacco, limiting N nutrition decreased SLA at comparable rates of photosynthesis (Fig. 7.1C). This was mainly due to increasing amounts of starch present in the leaves at N deficiency (cf. Fig. 7.2A). If SLA was corrected for starch the difference between N treatments largely disappeared (see Fichtner et al. 1993), although SLA still increased with decreasing photosynthesis.

The leaf area maintained per total plant dry weight (leaf area ratio, LAR) can be viewed as the integration of the partitioning of whole-plant biomass to root and shoot, (the latter primarily leaves in the case of the young tobacco plants), and of leaf biomass to leaf area. LAR, similarly to SLA (compare Fig. 7.1D and C), decreased with photosynthesis and increased with N supply. The range of LAR within the HN treatment was higher than the range of SLA because of the concomitant increase in shoot partitioning when photosynthates were limiting (Fig. 7.1B). The difference between the N treatments was larger for LAR than SLA due to the lower shoot/root patitioning at decreasing N supply. Thus, the leaf area maintained per total plant dry weight is enhanced when the carbon gain of the photosynthetically active leaf area is low, but is reduced by low N supply, which favors biomass partitioning to roots rather than leaves (e.g., Corre 1983).

7.2.3   Carbon and Nitrogen Storage in Relation to Photosynthesis

Carbon stored as starch in the leaves increased nonlinearly with photosynthesis and decreased with N supply at comparable rates of photosynthesis (Fig. 7.2A). Starch content increased with photosynthesis most markedly when growth did not respond, as in the LN treatment and the MN treatment at high photosynthetic rates (Fig. 7.1A). Similar observations were made in a more qualitative way when light (Waring et al. 1985; McDonald et al. 1986) or CO_2 (Raper et al. 1973; Allen et al. 1988) was increased. In general, the starch content increases with higher availability of photosynthates or lower N availability, i.e., when carbohydrate availability exceeds the demand for growth.

The carbon content in the leaves increased with the rate of photosynthesis, no matter whether it was altered genetically or by N nutrition (Fig. 7.2C). An elemental analysis revealed that this was accompanied by a decrease of the macroelements N and K at all levels of N supply, and S at MN and LN (cf. Fig. 7.2D and see Fichtner et al. 1993). This demonstrates a changed balance between the incorporation of carbon, taken up by the leaves, vs. nutrients, taken up by the roots, into biomass when photosynthesis declined. The decreased carbon content at low rates of photosynthesis will also reduce the tissue construction costs (Williams et al. 1987). This contributes further to a compensation of declining growth when carbon gain is reduced.

In the tobacco plants investigated, the leaf nitrate content increased when photosynthesis was inhibited genetically, but to a much lower degree when the external supply of ammonium nitrate was low as in MN and LN (Fig. 7.2B). The enhanced nitrate levels were accompanied by a decrease in nitrate reductase measured in vivo (data not shown). In these transgenic plants grown under uniform light conditions, there is no energy limitation even at low rates of photosynthesis (Quick et al. 1991b). Thus, the cause of the low nitrate assimilation might be the lower availability of photosynthates itself. In agreement with these results, Kaiser and Förster (1989) and Kaiser and Brendle-Behnisch (1991) found that leaf nitrate levels are higher and nitrate reductase activity is lower when plants are kept under conditions limiting photosynthesis, e.g., low CO_2.

Total leaf N concentration increased when photosynthesis was inhibited genetically and when N supply increased (Fig. 7.2D). The decrease of N with photosynthesis can be interpreted as a N dilution effect due to high CO_2 uptake. N dilution in biomass due to promotion of photosynthesis or growth can also be seen with increasing light (Waring et al. 1985), CO_2 (Tremblay et al. 1988; Hocking and Meyer 1991), or phosphate availability (Tremblay et al. 1988). Thus, a decrease in the N rich enzyme, Rubisco, was not followed by a decline in N allocation to leaves. Instead, the content of other proteins (Quick et al. 1991a,b, 1992) and marginally the content of amino acids (K. Fichtner, unpubl. results) increased. Additionally, when N and nitrate values are compared, it is obvious that especially at HN there is an increasing partitioning of N to nitrate when photosynthesis is inhibited.

7.2.4 The Tobacco System: Conclusions

The effect of the photosynthetic rate on growth depends on the N supply. While photosynthesis promotes growth in well-fertilized plants, there is no effect when N is deficient. When photosynthesis is decreased genetically in plants grown under uniform environmental conditions there is a strong decrease in biomass investment per unit leaf area, resembling changes in the light or CO_2 environment. The partitioning between above- and belowground biomass is much more strongly influenced by N nutrition than by photosynthesis when equivalent changes in photosynthetic rate are compared. Photosynthesis strongly affects biomass composition. With decreasing photosynthesis, carbon, and particularly starch, stores decline while N, and especially nitrate, stores increase.

7.3 Allocation in Relation to Shoot and Root Activity

The studies with tobacco indicated that the *relative* levels of carbon and N availability predicted carbon and N allocation to shoots, roots, and storage compounds better than did absolute levels of either resource. This is consistent with conceptual models that view allocation responses as tending to compensate for changes in resource availability ratios (Brouwer 1963; Davidson 1969; Schulze 1982). From this it would be expected that the ratio of shoot to root *activity* should predict shoot/root allocation better than photosynthesis or nitrogen acquisition rate alone. Here we test this idea, drawing on results obtained in studies examining relationships among photosynthesis, nitrate (NO_3^-) uptake, and allocation in wild radish plants (*Raphanus sativus* × *raphanistrum*) grown under different combinations of N and light availability (Koch et al. 1987, 1988; Küppers et al. 1988; G.W. Koch, unpubl. data).

7.3.1 Resource Ratios, Growth, and Allocation

Different light regimes and NO_3^- availabilities produced a similar range of RGR in radish, but through influences on different aspects of plant physiology. Whereas light primarily affected photosynthesis, N supply had a large influence on shoot/root allocation (expressed as the shoot fraction in Table 7.1; shoot fraction = 1 – root fraction). For a given N environment, daily light availability had only a modest influence on shoot allocation, while for a given light environment, varied N availability elicited a comparatively large change in allocation. The extremes of allocation occurred at extreme ratios of light to N availability; the greatest shoot allocation was obtained at low light and high N (8.6 mol quanta $m^{-2}d^{-1}$ and 100 mmol NO_3^- m^{-3}), while shoot allocation was lowest at high light and low N (25.2 mol quanta $m^{-2}d^{-1}$ and 0.01X NO_3^- treatment).

7.3.2 Photosynthesis, Specific Absorption Rate, and Allocation

The relationship between photosynthesis and shoot/root allocation in radish apparently depends on the cause of variation in photosynthesis. When

Table 7.1. Relative growth rate, allocation, and photosynthesis of wild radish (*Raphanus sativus* × *raphanistrum*) grown at different light and NO_3^- availabilities in solution culture (means ± SE, see Koch et al. 1987 and Küppers et al. 1988 for details). Daily quantum fluxes of 35.3, 25.2, 14.1, and 8.6 $mol\,m^{-2}\,d^{-1}$ were provided at photoperiod/PFD ($h/\mu mol\,m^{-2}\,s^{-1}$) combinations of 14/700, 10/700, 14/280, and 6/400, respectively. NO_3^- treatments designated by an "X" used relative addition of NO_3^- at different fractions of the projected optimal N demand ($[NO_3^-]$ generally below $5\,mmol\,m^{-3}$). At a daily quantum flux of $25.2\,mol\,m^{-2}\,d^{-1}$, photosynthesis per unit leaf area was estimated from the mean specific leaf area and the photosynthesis-leaf [N] relationship determined from the 5 and $100\,mmol\,m^{-3}$ NO_3^- treatments: photosynthesis ($nmol\,g^{-1}\,s^{-1}$) = 12.89 · $N(mg\,g^{-1})$ − 15.6, (r^2 = 0.85, p < 0.01, Koch, unpublished data). Ambient photosynthesis was measured on three to five plants in all other treatments. (Küppers et al. 1988)

Daily quantum flux $mol\,m^{-2}\,d^{-1}$	Solution $[NO_3^-]$ $mmol\,m^{-3}$	Relative growth rate $g\,g^{-1}\,d^{-1}$	Shoot fraction of total dry mass	Whole-plant organic N $mg\,g^{-1}$	Whole-shoot daily photosynthesis $mol\,C\,m^{-2}\,d^{-1}$
35.3	100	0.358 ± 0.008	0.74 ± 0.03	47.2 ± 2.9	1.01
35.3	10	0.271 ± 0.013	0.70 ± 0.02	34.5 ± 4.5	0.81
25.2	1000	0.302 ± 0.007	0.79 ± 0.01	49.2 ± 0.7	0.90
25.2	100	0.306 ± 0.010	0.79 ± 0.01	48.5 ± 1.1	0.95
25.2	50	0.268 ± 0.008	0.74 ± 0.01		
25.2	10	0.247 ± 0.007	0.69 ± 0.01	45.0 ± 0.6	0.76
25.2	5	0.206 ± 0.009	0.64 ± 0.01	31.6 ± 1.6	0.74
25.2	0.5X	0.237 ± 0.012	0.66 ± 0.03	37.3 ± 2.4	0.62
25.2	0.1X	0.172 ± 0.007	0.56 ± 0.02	35.4 ± 2.0	0.56
25.2	0.01X	0.135 ± 0.008	0.48 ± 0.03	31.1 ± 2.9	0.54
14.1	100	0.237 ± 0.011	0.79 ± 0.02	48.7 ± 1.2	0.45
14.1	10	0.223 ± 0.013	0.72 ± 0.03	39.5 ± 3.4	0.45
8.6	100	0.163 ± 0.005	0.81 ± 0.03	43.9 ± 1.0	0.24
8.6	10	0.145 ± 0.005	0.73 ± 0.03	38.8 ± 1.2	0.26

photosynthesis was varied by changing N availability at a given light level it was positively correlated with shoot allocation, while when varied by changing light availability across a given N regime, it bore no clear relationship to allocation (Fig. 7.3A). In the former case, N limitation elicited both a reduction in shoot N (and hence photosynthetic capacity) and in shoot growth relative to root growth, consistent with recent models of optimal shoot/root partitioning (Ågren and Ingestad 1987; Levin et al. 1989; Hilbert 1990). Under low light, however, photosynthesis is limited by light energy rather than N, and thus an increase in root allocation (i.e., decreased shoot fraction) is not beneficial.

Shoot/root allocation was more clearly related to the activity of the root system in radish than was photosynthesis (Fig. 7.3B). Over all treatments a curvilinear relationship was seen between the mean NO_3^- uptake rate per unit root mass and shoot allocation, with fractional shoot allocation saturating at high root activities. Although uptake rate and allocation were strongly and nearly linearly related at lower uptake rates, there was a broad range of higher uptake rates over which allocation showed little variation. In

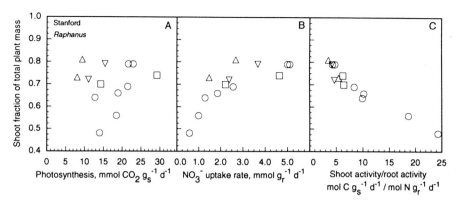

Fig. 7.3A–C. Shoot fraction as a function of **A** whole-shoot daily photosynthesis, **B** the mean specific NO_3^- uptake activity of the root system **C** the ratio of photosynthesis to NO_3^- uptake for wild radish (*Raphanus sativus × raphanistrum*) plants raised under the different combinations of light and NO_3^- availability described in Table 7.1. The mean root NO_3^- uptake activity in **B** was calculated from variables in Table 7.1. as: U = RGR $\cdot$ [N] $\cdot$ $1/F_r$, where RGR is the relative growth rate ($g\,g^{-1}\,d^{-1}$), [N] is whole-plant N concentration ($mmol\,g^{-1}$), and F_r is the root fraction of total plant mass (1 − shoot fraction). *Different symbols* depict different light treatments (mol quanta m^{-2}): *squares* 35.3; *circles* 25.2; *inverted triangles* 14.1; *triangles* 8.6

particular, the lowest light treatments had shoot fractions that tended to be higher for a given uptake rate than the high light treatments (i.e., the low light points in Fig. 7.3B are above the curve described by the 25.2 and 35.3 light treatments). Here N uptake was most likely regulated by the demand for N imposed by growth, which is low under low light, while at the same time shoot allocation was enhanced. In contrast, when external N supply was low but light, and thus N demand, was high, shoot allocation decreased to compensate for the low specific uptake activity of the root system.

The relative rates of photosynthesis and N uptake appear to describe most generally the compensatory changes in shoot/root allocation produced by variation in both light and N availability (Fig. 7.3C). Particularly among the NO_3^- availability treatments, which shared a common light level and photoperiod, the ratio of shoot to root activity was strongly and inversely related to shoot allocation. This ratio should be proportional to whole-plant C/N ratio, although the actual value of C/N would also depend on the root/shoot ratio and the daily C loss by shoot dark respiration and root respiration. At the physiological level, changes in the shoot/root activity ratio should be reflected by changes in the accumulation of C and N storage compounds, which in radish are primarily starch and NO_3^--N, respectively. For example, at high N/light availability ratios, N tends to accumulate in plant tissues as unreduced NO_3^--N which can be reallocated and used to support growth if external NO_3^- supply declines (Schulze et al. 1985; Koch et al. 1988).

7.3.3 The Radish System: Conclusions

The ratio of shoot to root activity appears to be the best predictor of allocation to shoots, roots, and specific storage compounds, there being no simple relationship between photosynthesis or root activity and allocation across different light and N environments. Variation in the relative availabilities of light and N can encompass conditions ranging from N acquisition being largely supply-limited at low N/light ratios to largely demand-driven at high N/light ratios. Whether N acquisition is demand-driven or supply-limited will vary with the plant factors (e.g., inherent growth potential, developmental stage) and environmental conditions (e.g., light, CO_2, or temperature) that determine photosynthesis and relative growth rate, and thus, plant N demand (Clarkson 1986; Garnier et al. 1989).

7.4 Storage as Related to Resource Availability

A further example of the tight coupling between N and carbon availabilities and stores, as demonstrated for tobacco and radish, is shown by recent experiments on *Phaseolus*. Specifically, these latter experiments were directed toward learning more about the role of N and carbon reserves versus stores. Reserve formation is the metabolically regulated synthesis of storage compounds, in competition for growth, in contrast to accumulation of stores that occurs when resource supply exceeds demands for growth and reproduction (Chapin et al. 1990).

These experiments utilized two genotypes of *Phaseolus lunatus*, the lima bean. One of these was a perennial wild type from the deciduous forests of western Mexico (Chamela, Jalisco). The second genotype was the lima bean cultivar, Burpee's Best. This genotype is annual, as are all cultivars of *P. lunatus*. The rationale for utilizing these two different genotypes was that it might be expected that the perennial would accumulate reserves, even at a loss of growth potential, since it must utilize stores to endure a long drought period and to regrow after its cessation. The annual, on the other hand, would not have such demands for stores, and hence, might not have to regulate reserve formation away from competitive growth potential.

We grew these two genotypes under differing supply rates of carbon (by altering the rate of photosynthesis by different PFD) and N. They were grown for 3 weeks under low ($100\,\mu mol\,m^{-2}\,s^{-1}$) or high ($400\,\mu mol\,m^{-2}\,s^{-1}$) PFD and under either low (0.5 mmol) or high (5.0 mmol) ammonium nitrate with other nutrients nonlimiting.

The results of these analyses (Mooney et al., submitted) indicate a strong and direct relationship between resource availability and growth, with storage playing an important role in buffering the effects of imbalances of resource availability and use. Specifically, it was shown that the concentration of N in

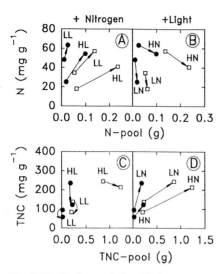

Fig. 7.4A–D. Interrelationship between resource supply and resource storage of nitrogen in two cultivars of *Phaseolus lunatus* (*circles* indicate wild type, and *squares* Burpee's Best pole bean) grown under high and low light (*HL* and *LL*) and high and low nitrogen availability (*HN* and *LN*). *Arrows* indicate the change in concentration and whole plant pool size of nitrogen and total nonstructural carbohydrate (*TNC*) when plants were given an increase in nitrogen availability when grown under either high or low light (**A, C**), or when they were given an increase in light availability when grown under high or low nitrogen (**B, D**). (Data from Mooney et al., submitted)

plants is positively related to the amount of N available and inversely to the amount of light available (Fig. 7.4A,B). The converse is the case in respect to total nonstructural carbohydrates TNC (Fig. 7.4C,D). Whole plant pool size of both N and TNC increased with resource availability.

These results indicate a fairly tight coupling between N and carbohydrate availabilities to support growth and the accumulation of either carbon or N when either of these resources is limiting. Further, these results would not support the concept of the storage of materials in a regulated manner (reserves) since resource accumulation was directly responsive to the balance between resource supply and demand. This was the case for both the perennial and annual bean genotypes.

7.5 Conclusions

We have demonstrated the direct effects of photosynthesis per se on plant growth rate, allocation, and storage. When N supply is ample, the relative growth rate is directly proportional to photosynthetic rate. When N is limiting, there is no effect of photosynthesis on growth. Allocation to leaf area increases as the *intrinsic* photosynthetic capacity is reduced, resulting in a carbon gain compensation similar to that which occurs under reduced light or CO_2. Under conditions where photosynthetic capacity is high, but N availability is low, starch accumulates, and conversely when there is high N availability and low photosynthetic capacity, N accumulates. The best predictor of allocation appears to be the ratio of photosynthesis to N uptake (shoot/root activity), rather than absolute rates of shoot or root activity or concentrations of carbon or N storage forms.

Acknowledgments. We gratefully acknowledge the Alexander von Humboldt Foundation and the National Science Foundation for support of the work summarized here.

References

Ågren GI, Ingestad T (1987) Root:shoot ratio as a balance between nitrogen productivity and photosynthesis. Plant Cell Environ 10: 579–586

Allen LH Jr, Vu JCV, Valle RR, Boote KJ, Jones PH (1988) Nonstructural carbohydrates and nitrogen of soybean grown under carbon dioxide enrichment. Crop Sci 28: 84–94

Bastow-Wilson J (1988) A review of evidence on the control of shoot:root ratio in relation to models. Ann Bot 61: 433-439

Brouwer R (1963) Some aspects of the equilibrium between overground and underground plant parts. Jaarb IBS 1963: 31–39

Chapin FS III (1980) The mineral nutrition of wild plants. Annu Rev Ecol Syst 11: 233–260

Chapin FS III, Schulze E-D, Mooney HA (1990) The ecology and economics of storage in plants. Annu Rev Ecol Syst 21: 423–447

Chiariello NR, Mooney HA, Williams K (1989) Growth, carbon allocation and cost of plant tissues. In: Pearcy RW, Ehleringer J, Mooney HA, Rundel PW (eds) Plant physiological ecology. Chapman and Hall, London, pp 327–365

Clarkson DT (1986) Regulation of the absorption and release of nitrate by plant cells: a review of current ideas and methodologies. In: Lambers H, Neetson JJ, Stulen I (eds) Fundamental, ecological and agricultural aspects of nitrogen metabolism in higher plants, Martinus Nijhoff, Dordrecht, pp 3–27

Corre WJ (1983) Growth and morphogenesis of sun and shade plants. II. The combined effects of light intensity and nutrient supply. Acta Bot Neerl 32: 277–294

Cure JD, Israel DW, Rufty TW Jr (1988) Nitrogen stress effects of growth and seed yield of nonnodulated soybean exposed to elevated carbon dioxide. Crop Sci 28: 671–677

Davidson RL (1969) Effect of root/leaf temperature differentials on root/shoot ratios in some pasture grasses and clover. Ann Bot 33: 561–569

Evans GC, Hughes AP (1961) Plant growth and the aerial environment. I. Effect of artificial shading on *Impatiens parviflora*. New Phytol 60: 150–180

Fichtner K, Schulze E-D (1992) The effect of nitrogen nutrition on growth and biomass partitioning of annual plants originating from habitats of different nitrogen availability. Oecologia 92: 236–241

Fichtner K, Quick WP, Schulze E-D, Mooney HA, Rodermel SR, Bogorad L, Stitt M (1993) Decreased ribulose-1,5-bisphosphate carboxylase-oxygenase in transgenic tobacco transformed with "antisense" *rbc*S V. Relationship between photosynthetic rate, storage strategy, biomass allocation and vegetative plant growth at three different nitrogen supplies. Planta 190: 1–9

Garbutt K, Williams WE, Bazzaz FA (1990) Analysis of the differential response of five annuals to elevated CO_2 during growth. Ecology 71: 1185–1194

Garnier E, Koch GW, Roy J, Mooney HA (1989) Responses of wild plants to nitrate availability: Relationships between growth rate and nitrate uptake parameters, a case study with two *Bromus* species, and a survey. Oecologia 79: 542–550

Hilbert DW (1990) Optimization of plant root:shoot ratios and internal nitrogen concentration. Ann Bot 66: 91–99

Hocking PJ, Meyer CP (1991) Effects of CO_2 enrichment and nitrogen stress on growth, and partitioning of dry matter and nitrogen in wheat and maize. Aust J Plant Physiol 18: 339–356

Ingestad T, McDonald AJS (1989) Interaction between nitrogen and photon flux density in birch seedlings at steady-state nutrition. Physiol Plant 77: 1–11

Kaiser WM, Brendle-Behnisch E (1991) Rapid modulation of spinach leaf nitrate reductase activity by photosynthesis. Plant Physiol 96: 363–367

Kaiser WM, Förster J (1989) Low CO_2 prevents nitrate reduction in leaves. Plant Physiol 91: 970–974

Koch GW, Winner WE, Nardone A, Mooney HA (1987) A system for controlling the root and shoot environment for plant growth studies. Environ Exp Bot 27: 365–377

Koch GW, Schulze E-D, Percival F, Mooney HA, Chu C (1988) The nitrogen balance of *Raphanus sativus* × *raphanistrum* plants. II. Growth, nitrogen redistribution and photosynthesis under NO_3^- deprivation. Plant Cell Environ 11: 755–767

Küppers M, Koch GW, Mooney HA (1988) Compensating effects to growth of changes in dry matter allocation in response to variation in photosynthetic characteristics induced by photoperiod, light, and nitrogen. Aust J Plant Physiol 15: 287–298

Levin SA, Mooney HA, Field C (1989) The dependence of plant root:shoot ratios on internal nitrogen concentration. Ann Bot 64: 71–75

Margolis HA, Vézina L-P (1988) Nitrate content, amino acid composition and growth of yellow birch seedlings in response to light and nitrogen source. Tree Physiol 4: 245-253

McDonald AJS, Ericsson A, Lohammar T (1986) Dependence of starch storage on nutrient availability and photon flux density in small birch (*Betula pendula* Roth). Plant Cell Environ 9: 433-438

Mohr H (1972) Lectures on photomorphogenesis. Springer, Berlin Heidelberg New York

Mooney HA, Winner WE (1991) Partitioning response of plants to stress. In: Mooney HA, Winner WE, Pell EJ (eds) Integrated responses of plants to stress. Academic Press, New York, pp 129–142

Mooney HA, Fichtner K, Schulze E-D (1993) Growth, photosynthesis, and content of carbohydrates and nitrogen in *Phaseolus lunatus* in relation to resource availability. Oecologia (submitted)

Poorter H, Remkes C (1990) Leaf area ratio and net assimilation rate of 24 wild species differing in relative growth rate. Oecologia 83: 553–559

Poorter H, Remkes C, Lambers H (1990) Carbon and nitrogen economy of 24 wild species differing in relative growth rate. Plant Physiol 73: 553–559

Potter JR, Jones JW (1977) Leaf area partioning as an important factor in growth. Plant Physiol 59: 10–14

Quick WP, Schurr U, Fichtner K, Schulze E-D, Rodermel SR, Bogorad L, Stitt M (1991a) The impact of decreased Rubisco on photosynthesis, growth, allocation and storage in tobacco plants which have been transformed with antisense rbcS. Plant J 1: 51–58

Quick WP, Schurr U, Scheibe R, Schulze E-D, Rodermel SR, Bogorad L, Stitt M (1991b) Decreased ribulose-1,5-bisphosphate carboxylase-oxygenase in transgenic tobacco transformed with "antisense" rbcS I. Impact on photosynthesis in ambient growth conditions. Planta 183: 542–554

Quick WP, Fichtner K, Schulze E-D, Wendler R, Leegood RC, Mooney H, Rodermel SR, Bogorad L, Stitt M (1992) Decreased ribulose-1,5-bisphosphate carboxylase-oxygenase in transgenic tobacco transformed with "antisense" rbcS IV. Impact on photosynthesis in conditions of altered nitrogen supply. Planta 188: 522–531

Raper CD Jr, Weeks WW, Downs RJ, Johnson WH (1973) Chemical properties of tobacco leaves as affected by carbon dioxide depletion and light intensity. Agron J 65: 988–992

Rodermel SR, Abbott MS, Bogorad L (1988) Nuclear-organelle interactions: nuclear antisense gene inhibits ribulose bisphosphate carboxylase enzyme levels in transformed tobacco plants. Cell 55: 673-681

Schulze E-D (1982) Plant life forms and their carbon, water and nutrient relations. In: Lange OL, Nobel PS, Osmond CB, Ziegler H (eds) Encyclopedia of plant physiology, vol 12B, Physiological plant ecology II. Springer, Berlin Heidelberg New York, pp 615–676

Schulze E-D, Koch GW, Percival F, Mooney HA, Chu C (1985) The nitrogen balance of *Raphanus sativus × raphanistrum* plants. I. Daily nitrogen use under high nitrate supply. Plant Cell Environ 8: 713–720

Sharkey TD, Vassey TL, Vanderveer PJ, Nierstra RD (1991) Carbon metabolism enzymes and photosynthesis in transgenic tobacco (*Nicotiana tabacum* L.) having excess phytochrome. Planta 185: 287–296

Stitt M, von Schwaewen A, Willmitzer L (1991) "Sink" regulation of photosynthetic metabolism in transgenic tobacco plants ecpressing yeast invertase in their cell wall involves a decrease of Calvin cycle enzymes and an increase of glycolytic enzymes. Planta 183: 40–50

Tremblay N, Yelle S, Gosselin A (1988) Effects of CO_2 enrichment, nitrogen and phosphorus fertilization during the nursery period on mineral composition of celery. J Plant Nutr 11: 37–49

Wardlaw IF (1990) The role of carbon partitioning in plants. New Phytol 116: 341–381

Waring RH, McDonald AJS, Larsson S, Ericsson T, Wiren A, Arwidsson E (1985) Differences in chemical composition of plants grown at constant relative growth rates with stable mineral nutrition. Oecologia 66: 157–160

Williams K, Percival F, Merino J, Mooney HA (1987) Estimation of tissue construction cost from heat of combustion and organic nitrogen content. Plant Cell Environ 10: 725–734.

8 Gas Exchange and Growth

J.S. Pereira

8.1 Introduction

Plant growth and ecosystem primary productivity are ultimately dependent on photosynthesis. Although it has long been recognized that it is the amount rather than the activity of the photosynthetic tissues that determines plant productivity in most cases (Monteith 1977, 1981; Jarvis and Leverenz 1983; Kriedemann 1986; Osmond 1987), there has been a great interest in the study of leaf photosynthetic rates during the last two decades because of the improvements in the instrumentation for gas exchange measurements and the discovery that the different biochemistry of photosynthesis in C_3 and C_4 plants resulted in different growth rates. At the same time, the installation of reductionist views in ecology and even in crop science led to the assumption that plant growth could be equated with leaf photosynthetic rates. The recognition that this was often a misinterpretation led to disenchanting contentions of the type, "photosynthetic rates play no role in crop yield". Moreover, most of the recent increases in agronomic productivity of cereal crops resulted largely from increases in the harvest index, sometimes even accompanied by a decrease in A (Evans 1976). It is obvious that for the purpose of growth and productivity studies, instantaneous rates of leaf photosynthesis (A) must be inserted in the adequate time and biological organization frameworks (Zelitch 1982; Osmond 1987). One solution for this is the use of simulation models. However, even though there are satisfactory models for canopy or whole-plant gas exchange (e.g., Norman and Campbell 1983; Wang and Jarvis 1990), it is still difficult to "translate" that into seasonal plant growth because of the gaps in our understanding of the mechanisms controlling carbon allocation (Gifford et al. 1984).

This chapter will review briefly the plant characteristics determining growth, namely plant metabolic rates vs. plant morphology and the role of assimilate partitioning in determining growth kinetics. In plant communities, light interception by the canopy largely determines the productivity, but the efficiency of conversion of solar radiation energy to biomass may vary with species and environmental factors. The influence of different phenologies and environmental factors on the relation between gas exchange rates and growth will be addressed as well.

8.2 How Plants Grow

Mathematical growth analysis has been the preferred methodology for more than 70 years to understand how plants grow and to study differences in growth among genotypes or in different environments. The key variable in this analysis is the growth rate expressed per unit of plant mass (W) or relative growth rate (R_w or RGR):

$$dW/dt * 1/W = R_w. \tag{1}$$

Of course, the relative growth rate is not constant throughout the plant life cycle, but decreases as maximum plant mass is reached. Accumulated growth can be expressed by any of a family of curves belonging to a generalized asymptotic function (Hunt 1982). This is important because growth comparisons must take into consideration development and how far a plant is from maximum R_w.

R_w is traditionally decomposed into two components, one more related to the rate of carbon assimilation and the second expressing plant structure. The first component is net assimilation rate (E_A or NAR):

$$E_A = 1/L_a * dW/dt, \tag{2}$$

with L_a as leaf area of the plant and W as whole plant mass. The second component of R_w is called leaf area ratio (F or LAR):

$$F = L_a/W. \tag{3}$$

The value of R_w is then

$$R_w = E_A * F. \tag{4}$$

It is worth noting that the time-averaged R_w is not equal to the product of averaged F e E_A over the same period because each of these variables follows different patterns of change with time. The variable E_A is essentially the balance between the daily integrals of photosynthesis and whole plant respiration expressed per unit of leaf area. It is not surprising, therefore, that E_A has been often correlated with net photosynthesis in a number of species (Könings 1989). However, the relationship between E_A and photosynthetic rate may be modified by changes in the proportion of respiration relative to photosynthesis. Respiration may account for up to 30 to 50% of the carbon assimilated daily (Lambers et al. 1990). Total respiration rates per unit of plant mass normally increase with R_w (Amthor 1989; Charles-Edwards et al. 1986). As discussed in Chapter 4, (this Vol), respiration is traditionally separated in growth respiration and maintenance respiration, but empirical quantification of these two components is difficult. The absolute growth rate, dW/dT, may be rewritten in terms of respiration as:

$$dW/dT = Y_G(Ap - mW), \tag{5}$$

where Y_G in the "growth conversion efficiency" given as $Y_G = \Delta S_T/(\Delta S_R + \Delta S_T)$, where S_T and S_R refer to the amounts of carbon substrate used as carbon skeletons for biomass accumulation and for respiration to provide energy for growth, respectively, A_p is the whole plant carbon assimilation, and m is the maintenance respiration coefficient (Amthor 1989 see also Chap. 4, this Vol.).

The so-called structural component of relative growth rate, i.e., F, is not directly related to gas exchange rates but gives information on the "leafiness" of the plant, i.e., the photosynthetic area per unit of plant dry weight. The value of F is, therefore, dependent on the partition of biomass between leaves and the total respiring biomass (leaf weight ratio, LWR) and on a foliage structure parameter, i.e., specific leaf area (SLA or σ).

Empirical data show that in many cases E_A decreases with increasing σ (Könings 1989). E_A will often decrease with increasing LWR. This means a compensation of any increase in F by a decrease in E_A, resulting in R_w being more conservative than either E_A or F.

In trying to relate growth and metabolic rates, we face the inherent difficulties of differences in scale. In fact, the instantaneous rates of gas exchange vary with leaf age and position and with time (of day, season), whereas growth integrates over space and time. In addition, the relationship is intrinsically complex because of the interplay of metabolic processes with structural characteristics in determining growth. The following equation is an attempt to relate growth with photosynthetic rate on a leaf area basis (A) and respiration (R) in simple herbaceous plants (Masle and Farquhar 1988; Farquhar et al. 1989),

$$dW/dt = (L_aA - R_r)*d - (R_r' + R_s')*n, \tag{6}$$

where the prime superscript denotes nighttime rates, d and n for the hours of light and dark and r and s for root and shoot, respectively. Equation (6) may be rewritten as $dW/dt = L_aAl*(1 - \phi)$, where ϕ is the proportion of carbon fixed that is respired and $1 = d/(d + n)$, i.e., light period as a proportion of the 24 h. The relative growth rate is then

$$R_w = dW/dt * 1/W = Al(1 - \phi)/\rho, \tag{7}$$

where ρ is the mass of carbon in the plant per unit leaf area, $W/(C_fL_a)$, where C_f is the conversion factor indicating the biomass/carbon ratio of recently produced dry matter. Notice that ρ is proportional to $1/F$ in a given plant or species. On the other hand, ρ is inversely related to C_f. Therefore high carbon concentration in the biomass (low C_f, reflecting, e.g., high lignin concentration) will influence growth rates negatively, other conditions being equal. A low lignin concentration is typical of fast-growing plants (Poorter 1989).

The ratio of carbon isotopes ($^{13}C/^{12}C$) in plant tissues is related to photosynthesis, namely the ratio of p_i (intercellular CO_2 partial pressure) and the partial pressure of CO_2 in the air, p_a (Farquhar et al. 1989, see also Chap. 19, this Vol.). The use of the discrimination for ^{13}C in plant tissues to study the relation between photosynthesis and growth has been limited. Masle and Farquhar (1988) found in wheat plants with different growth rates a strong negative correlation between ρ and the value of ^{13}C discrimination in leaf tissues. For a given increase in discrimination, the decrease in ρ was proportionally larger than the decrease in assimilation rate so that relative growth rate was positively correlated to ^{13}C discrimination. We also found that ^{13}C discrimination was negatively relate to ρ and positively related with growth and the quotient biomass production: solar radiation intercepted (Osório and Pereira, unpubl.). More work needs to be done in this area (see Chap. 19, this Vol.).

8.3 Photosynthesis and Growth Rates

One obvious case of differences in growth being ascribed to differences in metabolic rates is given by the generally higher rates of growth and pho-

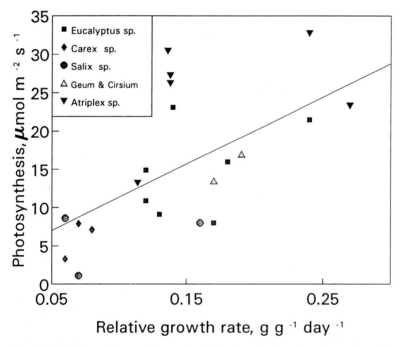

Fig. 8.1. Relation between R_W and leaf photosynthetic rates of woody and herbaceous species, *Eucalyptus* spp. (Mooney et al. 1978; Pereira et al. 1992a), *Carex* spp. (Könings 1989), *Salix aquatica* grown at different irradiances (Waring et al. 1985), and *Geum urbanum* and *Cirsium palustre* (Pons 1977). The regression line is $R_W = 2.65 + 87.13*A (r^2 = 0.25$, n.s.)

tosynthesis in C_4 than in C_3 plants. There are other examples of positive correlations between fast growth rates and high A (poplar clones, Ceulemans and Impens 1983) or low maintenance respiration rates (*Lolium perenne* genotypes, Wilson 1975). However, the relationship between A and R_w is quite variable and the correlation between the two is rather poor, as in Fig. 8.1.

It is unlikely that leaf photosynthesis or respiratory rates per se really explain most differences between fast and slow growers. Poorter (1989) reviewed growth and gas exchange data and concluded that only in a few experiments was photosynthetic rate on leaf area basis found to explain any significant part of the variation in R_w. In three *Eucalyptus globulus* clones with different inherent growth rates, photosynthetic rates were practically the same. The only important difference among these clones was a greater capacity to produce leaves in the fastgrowing plants when compared to the slow growers (Osório and Pereira, unpublished). On the other hand, there is no real ecological advantage to plants to have a high photosynthetic rate per se, whereas fast growth rates may be advantageous for ruderals as well as for some successful competitors (Grime 1979).

If A and R_w are poorly related, what are the differences between fast and slow growers? Even though Poorter (1989) found that A on a leaf area basis was unrelated with R_w, he also found that photosynthetic rates expressed on a leaf mass basis correlated well with R_w as a result of differences in σ. Slow growers had higher amounts of tissue "packed" per unit of leaf area, i.e., lower σ, than fast growers. Similar results have been reported in the literature, e.g., as in a comparison of 11 half-sib families of *Robinia pseudoacacia* showing a negative correlation coefficient between A and growth (Mebrahtu and Hanover 1991). Total dry weight accumulated was positively related to σ. Similar results are illustrated in Fig. 8.2 with data of the eucalypt trial shown in Table 8.1.

Because high g_s is often related with high photosynthesis, it has been assumed that high transpiration should be related with high growth rates. However, because the limitation of photosynthesis caused by stomata may vary widely with genotypes and environmental conditions, this relationship is weak. Typically, the gas phase (mainly stomatal) limitation represents ca. 30% of the total limitation to photosynthesis in C_3 plants at ambient CO_2 (Jones 1983). The gas phase limitation is considerably higher in C_4 plants.

Stomatal conductance varies diurnally according to well-described patterns (Tenhunen et al. 1986). This poses the methodological problem of defining the most meaningful A rate during a day. There is a great variety of criteria in the literature. The daily integral of A would provide the most useful information for comparison with growth [see Eqs. (6) and (7)]. However, for many porposes the daily maximum (A_{max}) is quite useful (Schulze and Hall 1982). Lack of standardization in these measures is certainly a source of noise in the type of relationships illustrated in Fig. 8.1. Nevertheless, in many circumstances, e.g., low soil water or high water vapor pressure deficit

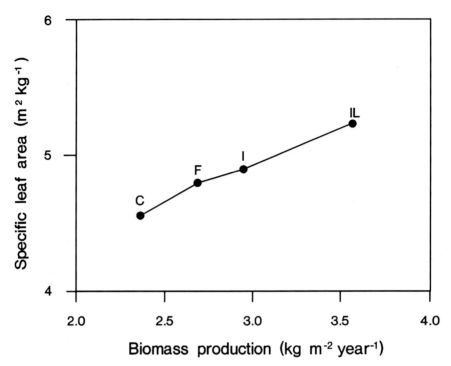

Fig. 8.2. Relationship between biomass productivity and specific leaf area (σ) in *Eucalyptus globulus* with different irrigation and fertiliser treatments (Araújo and Pereira, unpubl.; see text and Table 1 for details)

of the air, stomata may be a major cause for the decrease in photosynthetic carbon assimilation. Schulze and Hall (1982) estimated that on a daily basis the average value of A is at least 50% below the maximum value possible under ambient CO_2. Although in temperate climate regions low light may be the main reason for this difference, in arid regions stomatal closure may account for a substantial part of the difference between maximum A and daily average: for example 39 to 63% in *Prunus armeniaca* and *Hamada scoparia* respectively, in Israel.

As discussed below (see Sect. 8.8.1) primary productivity and evapo-transpiration are often correlated. However, the control that stomata exert on gas exchange becomes more complex as we move from the leaf level to the canopy. For example, stomatal control over transpiration depends on the coupling between the canopy and the atmosphere and may be rather coarse as, e.g., in short vegetation, which is virtually uncoupled from the atmosphere (Jarvis and McNaughton 1986).

8.4   The Importance of Allocation

An important feature invoked to explain the variation in inherent growth rates is the difference in the partitioning of photoassimilates between plant parts (see also Chap. 7, this Vol.). A major complication in the analysis of the relationships between growth and gas exchange rates is the marked difference in the timing of growth and gas exchange. Because maximum growth rates rarely coincide with maximum average photosynthetic rates, plants store carbon and nutrients as daily, long-term, and buffer reserves (Chapin et al. 1990; Geiger and Servaites 1991) which allow the plants to grow even when current assimilation rates are negligible or insufficient to maintain growth. The type and role of storage in growth in different time scales is exemplified in Fig. 8.3. It is clear that not all products of photosynthesis or nutrient uptake that may be used later for the buildup of structural or protection biomass result from a specific allocation to reserve storage, thus competing with the increase in structural biomass. Part of the carbon and nutrient storage may accumulate as a reflection of the difference between "resource supply over demand for growth and maintenance" (Millard 1988; Chapin et al. 1990). Usually, the greatest complexity in the relation between photosynthesis and growth may be encountered in

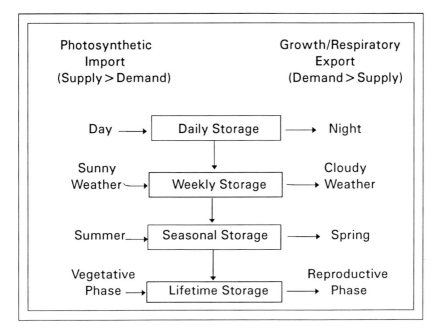

Fig. 8.3. Different time scales for storage of carbon and nutrients and causes for changes. (Chapin et al. 1990)

perennial woody species, where long-term storage is essential for the rapid growth rates in spring even before photosynthesis can provide enough carbon skeletons.

Growth of different plant parts is internally regulated and highly competitive for available carbohydrates. Renewal of fine roots, growth in diameter of perennial roots and shoots, or fruit growth all have different timing. For example, shoot elongation and canopy formation in many temperate zone woody plants occur early in the growth season largely at the expense of stored carbohydrates and nutrients rather than current season photosynthesis (Kozlowski and Keller 1966; Schulze 1982). In many evergreens of the temperate zone and regions with a mediterranean type of climate, carbon balance depends to some extent on winter photosynthesis, a period when growth does not occur (Larcher 1973; Pereira et al. 1986; Tenhunen et al. 1987). The use of stored carbohydrates leads obviously to the uncoupling of growth and carbon assimilation rates, a fact that is often ignored by modelers (see, for example, Fig. 8.4). In some species the storage of carbohydrates may actually compete with growth for current photosynthate availability (Chapin et al. 1990).

In the short term, whenever photosynthetic rates are in excess relative to carbohydrate use (slow growth, high photosynthetic rates) there is a tendency for the accumulation of carbohydrates, mainly starch, for later use. Very often, however, starch accumulation is negatively correlated with growth, as when low temperatures or nutrient supply reduce growth (Schulze 1982; McDonald et al. 1986; Schulze et al. 1991). It is possible that the need for stored carbohydrates is in part related with the fact that maximum growth rates do not coincide with maximum rates of gas exchange (Digby and Firn 1985) either in the long or in the short term. Storage of carbohydrates is essential for efficient growth, as suggested by data obtained with *Arabidopsis thaliana* mutants lacking the capacity to store starch (Schulze et al. 1991), which grew more slowly than the wild type. A better understanding of the mechanisms that determine carbon partition is needed,

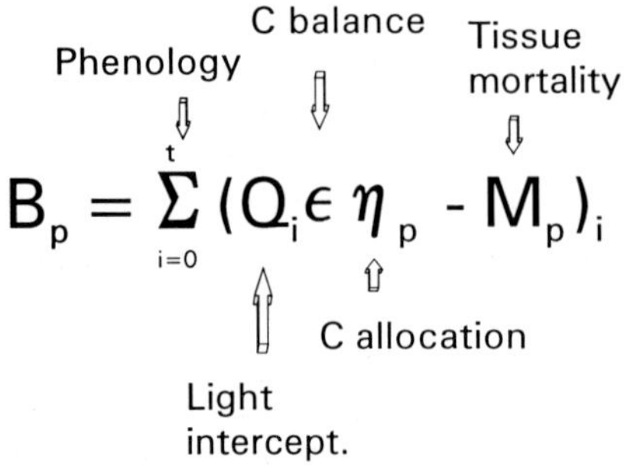

$$B_p = \sum_{i=0}^{t} (Q_i \epsilon \, \eta_p - M_p)_i$$

Phenology — C balance — Tissue mortality — C allocation — Light intercept.

Fig. 8.4. Plant growth model proposed by Charles-Edwards with indication of possible modules of a mechanistic model of plant growth. (After Reynolds et al. 1989)

taking into consideration that assimilation, phloem transport, and sink activity are not in phase even in the short term.

The kinetics of growth is largely dictated by how new photoassimilates are partitioned into "productive" (i.e., photosynthetic) or "support" (roots, shoots) biomass. A merely econometric approach would emphasize the importance of the rate of partitioning of photoassimilates into new leaf area as the component that would accelerate growth. For a given amount of net carbon increment, a plant will tend to grow relatively more in the following interval if partition of photoassimilates to new leaves is high, therefore "investing" more in new photosynthetic biomass, than plants "investing" less new carbon in leaves. In a number of cases R_w has been found well correlated with "leaf area partitioning", i.e., the relative increment in leaf area per unit increment in biomass (Potter and Jones 1977; Patterson et al. 1978).

Normally, the relative growth rates of two different components of the whole plant, are proportional. The constant of proportionality is called the "allometric constant". The allometric constants are often different from 1 and therefore the biomass ratios between plant parts change with time. For example, if the allometric constant of the root to shoot quotient is greater than 1, root biomass will increase faster than the shoot and, thus, whole-plant R_w will tend to decrease.

8.5 Do Growth Rates Influence Carbon Assimilation?

I have been discussing the influence that metabolic rates may have on growth. However, the inverse, i.e., to what extent growth activity may modify plant metabolism, is also pertinent. For example, fast growth rates increase respiration rates (Charles-Edwards et al. 1986; Amthor 1989). The answer is more complex for photosynthesis and it raises the question as to a possible feedback between "sink" and "source" of photoassimilates. This topic has been studied for some time (see Neales and Incoll 1968) and evidence for such a relationship between sink demand and photosynthesis has accumulated (King et al. 1967; Azcón-Bieto 1983; Lauer and Shibles 1987; Foyer 1988), although with some contradictory results (Geiger and Servaites 1991). Even though the results of some experiments to evaluate constraints on photosynthesis due to the accumulation of assimilates in leaves when growth or phloem transport has been restricted remain obscure, the biochemical evidence suggests the likelihood of a feedback relation between sink and source despite considerable variation among plants. This variation seems to be related in part with the capacity of leaves for starch synthesis and in part to metabolic regulation of sucrose synthesis (Huber 1981; Foyer 1987), which vary with genotype and environmental conditions. It has been shown that during periods when sucrose is produced more rapidly than it can be exported and accumulates in leaves, the concentrations of metabolites involved in determining the relative rates of starch and sucrose synthesis, especially the cytoplasmic concentrations of ortophosphate (Herold and Walker 1979) and of fructose 2, 6 bisphosphate (Stitt et al. 1984, 1987; Huber 1986), may change, leading to a decrease in photosynthetic rates. More recently, Krapp et al. (1991) showed that feeding leaves with glucose led to the long-term decrease in photosynthetic rate as a result of a decrease in Rubisco and other Calvin cycle enzymes and to the increase of respiratory rates. Although it is interesting to understand the short-term regulation of photosynthesis, these processes may be confounded by the fact that carbon assimilation and growth are not necessarily in phase.

8.6   Light Interception by Canopies and Plant Productivity

Biomass productivity of a plant community (B, e.g., in $t\,ha^{-1}\,year^{-1}$) may be estimated as a linear function of the amount of radiation intercepted by the canopy, Q_i (Monteith 1977, 1981; Jarvis and Leverenz 1983):

$$B = \Sigma(\varepsilon\,f\,Q_o), \tag{8}$$

where ε is the quotient biomass produced: solar radiation intercepted by the foliage (sometimes called "radiation use efficiency"), f is the mean intercepted fraction of solar radiation, and Q_o the accumulated amount of incident solar radiation over a certain period. The value of f that may be simply estimated as $[1 - \exp(-k\,L)]$ is a function of leaf area index (L) and light extinction coefficient (k). The product fQ_o over a certain period is equal to Q_i, which is therefore related not only with solar radiation during that period but also with the integral of leaf area index (L) over time that is leaf area duration (LAD or D). This term takes into account both the magnitude of photosynthetic area and its persistence in time and has been widely used in traditional growth analysis (Hunt 1982). A very rough approximation of biomass productivity is $B \approx D \times E_A$.

For a given species and geographical area, biomass productivity or yield is usually more dependent on variations in Q_i (or D) than on the variation in ε (Monteith 1977, 1981; Schulze 1982; Jarvis and Leverenz 1983; Russell et al. 1989; Cannell 1989). For example, the seasonal average value of ε for crops ranging from apples to cereals in Great Britain was approximately $1.5\,g\,MJ^{-1}$ given adequate water and nutrient supply. It was also shown that such a value of ε is consistent with a maximum A of $10.5\,\mu mol\,m^{-2}\,s^{-1}$ and a k value between 0.5 and 0.7, typical of a large variety of crops (Monteith 1981).

Table 8.1. Average estimated leaf area index (L), above-ground biomass productivity (B), mean fraction of the solar radiation intercepted by the foliage (f; with $Q_0 = 4.88\,GJ$ $m^{-2}\,year^{-1}$) and the quotient biomass production: radiation intercepted (ε) in the period of June 1988 to June 1989 (see text for details on treatments). (Tomé and Pereira 1991)

Treatments	L	B ($t\,ha^{-1}\,year^{-1}$)	f	ε ($kg\,GJ^{-1}$)
IL (Irrig. + fert.)	3.8	35.7	0.91	0.8
I (Irrig.)	3.0	29.5	0.78	0.8
F (Fert.)	2.8	26.9	0.77	0.7
C (Control)	2.3	23.7	0.71	0.7

As a consequence, the value of f, which depends on foliage area and canopy structure is a major factor in determining differences in productivity. This is shown in Table 8.1 with data from an irrigation and fertilization experiment conducted in Portugal with *Eucalyptus globulus*, where B was more closely related to f ($r^2 = 0.97$) than with ε. The treatments consisted of near optimal supply with irrigation and fertilizer (IL), with irrigation (I), with fertilizer added without irrigation (F) and a control (C) rainfed without fertilizer (Pereira et al. 1989; Madeira and Pereira 1990).

In this experiment the differences in long-term productivity were related with the rate of L and f increase. In treatment IL, the value of L was 3.2 ± 1.0 one year after plantation, whereas in C the value of L was only 1.6 ± 0.5. Significant differences in L were still found 3 years after planting (Table 8.1). These differences resulted from larger individual leaves, higher σ, but mostly from increased leaf numbers in the treatment plots in comparison with the control (Pereira 1990). Cannell (1989) also found that the primary reason why *Salix viminalis* produced more biomass than *Populus trichocarpa* in Scotland was a greater f (0.76 to 0.72 in *Salix* vs. 0.51 in *Populus*), and this resulted mainly from differences in leaf area development in the early summer, with *Salix* reaching a L of 2 about 15 days earlier than *Populus*.

The value of ε decreases with increasing k (Monteith 1981), because overall lower k values allow greater L for the same amount of intercepted light but the influence of canopy architecture in the mean fractional canopy interceptance is usually much less important than L in many crops. For example, in fast-growing trees the extinction coefficients vary between 0.4 and 0.6 for most of the growing season (Linder 1985; Cannell 1989; Gazarini et al. 1991) and in many cereal crops between 0.5 and 0.7 (Monteith 1981). In addition, canopy structure may influence the value of apparent ε because different amounts of radiation are intercepted by nonphotosynthetic biomass, namely branches in tree crowns. In a 3-year-old *E. globulus* plantation light interception due to branches was ca. 10% of that resulting from foliage (Gazarini et al. 1991).

The value of ε also depends on the way B and Q_i are calculated. For example, Cannell (1989) estimated ε for poplar and willow clones in Scotland as ca. $1.5 \, g \, MJ^{-1}$ for the whole plant. The difference between these and data of *E. globulus* shown in Table 8.1 results largely from considering only the aboveground biomass in the eucalypt data. However, an additional reason is the use of whole-year Q_o in the calculation of evergreen productivity instead of only growing season Q_o as in agricultural crops or in deciduous tree species. This results in the apparent decrease in ε of evergreens when compared to deciduous tree species or crops. In the 12 months values Q_o integrate periods when environmental stress (low winter temperatures and/or summer drought) decreases ε.

The short-term value of ε is quite variable and may be written as a function of whole-plant daily integrals of photosynthesis and respiration, A_p and R_p, respectively,

$$\varepsilon = (A_p - R_p)C_f/Q_i. \qquad (9)$$

The value of A_p (i.e., A for the whole canopy) depends on A at saturating light and quantum yield. Under natural conditions, daily average A is not greater than 70 to 30% of the maximum A at saturating light (A_{max}) because of stomatal closure and limiting light. The importance of A_{max} diminishes as incident light becomes more limiting. On the other hand, a decrease in quantum yield, caused for example by photoinhibition, may result in a lower A_p (Monteith 1981; Jarvis et al. 1989). However, the effect of a low quantum yield on A_p tends to be relatively smaller as incident irradiance increases.

The short-term ε is not necessarily correlated with dW/dt because it may happen, for example, that the same biomass increment coincides with either high irradiance (low ε) or with low irradiance (high ε). The fact that growth may continue with great variability in incident light and therefore in A_p, may be explained by mobilization of storage carbohydrates during cloudy periods (see Fig. 8.3). During the life of a crop, ε decreases as plants become older because R_w or dW/dt tend to zero. However, even average Q_i is not likely to decrease concomitantly and therefore a good correlation of R_w with short-term values of ε, cannot be expected.

Although essential for carbon fixation, incident light is seldom a limiting factor for production, but leaf area duration, D, is. As Monteith (1981) wrote, light is a determinant of plant production, not a discriminant. However, differences in total light intercepted over a season discriminate the productivity of different crops or plant communities. This means that most of the variation in crop or community growth rates are related to the impact of environmental factors such as temperature and water or nutrient availability in the value of D. One of the advantages of Eq. (8) is its easy use in "top down" physiological models of plant growth as illustrated in Fig. 8.4. However, the very simple relation between crop yields and Q_i on a seasonal basis has prompted a great deal of enthusiasm and some misinterpretations.

Demetriades-Shah et al. (1992) called the attention to some misuses of the short-term correlation between plant productivity and intercepted light especially in the case of remote sensing, but contributes very little to the understanding of its relevance in physiological terms because, as mentioned above, there are no physiological reasons to expect that short-term ε is related with R_w.

8.7   Phenology and Rates of Growth and Photosynthesis

Phenology and plant life-form often modify the relationship between growth and gas exchange. For example, ruderals and annual plants usually have high metabolic and growth rates. In normal conditions, this insures that the maximum number of viable propagules is produced during the favorable part of the year. However, competition and environmental stress may drastically change this simplistic generalization. For example, in the Mediterranean environment, C_3 annuals do best if they can grow and photosynthesize in the relatively low temperatures of the winter (when water is plentiful) even if these do not correspond to optimal temperatures. In woody perennials, fast growth rates have to be sacrificed to large partitioning coefficients to support perennial biomass. Usually, plants with greater longevity have slower growth rates because resistance of perennial biomass to decay (herbivory, disease, or saprophytic fungi) and to mechanical stress, result in additional respiratory costs and a high carbon concentration of the biomass (low C_f) related to the accumulation of lignin and other phenolic compounds. It is a low F and Y_G that will contribute to lower growth rates in long living woody perennials in comparison with herbaceous. The same is often true for the comparison between fast- and slow-growing trees and shrubs. The growth characteristics and physiology of different plant life-forms was reviewed comprehensively by Schulze (1982) and therefore I will focus on generalities.

It is often assumed that evergreens always have lower photosynthetic capacity than deciduous trees and most herbaceous plants. In a classical study, Schulze et al. (1977) showed that the evergreen *Picea abies* in central Germany compensated for the lower photosynthetic capacity of its needles with a longer photosynthetic season and larger foliage mass, when compared with the deciduous co-occurring *Fagus sylvatica* in terms of annual carbon budget. However, this generalization should not be taken too far. For example, Mediterranean evergreens often have photosynthesis rates of the same magnitude or greater than temperate zone deciduous woody plants (Körner et al. 1979). On the other hand, evergreens may differ a great deal in terms of leaf turnover, and that may change the photosynthetic capacity of the canopy.

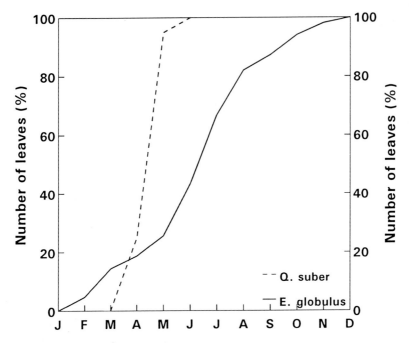

Fig. 8.5. Rates of leaf production in two evergreen tree species (*Quercus suber* and *Eucalyptus globulus*) growing under similar conditions in Portugal. (After Pereira et al. 1987 and Pereira unpubl.)

Normally, photosynthetic rates increase with leaf age up to full expansion, declining thereafter. In the whole canopy, the age structure of the leaf population may strongly influence the capacity of the plant to take up carbon. As shown in Fig. 8.5, two evergreen angiosperms with different growth rates in the same climatic conditions in Portugal, *Quercus suber*, and *Eucalyptus globulus*, have different leaf production phenologies. The slow-growing *Q. suber* produces a flush of shoot growth during the spring. The population of leaves produced will normally not be replaced until the following year. This results in the decline of the "photosynthetic capacity of the canopy" throughout the growing season as a result of leaf aging. As some leaves are shed during that period but are not replaced (compare Figs. 8.5 and 8.6), the amount of light intercepted by the canopy also declines. In *Q. suber* some leaves survive the first growth season whereas in *E. globulus* most leaves are replaced after 10–12 months (Fig. 8.6A). Better environmental conditions and higher growth rates seem to decrease the life expectancy of leaves and increase the turnover rate in the eucalypt (Fig. 8.6B). Similar findings have been reported for other species such as in *Erica tetralix* (Aerts 1989). Nutrient or water deficiency, however, decreases the rate of new leaf production and may increase the rate of leaf abscision (Harper

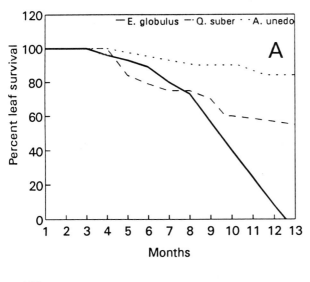

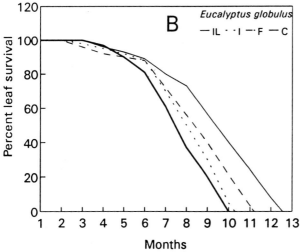

Fig. 8.6A,B. Survivorship curves for given cohortes of leaves of **A** Three concurring evergreen tree species in Portugal and **B** *Eucalyptus globulus* 1-year-old trees with different water and nutrient supply rates in Portugal. (Borralho and Pereira, unpubl.; see text and Table 1 for details on treatments)

1989; Metcalfe et al. 1990). In the fast-growing *E. globulus*, leaves are being produced for a much longer period, replacing older ones (compare Figs. 8.5 and 8.6) and maintaining a younger population of leaves mainly in the outer layers of the canopy. This will allow a high potential ε value to be maintained during most of the year.

8.8   Environmental Stresses Change the Relationship
Between Photosynthesis and Growth

A great deal of research has been devoted to the study of the effects of environmental factors on plant growth and metabolism. The traditional definition of environmental stress considers that stress conditions occur whenever growth rate, survival, or gas exchange are lower than maximum. It is virtually impossible to quantify these conditions. For example, in normal circumstances plants regulate their metabolism according to the resource or other limitations imposed, e.g., rates will slow down at temperatures lower than certain optimal values. On the other hand, different processes will respond differently to environmental factors. Nevertheless, I will use the term stress in this broad qualitative meaning. In terms of plant response to the environment three aspects are of importance: short-term responses, acclimation to new environmental conditions, and survival. Each of these implies very different metabolic and regulatory responses and the first two are most relevant for this study.

The econometric analogy sometimes used to study plant responses to what we may define as "resources" (water, nutrients) has some limitations. It often implies the assumption of optimization of resource use in relation to some "return" that may be growth, survival, or reproduction. It is very difficult to define goals at organism level, whereas at the ecological level the number or interactions are virtually endless. For similar reasons, the so-called efficiency terms, i.e., the ratios of inputs (e.g., nutrient uptake or water absorbed) vs. output (e.g., growth) should be interpreted with care. For example, a greater efficiency in resource use may not increase the fitness of the organism towards its environment. For example, in many cases ecological fitness may be more related with the efficacy of resource collection than with the efficiency of its use, especially when competition is involved, as argued for water use by Passioura (1982; see also Jones 1980).

We need to understand better the responses of physiological processes to environmental factors (e.g., foliage growth and how the whole organism integrates signals from the environment, or respiration) without being biased by a priori concepts about how the plants use resources. Presumably the most important response of plants when resources are limited is to slow down growth and adjust size to resources. Still, plants may not respond to resource availability by itself but to signals (hormonal or physical) that trigger responses in organs away from the sites where resource availability is sensed. The following sections will consider how water or nitrogen abundance and temperature affect growth and gas exchange. Most attention will be given to situations where growth limitation occurs without major deficiencies or tissue damage. Although of great importance, the interaction between factors, such as between temperature and water or nutrients in the soil, will be ignored.

8.8.1   Water Deficits

It is well known that growth at the cellular and organ levels is more sensitive
to water deficits than gas exchange (e.g., Bradford and Hsiao 1982). In the
study of the effects of water deficits there has been some confusion between
the effects of dehydration and water shortage per se. However, gas exchange and growth may be affected by low water availability without tissue
dehydration (Davies and Zhang 1991; see also Chap. 10, this Vol.).

In the long term, water deficits reduce the rate of leaf expansion, final
leaf size, and σ, and the rate of initiation of leaf primordia and the release
of axilary buds (Pereira 1989; Pereira and Chaves 1992, see also Chap. 10,
this Vol.). The relative importance of the effects of water deficits in each
of these processes may change with plant genotype and stage of plant development. In many cases, differences in leaf area may change growth
rates without important changes in net photosynthetic rates or in photosynthetic capacity, as shown by a reduction in the average biomass production rate between ca. 30 and 60% in three *E. globulus* clones grown in a soil
with one quarter of the field capacity of well-watered controls (Osório and
Pereira unpublished). These plants not only had similar photosynthesis
but also had the same leaf nitrogen concentration as the well-watered controls, thus ruling out the likelihood that water shortage creates nitrogen
deficiency. However, lower soil water content may interact with nutrient
uptake. Gollan et al. (1992) showed that the concentration of nitrate (and
phosphate) in the xylem sap decreased with decreasing soil water content in
plants kept fully turgid. Because growth is reduced, this could occur without
changes in the nitrogen concentrations in leaves but result in reduced growth
rates by interference with growth regulators such cytokinins (see below).
Nevertheless, studies in the field with almond trees failed to show any
decrease in cytokinin concentration in xylem sap with increasing water stress
under field conditions (Heilmeier, personal communication). Alternatively,
the response of the plant to ABA might be modulated by nitrate uptake, as
suggested by Schurr et al. (1992) for stomatal closure.

Root growth is often less affected by water deficits than shoots, and this
results in an increase in root/shoot ratio when water availability is low.
As in the case of shoots, hormonal control has been invoked together with
osmotic adjustment to explain the response of roots to water deficits. In the
case of roots, however, it has been suggested that increases in endogenous
abscisic acid ABA act differentially to maintain root elongation and inhibit
leaf elongation with water deficits (Sharp 1990).

After a period of water stress even if gas exchange rates are re-established,
whole-plant growth is reduced mainly due to the decrease in F [see Eqs. (3)
and (4)]. The reduction in water loss by stomatal closure is often not enough
to prevent tissue dehydration and this may cause a decrease in the photosynthetic capacity of leaf tissues. It has been shown that in many cases
photosynthetic capacity may remain unchanged at least until relative water

content decreases to around 70% (Chaves 1991). Stomata close at much less severe water deficits. However, the extent and nature of nonstomatal limitation of photosynthesis depend upon severity of dehydration, rate of imposition of stress, and the coincidence of water deficits and other stresses. There are, however, species differences in the response of photosynthetic capacity to water deficits (Quick et al. 1992).

The maintenance of the photosynthetic capacity at the chloroplast level until dehydration is too severe may be important for growth in many circumstances, especially in recovery from short dry spells. Indeed, leaf carbon assimilation may resume at high rates as soon as stomata open after rehydration (Chaves 1991). Recovery may become more complex if dehydration lasts for a relatively long period because then stored starch may be used up in the growth of strong sinks, e.g., reproductive growth in annuals (Chapin et al. 1990). In the short-term, partition of photoassimilates the accumulation of starch is sacrificed to maintain the levels of soluble carbohydrates near the well-watered values (Quick et al. 1992). Under the same circumstances there may also be a decrease in Rubisco leaf concentration (Jones 1973; Chaves 1991). In addition to the eventual decline in photosynthetic capacity, this would result in the decrease in nitrogen stored in the leaves for future growth. However, contradictory evidence was found for sunflower even after 4 weeks after withholding water (Fredeen et al. 1991). It is not clear how much of the reported decrease in photosynthetic capacity is the direct result of tissue dehydration or is due to leaf aging.

In addition to the more or less reversible effects of dehydration on leaves, water stress may result in accelerated acropetal leaf senescence, which may result in a decrease in whole-canopy photosynthetic capacity (see also Chap. 10, this Vol.). The importance of drought-induced leaf senescence and abscision is quite variable among species, and varies with the intensity and timing of water deficits (Jordan 1983; Pereira and Pallardy 1989; Ranney et al. 1990; Pereira and Chaves 1992).

Much less work has been done on respiratory rates under water stress, but again the rates of metabolism per se seem to be quite resistant to dehydration. However, both growth and maintenance respiration decrease because growth also decreases. For example, maintenance respiration in sorghum decreased from 50 to 30 mg $g^{-1} d^{-1}$ as the R_w decreased from 0.16 to $0.04 d^{-1}$ as a result of water stress (Wilson et al. 1980).

As mentioned above, biomass production is usually correlated with the amount of water transpired

$$B = wE, \qquad\qquad (10)$$

where w is the proportionality regression coefficient or water use efficiency and E the amount of water transpired. This is not surprising, given the leading role of stomatal aperture and leaf area in controlling both photosynthesis and transpiration. Net assimilation rate (E_A) of single plants has also been found correlated with the rate of transpiration on a leaf area

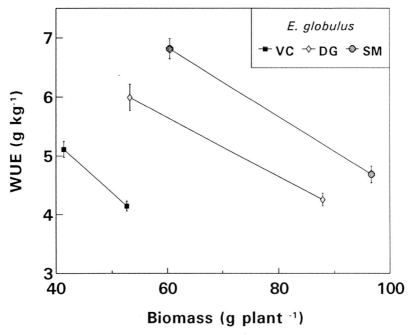

Fig. 8.7. Relationship between water use efficiency and biomass production in three *Eucalyptus globulus* clones (*VC*, *DG*, *SM*) well watered and with a reduced water supply. (Osório and Pereira, unpubl.)

basis (Könings 1989). On the other hand, E and Q_i are correlated because both depend upon incident radiation and leaf area index. However, the value w is highly variable with climate, especially with air humidity (Davies and Pereira 1992).

When plants are compared under strictly similar environments, a greater water use efficiency may be related with either higher growth and photosynthesis or with lower growth rates due to stomatal closure (Davies and Pereira 1992). As illustrated in Fig. 8.7, each of the three *E. globulus* clones under study had slightly different water use efficiencies with sufficient water supply (values at highest B on each line). However, water use efficiency increased as well as the discrimination for ^{13}C, when the plants were grown with water deficits, which resulted in stomatal closure, and was therefore negatively correlated with productivity (lowest B value on each line of Fig. 8.7).

It is worth noting, however, that even under similar conditions the B/E quotient is less that the instantaneous leaf photosynthesis/transpiration ratio due the respiratory losses in the transformation of the photoassimilated carbon into biomass and the nontranspiratory losses of water (Tanner and Sinclair 1983; Farquhar and Richards 1984). The reduction may be of the order of 20 to 50% according to Fisher and Furner (1978), or even greater.

We may use Eq. (7) to summarize how water deficits change the relationship between A and growth even though it does not allow any kinetic simulation of growth. From what was written above, water deficits will result in reduced growth because ρ increases and A decreases with water deficits. In many cases, changes in A result from stomatal closure decreasing maximum conductance and increasing the afternoon depression in conductance. In the long term, decrease in the photosynthesis of the whole canopy may occur, possibly due to aging of foliage and leaf shedding. The reasons for the changes in the respiration term of Eq. (7), ϕ, are not clear because only a few studies are available on respiration. It is likely, however, that in spite of the decline in the instantaneous rates of respiration, a greater percentage of carbon fixed goes to respiration under water stress, because of the decrease in photosynthetic rates. Changes in plant structure, leaf area, and the level of stored carbohydrates and nutrients will hamper the full recovery of growth rates after water deficits are alleviated even if gas exchange rates are restored. A great variety of responses may be expected, however, from much slower growth compared to well-watered controls to faster growth rates. The latter may result, for example, from previously water-stressed plants being smaller and having therefore less self-shading of foliage. In addition, the initiation of growth may create a greater sink capacity and the stimulation of A in the previously water-stressed plants.

8.8.2 Nitrogen Abundance

Nitrogen is often the major growth-limiting nutrient, and its effects on growth have been extensively studied. About 75% of the element in a C_3 plant leaf is allocated to the chloroplast (Chapin et al. 1987) and it is often found that photosynthetic rate of single leaves increases with nitrogen concentration in tissues (Field 1983; Field and Mooney 1986; Evans 1989, 1990; Sinclair and Horie 1989). However, the reverse may occur as well (e.g., Sheriff et al., 1986) if, for example, a lower proportion of tissue nitrogen concentration is allocated to the photosynthetic apparatus (thylakoids plus soluble protein) (Osmond 1987; Evans 1990).

Table 8.2. Gas exchanges (A, $\mu mol\,m^{-2}s^{-1}$; g_s $mmol\,m^{-2}s^{-1}$), photosynthetic capacity ($\mu mol\,O_2\,m^{-2}s^{-1}$), w_i water use efficiency ($mmol\,CO_2/mol\,H_2O$) and photosynthesis per unit N mass ($\mu mol\,mol^{-1}s^{-1}$) and nitrogen concentration in leaves [g (N) m^{-2}] of *E. globulus* grown with two levels of nutrient supply. Standard errors in parenthesis. (Pereira et al. 1992b)

Treatments	A	g_s	Photos. cap.	WUE	Photos. per unit of leaf N	[N]
High N	21.5 (0.9)	187.1 (22.5)	42.5 (4.4)	9.3 (1.5)	137.4 (35.3)	1.80 (0.12)
Low N	14.9 (0.8)	203.6 (27.9)	24.1 (4.2)	6.1 (1.1)	159.0 (39.1)	0.96 (0.17)

Table 8.2 shows the effects of nitrogen supply in gas exchange character-istics of *Eucalyptus globulus* plants grown in 10-l pots in the glasshouse with nutrient addition rates allowing for a high relative growth rate $(0.042\,d^{-1})$ and a low relative growth rate $(0.023\,d^{-1})$ (Pereira et al. 1992a). The higher growth rates of the high N plants were proportional to nitrogen concentra-tions in the leaf tissues and to significantly higher rates of net photosynthesis and photosynthetic capacity measured as O_2 evolution at CO_2 and light saturation. Stomatal aperture was unaffected and therefore water use ef-ficiency increased significantly with increasing N in the tissues. However, the different growth rates were also related with differences in biomass partitioning with an increase of ca. 48% in leaf area partitioning that re-sulted in a decrease (ca. 50%) in the root/shoot ratio after 56 days of growth. High N also resulted in a slight increase (20%) in the σ. All this resulted in a 34% increase in F in the high N plants. Similar morphological effects were also found to be dominant by McDonald et al. (1986), who showed that a reduction in the rate of nitrogen supply to birch seedlings decreased growth mainly because it reduced more the allocation of assim-ilates to new leaf area than the rate of photosynthesis per leaf area. The extra carbon in slow growing plants was accumulated as starch mainly in leaves.

The results shown in Table 8.2 contrast with those reported by Pereira et al. (1992a) for plants of the same species grown in the field with differ-ent rates of nutrients and water supply. In these, biomass production was positively related to leaf area (see Table 8.1) but the leaf photosynthetic capacity was higher in the control without fertilizers added (C) than in the optimal nutrition trees (IL), as shown in Table 8.3. However, as shown by the similar values of photosynthetic capacity expressed on a dry weight basis (Pereira et al. 1992b) or in nitrogen or chlorophyll concentration (see Table 8.3), the differences in photosynthetic capacity on a leaf area basis (in the first row of Table 8.3) result largely from differences in σ (see also Fig. 8.2). Nevertheless, the differences between treatments in nitrogen content of leaf tissues were modest and much smaller than in potted plants. The primacy of partition of N to foliage growth rather than to inreased leaf photosynthetic capacity on a leaf level in *E. globulus* (Pereira et al. 1992b) was also found by Cromer and Jarvis (1990) in *Eucalyptus grandis* seedlings grown with different nutrient addition rates under controlled laboratory conditions.

Low nitrogen supply reduces the rate of leaf expansion, final leaf size, and specific leaf area, as well as the release of axilary buds and the accumulation of starch (Dale 1982; Trewavas 1985; McDonald et al. 1986; Lambers et al. 1990; Pereira 1990; Fredeen et al. 1991; Schulze et al. 1991). Table 8.4 shows a clear loss of apical dominance (number of branches of higher orders inserted in lower order ones) in *E. globulus* plants grown in the field with high rates of nutrient supply (IL) in comparison to rain-fed controls without fertilizers added (C). This effect of high N was largely responsible for the increase in L and in growth rate in eucalyptus (see Table 8.1). It is possible

Table 8.3. Photosynthetic capacity on an area basis and on
N and chlorophyll concentration of leaves of two groups of
E. globulus grown without any fertilizers added and rain-fed
(C) and irrigated with "near optimal nutrition" (IL). (See
text and Table 1 for details.) Standard errors in parenthesis

	IL	C
Area (μmol m^{-2} s^{-1})	15.1 (1.9)	19.7 (2.2)
[N] (μmol mol^{-1} s^{-1})	165.2 (22.4)	148.4 (18.2)
[Chlorophyll] (μmol g^{-1} s^{-1})	32.0 (4.4)	34.0 (4.0)

Table 8.4. Number of third-order branches per unit second-
order branch, at three different levels of the crown in 3-year-
old trees of *E. globulus* grown with different nutrition levels
(see text and Table 1 for details). Standard errors in parenthesis. (After Pereira 1990)

	Treatments	
	IL	C
Percent of crown height		
0–60	9.0 (3.8)	7.0 (3.7)
60–80	8.3 (3.4)	5.8 (3.5)
80–100	7.4 (2.4)	3.7 (3.4)

that the loss in apical dominance related to high N nutrition may be related
with increased cytokinin production at high levels of nitrate in the root
medium (Samuelsson et al. 1991).

Nitrogen concentrations in leaf tissues decrease normally with leaf age,
with younger leaves having the highest concentrations and photosynthetic
rates. The increase in N concentration and photosynthetic capacity along a
plant or a branch may improve the whole-canopy carbon assimilation because
nitrogen retranslocated from older to younger (well exposed) leaves will
allow a higher photosyynthetic capacity and may lead to the optimization of
carbon gain by the whole canopy for a given nitrogen availability (Field
1983; Hirose and Werger 1987).

In field-grown *E. globulus*, the decrease in photosynthetic capacity with
nitrogen concentration (on an area basis) in leaves sampled along branch
length, from the proximal to the distal ends, was only ca. 32% in low
nutrients (C) and 19% in high nutrients (IL) (Pereira et al. 1992b). This is a
small decrease in comparison with other observations (see Field 1983; Evans
1989). We tested the impact of heterogeneity of leaf photosynthetic capacity
in carbon assimilation by the whole canopy by simulating with the MAESTRO
model parameterized for *E. globulus* juvenile foliage (Jarvis et al. 1989).

This impact was rather small because in spring days (when photosynthetic rates are highest) the amount of carbon fixed by the heterogeneous canopy was only 5 to 10% better than the homogeneous canopy (Pereira et al. 1992b). Hirose and Werger (1987), using a model parameterized for the actual distribution of leaf N in the canopy of *Solidago altissima*, which was more uniform than the "optimal" distribution, found that the heterogeneous canopy realized over 20% more A_p than the uniform N distribution and 4.7% less A_p than the "optimal" distribution. They also found that redeployment of N to the leaves at the top of the canopy with aging should be more effective in increasing A_p with high L than with low L values. This may also be the case of the modest variation in N concentration in *E. globulus* canopies which have typically low L values and little mutual shading of leaves. A decrease in light intensity or a modification in spectral composition as characterized by red/far-red ratio due to shade seem to be necessary to bring about an important nitrogen retranslocation and therefore a gradient in photosynthetic capacity in the canopy.

A greater nitrogen concentration in the foliage results normally in a decrease in the "photosynthetic nitrogen use efficiency" at leaf level (maximum A per unit of leaf nitrogen content per unit of area) as found, for example, in sunflower (Fredeen et al. 1991). In *E. globulus* there was a 16% decrease in "photosynthetic nitrogen use efficiency" in plants grown with high N supply (Table 8.2, differences significant $P < 0.05$). The same happened with the nitrogen use efficiency (NUE) calculated as the ratio of biomass produced/N uptake during a given period. In potted *E. globulus* plants, high N supply reduced NUE to 1/4 of the corresponding value in the low N plants. This reflects the observation that in many species an increase in nitrogen use efficiency is a phenotypic response to low nutrient availability (Millard and Proe 1991). In this case a greater proportion of N from older to new leaves in low N than in high N plants may explain these differences in part (Pereira et al. 1992a). On the other hand, in the field trial shown in Table 8.1, the NUE was not substantially different in high nutrition (IL) and the rain-fed without fertilizers added control (C). The same is shown for the instantaneous rates of carbon assimilation per unit of nitrogen mass in Table 8.3. This may be explained by the decrease in photosynthetic and growth efficacy due to summer water stress in the rain-fed control in comparison with the irrigated IL plots.

In conclusion, abundant nitrogen in the rooting medium leads to fast growth as a result of increases in F and/or A. In crops, Sinclair and Horie (1989) showed that increases in N concentration on a leaf area basis resulting in higher photosynthetic rate have a limited impact in ε unless the plants have deficiency symptoms. The value of N concentration that maximizes ε is lower than the optimal concentration for net photosynthesis. Before canopy closure (or at low L as in eucalypt stands), nitrogen abundance seems to be most important in canopy development bringing about increases in foliage area.

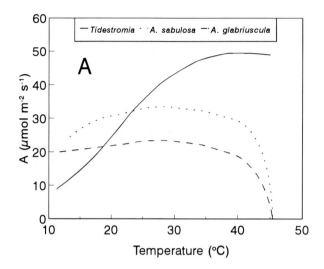

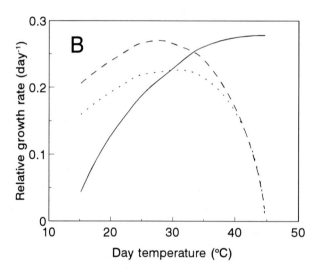

Fig. 8.8A,B. Responses to air temperature of A (measured at $1600\,\mu mol\,m^{-2}\,s^{-1}$ of photon flux density and 320 ppm CO_2) and R_w in three north-american shrubs, *Tidestromia oblongifolia*, *Atriplex sabulosa* and *Atriplex glabriuscula*. The leaves of *Tidestromia oblongifolia* were acclimated to high temperatures (grown at 40/30 °C day/night tempera-tures) and the two *Atriplex* species were acclimated at lower temperatures (grown at 15/11 °C day/night temperatures). (Berry and Raison 1981)

8.8.3 Temperature Effects

Air temperature influences growth and gas exchange differently (Fig. 8.8). Typically, the metabolic responses to temperature have an optimum value which not only varies with species but is different for different processes such as leaf and root growth, stomatal aperture, photosynthesis, or respiration (Berry and Raison 1981; Ong and Baker 1985). Traditionally, the decrease in growth at high temperatures is attributed to a higher optimum temperature for respiration than for net photosynthesis, thus decreasing the daily net carbon gain. For example, the importance of warm nights in reducing plant biomass production has been emphasized by several authors.

The decrease in A at temperatures above or below T_o results from decreases in quantum yield and light-saturated photosynthesis. At temperatures above T_o there is a "down-regulation" of Rubisco which is apparently dependent on changes in the activation state of the enzyme (Weis and Berry 1988). These responses are not static but may vary as a result of acclimation. For example, Slatyer (1977) showed ca. 10 °C change in temperature optima for photosynthesis due to acclimation to low and high temperatures in *Eucalyptus pauciflora*. In *Eucalyptus globulus*, Pereira et al. (1986) showed a shift in the temperature response of leaf respiration corresponding to seasonal changes in temperature, e.g., respiration rates at 20 °C were lower in hot summer months than in cooler winter and spring months (Fig. 8.9). The physiological and molecular aspects of acclimation of metabolism to changing temperatures are out of the scope of this work but they are under intense scrutiny in the present.

The temperature response of canopy growth does not coincide with metabolic responses. Canopy growth depends on the rate and duration of leaf expansion, the number of leaves and branches formed per unit time, and the death rate of leaves. For example, the rate of leaf expansion often has temperature optima (T_o) higher than the temperature that maximizes final leaf size because the duration of leaf expansion decreases with temperature (Ong and Baker 1985). Leaf production rates and branching also increase at higher temperatures (Quinby et al. 1973). By increasing the rate of canopy development, higher temperatures result in the increase in the value of D or f [Eq. (8)] even though the value of ε may not change (Marshall et al. 1991). However, as development is also accelerated, leaf longevity decreases at high temperatures. The end result may be that D (or green leaf area duration and consequently f) may not be improved at higher temperatures. This is illustrated by *Lupinus albus* plants grown at higher temperatures that produced 37% more leaves but had ca. 34% reduction in D in comparison to the plants grown at lower temperatures because of the accelerated rate of senescence in the warmer regime (M. Chaves pers. comm.). On the other hand, at temperatures substantially below the optimum for growth, plants

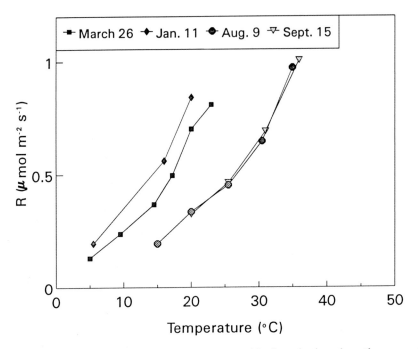

Fig. 8.9. Changes in the temperature response of leaf respiration along the season in *E. globulus* trees in Portugal. (Modified from Pereira et al. 1986)

tend to acquire characteristics of the "slow growers", e.g., low σ and leaf weight ratio (LWR) and accumulation of starch and soluble sugars.

The influence of temperature on growth through its effects on photosynthesis may be complicated because of the interaction of other environmental factors, especially light and water stress. Shaded leaves in the interior of the canopy may have their carbon balance reduced more at higher than at lower temperatures because of the differential in T_o for photosynthesis and respiration. However, more important is the increased tendency for the down-regulation of photosynthesis and photoinhibition at temperatures either below or above T_o (Ludlow and Bjorkman 1984; Baker et al. 1988). When high temperatures are superimposed on water deficits and high light, plants become even more susceptible to photoinhibition (Chaves et al. 1992), possibly because leaves with closed stomata are unable to dissipate absorbed solar radiation as latent heat and consequently leaf temperature increases further. Photoinhibition may reduce the carbon assimilation, A_p (see Sect. 8.6), although at high irradiances the decrease in A of the exposed leaves tend to be compensated by the enhanced contribution of leaves in the interior of the canopy not affected by photoinhibition.

Another complication that may arise in the relation between growth and gas exchange results from the effects of soil temperature on roots. Even if photosynthetic capacity does not change, root temperature may affect carbon partitioning coefficient to roots (changing root/shoot ratio and therefore F), root maintenance respiration, and water and nutrient absorption. All of these may directly or indirectly influence A and foliage growth.

8.9 Conclusions

Even though often equated, growth and net CO_2 fixation involve quite different processes and time scales and very seldom are leaf photosynthetic rates (on an area basis) correlated with R_W. Furthermore, growth rates can vary severalfold between plants (e.g., $40\times$ in data collated by Poorter 1989), whereas leaf photosynthetic rates (on an area basis) normally vary less ($10\times$ to $20\times$ according to Körner et al. 1979). Table 8.5 summarizes some of the traits commonly found when fast- and slow-growing plants are compared. The lack of correlation between growth and instantaneous gas exchange rates is obvious (see also Fig. 8.1).

In the comparison of growth and gas exchange there are methodological problems. Gas exchange is measured with the aim of expressing the instantaneous plant performance, whereas growth necessarily integrates along time and reflects morphological changes in the plants. It is redundant to state that dry matter accumulation shall equal the integral of photosynthetic carbon assimilation minus respiration. This is so if the period of integration is long enough, i.e., similar to what is used for growth estimates. Most important to understand the relation betweeen growth and gas exchange is not only the momentary carbon balance equation per se, but growth kinetics determined by the way new biomass is partitioned between productive

Table 8.5. Comparison of the characteristics most commonly found in fast- and slow-growing plants. (Idea from data collated by Lambers et al. 1990 and Poorter 1989)

	Fast growers	Slow growers
A (area basis)	=	=
A (dry mass)	+	−
σ (SLA)	++	− −
Leaf weight ratio	++	−
Leaf area partitioning	++	−
Respiration rate	+	−
Respir. as % Ap	−	+
Percent carbon in biomass	−	+
Starch in tissues	−	+

(photosynthetic), only respiratory, or passive (accumulation) compartments. This helps to explain the great variability in the relationship between growth and photosynthetic rates on a leaf area basis. In this context the regulation of photoassimilate partitioning and the formation of new organs (sources or sinks for assimilation of carbon) are points of major ignorance.

Whenever the rate of assimilation exceeds the rate of incorporation of carbon in plant structure or its use in respiration, carbon compounds are stored. The same may happen with macronutrients, namely nitrogen. Some of these stored compounds may be utilized at times when growth may occur at rates higher than those that could be supported by current photosynthetic rates. The lack of phase between growth and carbon assimilation (either on a daily or on a seasonal basis) is bound to make the relationship between growth and gas exchange rather complex. Furthermore, active growth may be occurring in some plant organs at the same time that storage occurs in other plant parts. The use of adequate simulation models and a better understanding of the mechanisms determining carbon and nitrogen allocation and internal cycling are essential tools to improve our understanding of the relationship with growth.

Environmental factors also modify the relationship between growth and gas exchange. Growth is normally reduced during acclimation to low nutrients, to water deficits, or to temperatures different from optimum. This may occur because of direct effects in metabolic rates (e.g., temperature) or through the whole-plant growth regulation (water or nitrogen) even without any negative effect on photosynthetic capacity. The end result of acclimation is the tendency towards the adjustment of plant size and development rate to resources available. In any case, carbon balance is negatively influenced by environmental stresses as a result of changes in plant structure (e.g., increase in nonphotosynthetic biomass relative to foliage) and stomatal closure, decreases in photosynthesis in part of the canopy, or increases in respiration rates. The changes in the rates of development resulting from the effects of environmental factors (e.g., leaf aging, flowering) may strongly modify the relationship between gas exchange and growth.

In recent years, our understanding of the biochemistry and biophysics of photosynthesis under different environmental conditions has progressed steadily, accompanied by the development of gas exchange equipment and chlorophyll a fluorometers usable in the field. However, the recognition of the importance of canopy expansion and carbon and nitrogen allocation in determining growth has not been accompanied by a research effort to understand the controling mechanisms, at least comparable to that which has been devoted to photosynthesis research.

Appendix: List of Symbols and Definitions

A – rate of net photosynthesis, $\mu mol(CO_2)\, m^{-2} s^{-1}$

A_{max} – rate of net photosynthesis at saturating light and ambient CO_2

A_P – integral of whole plant photosynthesis

B – biomass productivity (e.g., $t\, ha^{-1} year^{-1}$)

C_f – conversion factor indicating the biomass/carbon ratio of recently produced dry matter

D – LAD, leaf area duration, days

E_A – NAR, net assimilation rate $kg\, m^{-2} day^{-1}$

E – transpiration rate, $mmol\, m^{-2} s^{-1}$ (or $kg\, m^{-2} day^{-1}$)

f – fraction of solar radiation intercepted by a canopy over a given period

F – LAR, leaf area ratio, $m^2 kg^{-1}$

k – light extinction coefficient

L – leaf area index

L_a – leaf area of a plant

LWR – leaf weight ratio

m – maintenance respiration coefficient

NUE – nitrogen use efficiency, kg (dry weight)/g (N)

Q – accumulated solar radiation, i and o subscripts stand for intercepted and incident, respectively

R – rate of mitochondrial respiration, $\mu mol(O_2)\, m^{-2} s^{-1}$

R_P – integral of whole plant respiration

R_w – RGR, relative growth rate, day^{-1}

w – water use efficiency, g (dry weight)/kg (H_2O)

w_i – instantaneous water use efficiency, mmol CO_2/mol H_2O

W – plant biomass

Y_G – growth conversion efficiency, i.e., ratio of C substrate used for dry matter accumulation (ΔS_T) and total substrate used in growth respiration ($\Delta S_G = \Delta S_R - \Delta S_T$).

ε – radiation utilization coefficient, $g\, MJ^{-1}$

ϕ – the proportion of carbon fixed that is respired

ρ – the mass of carbon in the plant per unit leaf area

σ – SLA, specific leaf area, $m^2 kg^{-1}$

References

Aerts R (1989) The effect of increased nutrient availability on leaf turnover and above-ground productivity of two evergreen ericaceous shrubs. Oecologia 78: 115–120

Amthor JS (1989) Respiration and crop productivity. Springer Berlin Heidelberg New York

Azcón-Bieto J (1983) Inhibition of photosynthesis by carbohydrates in wheat leaves. Plant Physiol 73: 681–668

Baker NR, Long SP, Ort DR (1988) Photosynthesis and temperature, with particular reference to effects on quantum yield. In: Long SP, Woodward FI (eds) Plants and temperature. Company of Biologists Ltd, Cambridge, pp 347–375

Berry JA, Raison JK (1981) Responses of macrophytes to temperature. In: Lange OL, Nobel PS, Osmond CB, Ziegler H (eds) Physiological plant ecology. I. Responses to the physical environment. Springer Berlin Heidelberg New York, pp 277–338

Bradford KJ, Hsiao TC (1982) Physiological responses to moderate water stress. In: Lange OL, Nobel PS, Osmond CB, Ziegler H (eds) Physiological plant ecology. II. Water relations and carbon assimilation. Springer Berlin Heidelberg New York, pp 263–324

Cannell MGR (1989) Light interception, light use efficiency and assimilate partitioning in poplar and willow stands. In: Pereira JS, Landsberg JJ (eds) Biomass production by fast-growing trees. Kluwer, Dordrecht, pp 1–12

Ceulemans R, Impens I (1983) Net CO_2 exchange rate and shoot growth of young poplar (*Populus*) clones. J Exp Bot 34: 866–870

Chapin FS, Bloom AJ, Field CB, Waring RH (1987) Plant responses to multiple environmental factors. BioScience 37: 49–57

Chapin FS, Schulze E-D, Mooney HA (1990) The ecology and economics of storage in plants. Annu Rev Ecol Syst 21: 423–447

Charles-Edwards DA, Doley D, Rimmington GM (1986) Modelling plant growth and development. Academic Press Australia, Sydney

Chaves MM (1991) Effects of water deficits on carbon assimilation. J Exp Bot 42: 1–16

Chaves MM, Osório ML, Osório J, Pereira JS (1992) The photosynthetic response of *Lupinus albus* to high temperature is dependnet on irradiance and leaf water status. Photosynthetica 27: 521–528

Cromer RN, Jarvis PG (1990) Growth and biomass partitioning in *Eucalyptus grandis* seedlings in response to nitrogen supply. Aust J Plant Physiol 17: 503–515

Dale J (1982) The growth of leaves. Edward Arnold, London

Davies WJ, Zhang J (1991) Root signals and the regulation of growth and development of plants in drying soil. Annu Rev Plant Physiol Mol Biol 42: 55–76

Davies WJ, Pereira JS (1992) Plant growth and water use efficiency. In: Baker NC, Thomas H (eds) Crop photosynthesis: spatial and temporal determinants. (Topics in photosynthesis, vol 12. Elsevier, Amsterdam, pp 213–233

Demetriades-Shah TH, Fuchs M, Kanemasu ET, Flitcroft I (1992) A note of caution concerning the relationship between cumulated intercepted solar radiation and crop growth. Agric For Meteorol 58: 193–207

Digby J, Firn RD (1985) Growth substances and leaf growth. In: Baker NT, Davies WJ, Ong CK (eds) Control of leaf growth. Cambridge University Press, Cambridge, pp 57–76

Evans JR (1989) Photosynthesis and nitrogen relationships in leaves of C_3 plants. Oecologia 78: 9–19

Evans JR (1990) Photosynthesis – the dependence of nitrogen partioning. In: Lambers H, Cambridge ML, Konings H, Pons TL (eds) Causes and consequences of variation in growth rate and productivity of higher plants. SPB Academic Publ, The Hague, pp 159–174

Evans LT (1976) Physiological adaptation to performance as crop plants. Philos Trans R Soc Lond B 275: 71–83

Farquhar GD, Richards RA (1984) Isotopic composition of plant carbon correlates with water-use efficiency of wheat cultivars. Aust J Plant Physiol 11: 539–552

Farquhar GD, Ehleringer JR, Tand KT, Hubick (1989) Carbon isotope discrimination and photosynthesis. Annu Rev Plant Physiol Plant Mol Biol 40: 503–537

Field C (1983) Allocating leaf nitrogen for the maximization of carbon gain: leaf age as a control on the allocation program. Oecologia 56: 341–347

Field C, Mooney HA (1986) The photosynthesis-nitrogen relationship in wild plants. In: Givnish TJ (ed) On the economy of form and function. Cambridge University Press, Cambridge, pp 25–55

Fisher RA, Turner NC (1978) Plant productivity in the arid zone and semiarid zones. Annu Rev Plant Physiol 29: 277–317

Foyer CH (1987) The basis of source-sink interaction in leaves. Plant Physiol Biochem 25: 649–57

Foyer CH (1988) Feedback inhibition of photosynthesis through source-sink regulation in leaves. Plant Physiol Biochem 26: 483–92

Fredeen AL, Gamon JA, Field CB (1991) Responses of photosynthesis and carbohydrate-partitioning to limitations in nitrogen and water availability in field-grown sunflower. Plant Cell Environ 14: 963–970

Gazarini LC, Araújo MC, Borralho N, Pereira JS (1991) Plant area index in *Eucalyptus globulus* plantations determined indirectly by a light interception method. Tree Physiol 7: 107–113

Geiger DR, Servaites JC (1991) Carbon allocation and response to stress. In: Mooney HA, Wimmar WE, Pell EG (eds) Response of plants to multiple stresses. Academic Press, New York, pp 103–127

Gifford RM, Thorne JH, Hitz WD, Giaquinta RT (1984) Crop productivity and photo-assimilate partitioning. Science 225: 801–808

Gollan T, Schurr U, Schulze ED (1992) Stomatal response to drying soil in relation to changes in the xylem sap composition of *Helianthus annuus*. I. The concentration of cations, amions, amino-acids in, and pH of, the xylem sap. Plant Cell Environ 15: 551–559

Gowing DJG, Davies WJ, Jones HG (1990) A positive root-sourced signal as an indicator of soil drying in apple *Malus × domestrica*. Borkh J Exp Bot 41: 1535–1540

Grime JP (1979) Plant strategies and vegetation processes. Wiley, Chichester

Harper JL (1989) Canopies as populations. In: Russell G, Marshall B, Jarvis PJ (eds) Plant canopies. Their growth form and function. Cambridge University Press, Cambridge, pp 105–128

Herold A, Walker DA (1979) Transport across chloroplast envelopes – the role of orthophosphate. In: Giebish G, Tosteson DC, Ussing HH (eds) Handbook on transport Springer Berlin Heidelberg New York, pp 411–439

Hirose T, Werger MJA (1987) Maximizing daily canopy photosynthesis with respect to the leaf nitrogen allocation pattern in the canopy. Oecologia 72: 520–526

Huber SC (1981) Interspecific variation in activity and regulation of leaf sucrose phosphate syntheses. Z Pflanzenphysiol 102: 443–50

Huber SC (1986) Fructose 2,6-bisphosphate as a regulatory metabolite in plants. Annu Rev Plant Physiol 37: 233–46

Hunt R (1982) Plant growth curves. The functional approach to plant growth analysis. Edward Arnold, London

Jarvis P, Leverenz JW (1983) Productivity of temperate, deciduous and evergreen forests. In: Lange OL, Nobel PS, Osmond CB, Ziegler H (eds) Physiological plant ecology. IV. Ecosystem processes. Mineral cycling, productivity and man's influence. Springer Berlin Heidelberg New York, pp 233–280

Jarvis PG, McNaughton KG (1986) Stomatal control of transpiration: scaling up from leaf to region. In: MacFadyen A, Ford ED (eds) Advances in ecological research, vol 15. Academic Press, London, pp 1–49

Jarvis PG, Wang YP, Borralho NMG, Pereira JS (1989) Simulation of the roles of stress on radiation absorption, assimilation transpiration and water use efficiency of stands of *Eucalyptus globulus*. In: Pereira JS, Landsberg JJ (eds) Biomass production by fast-growing trees. Kluwer, Dordrecht, pp 169–179

Jones HG (1973) Moderate-term stress and associated changes in some photosynthetic parameters in cotton. New Phytol 72: 1095–1104

Jones HG (1980) Interaction and integration of adaptative responses to water stress: the implications of an unpredictable environment. In: Turner NC, Kramer PJ (eds) Adaptation of plants to water and high temperature stress. Wiley New York, pp 353–363

Jones HG (1983) Plants and microclimate. Cambridge University Press, Cambridge, UK

Jordan WR (1983) Whole plant response to water deficits: an overview. In: Taylor HM, Jordan WR, Sinclair TR (eds) Limiatations to efficient water use in crop production. Am Soc Agronomy Madison, Wisc (USA), pp 289–317

King RW, Wardlaw IF, Evans LT (1967) Effect of assimilated utilisation on photosynthetic rate in wheat. Planta 77: 261–276

Könings H (1989) Physiological and morphological differences between plants with a high NAR or a high LAR as related to environmental conditions. In: Lambers H, Cambridge ML, Konings H, Pons TL (eds) Causes and consequences of variation in growth rate and productivity of higher plants. SPB Academic Publ, The Hague, pp 101–123

Körner CH, Scheel JA, Bauer H (1979) Maximum leaf diffusive conductance in vascular plants. Photosynthetica 13: 45–82

Kozlowski TT, Keller T (1966) Food relations of woody plants. Bot Rev 32: 293–382

Krapp A, Quick WP, Stitt M (1991) Ribulose-1,5-bisphosphate carboxylase-oxygenase, other Calvin-cycle enzymes and chlorophyll decrease when glucose is supplied to mature spinach leaves via the transpiration stream. Planta 186: 58–69

Kriedemann PE (1986) Stomatal and photosynthetic limitations to leaf growth. Aust J Plant Physiol 13: 15–31

Lambers H, Freijsen N, Poorter H, Hirose T, van den Werf A (1990) Analysis of growth based on net assimilation rate and nitrogen productivity. Their physiological background. In: Lambers H, Cambridge ML, Konings H, Pons TL (eds) Causes and consequences of variation in growth rate and productivity of higher plants. SPB Academic Publ, The Hague, pp 1–17

Larcher W (1973) Oekologie der Pflanzen. Ulmer, Stuttgart

Lauer MJ, Shibles R (1987) Soybean leaf photosynthetic response to changing sink demand. Crop Sci 27: 1197–1201

Linder S (1985) Potential and actual production in Australian forest stands. In: Landsberg JJ, Parsons W (eds) Research for Forest Management CSIRO, Melbourne, Australia, pp 11–35

Ludlow MM, Bjorkman O (1984) Paraheliotropic leaf movements in *Sirartro* as a protective mechanism against drought-induced damage to primary photosynthetic reactions: damage by excessive light and heat. Planta 161: 505–518

Madeira M, Pereira JS (1990) Productivity, nutrient immobilization and soil chemical properties in an *Eucalyptus globulus* plantation under different irrigation and fertilization regimes. Water Air Soil Pollut 54: 621–634

Marshall B, Squire GR, Terry AC (1991) Effect of temperature of interception and conversion of solar radiation by stands of groundnut. J Exp Bot

Masle J, Farquhar G (1988) Effects of soil strength on the relation of water-use efficiency and growth to carbon isotope discrimination in wheat seedlings. Plant Physiol 86: 32–38

McDonald AJS, Lohamar T, Ericsson A (1986) Growth response to step-decrease in nutrient availability in small birch (*Betula pendula* Roth). Plant Cell Environ 9: 427–432

Mebrahtu T, Hanover JW (1991) Family variations in gas exchange, growth and leaf traits of black *locust half-sib* families. Tree Physiol 8: 185–193

Metcalfe J, Davies WJ, Pereira JS (1990) Leaf growth of *Eucalyptus globulus* seedlings under water deficit. Tree Physiol 6: 221–227

Millard P (1988) The accumulation and storage of nitrogen by herbaceous plants. Plant Cell Environ 11: 1–8

Millard P, Proe MF (1991) Leaf demography and the seasonal internal cycling in sycamore (*Acer pseudoplatanus* L.) seedlings in relation to nitrogen supply. New Phytol 117: 87–96

Monteith JL (1977) Climate and efficiency of crop production in Britain. Philos Trans R Soc Lond B 281: 277–294

Monteith JL (1981) Does light limit crop production? In: Johnson CB (ed) Physiological processes limiting plant productivity. Butterworths, London, pp 23–38

Mooney HA, Ferrar PJ, Slatyer RO (1978) Photosynthetic capacity and carbon allocation patterns in diverse growth forms of *Eucalyptus*. Oecologia 36: 103–111

Neales TF, Incoll LD (1968) The control of leaf photosynthetic rate by the level of assimilate concentration in the leaf: A review of the hypothesis. Bot Rev 34: 107–125

Norman JM, Campbell GS (1983) Application of a plant environment model to problems in irrigation. In: Hillel D (ed) Advances in irrigation. Academic Press, New York, pp 158–188

Ong CK, Baker NR (1985) Temperature and leaf growth. In: Baker NR, Davies WJ, Ong CK (eds) Control of leaf growth. Cambridge University Press, Cambridge, pp 175–200

Osmond CB (1987) Photosynthesis and carbon economy of plants. New Phytol 106 (Suppl): 161–175

Passioura JB (1982) Water in the soil-plant-atmosphere continuum. In: Lange OL, Nobel PS, Osmond CB, Ziegler H (eds) Physiological plant ecology. II. Water relations and carbon assimilation. Springer, Berlin Heidelberg New York, pp 5–33

Passioura JB (1988) Root signals control leaf expansion in wheat seedlings growing in drying soil. Aust J Plant Physiol 15: 687–693

Patterson DT, Meyer CT, Quinby PC (1978) Effects of irradiance on relative growth rates, net assimilation rates and leaf area partitioning in cotton and three associated weeds. Plant Physiol 62: 14–17

Pereira JS (1990) Whole-plant regulation and productivity in forest trees. In: Davies WJ, Jeffcoat B (eds) Importance of root-to-shoot communication in the response to environmental stress. BPGRG Monograph 21-1990. British Society for Plant Growth Regulation, Bristol, pp 237–250

Pereira JS, Chaves MM (1992) Plant water deficits in Mediterranean ecosystems. In: Smith JAC, Griffiths H (eds) Plant responses to water deficits – from cell to community. BIOS Scientific Publ (environmental Plant Biology Series), pp 237–251

Pereira JS, Pallardy S (1989) Water stress limitations to tree productivity. In: Pereira JS, Landsberg JJ (eds) Biomass production by fast-growing trees. Kluwer, Dordrecht, pp 37–56

Pereira JS, Tenhunen JD, Lange OL, Beyschlag W, Meyer A, David MM (1986) Seasonal and diurnal patterns in leaf gas exchange of *Eucalyptus globulus* trees growing in Portugal. Can J For Res 16: 177–184

Pereira JS, Beyschlag G, Lange OL, Beyschlag W, Tenhunen JD (1987) Comparative phenology of four mediterranean shrub species growing in Portugal. In: Tenhunen JD, Catarino FM, Lange OL, Oechel WC (eds) Plant response to stress. Functional analysis in Mediterranean ecosystems. Springer, Berlin Heidelberg New York, pp 503–514

Pereira JS, Linder S, Araújo MC, Pereira H, Ericsson T, Borralho N, Leal L (1989) Optimization of biomass production in *Eucalyptus globulus* plantations. A case study. In: Pereira JS, Landsberg JJ (eds) Biomass production by fast-growing trees. Kluwer, Dordrecht, pp 101–121

Pereira JS, Chaves MM, Carvalho PO, Caldeira MC, Tomé J (1992a) Carbon assimilation growth and nitrogen supply in *Eucalyptus globulus* plants. In: Roy J, Granier E (eds) Whole plant perspective on carbon nitrogen interactions. SPB Academic Publ, The Hague (in press)

Pereira JS, Chaves MM, Fonseca F, Araújo MC, Torres F (1992b) Photosynthetic capacity of leaves of *Eucalyptus globulus* (Labill.) growing in the field with different nutrient and water supplies. Tree Physiol 11: 381–389

Pons TL (1977) An ecophysiological study in the field layer of ash coppice. II Experiments with *Geum urbanum* and *Cirsium palustre* in different light intensities. Acta Bot Neerl 26: 29–42

Poorter H (1989) Interspecific variation in relative growth rate: on ecological causes and physiological consequences. In: Lambers H, Cambridge ML, Konings H, Pons TL (eds) Causes and consequences of variation in growth rate and productivity of higher plants. SPB Academic Publ, The Hague, pp 45–68

Potter JR, Jones JW (1977) Leaf area partitioning as an important factor in growth. Plant Physiol 59: 10–14

Quick WP, Chaves MM, Wendler R, David MM, Rodrigues ML, Passarinho JA, Pereira JS, Adcok MD, Leegood RC, Stitt M (1992) The effect of water stress on photosynthetic carbon metabolism in four species grown under field conditions. Plant Cell Environ 15: 25–35

Quinby JR, Hesketh JD, Voigt RL (1973) Influence of temperature and photoperiod on floral initiation and leaf number in sorghum. Crop Sci 13: 243–246

Ranney TG, Whitlow TH, Bassuk NL (1990) Response of fine temperate deciduous tree species to water stress. Tree Physiol 6: 439–448

Reynolds JF, Acock B, Dougherty RL, Tenhunen, JD (1989) A modular structure for plant growth simulation models. In: Pereira JS, Landsberg JJ (eds) Biomass production by fast-growing trees. Kluwer, Dordrecht, pp 123–134

Russell G, Jarvis PG, Monteith JL (1989) Absorption of radiation by canopies and stand growth. In: Russell G, Marshall B, Jarvis PJ (eds) Plant canopies. Their growth, form and function. Cambridge University Press Cambridge, pp 21–39

Samuelsson ME, Eliasson L, Larsson C-M (1991) Nitrate-regulated growth and cytokinin responses in seminal roots of barley. Plant Physiol 98: 309–315

Schulze ED (1982) Plant life-forms and their carbon, water and nutrient relations. In: Lange OL, Nobel PS, Osmond CB, Ziegler H (eds) Physiological plant ecology. II. Water relations and carbon assimilation. Springer, Berlin Heidelberg New York, pp 615–676

Schulze ED, Hall AE (1982) Stomatal responses, water loss and CO_2 assimilation rates of plants in contrasting environments. In: Lange OL, Nobel PS, Osmond CB, Ziegler H (eds) Physiological plant ecology. II. Water relations and carbon assimilation. Springer, Berlin Heidelberg New York, pp 181–229

Schulze ED, Fuchs M, Fuchs MI (1977) Spatial distribution of photosynthetic capacity and significance of the evergreen habit. Oecologia 30: 239–248

Schulze ED, Turner NC, Golan T, Shakel KA (1986) Stomatal responses to air humidity and to soil drought. In: Zeiger E, Farquhar GD, Cowan IR (eds) Stomatal function. Stanford University Press, Stanford, California, pp 311–321

Schulze W, Stitt M, Schulze ED, Neuhaus HE, Fichtner K (1991) A quantification of the significance of assimilatory starch for growth of Arabidopsis thaliana L. Heynh. Plant Physiol 95: 890–895

Schurr U, Gollan T, Schulze ED (1992) Stomatal response to drying soil in relation to changes in the xylem sap composition of Helianthus annuus. II. Stomatal sensitivity to abscisic acid imported from the xylem sap. Plant Cell Environ 15: 551–559

Sharp RE (1990) Comparative sensitivity of root and shoot growth and physiology to low water potentials. In: Davies WJ, Jeffcoat B (eds) Importance of root-to-shoot communication in the response to environmental stress. BPGRG Monograph 21-1990. British Society for Plant Growth Regulation, Bristol, pp 29–44

Sheriff DW, Nambiar EKS, Fife DN (1986) Relationships between nutrient status, carbon assimilation and water use efficiency in Pinus radiata (D. Don) needles. Tree Physiol 2: 73–88

Sinclair TR, Horie T (1989) Leaf nitrogen, photosyntesis and crop radiation use efficiency: a review. Crop Sci 29: 90–98

Slatyer RO (1977) Altitudinal variation in the photosynthetic characteristics of snow gum, Eucalyptus pauciflora Sieb. ex Spreng. III. Temperature response of material grown in contrasting thermal environments. Aust J Plant Physiol 4: 301–312

Stitt M, Kurzel B, Heldt HW (1984) Control of photosynthetic sucrose synthesis by fructose 2,6-bisphosphate. II. Partitioning between sucrose and starch. Plant Physiol 75: 548–553

Stitt M, Gerhardt R, Wilke I, Heldt HW (1987) The contribution of fructose 2,6-bisphosphate to the regulation of sucrose synthesis during photosynthesis. Physiol Plant 69: 377–386

Tanner CB, Sinclair TR (1983) Efficient water use in crop production: research or research? In: Taylor HM, Jordan WR, Sinclair TR (eds) Limitations to efficient water use in crop production. Am Soc Agronomy Madison, Wisc (USA), pp 1–27

Tenhunen JD, Pearcy RW, Lange OL (1986) Diurnal variations in leaf conductance and gas exchange in natural environments. In: Zeiger E, Farquhar GD, Cowan IR (eds) Stomatal function. Standford University Press, Stanford, California, pp 323–351

Tenhunen JD, Catarino FM, Lange OL, Oechel WC (1987) Plant response to stress. Functional analysis in Mediterranean ecosystems. Springer, Berlin Heidelberg New York

Tomé M, Pereira JS (1991) Growth and management of eucalypt plantations in Portugal. In: Ryan PJ (ed) Productivity in perspective. Third Australian Forest Soils and Nutrition Conference. Forestry Commission of NSW, Sydney, Australia, pp 147–157

Trewavas A (1985) A pivotal role for nitrate and leaf growth in plant development. In: Baker NR, Davies WJ, Ong CK (eds) Control of leaf growth. Cambridge University Press, Cambridge, pp 77–91

Wang YP, Jarvis PG (1990) Description and analysis of an array model – MAESTRO. Agric For Meteorol 51: 257–280

Waring RH, McDonald AJS, Larsson S, Ericsson T, Wiren A, Ardwidsson E, Ericsson A, Lohamar T (1985) Differences in chemical composition of plants grown at constant relative growth rates with stable mineral nutrition. Oecologia 66: 157–160

Weis E, Berry JA (1988) Plants and high temperature stress. In: Long SP, Woodward FI (eds) Plants and temperature. Company of Biologists Ltd, Cambridge, pp 329–346

Wilson D (1975) Variation in leaf respiration in relation to growth and photosynthesis of Lolium. Ann Appl Biol 80: 323–338

Wilson D (1975) Variation in leaf respiration in relation to growth and photosynthesis of Lolium. Ann Appl Biology 80: 323–338

Wilson DR, van Bavel CHM, McCree KJ (1980) Carbon balance of water-deficient grain sorghum plants. Crop Sci 20: 153–159

Zelitch I (1982) The close relationship between net photosynthesis and crop yield. Bio-Science 32: 796–802

Part B
Responses of Photosynthesis
to Environmental Factors

9 Internal Coordination of Plant Responses to Drought and Evaporational Demand

R. Lösch and E.-D. Schulze

9.1 Introduction

Higher plants balance between desiccation by transpiration during periods of CO_2 uptake through open stomata and starvation during periods of stomatal closure when preventing excessive water loss. Since fluxes of water vapor and CO_2 are linked by using the same pathway through the stomatal aperture, the question of how plants regulate their water status is also of significance to the ecophysiology of photosynthesis, the main theme of this Volume.

Major advances in the understanding of stomatal regulation have been made in the last decade. Initially, emphasis was given to the investigation of the environmental response of gas exchange (Schulze and Hall 1982); however, it has become apparent meanwhile (Schulze 1986; Zhang and Davies 1990) that internal regulations are just as important for understanding plant responses to drought in the soil and to evaporative demand in the atmosphere. This internal regulation includes hormonal root-shoot signals, as well as structural components in the leaf and in the stem.

In the following, we will briefly review stomatal responses to environmental stress, and we will then more explicitly explore physiological and structural interactions between root and shoot as part of an internal co-ordination of regulating plant water relations.

9.2 Environmental and Plant-Internal Influences on Transpiration

Plants are the interface in the steep gradient of matter and energy exchange within the soil-plant-atmosphere system. At lower levels of structural organization, poikilohydric plants depend passively upon their environment. In contrast, higher plants are generally decoupled from the physically driven fluxes of matter and energy so that permanent gradients of chemical substances and energy are established. By active metabolic processes, selectivity in ion uptake can be achieved, resulting in quite different ion concentrations in the soil and in plant compartments. By a highly water-impermeable cutinized epidermis the homoiohydric plants keep their water relations at more or less stable levels even in dry surroundings. By simultaneous root water uptake and controlled stomatal transpiration they keep a persistent but regulated water flow through the plant, driven by the water potential gradient between soil and atmosphere. The evaporation of this water in the leaves permits a partitioning of the energy absorbed by shoots and leaves from the incoming radiation into sensible and latent heat. As a consequence, shoot temperatures are at least partly decoupled from air temperatures and remain in a range which is advantageous for overall metabolism and water status (Lange 1959; Gates 1976). Ion distribution within the plant and energy consumption by evaporation are linked intimately to the water transport through the plant within the soil-plant-atmosphere pathway. Apparently, a control of this transpiration stream through the plant is more efficient at the interface where water molecules escape from the leaf, i.e., at the stomata, than a regulation of root water uptake (Meinzer and Grantz 1990). Therefore, a complicated system of stomatal responses to the environmental parameters has developed during the evolution of cormophytic plants which allows them to adapt their transpiration to the balance of supply and demand (Schulze and Hall 1982) and to grow in different environments, from humid to extremely arid sites.

Transpirational water diffusion through the stomatal pores is driven by the leaf-to-air vapor pressure difference and restricted by the variable diffusion resistance of the stomatal pores. The cuticle acts as a parallel resistor of high, constant resistance. The boundary layer resistance is connected in series with both the cuticular and the stomatal resistance. It is dependent on plant and leaf size, but not on stomatal responses to the environment. A variable stomatal resistance is mediated by the guard cell responses to factors such as visible and thermal radiation, ambient humidity, CO_2 content of the air, and sometimes various pollutants. Their effects have been studied in great detail (Hall et al. 1976; Schulze and Hall 1982), and successful attempts have been made to mathematically model stomatal conductances and transpiration as function of these external factors (e.g., Lösch and Tenhunen 1981; Küppers and Schulze 1985; Lösch et al. 1992).

Bulk plant water potential has been taken most often in such studies as a term integrating the effects of internal plant factors, particularly water shortage. Several findings, however, indicate that this parameter is a rather incidental one, not linked directly with the stomatal responses to drought. For example, use of this parameter alone often does not yield satisfactory results in simulation approaches (Rosa et al. 1991; Lösch et al. 1992). Long-term and short-term water stress effects were distinguished (Schulze and Küppers 1979), and with time it became evident that stomatal closure upon water shortage is a completely metabolically governed process. Therefore, the causal interpretation of drought effects upon stomatal apertures shifted from the assumption of simple hydraulic effects (Stålfelt 1956) to the detection of a sophisticated interaction between altered plant functions by water stress and hormonal signals, particularly abscisic acid (ABA). Under water stress, ABA is produced or liberated from storage sites in leaves, roots, or even the guard cells themselves, and transported to the site of its action, the guard cell plasmalemma. There it changes the membrane permeability by depolarization of the membrane potential, and enhances thereby the efflux of potassium ions (summaries of recent findings about the mechanisms: Lösch 1989, 1993b). Altered guard cell solute contents finally induce changed pore apertures as a result of altered turgor balances between the guard cells and the surrounding epidermis.

During the decade 1970–1980, wide knowledge was accumulated about the effects of plant environmental factors on stomatal behavior, particularly about the effects of air humidity (Lange et al. 1971; Hall et al. 1976; Lösch and Tenhunen 1981). During the 1980s, changes occurred in the appreciation of internal plant effects on stomatal regulation. Furthermore, the focus was extended to the coordination of morphological structure and functional processes within different plant parts. In this way, an integrated view of whole-plant responses to the environment becomes apparent which may allow an extension from whole-plant individuals to stands of vegetation.

9.3 Root-Leaf Signals Under Moisture Shortage Contribute to Drought Avoidance Responses of Leaves

Contradictory findings concerning a threshold value for stomatal closure under the influence of decreasing water potential (Hsiao 1973; Biscoe et al. 1976; Davenport et al. 1977) and hysteresis effects of stomatal behavior after relief of water stress ("after effects"; e.g., Fischer et al. 1970) provided the first indications that leaf conductance does not depend simply on immediate turgor hydraulics and leaf water potentials. Schulze and Küppers (1979) found decreasing conductances of *Corylus avellana* leaves under controlled conditions when plant water potentials were progressively lowered during a period of several days. In contrast, reduced water potentials during the course of a day did not negatively affect leaf conductances if leaf turgor values approached zero. Field measurements of barley leaf conductances during diurnal courses gave high values even at actual water potentials of less than $-2\,MPa$ and turgor values near zero (Lösch et al. 1992). Leaf conductances were reduced only during a prolonged drought when water deficits of the sandy soil in the barley field increased corresponding to a flag leaf predawn water potential of approximately $-1.6\,MPa$. Decreased maximal conductances with lower predawn water potentials at progression of seasonal drought, and continuously high conductances irrespective of low noon water potentials at sufficient soil water supply in plants from winter or summer rain climates (Lösch et al. 1982; Ullmann 1985) may similarly indicate a stomatal response to long-term but not to short-term decreases of plant water potential. Generally, long-term water stress is caused by decreases of soil water potential, while short-term changes in plants occur mainly on a daily basis in leaves.

Bates and Hall (1981), studying *Vigna unguiculata*, found that leaf conductance declined although total leaf water potential remained constant over a wide range of soil water deficits. Turner et al. (1985) and Gollan et al. (1985) extended these findings with experiments using *Helianthus annuus* and *Nerium oleander*. They found that leaf conductances and photosynthesis decreased when about two-thirds of the soil extractable water had been utilized. Küppers et al. (1988) found *Vigna* leaf water contents nearly unchanged when the soil extractable water content decreased from 60 to 0%. Maximal leaf conductance and assimilation values decreased dramatically once the amount of soil extractable water was reduced to 20%. Berard and Thurtell (1991) reported also reduced photosynthetic rates of maize as a consequence of decreased leaf conductances when soil moisture was low but leaf water potentials high in a humid atmospheric environment.

Gollan et al. (1986) experimentally decoupled leaf and soil water potentials. They compensated for decreased leaf water potential by applying pneumatic pressure within a modified pressure chamber and kept wheat and sunflower leaves turgid while their rooting substrate dried. Leaf conductances decreased under such treatments, when soil water content was reduced to 30% of field capacity. This suggests that information about root water relations is mediated to leaves without change in leaf water status.

Gowing et al. (1990) discuss three possibilities for signals from roots to shoots. (1) A negative message coming from turgid roots that promotes growth and stomatal opening. (2) A positive root signal produced from roots under water shortage which increases in intensity with increasing drought conditions in the root zone. (3) An altered transport rate for substances moving in the transpiration stream from roots to the shoot if the former are affected by dry soil conditions. Gowing et al. (1990) analyzed these possibilities using vegetatively propagated apple plants which were cultivated with a split root system. One half of the root system was droughted while the overall plant water potential remained unchanged because of water supply from the other half of the root system. Nevertheless, leaf growth was inhibited when one half of the root system was dried. This growth inhibition was alleviated after the dry roots were excised. This suggests that there is a positive inhibitor, produced by drying roots and transported to the shoot.

The experiment by Gowing et al. (1990) repeats an earlier split-root study using maize (Blackman and Davies 1985). Plants with half their root systems subjected to drought displayed the same water, solute, and turgor potentials, but significantly lower leaf conductances, than the completely watered controls. Also the bulk leaf ABA contents of treated and control plants did not differ. The stomatal closure of the insufficiently watered plants became reversed in discs which were removed from the leaves and floated on kinetin or zeatin solutions. It was concluded, therefore, that a continuous supply of cytokinins from the roots counteracted any stomatal closing tendencies. This is the case of a negative message from the roots in order to sustain maximal stomatal opening in well-watered plants. But Davies et al. (1986) emphasized how little is known, even currently, about the effects of phytohormones other than ABA on plant water relations. Only few data exist about the involvement of kinetin in plant water stress responses (e.g., Itai and Vaadia 1965, 1971; Bradford and Hsiao 1982), and also the findings of Blackman and Davies (1984, 1985) on the effects of root-produced cytokinins on stomatal opening have not been repeated in subsequent years. Fußeder et al. (1992) found no clear evidence of antagonistic effects of cytokinins and ABA in field-grown almond trees. Only when ABA was not limiting stomatal aperture, may cytokinins have affected stomatal opening in the early morning. Munns and King (1988) postulated that still other messenger substances, apart from ABA, are transported from roots under soil water shortage to the leaves and cause stomatal closure. They collected xylem sap from detopped pressurized wheat plants growing in dry soil and fed this xylem sap into detached wheat leaves through their cut edges. Stomatal closure occurred in these leaves even if ABA was removed from the xylem sap prior to application. However, it was shown later that this may have been due to specific experimental conditions, and that ABA fed into the xylem sap caused a dose response on stomatal aperture (Heckenberger 1993).

It has been further suggested (Gollan et al. 1992) that in addition to the amount of phytohormones, the concentrations of mineral ions such as Ca^{2+} and nitrate will be affected by a reduced water uptake and transport through the plant. Ions like Ca^{2+} are known to be involved in the regulation of stomatal apertures (e.g., De Silva et al. 1985; Atkinson et al. 1989). Such an effect would belong to the third category suggested by Gowing et al. (1990), the modification by drought of the rate of substances flowing in the transpiration stream.

From several investigations it has become clear meanwhile that higher amounts of ABA are produced in roots experiencing water stress (e.g., Rivier et al. 1983; Robertson et al. 1985; Lachno and Baker 1986; Wartinger et al. 1990). A long-distance transport of ABA from roots to shoots also has been documented in several studies (e.g., Hartung 1977; Prochazka 1982), and concentrations of ABA in the xylem sap have been measured (Munns and King 1988; Zhang and Davies 1989a; Schurr et al. 1992). However, attempts to correlate bulk leaf ABA levels with stomatal closure failed when the roots were exposed to drought (Beardsell and Cohen 1975; Blackman and Davies 1985; Zhang et al. 1987). This might be due to the fact that bulk leaf ABA contents were determined rather than epidermal contents. Large amounts of physiologically inactive ABA sequestered in the mesophyll (Heilmann et al. 1980) will mask the detection of local epidermal ABA changes which affect the guard cells.

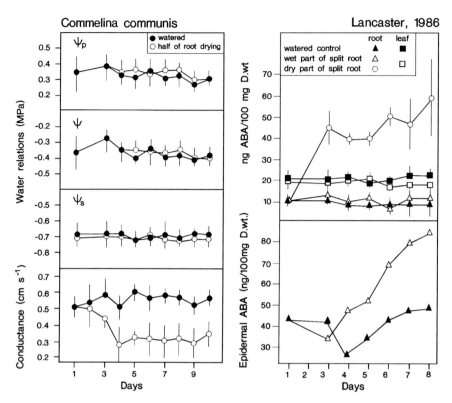

Fig. 9.1. Response of *Commelina* leaves from plants with split roots between two pots. Water was applied to both halves of the root system (*closed symbols*) or water was withheld from one half of the root system after day 1 (*open symbols*). *Left panel* Water relations and conductance; *right panel* ABA contents of leaves and shoots (*top*) and of the lower epidermis (*bottom*). (After Zhang et al. 1987).

Zhang et al. (1987) determined epidermal ABA contents of *Commelina* leaves which had high bulk turgor due to sufficient water supply from one half of the root system while the other half experienced drying soil (Fig. 9.1). As in other split-root experiments, leaf conductances decreased and showed only half the conductance of the controls following the fourth day of the drying cycle. Bulk water relations did not change nor did bulk ABA contents which remained at about 0.2 ng ABA mg^{-1} dw in leaves and 0.1 ng ABA mg^{-1} dw in roots. ABA contents of the lower epidermis, however, increased steadily with progression of soil drying from about 0.4 ng ABA mg^{-1} epidermal dry weight during the first four days of the investigation to nearly 0.9 ng ABA mg^{-1} dw by day 8 of the experiment. In plants with roots kept well watered, epidermal ABA contents remained in the range of 0.4–0.5 ng ABA mg^{-1} dw.

The increase in epidermal ABA appears to result from a greater ABA synthesis in the roots, and there is indirect evidence that this ABA could be transported to the leaves (Zhang and Davies, 1987). In isolated root tips of *Pisum* and *Commelina*, root tip water contents were reduced from 100 to 60% and the corresponding solute potential fell from -0.8 to -1.4 MPa. Turgor was reduced from 0.3 to 0.1 MPa. The root tip ABA content rose from 0.05 to more than 0.5 ng ABA mg^{-1} dw. ABA increase with water stress was even more pronounced in *Commelina* (from 0.2 to 1.2 ng ABA mg^{-1} dw). In both cases there was a threshold (0.2 MPa for pea and 0.3 MPa for *Commelina* roots) when the ABA content increased drastically with further tissue drought. In another study using maize as the experimental plant, Zhang and Davies (1989b) demonstrated an increase of root ABA contents progressing deeper into the soil as the profile dried down from the surface. Gollan et al. (1989) showed that with a drought-induced increase of root ABA content, a corresponding increase in xylem sap ABA concentration occurred, and Zhang and Davies (1989a) showed a log-linear relationship between the reduction of sunflower leaf conductance and the increase of ABA concentration in the xylem sap. Despite all this evidence of ABA effects on stomata, it must be made clear that the underlying mechanism is still more complicated. Wartinger et al. (1990) found no relation between ABA transport in the xylem and stomatal conductance during the course of a day, but a threshold response of maximum conductance (Fig. 9.2).

There is sufficient evidence to show that with soil drought a greater amount of ABA is produced by the roots and becomes transported in the transpiration stream to the leaf epidermis. There it may cause some degree of stomatal closure without necessarily involving the bulk leaf water potential. However, several factors may interact with a direct and instantaneous response. ABA may be (1) sequestered into the mesophyll, (2) metabolized in the mesophyll and in the epidermis (Heckenberger 1993), and (3) loaded into the phloem and thus transported back to the root (Schurr 1992a). (4) In addition, other substances being produced or taken up by the roots under drought might be transported by the transpiration stream (e.g., calcium ions, other phytohormones like kinetins) and act synergistically with ABA in regulating plant water loss in a coordinated fashion. Gollan et al. (1992) demonstrated that, in addition to Ca^{2+} and PO_4^{3-}, NO_3^- transport rate is reduced with soil water shortage (Fig. 9.3). This might increase the acropetal transport of ABA by changing the xylem pH (Schurr 1992a,b).

In addition, one cannot exclude the fact that turgor reduction of the leaves will affect stomatal apertures hydraulically and that ABA synthesis or release from storage compartments of the aerial parts of the plant occurs. Zhang and Davies (1989a) observed that old leaves lost turgor first in a drying cycle and that upon their wilting, xylem sap ABA concentrations rose sharply. Thus, ABA coming from both roots and older leaves could intensify stomatal closure under drought. Finally, even a local feedback accumulation of active ABA must be considered, since the guard cell acidification by external ABA acting upon the plasmalemma could effect a

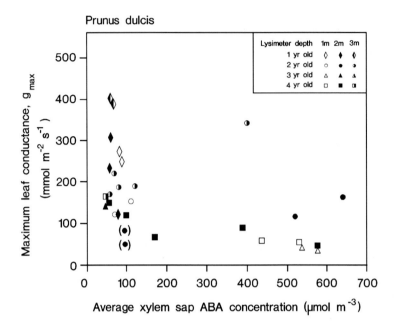

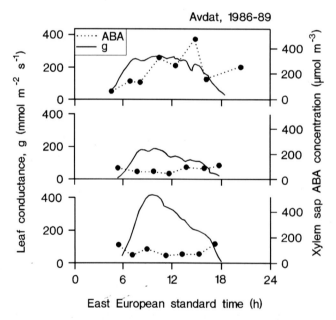

Fig. 9.2. Stomatal response of *Prunus dulcis* to ABA. *Lower panel* Different daily courses of stomatal conductance and ABA. *Top* 2-year-old tree in a 3-m-deep lysimeter, June 16; *middle* 2-year-old tree in a 3-m-deep lysimeter, September 5; *bottom* 1-year-old tree in a 3-m-deep lysimeter, September 15. *Upper panel* Maximum leaf conductance as related to average xylem sap ABA concentration. (After Wartinger et al. 1990)

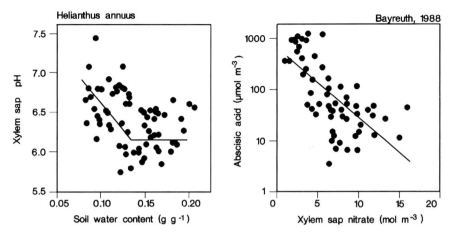

Fig. 9.3. Change of xylem sap pH with soil water content (*left*) and the relation between ABA and nitrate concentration in xylem sap during soil drying (*right*). Highest ABA values occur in dry soil. (After Gollan et al. 1992; Schurr et al. 1992)

release of sequestered ABA from guard cells themselves (Behl and Hartung 1986).

A quantitative determination of the effects of ABA coming from different sources in the plant is rather difficult. Wolf et al. (1990) attempted compartmental ABA determinations and flux calculations to quantify rates of long-distance transport of ABA, its biosynthesis and degradation in salt-stressed lupin plants. They emphasized that ABA is produced in leaves and roots. It is distributed basipetally by the phloem stream and acropetally by the transpiration stream within the xylem. Schurr (1992b) in fact measured a considerable recirculation of ABA in *Rhizinus communis*. In well-watered plants almost 100% of the ABA transported by the xylem water flow from roots to the shoot were recirculated from the aerial parts of the plant back to the root. In the study of Wolf et al. (1990), ABA transport increased considerably in both phloem and xylem under NaCl-induced water stress; but the amounts transported acropetally in the xylem became much greater, resulting from an increase of ABA synthesis in the roots. These calculations of ABA fluxes confirm the important role of root ABA production under stress conditions. At the same time, they reconcile findings about ABA production primarily in the shoots (Jackson et al. 1988), which occurs particularly under stress-free conditions. Wolf et al. (1990) did not differentiate between a compartmentation of the ABA fluxes to the leaf mesophyll and the epidermis. Other studies (e.g., Wright and Hiron 1969; Zhang and Davies 1989b) showed that not only epidermal but also bulk leaf ABA content will increase under prolonged drought.

The ABA increase in leaves is not proportional to the xylem input because ABA is not only recirculated via the phloem but is also rapidly

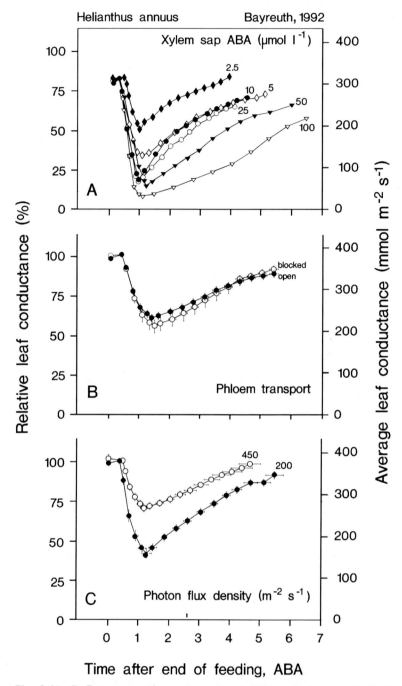

Fig. 9.4A–C. Response and recovery of stomatal conductance of *Helianthus annuus* following feeding of ABA into the xylem. The figures show the response of stomata following a 30 minute ABA addition. **A** Feeding of different ABA concentrations (μM: as indicated by the numbers at the response curves) terminated at time 0. An initial closing response and a slower recovery were related to the ABA concentration which was applied. **B** Feeding of 10 μM ABA into leaves in which the phloem was functional or blocked by cooling of the petiole to 0 °C. **C** Feeding of 5 μM ABA at different light intensities shows that the initial response and the recovery is light-dependent. (After Heckenberger 1993)

metabolized in the leaf. Feeding ABA into the xylem of sunflower leaves at a known dose caused a transient depression of conductance (Fig. 9.4) and a rapid recovery which was not caused by ABA export (as shown by phloem blockage). Recovery was related to the light-dependent metabolization of ABA in the leaf (Heckenberger 1993). The light-dependent metabolization of ABA may also explain the significant relation between maximum conductance in the morning and xylem ABA and the contrasting lack of such correlation during the course of a day (see Fig. 9.2; Wartinger et al. 1990).

9.4   Leaf Anatomy, Canopy Structure, and Stomatal Function

Water stress reduces cell and tissue growth by immediate hydraulic effects as well as by phytohormones such as ABA (Bradford and Hsiao 1982). It has been speculated that such nonhydraulic effects of soil drying reduce leaf expansion. An obvious reduction in leaf growth of cereal species without apparent change of leaf water potential was observed when roots were exposed to soil compaction (Masle and Passioura 1987) or soil drying (Passioura 1988; Saab and Sharp 1989) due to possible nonhydraulic root signals. Possible differences in water potential between the mature leaf tissue and the basal meristematic region of the gramineoid leaves of these species were not considered. However, Michelena and Boyer (1982) already showed that osmotic adaptation occurs preferentially in the elongating region of maize leaves, so that turgor is maintained there even at low water potentials. However, also in these cases a decrease of leaf elongation during a drying cycle was observed which apparently was due to a nonhydraulic effect. Zhang and Davies (1990) found a negative log-linear relationship between the rate of leaf growth and the concentration of ABA in the xylem of maize and sunflower. They emphasized that the higher ABA concentrations under drought treatments did not result simply from a reduced amount of water as solvent for the acropetally transported ABA. Instead, extra ABA resulting from soil drying is present in the xylem sap and is probably responsible for the reduced leaf elongation.

Additional experiments are needed with dicots in which the leaf development differs from that in gramineoid leaves. However, sufficient evidence emerges of morphogenetic effects of root messengers produced during soil drought stress and possible influences of ABA on cell wall elasticity and membrane conductivity (Glinka and Reinhold 1971; Kutschera and Schopfer 1986a,b). It is suggested that cell growth will be modified by phytohormonal effects without changes in turgor resulting in greater or lesser rates of cell and tissue enlargement. The mechanisms of these processes were reviewed, e.g., by Aspinall (1986) and Barlow (1986).

Apparently, these growth processes result in plant foliage with a surface area and a structure that correspond with the water availability and supply capacity of the plant's root and shoot system. An example for such a morphogenetic, structural coordination may be the case of hazel (*Corylus avellana*) shrubs growing in a hedgerow in North Germany: from porometer measurements of diurnal courses throughout the season by Linnenbrink (unpublished) south-facing leaves had 1.4 ($\pm$0.18 SE) greater transpirational water loss per unit leaf area than north-facing ones. Transpiration of leaves in the center of the shrub was still less. But the greater water loss of south-facing leaves under more severe evaporation conditions is nearly equal to the sum of transpirational water loss of all the leaves calculated for each canopy compartment. Based on differences in the specific leaf area, an outer or more south-facing, an inner or more north-facing crown, and an inner, shaded portion of the canopy can be distinguished. Since only one-fifth of the foliage belongs to the sun-exposed part of the canopy and more than half of the leaf area is in the shadow of the shrub interior, on average the sun foliage has only a 12% higher gross transpiration than the latter (Lilienfein et al. 1991). Similarly, differences exist between transpiration rates of leaves in different canopy positions and canopy-related water loss from the periphery and the interior of co-occurring *Alnus* and *Sambucus* shrubs (Weisheit, Linnenbrink, unpubl. data). Xylem sap flow rates differ considerably between north- and south-facing branches of *Carya illinoensis*, resulting in a 40% smaller amount of water transported to the more shaded parts of the crown (Steinberg et al. 1990). Normalized in relation to the supplied partial leaf area, however, flow rates in all parts of the tree are in the same order of magnitude. This points to a versatile allocation of trunk water supply to the different parts of a canopy, according to the respective actual demands (Hinckley and Ritchie 1970). Similarly, Cregg et al. (1990) determined biomass development, leaf conductance, and predawn xylem potentials of a thinned *Pinus taeda* stand. Also in this case, leaf area, transpiration, and water potential interact in such a way that homeostasis of water relations results.

A relationship exists between leaf area and maximal stomatal conductance of *Saccharum* so that transpiration from a plant does not exceed a certain maximum value even if total leaf area increases (Meinzer and Grantz 1990). This results from a decrease in area-related maximal conductance once leaf area has exceeded a value of $0.2\,m^2$ plant^{-1}. Upon partial removal of the transpiring leaves, a rapid stomatal adjustment occurred, reestablishing the relationships between maximum conductance and remaining leaf area to the previous level. Leaf water balance at maximal transpiration was independent of leaf area, e.g., leaf water potentials remained constant while transpiration per plant increased. The ratio of transpiration to the soil-xylem water potential difference, taken as the hydraulic conductivity, decreased with increasing total leaf area if unit-area related, and came to a constant maximal value at leaf areas greater than $0.4\,m^2$ plant^{-1} if related to the total leaf area of a plant. This is the same pattern as that exhibited by stomatal conductance. Extending the earlier "pipe" model of Waring et al. (1982), which proposed a relation between leaf area index and xylem sapwood area, Meinzer and Grantz (1990) concluded that a coordination exists between the vapor phase conductance and the liquid flow conductivity of the sugarcane plant. They speculate about root metabolites which adjust vapor exchange rates to the water supply capacity of the plant. This capacity is determined not only by the ability for water uptake of the roots but also by the transport efficiency of the xylem pathway through the shoots. The coordination between root water gain, transpiring leaf area, stomatal control of transpiration, and an appropriate structure of the vessel system will be important for plant water balance.

9.5 Xylem Conductivity and Leaf Conductance

In the sugarcane plants studied by Meinzer et al. (1992), shoot hydraulic conductivity per leaf area ("specific conductivity") decreases in a similar manner as unit-area-related leaf conductance and transpiration once a critical leaf area of $0.2\,m^2$ plant^{-1} is exceeded. The parallel change of the three parameters prevents steep water potential gradients along the stalks, which could cause cavitation of vessel elements. The possibility of vessel embolism can result from the strong tension under which water exists in the xylem ducts according to the well-established cohesion theory of sap ascent in plants (Dixon 1914). Cavitation occurs when one of the continuous strands of water through the conducting elements becomes interrupted by the sudden expansion of a submicroscopic vapor bubble produced by nucleation at hydrophobic cracks or by meniscal failure at a pore in the cell wall. Such an event is accompanied by the acoustic emission of energy which results from the abrupt enlargement of the gas volume within the conducting element. It can be monitored by microphones and a suitable signal amplification, both in the low-frequency, audible range and also at ultrasonic frequencies (Milburn and Johnson 1966; Tyree and Dixon 1983). Tyree and Sperry (1989) emphasized that cavitation is a common phenomenon in nature, and occurs particularly often with winter freezing and under water stress. Since each cavitation of a tracheid or a vessel will block the water transport through this particular strand, it will additionally aggravate the risk of drought damage. Following cavitation, fewer functional vessels have to sustain the transpiration stream, which comes under even stronger tension. A repair of embolism is possible only at very high water potentials after rain or by root pressure during overnight replenishment of the plant water reserves. Tyree and Sperry (1989) and Field and Holbrook (1989) proposed the hypothesis that plants very often operate on the verge of a catastrophic breakdown of the water transport system caused by a runaway embolism. Stomatal regulation may function just to keep transpirational flow at a rate which allows sufficient carbon gain (Cowan 1982) but prevents breakdown of the plant's hydraulic system. A feedback regulation responding to bulk water shortage in the aerial parts of a plant might not be sensitive enough to fulfill such a requirement. Again, direct signals from the water uptake interface, the roots, to the water loss interface, the stomata, could be a suitable solution for such control.

In order to be efficient, such a regulation needs information about the water transport capacity of the conducting pathway. Knowledge is scarce about the means of such a communication. From parallel increases of water uptake and transpiration not accompanied by greater water potential gradients, a variable hydraulic conductivity has been inferred in several studies (e.g., Aston and Lawlor 1979; Meinzer et al. 1988). Such variable whole-plant conductivities are usually attributed to altered root resistances for water uptake (Steudle 1993). They are short-term effects, and the calculated changes of conductivity mostly result in an improvement of the acropetal water transport capacity. In the case of xylem embolism, this will deteriorate, and stomata have to respond by closure. Sperry (1986) has demonstrated such a sequence of events as occurring in the palm, *Rhapis excelsa*. Jones and Sutherland (1991) modeled the stomatal responses to xylem embolism and emphasized that optimal stomatal control of plant water relations required information about leaf and soil water potentials.

The species-specific balance between average leaf conductances and shoot conductivities will be developed during ontogeny. A positive correlation between conductance and conductivity has been shown to exist in several species. For example, Küppers (1984) reported that Central European hedgerow shrubs differed in the efficiency of their conducting sapwood area. He suggested that a relationship exists between the average hydraulic conductivity calculated as the seasonal mean of transpiration change per change of leaf water potential and the ratio of sapwood area/supplied leaf area. A low water supply capacity due to a peculiar xylem anatomy might also influence the occurrence of a species in the gradient between humid and arid habitats. This can be deduced from comparative studies on the water relations of Canarian laurel forest trees (Lösch 1993a). The species which compose this vegetation differ among each other with respect to plasmatic drought tolerance, stem hydraulic conductivity, and water loss avoidance. The latter is quantified as percentage water loss per time from desiccating saturated leaves under controlled conditions. All laurel forest species turned out to be much less efficient with respect to all these three parameters when compared with Mediterranean sclerophyllous species. It can be calculated from these data and the maximum leaf conductances of the laurophyllous species (unpubl.) that in nearly half of the species leaf water contents will fall below critical values if the saturation deficit of the ambient air decreases below $10 \, \mathrm{g \, m^{-3}}$ for prolonged periods.

Coordinated growth of trees was demonstrated for *Nothofagus*, in which the maximum transpiration rate per tree was linearly related to the sap-wood area and the circumference of the tree (Fig. 9.5). The slope of this relation may be taken as an indication for the requirement for stem growth

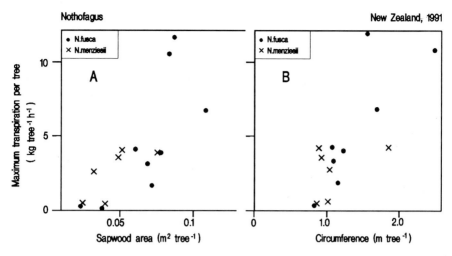

Fig. 9.5A,B. Maximum transpiration of *Nothofagus* trees as related to sapwood area and tree circumference. (Köstner et al. 1992)

for water transport into an increasing size crown. Sapwood increased by $0.01 \, m^2$ or by $0.06 \, m$ of circumference when maximum transpiration per tree increased by $1 \, l \, h^{-1}$. This may set the requirement for stem growth during expansion of the tree crown (Köstner et al. 1992).

9.6 Conclusions

Water flow through the plant compartment of the soil-plant-atmosphere continuum is a physiologically controlled process.

– The regulation occurs at the plant-air interface, where the transition from liquid water to vapor occurs and the steepest gradient of water potential exists. Stomata respond immediately to changes of this gradient in a complicated cellular feedback regulation which acts as if it were a feed-forward response. This was not discussed in the present chapter in detail (see review by Schulze 1993).
– Root-epidermis signals about water shortage at the soil/plant interface, bypassing the bulk shoot and leaf water relations, are another feedforward effect which contributes to balanced water relations within the plant compartment of the soil-plant-atmosphere continuum.
– Water transport capacity of the shoots insures that range of this water flow which is coupled with energy and substance exchanges between the plant and its environment. Feedback responses may not be sufficient to keep homeostasis in the homeohydric plant which is interconnected

in the soil-plant-atmosphere continuum with two compartments of highly variable water relations.

- One can speculate whether it is a basic attribute of living organisms that they attempt optimally not only to react by feedback regulation to changes in their environment but also by adapting themselves, by means of feedforward mechanisms, to a variety of circumstances. This will allow them to cope much more effectively with forthcoming or locally heterogeneous situations.
- The combination of feedforward and feedback mechanisms acting at the same regulatory point may give the necessary flexibility to respond in a variable environment.
- The coordination at the stomatal level is not sufficient to stabilize plant water status. The same signals appear to affect growth also. Thus, coordination in plant transpiration and structure is achieved. Detailed information is still lacking, however, as to how this goal is achieved functionally.

References

Aspinall D (1986) Metabolic effects of water and salinity stress in relation to expansion of the leaf surface. Aust J Plant Physiol 13: 59–73

Aston MJ, Lawlor DW (1979) The relationship between transpiration, root water uptake, and leaf water potential. J Exp Bot 30: 169–181

Atkinson CJ, Davies WJ, Mansfield TA (1989) Changes in stomatal conductance in intact ageing wheat leaves in response to abscisic acid. J Exp Bot 40: 1021–1028

Barlow EWR (1986) The water relations of expanding leaves. Aust J Plant Physiol 13: 45–58

Bates LM, Hall AE (1981) Stomatal closure with soil water depletion not associated with changes in bulk leaf water status. Oecologia 50: 62–65

Beardsell MF, Cohen D (1975) Relationships between leaf water status, abscisic acid levels and stomatal resistance in maize and sorghum. Plant Physiol 56: 207–212

Behl R, Hartung W (1986) Movement and compartmentation of abscisic acid in guard cells of *Valerianella locusta*: effects of osmotic stress, external H^+-concentration and fusicoccin. Planta 168: 360–368

Berard RG, Thurtell GW (1991) The interactive effects of increased evaporative demand and soil water on photosynthesis in maize. Can J Plant Sci 71: 31–39

Biscoe PV, Cohen Y, Wallace JS (1976) Daily and seasonal changes of water potential in cereals. Philos Trans R Soc Lond B 273: 565–580

Blackman PG, Davies WJ (1984) Modification of the CO_2 responses of maize stomata by abscisic acid and by naturally occurring and synthetic cytokinins. J Exp Bot 35: 174–179

Blackman PG, Davies WJ (1985) Root-to-shoot communication in maize plants of the effects of soil drying. J Exp Bot 36: 39–48

Bradford KJ, Hsiao TC (1982) Physiological responses to moderate water stress. In: Lange OL, Nobel PS, Osmond CB, Ziegler H (eds) Physiological plant ecology II, Encycl Plant Phys, NS 12B. Springer, Berlin Heidelberg New York, pp 263–324

Cowan IR (1977) Stomatal behavior and environment. Adv Bot Res 4: 117–288

Cowan IR (1982) Regulation of water use in relation to carbon gain in higher plants. In: Lange OL, Nobel PS, Osmond CB, Ziegler H (eds) Physiological plant ecology II, Encycl. Plant Phys NS 12B. Springer, Berlin Heidelberg New York, pp 589–614

Cowan IR, Farquhar GD (1977) Stomatal function in relation to leaf metabolism and environment. Symp Soc Exp Biol 31: 471–505

Cregg BM, Hennessey TC, Dougherty PM (1990) Water relations of loblolly pine trees in southeastern Oklahoma (USA) following precommercial thinning. Can J For Res 20: 1508–1513

Davenport TL, Jordan WR, Morgan PW (1977) Movement and endogenous levels of abscisic acid during water-stress-induced abscission in cotton seedlings. Plant Physiol 59: 1165–1168

Davies WJ, Metcalfe J, Lodge TA, da Costa AR (1986) Plant growth substances and the regulation of growth under drought. Aust J Plant Physiol 13: 105–125

De Silva DLR, Hetherington AM, Mansfield TA (1985) Synergism between calcium ions and abscisic acid in preventing stomatal opening. New Phytol 100: 473–482

Dixon HH (1914) Transpiration and the ascent of sap in plants. Macmillan, London

Field CB, Holbrook NM (1989) Catastrophic xylem failure: tree life at the brink. TREE 4: 124–125

Fischer RA, Hsiao TC, Hagan RM (1970) After-effect of water stress on stomatal opening potential. J Exp Bot 21: 371–385

Fußeder A, Wartinger A, Hartung W, Schulze E-D, Heilmeier H (1992) Cytokinin in the xylem sap of desert grown almond (*Prunus dulcis*) trees: daily courses and their possible interactions with abscisic acid and leaf conductance. New Phytol 122: 45–52

Gates DM (1976) Energy exchange and transpiration. In: Lange OL, Kappen L, Schulze E-D (eds) Water and plant life. Ecological Studies, vol 19. Springer, Berlin Heidelberg New York pp 137–147

Glinka Z, Reinhold L (1971) Abscisic acid raises the permeability of plant cells to water. Plant Physiol 48: 103–105

Gollan T, Turner NC, Schulze E-D (1985) The responses of stomata and leaf gas exchange to vapour pressure deficits and soil water content. Oecologia 65: 356–362

Gollan T, Passioura JB, Munns R (1986) Soil water status affects the stomatal conductance of fully turgid wheat and sunflower leaves. Aust J Plant Physiol 13: 459–464

Gollan T, Davies WJ, Schurr U, Zhang J (1989) Control of gas exchange: evidence for root-shoot communication on drying soil. Ann Sci For 46: 393s–400s

Gollan T, Schurr U, Schulze E-D (1992) Stomatal response to drying soil in relation to changes in the xylem sap composition of *Helianthus annuus*. I. The concentration of cations, anions, amino acids in, and pH of, the xylem sap. Plant Cell Environ 15: 551–559

Gowing DJG, Davies WJ, Jones HG (1990) A positive root-sourced signal as an indicator of soil drying in apple, *Malus x domestica* Borkh. J Exp Bot 41: 1535–1540

Hall AE, Schulze E-D, Lange OL (1976) Current perspectives of steady-state stomatal responses to environment. In: Lange OL, Kappen L, Schulze E-D (eds) Water and plant life. Ecological Studies, vol 19. Springer, Berlin Heidelberg New York, pp 169–188

Hartung W (1977) Der Transport von [2-^{14}C] Abscisinsäure aus dem Wurzelsystem intakter Bohnenkeimlinge in die oberirdischen Teile der Pflanze. Z Pflanzenphysiol 83: 81–84

Heckenberger U (1993) Stomatareaktion auf kontrollierte Veränderungen des Xylemsaftes mit Abscisinsäure. Diplom Thesis, Bayreuth

Heilmann B, Hartung W, Gimmler H (1980) The distribution of abscisic acid between chloroplasts and cytoplasm of leaf cells and the permeability of the chloroplast envelope for abscisic acid. Z Pflanzenphysiol 97: 67–78

Hinckley TM, Ritchie GA (1970) Within-crown patterns of transpiration, water stress, and stomatal activity in *Abies amabilis*. For Sci 16: 490–492

Hsiao TC (1973) Plant responses to water stress. Annu Rev Plant Physiol 24: 519–570

Itai C, Vaadia Y (1965) Kinetin-like activity in root exudate of water-stressed sunflower plants. Physiol Plant 18: 941–944

Itai C, Vaadia Y (1971) Cytokinin activity in water-stressed shoots. Plant Physiol 47: 87–90

Jackson MB, Young SF, Hall KC (1988) Are roots a source of abscisic acid for the shoots of flooded pea plants? J Exp Bot 39: 1631–1637

Jones HG, Sutherland RA (1991) Stomatal control of xylem embolism. Plant Cell Environ 14: 607–612

Köstner BMM, Schulze E-D, Helliher FM, Hollinger DY, Byers JN, Hunt JE, McSeveny TM, Meserth R, Weir PL (1992) Transpiration and canopy conductance in a pristine broad-leaved forest of Nothofagus: an analysis of xylem sap flow and eddy correlation measurements. Oecologia 91:350–359

Küppers BIL, Küppers M, Schulze, E-D (1988) Soil drying and its effect on leaf conductance and CO_2 assimilation of Vigna unguiculata (L.) Walp. Oecologia 75: 99–104

Küppers M (1984) Carbon relations and competition between woody species in a Central Europaean hedgerow. II. Stomatal responses, water use, and hydraulic conductivity in the root/leaf pathway. Oecologia 64: 344–354

Küppers M, Schulze E-D (1985) An empirical model of net photosynthesis and leaf conductance for the simulation of diurnal courses of CO_2 and H_2O exchange. Aust J Plant Physiol 12: 513–526

Kutschera U, Schopfer P (1986a) Effect of auxin and abscisic acid on cell wall extensibility in maize coleoptiles. Planta 167: 527–535

Kutschera U, Schopfer P (1986b) In-vivo measurement of cell-wall extensibility in maize coleoptiles: effects of auxin and abscisic acid. Planta 169: 437–442

Lachno DR, Baker DA (1986) Stress induction of abscisic acid in maize roots. Physiol Plant 68: 215–221

Lange OL (1959) Untersuchungen über Wärmehaushalt und Hitzeresistenz mauretanischer Wüsten- und Savannenpflanzen. Flora 147: 595–651

Lange OL, Lösch R, Schulze E-D, Kappen L (1971) Responses of stomata to changes in humidity. Planta 100: 76–86

Lilienfein U, Linnenbrink M, Weisheit K, Lösch R, Kappen L (1991) Licht- und Temperaturklima, Blattflächenverteilung im Kronenraum und Transpiration der Hasel in einer schleswig-holsteinischen Wallhecke. Verh Ges Ökol 20: 189–196

Lösch R (1989) Plant water relations. Prog Bot 50: 27–49

Lösch R (1993a) Water relations of Canarian laurel forest trees. In: Borghetti M, Grace J, Raschi A (eds) Water transport in plants under climatic stress. Cambridge University Press pp 243–246

Lösch R (1993b) Plant water relations. Prog Bot 54: 102–133

Lösch R, Tenhunen JD (1981) Stomatal responses to humidity-phenomenon and mechanism. In: Jarvis PG, Mansfield TA (eds) Stomatal physiology. Society for Experimental Biology Seminar Series, vol 8 Cambridge University Press, Cambridge, pp 137–161

Lösch R, Tenhunen JD, Pereira JS, Lange OL (1982) Diurnal courses of stomatal resistance and transpiration of wild and cultivated Mediterranean perennials at the end of the summer dry season in Portugal. Flora 172: 138–160

Lösch R, Jensen CR, Andersen MN (1992) Diurnal courses and factorial dependencies of leaf conductance and transpiration of differently potassium-fertilized and watered field-grown barley plants. Plant Soil 140: 205–224

Masle J, Passioura JB (1987) The effect of soil strength on the growth of young wheat plants. Aust J Plant Physiol 14: 643–656

Meinzer FC, Grantz DA (1990) Stomatal and hydraulic conductance in growing sugarcane: stomatal adjustment to water transport capacity. Plant Cell Environ 13: 383–388

Meinzer FC, Sharifi MR, Nilsen ET, Rundel PW (1988) Effects of manipulation of water and nitrogen regime on the water relations of the desert shrub Larrea tridentata. Oecologia 77: 480–486

Meinzer FC, Goldstein G, Neufeld HS, Grantz DA, Crisosto GM (1992) Hydraulic architecture of sugarcane in relation to patterns of water use during plant development. Plant Cell Environ 15: 471–477

Michelena VA, Boyer JS (1982) Complete turgor maintenance at low water potentials in the elongating region of maize leaves. Plant Physiol 69: 1145–1149

Milburn JA, Johnson RPC (1966) The conduction of sap. II. Detection of vibrations produced by sap cavitation in *Ricinus* xylem. Planta 66: 43–52

Munns R, King RW (1988) Abscisic acid is not the only stomatal inhibitor in the transpiration stream of wheat plants. Plant Physiol 88: 703–708

Passioura JB (1988) Root signals control leaf expansion in wheat seedlings growing in drying soil. Aust J Plant Physiol 15: 687–693

Prochazka S (1982) Translocation of ^{14}C-abscisic acid from roots into the aboveground part of pea (*Pisum sativum* L.) seedlings. Biol Plant 24: 53–56

Rivier L, Leonard JF, Cottier JP (1983) Rapid effect of osmotic stress on the control and exodiffusion of abscisic acid in Zea mays roots. Plant Sci Lett 31: 133–137

Robertson JM, Pharis RP, Huang YY, Reid DM, Yeung EC (1985) Drought-induced increases in abscisic acid levels in the root apex of sunflower. Plant Physiol 79: 1086–1089

Rosa LM, Dillenburg LR, Forseth IN (1991) Responses of soybean leaf angle, photosynthesis and stomatal conductance to leaf and soil water potential. Ann Bot 67: 51–58

Saab IN, Sharp RE (1989) Non-hydraulic signals from maize roots in drying soil: inhibition of leaf elongation but not stomatal conductance. Planta 179: 466–474

Schulze E-D (1986) Carbon dioxide and water vapor exchange in response to drought in the atmosphere and in the soil. Annu Rev Plant Physiol 37: 247–274

Schulze E-D (1993) The regulation of transpiration: Interactions of feedforeward, feedback and futile cycles. In: Schulze E-D (ed) Flux control in biologial systems; from the
enzyme to the population and ecosystem level. Academic Press (in press)

Schulze E-D, Hall AE (1982) Stomatal responses, water loss and CO_2 assimilation rates of plants in contrasting environments. In: Lange OL, Nobel PS, Osmond CB, Ziegler H (eds) Physiological plant ecology II. Encycl Plant Phys, NS 12B. Springer, Berlin Heidelberg New York, pp 181–230

Schulze E-D, Küppers M (1979) Short-term and long-term effects of plant water deficits on stomatal response to humidity in *Corylus avellana* L. Planta 146: 319–326

Schurr U (1992a) Stomatal response to drying soil in relation to changes in the xylem sap composition of *Helianthus annuus*. II. Stomatal sensitivity to abscisic acid imported from the xylem sap. Plant Cell Environ 15: 561–567

Schurr U (1992b) Die Wirkung von Bodenaustrocknung auf den Xylem- und Phloemtransport von *Ricinus communis* und deren Bedeutung für die Interaktion zwischen Wurzel und Sproß. Doktorarbeit, Bayreuth

Schurr U, Gollan T, Schulze E-D (1992) Stomatal response to drying soil in relation to changes in the xylem sap composition of *Helianthus annuus*. II. Stomatal sensitivity to abscisic acid imported from the xylem sap. Plant Cell Environ 15: 561–567

Sperry JS (1986) Relationship of xylem embolism to xylem pressure potential, stomatal closure, and shoot morphology in the palm *Rhapis excelsa*. Plant Physiol 80: 110–116

Stålfelt MG (1956) Die stomatäre Transpiration und die Physiologie der Spaltöffnungen. In: Ruhland W (ed) Handbuch der Pflanzenphysiologie, Bd 3. Springer, Berlin Göttingen Heidelberg, pp 351–426

Steinberg SL, McFarland MS, Worthington JW (1990) Comparison of trunk and branch sap flow with canopy transpiration in pecan. J Exp Bot 41: 653–659

Steudle E (1993) The regulation of plant water at the cell, tissue and organ level: role of active processes and of compartimentation. In: Schulze E-D (ed) Flux control in biological systems; from the enzyme to the population and ecosystem level. Academic Press (in press)

Turner NC, Schulze E-D, Gollan T (1985) The responses of stomata and leaf gas exchange to vapour pressure deficits and soil water content. II. In the mesophytic herbaceous species *Helianthus annuus*. Oecologia 65: 348–355

Tyree MT, Dixon MA (1983) Cavitation events in *Thuja occidentalis* L. Ultrasonic acoustic emissions from the sapwood can be measured. Plant Physiol 72: 1094–1099

Tyree MT, Sperry JS (1989) Vulnerability of xylem to cavitation and embolism. Annu Rev Plant Physiol Mol Biol 40: 19–38

Ullmann I (1985) Tagesgänge von Transpiration und stomatärer Leitfähigkeit sahelischer und saharischer Akazien in der Trockenzeit. Flora 176: 383–409

Waring RH, Schroeder PE, Oren R (1982) Application of the pipe model theory to predict canopy leaf area. Can J For Res 12: 556–560

Wartinger A, Heilmeier H, Hartung W, Schulze E-D (1990) Daily and seasonal course of leaf conductance and abscisic acid in the xylem sap of almond trees (*Prunus dulcis* (Miller) D.A. Webb) under desert conditions. New Phytol 116:581–587

Wolf O, Jeschke WD, Hartung W (1990) Long-distance transport of abscisic acid in NaCl-treated intact plant of *Lupinus albus*. J Exp Bot 41: 593–600

Wright STC, Hiron RWP (1969) (+)-abscisic acid, the growth inhibitor in detached wheat leaves following a period of wilting. Nature 224: 719–720

Zhang J, Davies WJ (1987) Increased synthesis of ABA in partially dehydrated root tips and ABA transport from roots to leaves. J Exp Bot 38: 2015–2023

Zhang J, Davies WJ (1989a) Sequential response of whole plant water relations to prolonged soil drying and the involvement of xylem sap ABA in the regulation of stomatal behaviour of sunflower plants. New Phytol 113: 167–174

Zhang J, Davies WJ (1989b) Abscisic acid produced in dehydrating roots may enable the plant to measure the water status of the soil. Plant Cell Environ 12: 73–81

Zhang J, Davies WJ (1990) Does ABA in the xylem control the rate of leaf growth in soil-dried maize and sunflower plants? J Exp Bot 41: 1125–1132

Zhang J, Schurr U, Davies WJ (1987) Control of stomatal behaviour by abscisic acid which apparently originates in the roots. J Exp Bot 38: 1174–1181

10   As to the Mode of Action of the Guard Cells in Dry Air

I.R. Cowan

10.1   Introduction

The first observation that stomata tend to close in dry air seems to have been made by one whose name is not usually found in reviews of stomatal physiology. Joseph Banks (1805) wrote that the pores in the stems of wheat, "which exist also on the leaves and glumes", are shut in dry weather and open in wet. The context is an essay on the cause of mildew in wheat, in which Banks supposes that the infection takes place through the stomata. He presumed the pores are "a provision intended no doubt to compensate, in some measure, the want of locomotion in vegetables" and remarked that "A plant cannot when thirsty go to the brook and drink: but it can open innumerable orifices for the reception of every degree of moisture, which either falls in the shape of rain or dew, or is separated from the mass of water always held in solution in the atmosphere. . . ."

Nearly a century later, Francis Darwin (1898) speculated on the role of stomatal response to humidity in the following passage, from which the title of this article is taken. "Some interesting questions arise as to the mode of action of the guard cells in dry air. Can we look at guard cells as sense-organs which, when the leaf is threatened by want of water, perceive the danger before the rest of the leaf? This idea is not wholly fanciful. Stahl has given some reason to believe that the transpiration of the guard cells is especially active; in dry air, or with a diminished water-supply, they should therefore be the first to suffer." In the work to which Darwin refers, Stahl (1894) suggested that the most important step in the "Emancipation" of plants from high atmospheric humidity lay in the capability of stomata to close. Ernst Stahl, who qualified as lecturer in Würzburg before moving to Jena, was an accomplished ecophysiologist with an ingenuity that will be indicated later.

The idea, then, that stomatal response to atmospheric conditions promoting transpiration is adapted to avert impending danger is quite an old one. But the credit for the first systematic study of the ecological implications of the humidity response surely goes to Otto Lange and his co-workers, who, with cultivated and wild species in the Negev Desert, examined the way in which the compromise between the needs to take up CO_2 and minimize water loss is influenced by atmospheric humidity deficit. It was a stimulus for theory of optimal stomatal action in which the responses to light and dryness of the atmosphere are equally important – the first in promoting photosynthesis[1] and the second in prolonging the conditions in which photosynthesis can occur. However, this chapter is not so much to do with the effect of the response to humidity on plant carbon and water economy as with the mechanism; a mechanism which so far has defied conclusive explanation. In this matter, also, the work of the Würzburg group is of the utmost importance.

10.2 Two Seminal Experiments

Lange et al. (1971) caused stomata to close and open by decreasing and increasing the humidity of air flowing over the outer surface of strips of the lower epidermis of *Polypodium vulgare* and *Valerianella locusta*. They were able to induce movements in a single stoma or a group of stomata by the use of small jets of air of varying humidity. They found also that the reaction of the stomata was influenced by the humidity of the air to which the inner surface of the epidermis was exposed. When the inner surface of the epidermis was put in contact with liquid water over its whole length, the stomata did not respond to variation in ambient humidity. When a small subepidermal space was created by introducing an air bubble only 2 mm in diameter, then the stomata above reacted to change in ambient humidity. The species that were used in this experiment are unusual in that, in intact leaves, the lower epidermis is attached to the remainder of the leaf only at the margin and main veins; the stomatal apparatus is particularly well suited to act as a humidity sensor because there is minimal hydraulic contact with the mesophyll.

The second experiment (Schulze et al. 1972) was one of a series with cultivated and wild plants in the Negev Desert (Lange et al. 1975). The data in Fig. 10.1 relate to single attached twigs of *Prunus armeniaca* enclosed in a naturally illuminated cuvette in which ambient temperature was controlled so as to keep leaf temperature constant. As the difference in humidity between leaves and air was successively increased by decreasing ambient humidity, the conductance of the stomata to vapor diffusion was caused to decrease and the net rate of assimilation to decrease. In this respect, the observations are unremarkable. What is remarkable is that each decrease in conductance was so great that it caused the rate of transpiration to decrease despite the increase in humidity difference between leaf and air.

---

[1] If the sensitivity to light is described as a tendency for stomata to close when light intensity is inadequate to sustain rapid photosynthesis, it becomes evident that it, also, is an adaptation that conserves water.

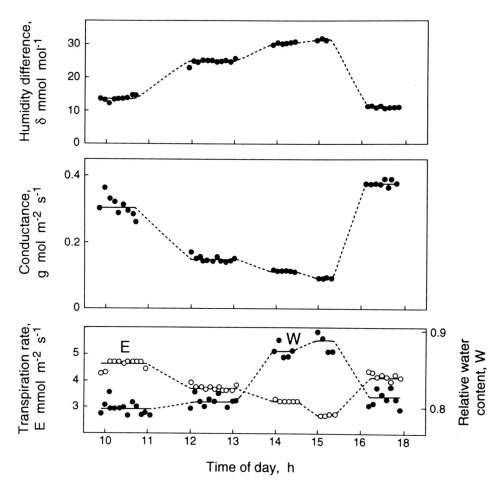

Fig. 10.1. Conductance to vapour transfer, g, transpiration rate, E, and relative water content, W, in *Prunus armeniaca* at various differences in humidity between leaf and air, δ. Leaf temperature was kept constant. Lighting was natural. (After Schulze et al. 1972)

The reduction in rate of transpiration was accompanied by increase in leaf water content. When, finally, ambient humidity was increased, then rate of transpiration also increased, and leaf water content declined. Similar results were obtained with the wild desert plants *Hammada scoparia* and *Zygophyllum dumosum*.

It was assumed, and later shown (see Hall et al. 1976), that stomatal response of this kind is a result of sensitivity to the difference between the humidities in the air space of the leaf and in the ambient air, the first being taken as saturation humidity at leaf temperature. Therefore it is a response to the environmental conditions promoting transpiration rather than humidity per se. However, for the sake of brevity, it is referred to hereafter as the direct humidity response. Actually, the term direct is not entirely apposite either, as will be understood later, but the reason for using it was well put by Schulze et al. "When the stomata close at low air humidity the water content of the leaves increases. The stomata open again at high air humidity in spite of a decrease in leaf water content. This excludes a reaction via the water potential in the leaf tissue and proves that the stomatal aperture has a direct response to the evaporative conditions in the atmosphere". Schulze et al. implied that the mechanism might be associated with peristomatal transpiration; that is to say, loss of water from the external surface of the leaf in the neighborhood of the stomatal apparatus taking place at a rate sufficient to affect the turgor of the guard cells. Raschke (1970) had been drawn to the same conclusion by observations made with detached leaves of *Zea mays*.

Numerous experiments have now confirmed the existence of the direct humidity response (Schulze and Hall 1982). The argument of Schulze et al. relating to mechanism seemed to be compelling (Cowan 1977; Farquhar 1978). It conflicted with the view, widely held at that time, that the response of stomata to ambient evaporative conditions is indirect, a feedback process depending on the leaf-internal consequences of variation in rate of transpiration. However, I shall try to show that what had been observed in Würzburg and in the Negev is not inconsistent with the feedback hypothesis; that the so-called direct humidity response could be brought about indirectly and that the explanation might rest with a generally unsuspected mechanical property of guard cells. Before doing so, I shall draw attention to some relevant studies of stomata that preceded the two that have been described, and some that followed.

10.3 Some Relevant Observations

10.3.1 On Stomatal Mechanics

It was von Mohl (1856) who, having made observations of stomatal movements in separated epidermis bathed in sugar solution of varied concentration, concluded that increase in guard cell volume and pore aperture resulted from increase in hydraulic pressure in the guard cell. Von Mohl also found by puncturing guard cells and subsidiary cells that the turgor pressure of the latter tended to diminish stomatal aperture. He described the interaction of guard cells and epidermal cells as one of "antagonism". A well-known drawing (Fig. 10.2) by Schwendener (1881) serves to illustrate the anatomy and change in anatomy associated with these movements in stomata having "kidney-shaped" guard cells, i.e., stomata in most dicotyledonous plants. Increase in pressure in the guard cell, so it is thought, causes the relatively thin dorsal walls to stretch and protrude into the

(a)

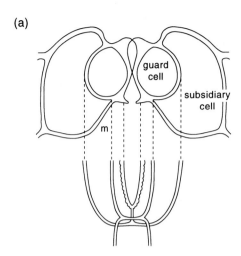

(b)

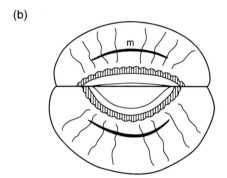

Fig. 10.2a,b. Stoma of *Sprekelia* (*Amaryllis*) formosissima. **a** Median transverse and surface views, **b** surface view of guard cell with pore closed, and with pore open. (After Schwendener 1881). Superimposed on **b** is an indication of the orientation of cell wall microfibrils

neighboring cells, with a concomitant bending of the ventral wall and opening of the pore. The mechanical basis of opening became more obvious when Ziegenspeck (1938, 1955) showed that the cellulose microfibrils of the guard cell walls were oriented in the way superimposed in Fig. 10.2b, suggesting that the length of the guard cell is more readily extensible than the transverse perimeter.

Three other important features of stomatal opening are to be found in Fig. 10.2b, two of them obvious, the other less so. Although the guard cell with the wider pore has thickened and bulged at its poles, the central lengthwise axis of the apparatus has not elongated. Secondly, the length of the pore, as defined by the thickened and ridged upper parts of the ventral walls, has shortened. Thirdly, something the reader may wish to verify by measurement, the perimeter of the pore, defined in the same way, has changed very little, if at all. In most descriptions and models of movement in stomata having kidney-shaped guard cells (Raschke 1979; Sharpe et al. 1987), the major axis of the pore is constant in length and opening of the pore results from a stretching of the ventral walls. That the walls are stretched in widely open stomata is undoubtedly true, and it is perhaps placing an unreasonable burden on Schwendener's drawing to conclude that they are not in the early stages of opening also. However, the drawing does serve to emphasize that what appears to be a rather insignificant diminution of length is sufficient to accommodate a visually much more impressive increase in width of the pore without stretching of the walls.

Most investigations of the way in which stomatal aperture, guard cell volume, and guard cell turgor are interrelated have used epidermal strips exposed to solutions of varying osmotic pressure. All of them, including those with epidermal strips in which the subsidiary cells as well as guard cells have been maintained intact (e.g., see MacRobbie 1987), have yielded results which have been interpreted as a monotonic increase of aperture and volume with pressure. (I use the word interpreted, because sometimes the number of data points has been inadequate to demonstrate any conflict there may have been with what has been a common assumption.) There have, however, been at least two suggestions that the state of affairs may sometimes be more complex.

Stålfelt (1927), on the basis of observations of the time course of widths of the stomatal apparatus as a whole and of the pore in *Vicia faba* following illumination, distinguished two phases of stomatal movement. In the initial *Spanniungsphase*, the length of which depends on the intensity of illumination, the guard cells thicken but the pore does not open. Just before the pore is observed to open, the cells narrow somewhat. Once opening is under way the guard cells thicken again, and the width of the whole apparatus increases in concert with increase in the width of the pore during what Stålfelt called the *motorische Phase*. Stålfelt explained the events he saw in terms of pressure in the guard cell thus:

> "Es genügt hierbei nicht, daß der Turgor den Druckwert der Nebenzellen erreicht. Erst wenn der Druck der Schließzellen einen gewissen Mehrbetrag erreicht hat, setzt die Bewegung ein. Sie kann dann in einer Zeit von viel geringeren Drucken weitergetrieben werden, d.h. die Bewegung muß sich unter vermindertem Drucke fortsetzen."[2]

Some support for the idea that guard cell pressure decreases at the onset of the *motorische Phase* was provided by determinations of sugar concentrations

---

[2] It is not sufficient that the magnitude of the turgor pressure of the subsidiary cells be attained. The movement first begins when the pressure of the guard cells exceeds (that of the subsidiary cells) by a certain amount. Then the cells can be driven apart during a period of much smaller pressure, i.e., the movement continues under reduced pressure.

required to plasmolyze guard cells in epidermal samples taken at various intervals. Unfortunately, sameples were allowed to equilibrate for 40 min with the sugar solutions, a period in which considerable solute exchange may have taken place. Although Stålfelt, aware of the problem, argued that the determinations would provide a relative indication of guard cell pressure at the time the samples were taken, the evidence must be regarded as suspect.

Meidner and Edwards (1975), also, had the idea that larger pressure in the guard cells may be required to start the process of bending the ventral walls than to continue the process. The context was a discussion of their experiments (see also Meidner 1982) in which capillaries were inserted into guard cells in intact leaves of *Tradescantia virginiana* and used to manipulate turgor pressure. When a capillary was inserted into a guard cell of an open stoma, turgor was lost and the pore closed partially, apparently in association with penetration of cytoplasm into the capillary. On application of hydraulic pressure in the capillary, the pore could be made to reopen up to, and beyond its original width. However, it was impossible to open an initially closed stoma by applying a pressure in one of its guard cells of 10 bar, the maximum of which the apparatus was capable. An opening movement could be achieved only with stomata in which the pore was already at least very slightly open. The observations indicate, as Stålfelt implies, that stomata may be capable of existing in either of at least two states, one closed and the other partially open, at the one level of turgor pressure in the guard cells. The contingency that this is due to an intrinsic mechanical property of stomata and not to temporally varying elasticity of the guard cell wall (an alternative suggestion by Meiduer and Edwards) is the keystone of the hypothesis that will be advanced about stomatal responses to humidity.

10.3.2   Signals and Responses

Joseph Banks' observation that stomata close in dry air was noted by von Mohl and contrasted with evidence that stomata were open in leaves free of dew and in bright sun, conditions tending to promote rapid transpiration. However, then, and for some time after, it was not fully understood that stomata could respond to many different stimuli. There were those, such as Schwendener, who held that light was the only stimulus and others, such as Leitgeb (1886), who believed that change in leaf water status was the predominant influence. The articles by Stahl and F. Darwin were perhaps the first in which there emerged a more balanced concept of stomatal action. With regard to the influence of atmospheric humidity, Darwin found that stomata in plants transferred from a moist to a drier atmosphere closed "without there being the slightest appearance of flaccidity in the leaves", having taken pains to ensure that his observations were not confounded by change in light intensity. However, Darwin had difficulty in understanding how the guard cells could become "more flaccid than the rest of the leaf", given some observations that suggested to him they must "share in the general turgor of the leaf". He concluded his discussion of stomatal mechanism with the following remarks. "There is another possibility in which I am inclined to believe, although I can give no positive evidence for it. The guard cells may lose turgor spontaneously, i.e., not by simple evaporation, but in response to a stimulus. And this may be the slight flaccidity of the rest of the leaf."

Evidence for Darwin's idea was provided by Stålfelt (1929) when he observed that stomata in *Vicia faba* began to close some 13 min after leaf water content had been reduced by a particular small amount, and that the movement, once begun, was not halted by resupplying the leaf with water. He suggested that the movement was associated with loss of solute from guard cells and called it "hydroactive" – as distinct from the more rapid, and opposing movements caused by fluctuations of turgor in the neighbouring epidermal cells which he called, somewhat inappropriately, "passive". Later still, a possible stimulus of hydroactive movements was identified. Little and Eidt (1968) and Mittleheuser and van Stevenink (1969) demonstrated that abscisic acid could cause stomata to close, and Wright and Hiron (1969) showed that abscisic acid was formed in wilting leaves. The role of abscisic acid as a messenger sensitizing guard cell metabolism to plant water stress has been reviewed by Raschke (1987). More recently, the existence of yet another signal, triggered by hydraulic conditions but not itself hydraulic, causing stomatal movement has been established, although its character has yet to be identified. It has been shown that the response of stomata to declining soil water content is to some extent independent of the effect of soil water content on the state of water in leaf tissue (Bates and Hall 1981; Blackman and Davies 1985; Turner et al. 1985; Gollan et al. 1986).

The importance of these findings, in the context of this chapter, is that stomata may respond to the water relations of plant tissues remote from the guard cells without necessarily being in close hydraulic communication with those tissues; and to the extent that the state of water in their own hydraulic microenvironment is different from that in other parts of the plant, being particularly affected by transpiration, they retain the capability of being particularly sensitive to conditions promoting transpiration. This is rather different from the suggestion by Darwin that a stimulus originating elsewhere in the leaf might account for the closure of stomata in dry air.

An extreme possibility is that the mechanism of the direct humidity response is isolated from the hydraulic continuum within the plant altogether; that the configuration of the guard cell wall is mechanically affected by the relative humidity of the external microenvironment. It is an idea engendered by observations (Ball et al. 1987) that the responses of stomata to variations in vapor pressure and temperature can be more coherently treated as though they were both manifestations of a sensitivity to relative humidity rather than difference in absolute humidity between leaf and air. The principle involved needs no discussion partly because it must be evident to those in a discipline that has made practical use, extending from the yucca hygrometer to the hair hygrograph, of the influence of absorbed water on strain in biological materials; but more particularly because a recent experiment has disproved the idea. Mott and Parkhurst (1991) broke the nexus between humidity and transpiration rate by measuring stomatal conductance at different ambient vapor pressures in air and in helox (21% O_2 and 79% He, in which water vapor diffuses 2.3 times more redily than in air). They showed that stomata were not at all affected by ambient vapor pressure per se, but were very much affected by the evaporative potential – this being the product of the relevant diffusion coefficient and the difference in humidity across the leaf epidermis and boundary layer. The experiments are made the more significant by the fact that one of the species used, *Phaseolus vulgaris*, exhibited the direct humidity response; that is to say, rate of transpiration declined with increase in evaporation potential.

It seems, then, that the mechanism of the direct humidity response must share in the general turgor of the leaf to some extent, and indeed the sequence of events that takes place when conditions affecting transpiration in a leaf are varied is similar to that occurring after a sudden disturbance of the water supply to the leaf, as Darwin observed. When rate of transpiration is stimulated, there is often a transient opening – a "passive" movement due to decrease in the "antagonism" of the epidermal cells. There is then a closing movement, sometimes followed by fluctuations. The pattern shown in Fig. 10.3 is very much like that seen by Raschke (1970) when the pressure of water supplied to a detached leaf of *Zea mays* was suddenly reduced. It is the nature of the closing movement that is little understood. When the potential of water in the epidermis is caused to vary, there must undoubtedly result some efflux or influx of water in the guard cells. Are the consequent volume changes sufficient, in normal circumstances, to cause substantial stomatal movement? And if additional water movement is brought about as a result of solute transport and change in the content of osmotica in the guard cell, is that solute transport instigated directly by the change in the state of water, as with a turgor pressure sensing mechanism for example (Gutknecht 1968; Coster et al. 1970)? Or is it a response to the changes in concentrations of ions in the guard cell, following the initial hydromechanical volume change? It is perhaps worth noting that Stålfelt (1955), in reexamining "hydroactive" closure of stomata, was uncertain whether loss of solute preceded or followed the initial volume changes in the guard cells, and, indeed, remarked on evidence that closure might sometimes take place without any loss of solute at all. To the extent that phytohormones initiate closure, effects on metabolism or membrane permeability come first. But when closure is initiated by change in the local state of water we remain as uncertain today as was Stålfelt.

There are two reasons for thinking that the direct humidity response is hydromechanical in origin. First, the response can be fast. Fanjul and Jones

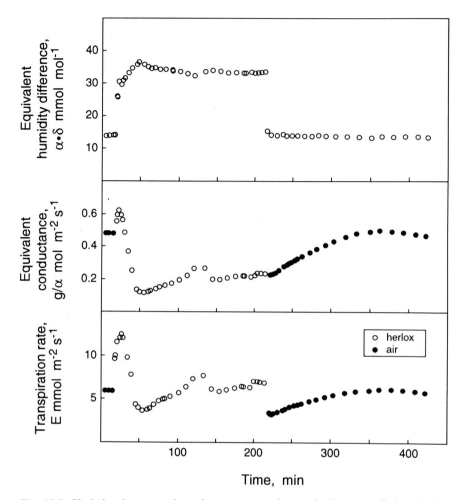

Fig. 10.3. Variation in stomatal conductance, g, and transpiration rate, E, in a leaf of *Vicia faba* in helox (O) and air (●). As water vapor diffuses 2.33 times faster in helox than in air, g/α, with α being 1 with air and 2.33 with helox, represents the variation in conductance that would have occurred in response to a variation in humidity difference $\alpha \cdot \delta$ in air. (After Mott and Parkhurst 1991)

(1982) showed that marked closing movements of stomata in apple due to decrease in ambient humidity were largely completed within a minute. The speed is similar to that of hydraulic equilibration of guard cells in epidermal strips with an external solution (Fischer 1973); generally responses to stimuli affecting guard cell metabolism, such as light, are much slower. Secondly, Lösch and Schenk (1978) in Würzburg found that closing and opening of stomata in *Valerianella locusta* due to variation in ambient humidity was not associated with variation in potassium content of guard cells. Changes in the content of potassium, thought to be the cation mainly responsible for osmotic pressure, did take place, but followed, rather than accompanied, stomatal movement. Of course, it seems inevitable that the processes affecting control of solute content in guard cells will be influenced by a change in solute concentration consequent on change in guard cell volume (see MacRobbie 1987), and they might well cause a readjustment enhancing and prolonging the initial movements. Responses to humidity sometimes, and perhaps usually, extend over longer periods of time than those observed by Fanjul and Jones. In particular, reopening at high humidity following closure at low humidity has been shown to be slow, as in Fig. 10.3 (see also Kappen et al. 1987; Kappen and Haeger 1991), and it is difficult to imagine that metabolism is not somehow involved.

10.3.3 Hydrology of the Epidermis

The evidence due to Stahl that the transpiration from the guard cells is especially active was obtrained by using, as a tracer of water movement in leaves, thallium supplied in solution to cut petioles as sulfate, and subsequently precipitated as thallium chloride following the supply of sodium chloride. After short periods of uptake, the black crystals of thallium chloride could be seen only in the guard cells. With successively longer periods they became apparent in the epidermal cells neighboring the guard cells, and then throughout the epidermis. The technique was a neat modification of one employed by Schimper (1890) – the use of thallium sulfate as a reagent to locate sodium chloride taken up by leaf tissue.

To explain the direct humidity response on the basis of evaporation from the region of the guard cell, it is necessary to assume that the distributions of the sites of evaporation and conductivities to water in the epidermis are such that the draw-down in water potential is sufficiently greater in the guard cells than the subsidiary cells in order that the "passive" action of the subsidiary cells be overcome. There would seem to be no conceptual impediment to that; nor to a difference in time constants for release or uptake of water by the cells such that the initial response of the stomata to increase in rate of transpiration is to open. All of this can be mimicked by an electrical analog consisting of resistors and two capacitors representing the pressure:volume relations of the cells (e.g., Cowan 1972). The critical question is whether evaporation from the neighborhood of the guard cells is under stomatal control or not.

Seybold (1962) supposed that transpiration in the vicinity of the stomatal complex, peristomatal transpiration as he called it, takes place through the external epidermal cuticle. Maercker (1965a,b, and as Maier-Maercker 1979) has provided evidence, and summarized that of others showing that rate of water loss near the stomata is greater than that from other regions of the epidermis, and has argued that it plays a decisive role in the stomatal control system. However, there is a lack of quantitave data to support the argument. Also, it might be thought surprising if a stomatal response that is presumably adapted to conserve water were to depend for its realization on a continuous and uncontrolled loss of water that is not coupled to the accession of CO_2 by the leaf.

With respect to this last comment, and for other reasons, the idea, espoused in particular by Meidner (1986) and his associates, that the stomatal response to humidity is due to transpiration from the inner walls of the guard cell has much to recommend it. Edwards and Meidner (1978) showed that cuticle is absent on the inner paradermal walls and innermost parts of the ventral walls of guard cells in *Polypodium vulgare*, whereas it is thick on the outer walls of the epidermis and the outer parts of the ventral walls, and present to varying degrees on the inner walls of subsidiary cells and other epidermal walls bordering leaf internal air space. It is easy to prove (Cowan 1977) that the inner guard cell walls, provided they are wet, must sustain a major proportion of the water loss by evaporation within the leaf, simply due to their proximity to the stomatal pore, even if all the other cell walls abutting air space within the leaf are also wet. It seems likely that the observations of Lange et al. with *P. vulgare* relate to evaporation from the inner wall of the guard cell, the absence of a humidity response when the inner walls were in contact with liquid water having less to do with supply of water to the cell than prevention of water loss from it.

In contrast to the observations with *P. vulgare*, Nonami et al. (1990) found that the inner walls of guard cells in *Tradescantia virginiana* support cuticle, and argued that transpiration from these surfaces must be very small, just as they proved it to be from the external cuticle of the leaf. However, of course, the question is not simply to do with the amounts of transpiration from the guard cell and subsidiary cell, but the amounts relative to the hydraulic conductances from epidermal cell to subsidiary cell, and subsidiary cell to guard cell. Also, it is to be doubted that thickness is a reliable indicator of the permeability of a cuticle to water. Permeability may be influenced by the amount and nature of waxes embedded within the cutin; and it is possible, too, that the permeability of the external cuticle of leaves is diminished by incipient drying (Schönherr 1976; Meidner 1986). Indeed, Meidner (1976) has demonstrated that the inner walls of the epidermis of *Tradescantia virginiana* can transpire and be continually replenished with water via the epidermis at rates which are a significant proportion of rates of transpiration in intact leaves. Water loss from the outer surface of the epidermis was shown to be much smaller.

There is a weightier objection to the notion that transpiration from the inner walls of guard cells is responsible for stomatal sensitivity to ambient humidity. Whatever the proportion the guard cell may sustain of the total amount of water evaporated within the substomatal cavity and lost through the stomata, that proportion is unlikely to increase as the stomatal aperture diminishes (indeed, the reverse is probable). How, then, can closure of the stomata be associated with water loss from the inner wall of the guard cell when closure causes the total rate of transpiration and, therefore, rate of water loss from the inner wall of the guard cell to decrease? That is the problem to be addressed if the observations of Lange and his associates are to be explained without recourse to the hypothesis of peristomatal transpiration.

10.4 Hypothesis

10.4.1 Feedback

If it is to be rate of transpiration from the leaf, or a proportion of that rate, that influences stomatal function when stomatal aperture responds to ambient humidity, then it is appropriate to replot the data of Schulze et al. in the form of Fig. 10.4a. As rate of transpiration is the product of leaf conductance and humidity difference, humidity differences are represented in the figure as the inverse slopes of straight lines from origin to data points. When humidity difference is increased, conductance decreases. At first rate of transpiration increases but eventually, when humidity difference is large, rate of transpiration decreases also.

That the potential of water in the leaf declines with increase in rate of transpiration may be inferred from the variation in leaf water content (Fig. 10.4b). Presumably the potential of water in the guard cells also declines – although, for reasons that have been discussed, not necessarily to the same extent. That is why the transpiration axis in Fig. 10.4 is oriented as it is; it is taken to represent decreasing water potential in the guard cells, Ψ, as indicated at the base of the figure. The result is what may seem a very strange variation of stomatal aperture with potential, one in which there is a phase in which aperture increases while potential decreases. Of course, increase in the rate of transpiration will cause decrease in potential of water in the subsidiary cells also, and that will tend to open the stomatal pores. However, while it may be that this enhances the peculiar characteristic of Fig. 10.4a, I do not think it can be responsible for it.

Let it be supposed, instead, that what underlies the relationship of conductance to potential in Fig. 10.4a is a relationship between guard cell volume, V, and turgor pressure, P. To deduce the second from the first would require more information than is available but, provided conductance is a monotonically increasing function, $g = g(V)$, of volume, the general characteristics of its shape are readily deduced. The transformation of the potential axis in Fig. 10.4 to one representing pressure is given by

$$P = \Psi + \Pi,$$

where Π is the osmotic pressure within the guard cell. We do not know to what extent solute content of the guard cells may have readjusted to the

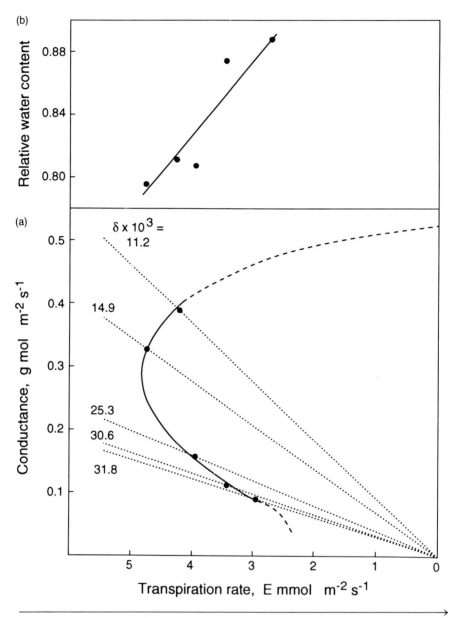

Fig. 10.4. Data of Schulze et al. (see Fig. 10.1) rearranged to show dependence of conductance, g, and relative water content, W, on rate of transpiration, E. Humidity differences between leaf and air, δ, are represented as the *inverse slope of the straight lines*. The *lower abscissa* indicates the presumed effect of rate of transpiration on water potential in the guard cells of the stomata, Ψ

change in guard cell volume while the stomata were first caused to close and later to open. If adjustment took place to the extent that Π returned to the same value after each disturbance, then the shape of the relationship between V and P that we seek would differ from that between g and V only insofar as the function g(V) may be curvilinear. If, at the other extreme, no adjustment took place at all so that Π varied inversely with volume, then

$$\frac{dP}{dV} = \frac{d\Psi}{dg} \cdot \frac{dg}{dV} - \frac{\Pi}{V},$$

from which it is seen that dP/dV is negative over a range of g encompassing and extending beyond that in which dΨ/dg is negative. Nevertheless, it may be assumed there is a limit to this trend at large volume, the forces resisting further distension of a greatly swollen guard cell becoming sufficiently large for dP/dV to be positive. It appears that the V(P) relationship must have the same essential features of shape as the g(Ψ) relationship in Fig. 10.4a.

If guard cell expansion has the characteristics described, it poses novel questions about the function and structure of the stomatal apparatus.

10.4.2   Of Bubbles and Balloons

Reproduced in Fig. 10.5 is the volume:pressure relationship of a rubber balloon (Bini Balloons, made in Denmark). I had been given to think of its relevance as an analogy to the functioning of guard cells 15 years ago (Cowan 1977) when I wrote of the observations of Meidner and Edwards using a pressure probe to manipulate guard-cell turgor that they, "... might be thought to indicate that the initial expansion of the guard cell is essentially an irreversible process – irreversible in the thermodynamic sense, that is – the internal pressure decreasing with increase in volume as with the inflation of a rubber balloon". I further envisaged, in the same article, that decline of conductance with increase of water potential in guard cell might have something to do with the experiments of Schulze et al. However, I rejected the idea on the grounds that the stomatal control system would then be unstable. That was a mistake: stomata may sometimes be unstable, and the dynamics of balloons sometimes are not.

Figure 10.5 quantifies a matter of common practical experience. There is a *Spannungsphase* in which one needs to exert a greater pressure to initiate inflation than to continue inflation once a certain volume has been achieved. It is succeeded by a *motorische Phase*, increase in volume being accompanied by decline in pressure. Eventually, at large volume, the balloon reverts to a "normal" characteristic, further increase in volume requiring increase in pressure up to, and beyond the initial, threshold pressure. Remarkably, the characteristic of the balloon in the *motorische Phase* is somewhat like that of a soap bubble; pressure varies approximately as the

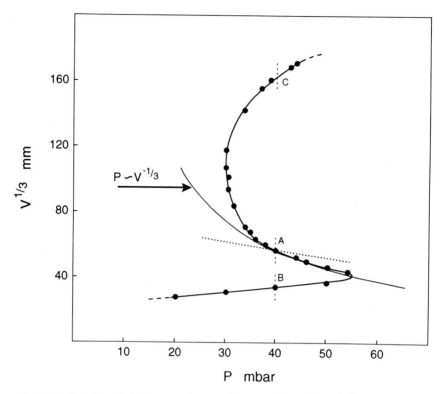

Fig. 10.5. Relationship between volume and pressure in a rubber balloon

inverse cube root of volume, as indicated by the rectangular hyperbola in Fig. 10.5. While in this phase the balloon is unstable in the following sense. If it were connected, when in the state represented by A in Fig. 10.5, to an effectively infinite source of air having the same pressure, then the balloon would almost immediately deflate to the point B or inflate to the point C. The direction of the change would not be predeterminate. Insofar as guard cells behave like this, their states as observed by Meidner and Edwards with closed and open stomata would be represented by B and C; and it would have been the magnitude of the pressure barrier to the right that prevented transition from the one state to the other, rather than the absence of a "loosening" factor affecting the intrinsic properties of cell walls.

However, the balloon can be maintained in state A if it is closed off. It is then made stable because the pressure of air within it varies inversely as V, whereas the pressure required for inflation varies inversely as $V^{1/3}$. That suggests the means of stabilizing a balloon in the *motorische Phase* by controling the pressure of air supplied to it. The increase or decrease in pressure of the air supplied with decrease or increase in volume must exceed that of the internal pressure – as illustrated by the sloping line intersecting point A in Fig. 10.5. The dynamics of control also need consideration. If one attempts to maintain the volume of a balloon constant, during the process of inflating it in the normal way, the result is far from perfect; it is almost impossible to prevent erratic fluctuations. The task becomes much easier if the passageway into the balloon is constricted so that the influx or efflux of air encounters an increased resistance. Herein lies an indication of the second requirement for stability. The speed with which external pressure is caused to vary with reference to observations of volume must exceed the speed of volume change in response to change of external pressure.

As with air pressure and a balloon, so perhaps with water potential and the guard cell. The source of water is in the guard cell wall, and control of water potential is provided by the influence of stomatal aperture on rate of evaporation in the parts of the wall bordering the substomatal cavity. Stability requires that the steady-state change in potential of water in the wall due to a given change in aperture should exceed the change in potential of water in the guard cell associated with that change in aperture. This is fulfilled by the data in Fig. 10.4a; the inverse slope of each of the straight lines is greater than the inverse slope of the curve at the point of intersection. The second requirement is that the speed with which potential of water in the wall responds to variation in stomatal aperture should exceed the speed with which aperture responds to variation in potential in the wall. Insofar as stomatal aperture became steady at each level of ambient humidity (see Fig. 10.1), it would seem that this condition was met also.

Through these arguments it can be appreciated that the direct humidity response might satisfactorily be explained by a feedback control system, provided the guard cells have the property that internal water potential tends to increase with decrease in guard cell volume and stomatal aperture. I am inclined to put the matter more positively and say that the observations of Schulze et al. (1990) are evidence that guard cells do have that property. To carry the description of the control system envisaged further really requires mathematical treatment, which I hope to present in another article. However, certain aspects, particularly the roles of solute and solute regulation in the guard cell, demand at least a qualitative commentary here.

If solute were conserved in the guard cell, so that osmotic pressure varied inversely with volume, it would gravitate against $dg/d\Psi$ being negative, as an examination of the equation in the preceding section makes clear. Indeed, if the guard cell were to have the characteristic of Fig. 10.5, it would be made intrinsically stable for the same reason that a closed, air-filled balloon is stable, and the $g(\Psi)$ relationship in Fig. 10.4 could not be explained in terms of the mechanical properties of the guard cell. To explain the relationship it would be necessary that the relative decrease in pressure with relative increase in volume exceed unity (whereas it is approximately 1/3 in a balloon). That would be achieved by an air-filled system of a balloon connected to a considerably larger rigid container. The image need not offend our sensibility of the shape of guard cells for, after all, the balloon is an analogy and not an explanation of guard cell mechanics. It serves the purpose of emphasizing that there are some rather demanding requirements relating to mechanics and structure if solute is conserved in the guard cell and the direct humidity response is to be explained along the lines I postulate.

I am led to think, partly for that reason but also because of the protracted nature of responses to humidity, that the guard cell movement is associated with (though not driven by) active control of osmotic pressure. A simple model is represented by the relationship $d(\Pi V)/dt \sim \Pi_* - \Pi$, expressing a continuing tendency for Π to return to a magnitude Π_* which is dependent on light and other factors affecting active solute transport and/or membrane permeabilities but independent of guard cell volume. With control of this nature, water potential would tend to vary as turgor pressure and therefore solute would have no influence whatever on the shape of the steady-state relationship between g and Ψ such as that in Fig. 10.4. However, osmotic pressure would determine the position of the curve with respect to the Ψ axis; the greater Π_*, the farther the curve would be placed to the left, and therefore the larger the stomatal conductance corresponding to any given humidity difference. And the dynamics of solute control would influence the dynamics of change from one steady state to another. Following a step change, a decrease say, in humidity difference, there would be a rapid loss of water from the cells of the stomatal complex culminating in a quasi-steady state in which stomatal aperture might be increased due to reduction in the "antagonism" of the subsidiary cells. There would then be a slower loss of water from the guard cells, limited in speed but not in magnitude, by the rate at which the solute control system operates. The associated closure of the stomata in this phase would be assisted by increased turgor in the subsidiary cells.

The assumption that osmotic pressure is maintained more nearly constant than it would be if solute were conserved relieves us of some difficulties in interpreting Fig. 10.4a in terms of guard cell mechanics; but the necessity of explaining how dP/dV might be negative remains. To invoke elastic properties such as those of rubber will not suffice for several reasons, one of which has to do with stability of guard cell shape. Although it might be possible to construct a balloon which has the shape of a deflated guard cell when deflated, and that of an inflated guard cell when fully inflated, it would not retain the end-to-end symmetry of a guard cell during the phase when pressure decreases with volume. Even a spherical balloon does not stretch uniformly during inflation, as may readily be shown by drawing a grid on its surface.

10.4.3 Piers and Vaults

The evolution, in the 12th and 13th centuries, of Gothic cathedrals of successively lighter construction but wider vaults involved a development of

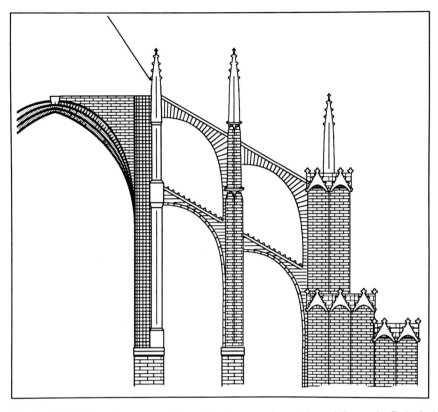

Fig. 10.6. Vaulting, clerestory wall, and buttresses of a section of Beauvais Cathedral, which collapsed in 1284. (After Mark 1982)

increasingly elaborate buttresses to counter the outward forces imposed on the central piers by the weight of vaulting and its surcharge (Fig. 10.6). The concept I have of stomatal opening has something in common with this trend. It has to do with a varying balance between forces tending to drive the ventral walls of the pore apart by pressing inwards from the poles of the guard cells and others which act to push the walls together. It is essentially two-dimensional in that only forces parallel to the epidermal plane are considered; the ventral and dorsal walls of the guard cell are strictly normal to the epidermis and constant and uniform in depth, and bulging of the paradermal walls is ignored.

Consider the representation of the stomatal apparatus with closed pore in Fig. 10.7a. Each ventral wall, of unit length say, is connected, through radiating microfibrils in the paradermal walls, to the dorsal wall, which has a projected length parallel to the pore of m. It will be assumed that the polar walls of the guard cell are constrained by the structure of the epidermis from bulging in the longitudinal direction but are nevertheless readily stretched so

(a)   (b)

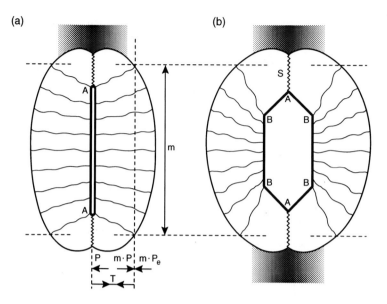

Fig. 10.7a,b. Representations of the stomatal apparatus. **a** With uniformly distributed radiating microfibrils connecting ventral wall of unit length to dorsal walls having projected lengths m; P is pressure in the guard cells, P_e the effective pressure of the epidermis on the dorsal walls and T the tensile force in the microfibrils. **b** With microfibrils connected to central sections, B-B, of ventral walls which bend at B and are hinged at A. The common walls of the cells, S, are extensible

that they do not appreciably impede an outward movement of the dorsal walls. Therefore the microfibrils sustain a tensile force normal to the pore of $T = m(P - P_e)$, P_e being the effective pressure exerted on the dorsal walls by the epidermis. It follows that the outward force on the ventral wall is positive only if $m(P - P_e) > P$, or, with $m = 1.25$ as in Fig. 10.7a, $P > 5P_e$. As the effective pressure, P_e, is likely to exceed the pressure in the subsidiary cells (because the subsidiary cells are deeper than the guard cells), this condition implies that P would need to be very large indeed for the pore to begin to open.

However, it is improbable that the forces acting to open the stomatal pore are uniformly distributed along the ventral walls, or indeed that the elastic properties of the ventral walls are uniform either. In the illustration in Fig. 10.7b the arrangement of microfibrils is such that the whole tensile force is exerted on the central sections B-B, one half the total length of the ventral walls. The walls are allowed to bend only at the points B, and are hinged at the apices of the pore, A. It is easy to show that, with the pore being closed, the moments of forces about the points B are positive – that is to say, there is a tendency for the pore to open – if $m(P - P_e) > 3P/4$. With $m = 1.25$, as in the illustration, the condition is $P > 2.5P_s$.

What now happens when the pore opens a little, and narrow vaults are formed by the divergence of the wall members A-B? It will be assumed that the paradermal walls at the poles of the guard cells and the common walls of the guard cells impede the movement very little, if at all, at first. This is reasonable, since one can imagine that the apices of the pore are pressed polewards in the *Spannungsphase*, i.e., in the phase in which the pore is closed and the pressure is less than that required to initiate opening. Therefore, the action of the polar walls in likened to that of springs, S in Fig. 10.7, in which the tensions are negligible within a small range of extensions but increase rapidly beyond. For small apertures, then, the components of force directed parallel to the major axis of the pore are predominantly due to the pressure, P, acting on the vault members A-B. The moments of these forces about the points B increase disproportionally with increase in the angle of bending and the width of the pore. If the pressure sufficient to initiate opening were maintained, the pore would continue to widen until restrained by the tension developed in S. Greater apertures would require increase in P.

The spontaneous opening described, and the associated increase in guard cell volume, could be arrested by decrease in the pressure P, causing the

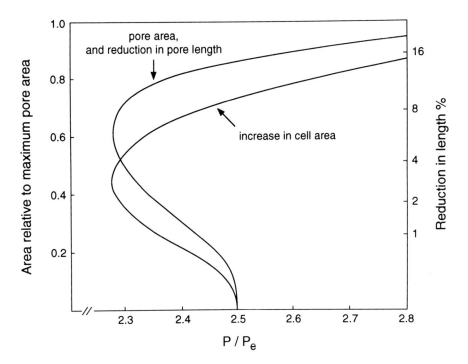

Fig. 10.8. Calculations based on Fig. 10.7b: pore area, increment in guard cell area, and pore length as functions of pressure in the guard cells, *P*, relative to epidermal pressure, *P_e*

tension in the radiating microfibrils to decrease and the net forces on the walls B-B to be directed inwards. Even so, the system would remain unstable in the absence of continuous adjustment of pressure to maintain the balance between the forces parallel to, and normal to the axis of the pore. Figure 10.8 shows an example of variations of pore area and guard cell volume with balancing pressure, estimated with particular assumed elastic properties relating to bending at B and wall extension at S, and with P_e taken as constant. Increment in guard cell area (equivalent to cell volume in a two-dimensional representation of stomatal movement) has been calculated on the basis that the increment of cell + pore area is m × (length of closed pore) × (width of pore).

What has been described here is a grossly simplified model of stomatal action. Nevertheless it serves to demonstrate that it is mechanically possible for an increase in stomatal aperture and guard cell volume to be associated with decrease in the pressure in the guard cells. The relationships in Fig. 10.8 have the characteristics I have postulated in this attempt to account for the direct humidity response as a manifestation of an indirect, feedback process.

10.5 Conclusions

Three ideas have been advanced. The most general of them is that the direct humidity response is due to a negative feedback system with a rather special property: the feedback gain sometimes exceeds unity and yet the system remains stable because it contains an element, a subsidiary, positive feedback loop which, in isolation, is unstable. This, if correct, is important, for it suggests that the response is susceptible to a variety of explanations of a kind not previously considered. It could hardly fail to have implications for other aspects of stomatal function, such as the response to light. The possibility that the system as a whole sometimes becomes unstable, control then having an "on-off"characteristic, might be the basis of heterogeneity and patchiness of stomatal aperture.

The second idea is to do with the specific nature of the internal, intrinsically unstable element. That the instability is due to decrease in pressure with increase in volume in the guard cell, perhaps allied to a tendency for osmotic pressure to be maintained more nearly constant than it would be if solute were conserved, is not the only suggestion possible. However, it seems more plausible than any other I have been able to devise. There is at least some slight support for it in the work of Stålfelt, and Meidner and Edwards.

Third is the idea that decrease in pressure in the guard cell with increase in volume and stomatal aperture is associated with increase in a bending moment due to forces directed inwards from the poles of the guard cells. To develop this notion further in the form of a realistic model of guard cell mechanics would require more knowledge of the elastic properties of component parts of a guard cell than we have, or are likely to acquire by direct means. Support for the idea is more likely to emerge from the reverse procedure: deconvolution of data relating to turgor pressure, cell volume, and cell shape.

Acknowledgments. Without the opportunity afforded me by Otto Lange to think and work in his laboratory, the ideas I have described here, such as they are, would not have come to fruition. Without his contribution to the study of stomata, involving, as did the work of Darwin and Stahl in the last century, meticulous and ingenious experiment informed by speculation about the adaptive function of stomata, they would not have been conceived.

References

Ball JT, Woodrow I, Berry JA (1987) A model predicting stomatal conductance and its contribution to the control of photosynthesis under different environmental conditions. In: Biggins J (ed) Progress in photosynthesis research, vol IV. Martinus Nijhoff, Dordrecht, pp 221–224

Banks J (1805) A short account of the causes of the diseases in corn, called by farmers the blight, the mildew, and the rust. In: Curtis LW (ed) Practical observations on the British grasses, especially such as are best adapted to the laying down or improving meadows and pasture, likewise an enumeration of the British grasses, 6th edn, 1824. Sherwood, Jones and Co, London, pp 151–166

Bates LM, Hall AE (1981) Stomatal closure with soil water depletion not associated with changes in bulk leaf water status. Oecologia 50: 62–65

Blackman PG, Davies WT (1985) Root to shoot communication in maize plants of the effects of soil drying. J Exp Bot 36: 39–48

Coster HGL, Steudle E, Zimmermann U (1970) Turgor pressure sensing in plant cell membranes. Plant Physiol 58: 636–643

Cowan IR (1972) Oscillations in stomatal conductance and plant functioning associated with stomatal conductance: observations and a model. Planta 106: 185–219

Cowan IR (1977) Stomatal behaviour and environment. Adv Bot Res 4: 117–228

Darwin F (1898) Observations on stomata. Philos Trans R Soc Ser B 190: 531–621

Edwards M, Meidner H (1978) Stomatal responses to humidity and the water potentials of epidermal and mesophyll tissue. J Exp Bot 29: 771–780

Fanjul L, Jones HG (1982) Rapid stomatal responses to humidity. Planta 154: 135–138

Farquhar GD (1978) Feedforward responses of stomata to humidity. Aust J Plant Physiol 5: 787–800

Fischer RA (1973) The relationship of stomatal aperture and guard-cell turgor pressure in *Vicia faba*. J Exp Bot 24: 387–399

Gollan T, Passioura JB, Munns R (1986) Soil water status affects the stomatal conductance of fully turgid wheat and sunflower leaves. Aust J Plant Physiol 13: 459–464

Gutknecht J (1968) Salt transport in *Valonia*: inhibition of potassium uptake by small hydrostatic pressures. Science 160: 68–70

Hall AE, Schulze E-D, Lange OL (1976) Current perspectives of steady-state stomatal response to environment. In: Lange OL, Kappen L, Schulze ED (eds) Water and plant life. Ecological studies 19. Springer, Berlin Heidelberg New York, pp 169–185

Kappen L, Haeger S (1991) Stomatal responses of *Tradescantia albiflora* to changing air humidity in light and in darkness. J Exp Bot 241: 979–986

Kappen L, Andresen G, Lösch R (1987) In situ observations of stomatal movements. J Exp Bot 38: 126–141

Lange OL, Lösch R, Schulze E-D, Kappen L (1971) Responses of stomata to changes in humidity. Planta 100: 76–86

Lange OL, Schulze E-D, Kappen L, Buschbom U, Evenari M (1975) Photosynthesis of desert plants as influenced by internal and external factors. In: Gates DM, Schmere RB (eds) Perspectives of biophysical ecology. Springer, Berlin Heidelbery New York, pp 121–143

Leitgeb H (1886) Beiträge zur Physiologie der Spaltöffnungsapparate. Mittheilungen aus dem Botanischen Institute zu Graz Bd I

Little CHA, Eidt DC (1968) Effect of abscisic acid on budbreak and transpiration in woody species. Nature 220: 498–499

Lösch R, Schenk B (1978) Humidity responses of stomata and the potassium content of guard cells. J Exp Bot 29: 781–787

MacRobbie EAC (1987) Ionic relations of guard cells. In: Zeiger H, Farquhar GD, Cowan IR (eds) Stomatal function. Stanford University Press, Stanford, pp 125–162

Maercker U (1965a) Mikroautoradiographischer Nachweis tritiumhaltigen Transpiration wassers. Naturwissenschaften 52: 15

Maercker U (1965b) Zur Kenntnis der Transpiration der Schließzellen. Protoplasma 60: 61–78

Maier-Maercker U (1979) "Peristomatal transpiration" and stomatal movement: a controversial view I. Additional proof of peristomatal transpiration by hydrophotography and a comprehensive discussion in the light of recent results. Z Pflanzenphysiol 91: 25–43

Mark R (1982) Experiments in Gothic architecture. MIT Press, Cambridge, Mass

Meidner H (1976) Vapour loss through stomatal pores with the mesophyll tissue excluded. J Exp Bot 27: 172–174

Meidner H (1982) Guard cell pressures and wall properties during stomatal opening. J Exp Bot 33: 355–359

Meidner H (1986) Cuticular conductance and the humidity response of stomata. J Exp Bot 37: 517–525

Meidner H, Edwards M (1975) Direct measurements of turgor pressure potentials of guard cells, I. J Exp Bot 26: 319–330

Mittleheuser CJ, van Steveninck RFM (1969) Stomatal closure and inhibition of transpiration induced by (RS)-abscisic acid. Nature 221: 281–282

Mohl Hv (1856) Welche Ursachen bewirken die Erweiterung und Verengung der Spaltöffnungen? Bot Zeit 14: 697–704

Mott KA, Parkhurst DF (1991) Stomatal responses to humidity in air and helox. Plant Cell Environ 14: 509–515

Nonami H, Schulze E-D, Ziegler H (1990) Mechanisms of stomatal movement in response to air humidity, irradiance and xylem water potential. Planta 183: 57–64

Raschke K (1970) Stomatal responses to pressure changes and interruptions in the water supply of detached leaves of *Zea mays* L. Plant Physiol 45: 415–423

Raschke K (1979) Movements of stomata. In: Haupt W, Feinleib ME (eds) Physiology of movements. Encyclopedia of plant physiology 7. Berlin Springer Berlin Hedelberg New York, pp 383–441

Raschke K (1987) Action of abscisic acid on guard cells. In: Zeiger E, Farquhar GD, Cowan IR (eds) Stomatal function. Stanford University Press, Stanford, pp 253–279

Schimper AFW (1890) Pflanzen-geographlz anf physiologischer fremdlap. Gustav Fischer Unrlc, Jena, 875 pp.

Schönherr J (1976) Water permeability of isolated cuticular membranes: the effect of cuticular waxes on diffusion of water. Planta 131: 159–164

Schulze E-D, Hall AE (1982) Stomatal responses, water loss and CO_2 assimilation rates of plants in contrasting environments. In: Lange OL, Nobel PS, Osmonol CB,

Ziegler H (eds) Physiological plant ecology 2. Encyclopedia of plant physiology 12B. Springer, Berlin Heidelberg New York, pp 181–230

Schulze E-D, Lange OL, Buschbom U, Kappen L, Evenari M (1972) Stomatal responses to changes in humidity in plants growing in the desert. Planta 108: 259–270

Schwendener S (1881) Über Bau und Mechanik der Spaltöffnungen. Monatsber Kgl Acad Wiss Berl 46: 833–867

Seybold A (1962) Ergebnisse und Probleme pflanzlicher Transpirationsanylesen. Jahresh Heidelb Akad Wiss 1961/62: 5–8

Sharpe PJH, Wu H, Spence RD (1987) Stomatal mechanics In: Zeiger E, Farquhar GD, Cowan IR (eds) Stomatal function. Stanford University Press, Stanford, pp 91–114

Stahl E (1894) Einige Versuche über Transpiration und Assimilation. Bot Zeit 52: 117–145

Stålfelt MG (1927) Die photischen Reaktionen im Spaltöffnungsmechanismus. Flora NF 21: 236–272

Stålfelt MG (1929) Die Abhängigkeit der Spaltöffnungsreaktionen von der Wasserbilanz. Planta 8: 287–340

Stålfelt MG (1955) The stomata as a hydrophotic regulator of the water deficit of the plant. Physiolgial Plant 8: 572–593

Turner NC, Schulze E-D, Gollan T (1985) The responses of stomata and leaf gas exchange to vapour pressure deficits and soil water content. II. In the mesophytic herbaceous species Helianthus annuus. Oecologia 65: 348–355

Wright STC, Hiron RWP (1969) (+)-abscisic acid, the growth inhibitor induced in detached wheat leaves by a period of wilting. Nature 224: 719–720

Ziegenspeck H (1938) Die Micellierung der Turgeszenzmechanismen. I. Die Spaltoffnungen (mit phylogenetischen Ausblicken). Bot Arch 39: 268–309

Ziegenspeck H (1955) Das Vorkommen von Fila in radialer Anordnung in den Schliesszellen. Protoplasma 44: 385–388

11 Direct Observations of Stomatal Movements

L. Kappen, G. Schultz, and R. Vanselow

11.1 Introduction

Stomata, the main channels for CO_2 and water vapor exchange, are well known as systems responding to internal factors mediated by the leaf tissues, and to external factors resulting from ambient conditions acting directly or indirectly. This was well documented by Raschke (1979) and has been discussed and reviewed in more recent times (e.g., Schulze and Hall 1982; Zeiger et al. 1987; and others).

The ecophysiological parameters for stomatal movements are depicted in Fig. 11.1. The relationships between these parameters and the stomatal responses are shown as far as they are relevant for the present study. For further understanding of stomatal functions and plant water relations, the reader is referred to Lösch and Schulze (Chap. 9, this Vol.). Our considerations are based on the following concept.

Light is known to be one of the principal parameters causing opening of the stomatal apparatus. Light acts directly (Sharkey and Ogawa 1987; Poffenroth et al. 1992) and indirectly as it controls photosynthetic CO_2 assimilation and therefore induces a CO_2 gradient between ambient and substomatal CO_2 concentration (Morison 1987). The CO_2 gradient dependence is obviously quite strong as it causes closing of the stomata of CAM plants in light (Ting 1987). Nevertheless, stomata of various C_3 plant species are not always totally closed in darkness (Meidner and Mansfield 1968). Several findings demonstrate that opening of stomata is caused by CO_2-free air or nitrogen (Louguet 1972) in darkness.

Because CO_2 uptake and water loss occur concomitantly, a conflict may arise between water stress and carbon assimilation. An optimalization of the relation between CO_2 gain and water loss was, for instance, modeled by Cowan and Milthorpe (1968).

Evidence is given (Stalfelt 1962) that ambient heat influences stomatal aperture not only indirectly by changing evaporative conditions and CO_2 metabolism but also directly by the fact that stomata are wide open at superoptimal temperatures even in darkness (Brunner and Eller 1974).

Water stress resulting from reduced soil water supply to the roots is transmitted to the stomatal apparatus by a decrease of the xylem water

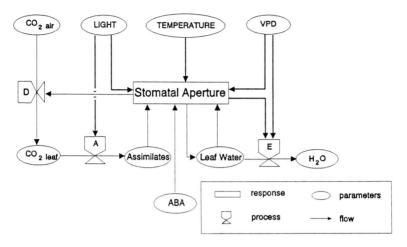

Fig. 11.1. Stomatal aperture in response to environmental parameters (*light, temperature, VPD, CO₂ air*) and to leaf or plant internal parameters (*leaf water, ABA, CO₂ leaf*). *VPD* Water vapor pressure deficit between leaf and air; *E* evaporation; *A* assimilation; *D* diffusion; *leaf water* leaf water potential and water content; *ABA* abscisic acid; *boldface arrows* relevant in this study

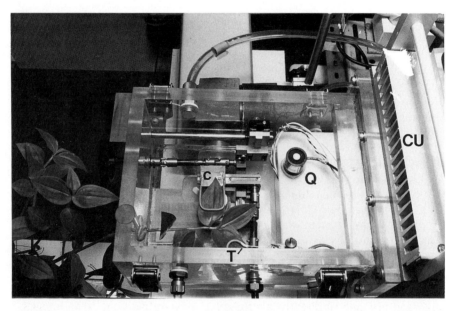

Fig. 11.2. Gas exchange chamber with a microscope gliding stage. A leaf of *Tradescantia albiflora* is fixed by a clamp (*c*) above the microscope objective lens. Leaf temperature is measured by a thermocouple (*T*) fixed by a second clamp. A quantum flux sensor (*Q*) is installed near the conditioning unit (*CU*). The conditioning unit is attached to the cuvette (*right-hand side* of the picture)

potential (Schulze 1986) or by a direct ABA signal from the root (Zhang and Davies 1989; Schurr 1992). Water stress response can be induced in many plant species by a decrease in the water content in the ambient air (Lange et al. 1971; Schulze et al. 1987; Grantz 1990). Stomata respond promptly to air humidity changes and therefore experiments based on humidity alterations are very useful in testing stomatal movements.

In most cases, stomatal movement or conductance has been estimated by calculations using water relations, transpiration, and temperature as parameters (Ball 1987; Von Caemmerer and Farquhar 1981). Rarely, stomatal aperture has been observed directly (Elkins and Williams 1962; Lösch 1977; Omasa et al. 1983), but it was not possible to observe stomatal responses in darkness. In the present chapter, we demonstrate to what extent visual inspection of stomatal responses in situ, coupled with simultaneous measurements of CO_2 and water fluxes, are capable of giving direct evidence of the hitherto reported stomatal responses to plant internal and external factors. Moreover, it is demonstrated that our method allows more insight into the mechanism of opening and closing of the stomata. So, for instance, a transient phase in the response to changed air humidity clearly indicates the role of a hydraulic component in the stomatal response mechanism.

11.2 The Methodical Approach

The methods used here were described in detail by Kappen et al. (1987) and Kappen and Haeger (1991). Principally, the system consists of a conditioned gas exchange cuvet ($7\,dm^3$) with CO_2 and H_2O analysis units (H. Walz, Effeltrich, Germany). The system operates with an open gas stream and records transpiration, net photosynthesis, and dark respiration. The leaf, still attached to the intact plant, is sealed within the gas exchange cuvette and fixed by a clamp that can be moved by a microscopic gliding stage (Kappen et al. 1987). This is hand-operated by three external adjustment knobs. Motor-driven remote control was mounted recently and will be used in future experiments (see Omasa et al. 1983).

The conditioning unit of the gas exchange cuvette is laterally mounted so that the microscope device could be inserted through the bottom part of the chamber (Kappen et al. 1987; Fig. 11.2). Images of the stomata are taken either in light from incandescent lamps illuminating the gas exchange cuvette or in physiological darkness by means of IR light (780–1200 nm) that is directed through the microscope objective to the leaf surface. Images are taken by a TV camera and are visible on a computer monitor. Instead of the previously described image-analysis method (area determination from photographic pictures), we are now using digital picture analysis as described by Omasa and Onoe (1984) and Van Gardingen et al. (1989). Our system now includes an AT 286 computer. The image analysis is performed on a RGB computer monitor connected to a digitizing board (Bartscher Elektronik, Eschwege, Germany). For the documentation of images a streamer is used. Further details about hard- and software will be published elsewhere. The CO_2 exchange cuvette is installed in a walk-in environmental chamber, and the electronic control and computer devices are in an adjacent dry room.

Changes in stomatal aperture are recorded over time periods of 40 s. Each image is stored for later analysis. The contrast of the images is intensified by gray color shifting. Although it is well established that stomatal width better describes the stomatal conductance (Meidner and Mansfield 1968), we are at present using area evaluation for comparative terms. Changes in stomatal aperture are due not only to changes in the width alone (e.g., in *Vicia faba*), they also result from an overall change in elliptic parameters, including the long axis, as in *Tradescantia albiflora* (Kappen and Haeger 1991). Thus, the parameters taken as proportional for stomatal response depend on the species observed.

11.3 General Aspects

The data presented in this study illustrate not only the capacity and potentials but also the limitations of a newly developed method. The advantage lies in directly measuring stomatal movements in response to any trigger. These measurements are successful only if the real pore width can be depicted. Stomatal width or aperture can be used as a proportional measure if related to the same stoma in a time series or, with restrictions, to a repeatedly observed identical population of 20 to 50 stomata in the same experiment. Otherwise, relative degrees of opening may be used in order to compare different plant specimens or species.

Plant species can be investigated by our method only if their stomata are large enough (the central aperture must be easily visible in the optical system). A screening test run with leaves of 150 plant species (Schultz 1990) revealed that 43 taxa fulfilled these conditions. Of 61 woody plant species, 13 were suited for further investigations (e.g., *Euonymus europaea*, *Betula pubescens*, *Juglans regia*, *Populus nigra*, *Sambucus nigra*, *Vitis vinifera*) and of 25 hemicryptophytic herbaceous plant species, 12 were suitable (e.g., *Aegopodium podagraria*, *Atropa belladonna*, *Ranunculus repens*). Only a few grass species proved useful (*Avena sativa*, but not *Zea mays*).

Figure 11.3 illustrates the response time needed for stomatal aperture to reach steady state after an experimental increase of the water vapor pressure deficit (VPD) air to leaf from 5 mbar bar^{-1} to 11 mbar bar^{-1} (irradiance was 550 µmol photons m^{-2}s^{-1} PAR). An initial oscillation of the stomatal aperture is eliminated in this graph. The response time varied between 25

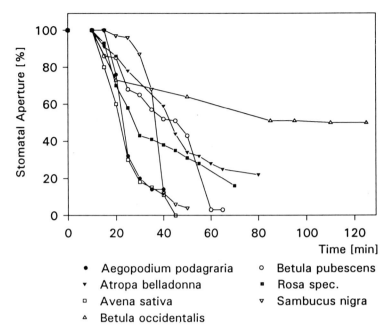

Fig.11.3. Stomatal response (pore width) of seven plant species to changes of VPD (5 to 11 mbar bar^{-1}) in light. Values in % of opening before the experiment

and 80 min according to the plant species. *Avena sativa* and *Sambucus nigra* showed a quick adjustment, while *Atropa belladonna* and *Betula occidentalis* reacted slowly. The degree of closing was very different: high in the first-mentioned species and low in the latter two. This indicates a species-specific sensitivity to decreasing air humidity. The differences may also be due to seasonal variation of the sensitivity, as occurred in a study with *Betula pubescens* and *Sambucus nigra* (B. Potrykus unpubl.). Stomata of these species showed very little closure in late summer and fall.

11.4 Stomatal Responses

11.4.1 Air-Humidity Response

If stomatal movements in response to changing air humidity are observed in more detail, all tested plant species showed a characteristic transient phase before they reached steady state. The transient phase in response to in-creased VPD consisted of an oscillation of the stomatal aperture starting opposite to the stimulus with a further opening and, after 5–30 min, a settling to a closure status proportional to the reduced humidity. The same

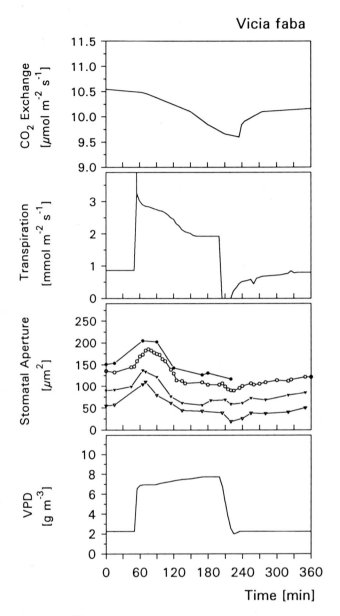

Fig.11.4. CO_2 exchange and response of four single stomata to changes of VPD in light (360 μmol photons m^{-2} s^{-1} PAR), air temperature 16 °C. (After Kappen et al. 1987)

was demonstrated in the reverse direction if VPD was decreased again (see Fig. 11.4). Such a transient phase was found to a greater or lesser extent in all species investigated so far (*Aegopodium podagraria, Atropa belladonna, Avena sativa, Betula pubescens, B. occidentalis, Rosa* sp., *Sambucus nigra,*

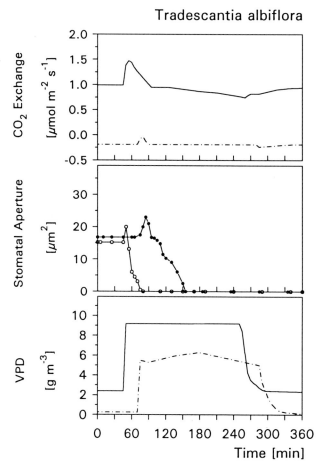

Fig.11.5. CO_2 exchange and response of the stomatal apparatus in the white area of a variegated leaf in light (—— and –○–), (360 μmol photons m^{-2} s^{-1} PAR, temperature 20 °C) and of a green leaf in darkness (–·– and –●–) (temperature 15 °C) to changes in VPD. (After Kappen and Haeger 1991)

Tradescantia albiflora). It also occurred in darkness, if the leaves were subjected to an increase of VPD (Fig. 11.5).

The transient phase of the stomatal apparatus is not always obvious from the net photosynthesis curves (Figs. 11.4, 11.5, 11.6). It also cannot be identified in the curves depicting the stomatal conductance and transpiration, because their oscillations are mainly due to changes in the experimental moisture-control system. However, even in cases when experimental moisture regulation did not temporarily deviate (*Rosa* sp., G. Schultz unpubl.), the transient response was obvious by image analysis.

If the humidity response is combined with soil water stress, it becomes apparent that the oscillations of the transient phase observed in the well-

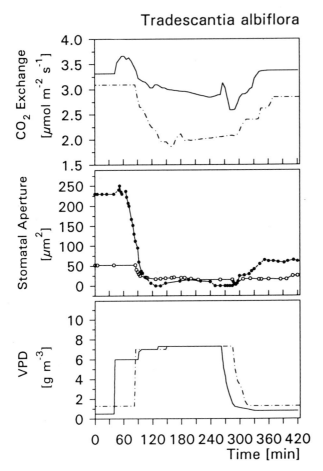

Fig.11.6. CO_2 exchange and response of the stomatal apparatus from a green leaf in light (350 μmol photons m^{-2} s^{-1} PAR, temperature 20 °C) to changes in VPD and at different soil water status: well-watered (—— and –●–), 7 days slightly and the last 2 days before the experiment not watered (–·– and –○–). (After Kappen and Haeger 1991)

watered plant cease with increasing stress (Fig. 11.6). A weak oscillation was seen if the plant was subjected to water stress for 2 days before the experiment. The oscillation disappeared if the plant was slightly watered for 1 week and then received no water at all 2 days before the experiment (Kappen and Haeger 1991).

11.4.2 Response to Changing CO_2 Concentrations of the Air

If air CO_2 concentration was decreased to about half its normal value, stomata of *Vicia faba* responded significantly in light (400 μmol photons

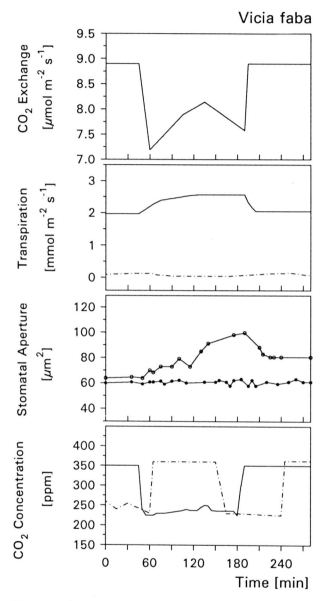

Fig. 11.7. CO_2 exchange and stomatal response (pore width) to changes in air CO_2 concentration in light (360 µmol photons $m^{-2}s^{-1}$ PAR) (—— and –o–) (VPD = 13 mbar bar^{-1}) and in darkness (–·– and –•–) (VPD = 5 mbar bar^{-1}). Dark respiration (constantly -0.3 µmol CO_2 $m^{-2}s^{-1}$) is not shown. Temperature 21 °C. (After Kappen et al. 1987)

m^{-2}s^{-1} PAR) but not in darkness (Fig. 11.7). Even though the leaves were subjected to a reduced air humidity that would normally cause significant stomatal closing, opening of the stomata was observed instead, most likely because photosynthesis created a CO_2 sink, and CO_2 deficiency in the stomatal cavity resulted. It was remarkable to observe a correlation between stomatal aperture and period of CO_2 deficiency. As soon as the CO_2 concentration had returned to normal, the stomata closed but to a little lesser extent than at the beginning of the experiment. The curve in Fig. 11.7 shows a time sequence of only one stomatal apparatus, but other stomata in the same experiment responded identically (Kappen et al. 1987).

CO$_2$ exchange mirrors well the metabolic response to varying CO_2 concentration in the air. Net photosynthesis decreased significantly with de-

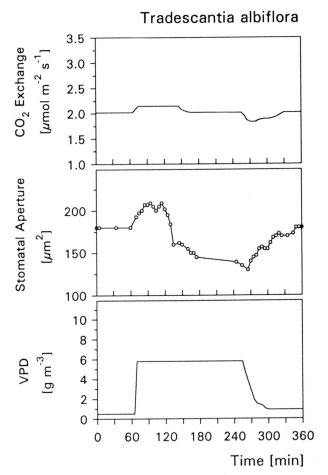

Fig. 11.8. CO$_2$ exchange and response of the stomatal apparatus of a green leaf in light (390 μmol photons m^{-2} s^{-1} PAR, temperature 20 °C) to changes in VPD, but air containing 177 ppm CO$_2$. (After Kappen and Haeger 1991)

creasing CO_2 supply, although stomatal aperture increased. In darkness the respiratory rate was not affected by the decrease in ambient CO_2 concentration. Transpiration and stomatal conductivity, however, were positively correlated with stomatal aperture (Fig. 11.7).

Figure 11.8 illustrates to what extent CO_2 deficiency overrules other influences such as low air humidity. At normal CO_2 concentration, stomata closed significantly if the water vapor pressure difference between ambient and leaf was increased (Fig. 11.6). At a CO_2 concentration of 177 ppm, stomata were closed only very slightly and showed a distinct transient phase with at least two oscillations (Fig. 11.8). These were also apparent after the CO_2 concentration returned to normal. The stomatal response had little effect on the CO_2 uptake rate of the leaf (Fig. 11.8) but transpiratory water loss may have been maintained at a high level (Fig. 11.7). Thus, CO_2 deficiency causes an unfavorable P/T ratio.

11.4.3 Response to Heat

Stomatal aperture and evaporation conditions can independently cause variations in transpiration rates (Schulze et al. 1972). Another factor is temperature. Our direct method confirms that stomata have a tendency to open wider at high leaf temperatures. At superoptimal temperatures (in *Vicia faba* above 35 °C), stomatal aperture was significantly increased, as was transpiration rate. In Fig. 11.9, stomatal conductance is calculated using the formula of Parlange and Waggoner (1970) based on the geometric changes of the stomatal aperture (length, width, and deepness of the stomatal pore). The values correspond well to the measured transpiration rate, but not to leaf conductance calculated from water relations in the chamber by means of the formula given by Ball (1987). The discrepancy between these different ways of determining stomatal conductance was also demonstrated in an experiment with 50 stomata (Vanselow 1990). These observations may stimulate further investigation of the parameters used in the Ball formula.

11.4.4 The Transient Phase and Other Pecularities of the Stomatal Response

The transient phase as a stomatal response "in a direction opposite to the final response" (Raschke 1970) was early recognized in experiments with leaves or leaf sections in osmotic solutions. Mohl (1856) already interpreted it as an antagonism between guard cells and epidermal cells, and Stalfelt (1929) termed it a hydropassive response. Hydropassive movements in response to air humidity changes were considered to be induced by extremely rapid humidity changes (Maier-Maerker 1979b) and thus to be

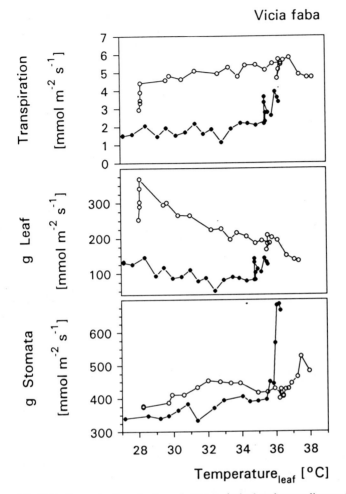

Fig.11.9. Transpiration, leaf conductance (calculated according to Ball 1987) and stomatal conductance (calculated according to Parlange and Waggoner 1970) in steady state with temperatures ranging from 27 to 39 °C in light (–o–) (550 µmol photons m^{-2} s^{-1} PAR) and in darkness (–●–)

abnormal. A weak transient response is visible in one of Hall's and Hoffmann's (1976) experiments with *Helianthus annuus* leaves exposed to an increased water vapor pressure difference; however, these authors did not mention the phenomenon. Our in situ experiments have clearly demonstrated transient phases in the stomatal response of various plant species. The transient phases are clearly a specific mode of response to moisture changes either in the osmoticum or in the air. They did not occur if the plant leaves were exposed to changes in light or in air CO_2 concentration, and they were not obvious if temperature was changed.

Based on his measurements of the cuticular conductance of the leaf, Meidner (1986) predicted transient phases in the stomatal movement induced by VPD changes. A stomatal response to an increase in VPD should indicate a feedback mechanism through an induced imbalance that causes turgor loss in the epidermal cells and an initial opening of the guard cells before a new steady state is reached. Such a hydropassive response is confirmed by the observations that the subsidiary cells of *Tradescantia* become visibly shriveled (Maier-Maerker 1979b; Kappen and Haeger 1991). Nonami et al. (1990), although not referring directly to transient phases, gave evidence by pressure probe neasurements that the stomatal opening is mediated by cellular water transport from the mesophyll to the epidermis and to the guard cells of *Tradescantia virginiana*. They also supported Meidner's (1986) finding that the sensor for the humidity response is not peristomatal transpiration as was postulated by Maier-Maerker (1979a). Nonami et al. (1990) suggest that the hydraulic system is controlled by the transpiratory water vapor loss from the mesophyll cells bordering the stomatal cavity. Therefore one must conclude that such a feedback control functions only when stomata are open; but if they close completely, they cannot react to changed humidity. The transient phase as a part of stomatal response was included by Assmann and Gershenson (1991) in their model describing the water vapor responses of stomata.

The reduction or disappearance of the transient-phase oscillations in response to changes of air humidity for a plant subjected to soil water stress may be due to an accumulation of abscisic acid (Zhang and Davies 1989; Gollan et al. 1989) and to a general loss of turgor.

Our observations show that the tendency of stomata to close is generally strong in darkness, but a certain percentage of stomata remain open or are only slightly closed (*Tradescantia albiflora*: 20%). An increase in VPD forces these open stomata to close in darkness. The pronounced stomatal oscillation during the transient phase demonstrates that these movements are only turgor-related and independent of metabolic processes. Once stomata were closed in darkness, they could not be opened by air humidity changes or even by a decrease in CO_2 concentration (177 ppm CO_2) of the ambient air. This observation and similar observations on stomata of white regions of variegated *Tradescantia* leaves in light lead to the conclusion that photosynthetic energy production is necessary to move the guard cells. This energy may be provided also by the mesophyll.

The strong influence of CO_2 deficiency in the air can be well demonstrated by our method. In light it overrides the humidity response of the stomata and thus may cause a dangerous water loss while the CO_2 uptake is low. The pronounced transient phases with several oscillations show the extent to which humidity and CO_2 effects interact with each other, so the humidity response is not absent, as was concluded by Raschke and Kühl (1969) from their experiment with maize. In darkness, stomata were not forced to open by a decrease in the CO_2 concentration to around 200 ppm (see also Willmer 1983) – that is equal to the air CO_2 concentration during the Ice Ages (Peel 1983) – even though air humidity was high. Stomata obviously react differently if they are subjected to CO_2-free air or pure nitrogen (Louguet 1972; Laffray et al. 1984). The question remains, however, whether other response mechanisms exist. If we assume that stomata open because they react to a CO_2 gradient that is caused by an assimilatory sink, as is quite obvious by stomatal response to low CO_2 concentrations in light, we have little reason to postulate stomatal opening in darkness. The CO_2 concentration in the tissue region is increased by respiration and consequently a gradient to a lower ambient CO_2 concentration may instead cause a tendency to close. If pure nitrogen is supplied, the lack of oxygen would principally change the system (Louguet 1972).

During evaluation of stomatal response to increasing temperature, it was interesting that leaf conductance as calculated by the moisture relations in the gas exchange cuvette differed from that calculated from the observed changes in stomatal aperture dimensions. Unless these differences can be explained by other experiments, they may indicate that the calculation of conductance using the chosen parameters (VPD, temperature) is insufficient, and that other parameters like convection have to be considered as well.

11.5 Conclusions

It is important to state that in all our experiments with herbaceous plants (and in general also with woody plants) the stomata respond in the same way to influences on the whole leaf. The stomatal response that we investigated was also in general accordance with transpiration and CO_2 uptake. Consequently, it is hard to believe that we have seen only an exceptional area of the leaf that accidentally corresponded with the gas exchange signals. The observed stomatal aperture cannot, however, be taken as an absolute measure because steady-state apertures vary to a large degree. This variation is apparently greater in leaves of woody plants than in those of herbaceous plant species (Kappen et al. in preparation). An unequal distribution of stomatal opening (patchiness) was demonstrated in mediterranean sclerophyllous leaves (Beyschlag and Pfanz 1990) and we suggest that it may be due to different steady-state apertures of stomata rather than to a pattern of patches with divergently reacting stomata. However, for the investigation of sclerophyllous plants, we must respect the technical limitations of our method because size and anatomy of these stomata do not allow direct observation.

Acknowledgment. The authors wish to thank Dr. R.J. Palmer Jr. for editing the English version of the manuscript.

References

Assmann SM, Gershenson A (1991) The kinetics of stomatal responses to VPD in *Vicia faba*: electrophysiological and water relations models. Plant Cell Environ 14: 455–465

Ball JT (1987) Calculations related to gas exchange. In: Zeiger E, Farquhar GD, Cowan IR (eds) Stomatal function. Stanford University Press, Stanford, pp 445–476

Beyschlag W, Pfanz H (1990) A fast method to detect the occurrence of nonhomogenous distribution of stomatal aperture in heterobaric plant leaves. Oecologia 82: 52–55

Brunner U, Eller BM (1974) Öffnen der Stomata bei hohen Temperaturen im Dunkeln. Planta 121: 293–302

Cowan IR, Milthorpe FL (1968) Plant factors influencing the water status of plant tissues. In: Kozlowski TT (ed) Water deficits and plant growth, vol I. Academic Press, London, pp 137–193

Elkins CB, Williams GG (1962) Still and timelapse photography of plant stomata. Crop Sci 2: 164–166

Gollan T, Davies WJ, Schurr U, Zhang J (1989) Control of gas exchange: evidence for root-shoot communication on drying soil. Ann Sci For (Paris) 46: 393–400

Grantz DA (1990) Plant response to atmospheric humidity. Plant Cell Environ 13: 667–679

Hall AE, Hoffmann GJ (1976) Leaf conductance response to humidity and water transport in plants. Agron J 68: 876–881

Kappen L, Haeger S (1991) Stomatal responses of *Tradescantia albiflora* to changing air humidity in light and in darkness. J Exp Bot 42: 979–986

Kappen L, Andresen G, Lösch R (1987) In situ observations of stomatal movements. J Exp Bot 38: 126–141

Laffray D, Vavasseur A, Garrec JP, Louguet P (1984) Moist air effects on stomatal movements and related ionic content in dark conditions. Study on *Pelargonium × hortorum* and *Vicia faba*. Physiol Veg 22(1): 29–36

Lange OL, Lösch R, Schulze E-D, Kappen L (1971) Responses of stomata to changes in humidity. Planta 100: 76–86

Lösch R (1977) Responses of stomata to environmental factors – experiments with isolated epidermal strips of *Polypodium vulgare*. I. Temperature and humidity. Oecologia 29: 85–97

Louguet P (1972) Influence de la pression partielle d'oxygène sur la vitesse d'ouverture et de fermeture des stomates de *Pelargonium × hortorum* à l'obscurité. Physiol Veg 10(3): 515–528

Maier-Maerker U (1979a) Additional proof of peristomatal transpiration by hygrophotography and a comprehensive discussion in the light of recent results. Z Pflanzenphysiol 91: 25–43

Maier-Maerker U (1979b) "Peristomatal transpiration" and stomatal movement: a controversial view. III. Visible effects of peristomatal transpiration on the epidermis. Z Pflanzenphysiol 91: 225–238

Meidner H (1986) Cuticular conductance and the humidity response of stomata. J Exp Bot 37:517–525

Meidner H, Mansfield TA (1968) Physiology of stomata. McGraw-Hill, London

Mohl von H (1856) Welche Ursachen bewirken die Erweiterung und Verengung der Spaltöffnungen? Bot Ztg 14: 697–721

Morison JIL (1987) Intercellular CO_2 concentration and stomatal response to CO_2. In: Zeiger E, Farquhar GD, Cowan IR (eds) Stomatal function. Stanford University Press, Stanford, pp 229–252

Nonami H, Schulze ED, Ziegler H (1990) Mechanisms of stomatal movement in response to air humidity, irradiance and xylem water potential. Planta 183: 57–64

Omasa K, Onoe M (1984) Measurement of stomatal aperture by digital image processing. Plant Cell Physiol 25(8): 1379–1388

Omasa K, Hashimoto Y, Aiga I (1983) Observations of stomatal movements of intact plants using an image instrumentation system with a light microscope. Plant Cell Physiol 24(2): 281–288

Parlange JY, Waggoner PE (1970) Stomatal dimensions and resistance to diffusion. Plant Physiol 46: 337–342

Peel DA (1983) Antarctic ice: the frozen time capsule. New Sci 98: 477

Poffenroth M, Green DB, Tallman G (1992) Sugar concentrations in guard cells of *Vicia faba* illuminated with red or blue light. Plant Physiol 98: 1460–1471

Raschke K (1970) Stomatal responses to pressure changes and interruptions in the water supply of detached leaves of *Zea mays* L. Plant Physiol 45: 415–423

Raschke K (1979) Movements of stomata. In: Haupt W, Feinleib ME (eds) Encyclopedia of plant physiology (NS), vol 7. Physiology of movements. Springer, Berlin Heidelberg New York, pp 383–441

Raschke K, Kühl U (1969) Stomatal responses to changes in atmospheric humidity and water supply: experiments with leaf sections of *Zea mays* (L.) in CO_2-free air. Planta 87: 36–48

Schultz G (1990) Möglichkeiten und Grenzen der in situ Beobachtung von Spaltöffnungen mit der Bildschirmmethode. Diplomarbeit, Universität Kiel, 110 pp

Schulze ED (1986) Carbon dioxide and water vapour exchange in response to drought in the atmosphere and in the soil. Ann Rev Plant Physiol 37: 247–274

Schulze ED, Hall AE (1982) Stomatal responses, water loss and CO_2 assimilation rates of plants in contrasting environments. In: Lange OL, Nobel PS, Osmond CB, Ziegler H (eds) Encyclopedia of plant physiology (NS), vol 12B. Physiological plant ecology II. Water relations and carbon assimilation. Springer, Berlin Heidelberg New York, pp 181–230

Schulze ED, Lange OL, Buschbom U, Kappen L, Evenari M (1972) Stomatal responses to changes in humidity in plants growing in the desert. Planta 108: 259–270

Schulze ED, Turner NC, Gollan T, Shackel KA (1987) Stomatal responses to air humidity and to soil drought. In: Zeiger E, Farquhar GD, Cowan IR (eds) Stomatal function. Stanford University Press, Stanford, pp 311–322

Schurr U (1992) Stomatal response to drying soil in relation to changes in the xylem sap composition of *Helianthus annuus*. II. Stomatal sensitivity to abscisic acid imported from the xylem sap. Plant Cell Environ 15: 561–567

Sharkey TD, Ogawa T (1987) Stomatal responses to light. In: Zeiger E, Farquhar GD, Cowan IR (eds) Stomatal function. Stanford University Press, Stanford, pp 195–208

Stalfelt G (1929) Die Abhängigkeit der Spaltöffnungsreaktion von der Wasserbilanz. Planta 8: 287–340

Stalfelt MG (1962) The effect of temperature on opening of the stomatal cells. Physiol Plant 15: 772–779

Ting IP (1987) Stomata in plants with crassulacean acid metabolism. In: Zeiger E, Farquhar GD, Cowan IR (eds) Stomatal function. Stanford University Press, Stanford, pp 353–366

Van Gardingen PR, Jeffree CE, Grace J (1989) Variation in stomatal aperture in leaves of *Avena fatua* L. observed by low-temperature scanning electron microscopy. Plant Cell Environ 12: 887–897

Vanselow RU (1990) Temperaturabhängigkeit der stomatären Reaktion von *Vicia faba*. Diplomarbeit, Universität Kiel, 90 pp

Von Caemmerer S, Farquhar GD (1981) Some relationships between the biochemistry of photosynthesis and the gas exchange of leaves. Planta 153: 376–387

Willmer CM (1983) Stomata. Longman, London, 166 pp

Zeiger E, Farquhar GD, Cowan IR (eds) (1987) Stomatal function. Stanford University Press, Stanford, 503 pp

Zhang J, Davies WJ (1989) Abscisic acid produced in dehydrating roots may enable the plant to measure the water status of the soil. Plant Cell Environ 12: 73–81

12 Carbon Gain in Relation to Water Use: Photosynthesis in Mangroves

M.C. Ball and J.B. Passioura

12.1 Introduction

"Mangrove" is an ecological term referring to a taxonomically diverse association of woody trees and shrubs that form the dominant vegetation in tidal, saline wetlands along tropical and subtropical coasts (Tomlinson 1986). The photosynthetic characteristics of mangroves are clearly those of plants utilizing C_3 photosynthetic biochemistry (Ball 1986). There is, however, a remarkable feature of the gas exchange characteristics of mangroves. Despite growing in environments with an abundant water supply, they transpire slowly and maintain high water use efficiencies for C_3 plants (Ball and Farquhar 1984a,b). These water use characteristics become increasingly conservative with increase in the salinities in which the plants are grown and with increase in the salt tolerance of the species (Ball 1988a). Conservative water use may have adaptive significance for survival in saline environments, but such behavior has far-reaching consequences for plant functioning. Maximizing carbon gain relative to water use is a whole-plant phenomenon involving a complex balance between several levels of plant function: stomatal behavior in relation to photosynthesis, variation in leaf properties in relation to light interception and evaporative demand, and the partitioning of carbon between structures supplying and those consuming carbon-based assimilates (Cowan and Farquhar 1977; Cowan 1986). Differences in water use characteristics thus find expression at all levels of plant form and function, and indeed are major determinants of mangrove forest structure along natural salinity gradients (Ball 1988a).

12.2 Water Relations: Why Be Conservative?

Several possible explanations come to mind for the mangroves' conservatism in water use, which are best discussed against a background of the essentials of plant water relations. The roots of mangroves in coastal environments are bathed in salt water whose osmotic potential is of the order of seawater, namely, $-2.5\,MPa$, but which may at times be much saltier. At low elevations in the intertidal zone, the soil is typically saturated with water so that the suction in the soil is close to zero and the soil water potential is therefore essentially equal to the osmotic potential. During transpiration, which, if it is to be sustained, requires the flow of water from roots to leaves, the leaf water potential must be lower than the soil water potential, and the osmotic potential of the leaves must be lower still if the leaves are to maintain turgor. The frictional losses of water potential associated with the passage of the transpiration stream through the plant may lower the water potential of the leaves to several hundred kilopascals below that of the soil. Typical values of leaf water potential in field-grown mangroves range from -2.5 to -6 MPa (Scholander et al. 1964; Scholander 1968; Rada et al. 1989; Smith et al. 1989). These water potentials are necessarily much lower than those of well-watered plants growing in freshwater, and possibly pose a threat of embolism in the plants' xylem vessels, with a consequent runaway increase in the overall hydraulic resistance, a point to which we will return.

The roots of the plants must, when they extract water from the soil, exclude most of the salt from the transpiration stream, for otherwise there would be a rapid and lethal buildup of salt at the evaporating surfaces in the leaves. Some mangrove species secrete salt from their leaves. These species allow more salt into the xylem than do the nonsecretors, but even secretors typically exclude 90% of the salt in the soil water as it enters the root (Scholander et al. 1962, 1966; Moon et al. 1986; Ball 1988b). The salt that does gain entry largely accumulates (if not secreted) in the vacuoles of the leaf cells, where it is the main contributor to the low osmotic potentials that are required there (Popp 1984a,b; Popp et al. 1984). The great longevity of the leaves of many mangrove species (Saenger and Moverley 1985) implies that the salt balance of the leaves is well controlled.

Given this background of the essentials of the salt and water relations of the plants, what are the processes that seemingly require the plants to have low transpiration rates despite the abundance of albeit salty water, and despite the presumed advantage accruing to a species that could transpire faster, and thereby gain carbon faster, than its conservative cousins?

One possibility is that the filters in the roots that exclude most of the salt from the transpiration stream may be so constituted that they present a very large resistance to the flow of water through them. Rapid transpiration would then induce such a low water potential in the leaves that an impossibly high concentration of solutes in the cells would be required to maintain turgor. However, the root density of mangroves is typically very high (Komiyama et al. 1987) and increases with increasing salinity, albeit at the expense of shoot growth (Ball 1988b). Such increase in root mass per unit leaf area concomitant with decrease in transpiration rates implies that the demands for water by the shoot are met by taking up water slowly over a large root system (Ball 1988b).

A second possibility, related to the first, is that water potentials below about 6 MPa may so tax the integrity of the xylem sap that embolism occurs

(Sperry and Tyree 1988). A low stomatal conductance may have evolved that keeps the water potential above a value that is on the verge of inducing substantial embolism (Tyree and Sperry 1988; Jones and Sutherland 1991). Sperry et al. (1988), working with *Rhizophora mangle*, showed that the stems lost little hydraulic conductivity at xylem pressures less than −6 MPa, but that the loss of conductivity was complete at a xylem pressure of −7 MPa.

A third possibility is that rapid transpiration may induce an excessive flow of salt to the leaves with the result that life of the leaves is excessively short. For a long time, apoplastic uptake of from 1 to 5% of the external salt solution was thought to account for the rates of salt transport to the shoot (Pitman 1977). If this were true, then the concentration of ions in the xylem should remain nearly constant with variation in volume flux, and salt uptake would be coupled with water uptake. However, salt concentrations in the

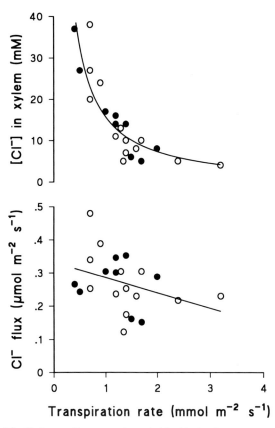

Fig 12.1a–c. Concentration of chloride in the xylem, and chloride flux to the leaves, as a function of transpiration rate, for *Aegiceras corniculatum* (○) and *Avicennia marina* (●). The flux was calculated from the product of concentration and transpiration rate in plants growing in nutrient solution containing 500 mM NaCl. (After Ball 1988b)

xylem decrease hyperbolically with increase in volume flux such that the flux of salt to the leaves does not increase with increase in transpiration rates (Fig. 12.1); indeed, the salt flux to leaves can even decrease at very high rates of water loss (Munns 1985; Ball 1988b). Constancy in salt flux contributes to maintenance of sustainable ion concentrations in leaves. Failure to maintain such ion concentrations would cause premature death of old leaves, and eventually affect growth of new leaves (Munns and Termaat 1986). The evidence on balance implies a high degree of control of salt

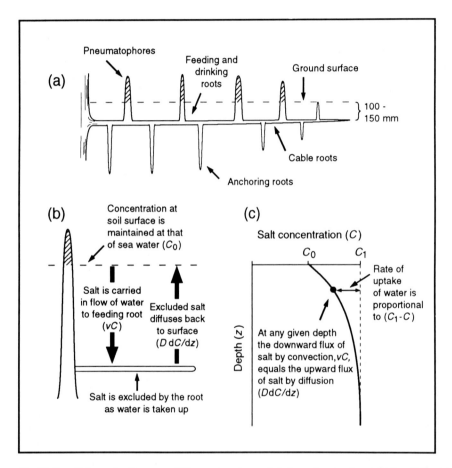

Fig 12.2. a Schematic diagram of the root system of *Avicennia germinans* (After Gill and Tomlinson 1977). The feeding and drinking roots form a dense mat in the top 100–150 mm of the soil. **b** Flow of salt in the zone of the feeding roots. **c** Schematic portrayal of salt concentration vs. depth when a quasi-steady state has been reached in which the downward convection of salt at a given depth (vC) is balanced by the upward diffusion ($D\,dC/dz$), where v is the (depth-dependent) velocity of soil water, and D is the diffusion coefficient of the salt. It is assumed that the rate of water uptake by the roots falls to zero when C equals C_1, which corresponds to the soil potential being equal to the leaf water potential. (After Passioura et al. 1992)

transport in relation to the expansion rate of growing tissues (Ball and Munns 1992).

A fourth possibility concerns the increase in soil salinity that arises when the roots extract water but exclude the salt. In tidal wetlands, where mangroves grow, the soils are typically poorly drained, indeed saturated, owing to their fine texture, the flatness of the landscape, and their regular inundation. As a result, the soils are not well flushed by the ebb and flow of tidal water. When roots extract water from saturated soil, the water must flow downwards from the soil surface towards the roots. Salt is carried, by convection, in this flow of water, but is excluded by the roots as they absorb the water. Thus, the salinity must rise in the soil occupied by the roots. It must keep rising until a sufficiently large concentration gradient develops to diffuse the salt back to the soil surface (where the concentration remains at approximately that of tidal water) as fast as it enters the soil by convection (Fig. 12.2). Calculations based on reasonable assumptions about the distribution and activity of the roots and about the transport characteristics of the salt in the soil suggest that the salt is likely to become so concentrated that it will severely limit the rate of water uptake by the roots (Passioura et al. 1992). The calculated limiting rate is about $1 \, mm \, day^{-1}$, which is but a small fraction of typical potential evaporation rates, and accords with measured values (Miller 1972; Wolanski and Ridd 1986).

Of these possible explanations for the generally low transpiration rate of mangroves, the second and fourth, concerning the danger of runaway embolism and the inhibitory buildup of salt around the roots, seem the most likely. But whatever the explanation, conservative water use has considerable consequences for carbon gain, as we elaborate below.

12.3 Implications of Conservative Water Use for Plant Function

Interspecific differences in water-use characteristics among the Rhizophoraceae are apparent from differences in the slope of assimilation rate with respect to stomatal conductance in response to variation in leaf temperature, irradiance, and vpd (Table 12.1). The slope, dA/dg, increases with increasing salinity tolerance of the species which, in turn, is correlated with a more frugal or conservative use of water. Thus, for a wide range of environmental factors affecting photosynthesis over the course of a day, stomatal conductance at a given assimilation rate is lower (and water use is more conservative) the greater the salinity tolerance of the species (Ball et al. 1988).

Conservative water use has implications for leaf functioning under natural field conditions. Salinity varies both temporally and spatially in a mangrove swamp, but the soil salinity around the roots changes more slowly than does

Table 12.1. Interspecific variation in gas exchange characteristics and in the display and properties of exposed leaves in relation to salinity tolerance in the Rhizophoraceae. Species are listed in order of increasing salinity tolerance, with *Bruguiera gymnorrhiza* being the least salt-tolerant. Values for dA/dg were calculated by linear regression of the assimilation rate (μmol CO_2 m^{-2}s^{-1}) as a function of stomatal conductance to water vapor (mmol m^{-2}s^{-1}) with variation in irradiance, leaf temperature, and leaf-to-air vapor pressure difference. dw dry weight. (After Ball 1988a)

Species	dA/dg (μmol/mmol)	Rosette area (cm^2)		Area per leaf (cm^2)		Specific leaf weight (g dw m^{-2})	Succulence (g water m^{-2})
		Total	Projected	Total	Projected		
Bruguiera gymnorrhiza	0.072	635	356	58	32	133.1	262.5
Rhizophora apiculata	0.096	553	196	69	25	148.8	348.4
Rhizophora stylosa	0.101	419	126	44	13	169.3	387.9
Ceriops tagal	0.113	102	39	8	3	189.2	463.2

the microclimate surrounding the leaves, which has a direct and immediate affect on diurnal water use in relation to carbon gain. The transpiration rate depends on both the leaf conductance to water vapor and the vapor pressure gradient between the leaf and air (vpd). Diurnal variation in vpd is caused mainly by variation in leaf temperature because ambient vapor pressure changes little over the course of a day. The smaller the difference between leaf temperature and air temperature, the more closely will the evaporative demand of the leaf reflect the saturation vapor deficit of the air.

Optimal leaf temperatures for photosynthesis in mangroves are very close to the average air temperatures in the tropical and subtropical environments in which the plants are grown. Both assimilation rate and stomatal conductance are maximal at leaf temperatures ranging from 25 to 30 °C, and decline precipitously with increase in leaf temperature above 35 °C (Moore et al. 1972, 1973; Andrews et al. 1984; Andrews and Muller 1985; Ball et al. 1988). The transpiration rates, even at optimal leaf temperatures, are not sufficient to prevent heating of the leaves above ambient air temperatures during periods of intense insolation. Leaves operating with high transpiration rates can take advantage of the high irradiances required for maintenance of high photosynthetic rates with minimal increase in leaf temperature over air temperature. In contrast, leaves with more conservative water use must avoid high irradiances if leaf temperatures are to be kept within physiologically acceptable limits. In such leaves, avoidance of high light intensities in the middle of the day, when the heat load on the leaf is greatest, would allow the leaves to maintain fairly constant, but low, assimilation rates throughout the day, thus achieving a greater net gain of carbon than if the leaves were horizontal and subject to temperature-dependent inhibition of photosynthesis for extended periods (Cowan 1982).

It follows that increase in leaf angle is a compromise between requirements for illumination and for maintenance of favorable leaf temperatures with minimal evaporative cooling. For example, when exposed canopy leaves of *Rhizophora apiculata* were held in a horizontal position, leaf temperatures increased from 4 to 11 °C above ambient air temperatures of approximately 30 °C, while incident irradiation increased from 1430 to 2085 $\mu mol\,m^{-2}\,s^{-1}$. In contrast, leaves left in their natural, almost vertical, orientation avoided the maximum heat load during midday when irradiance and air temperatures are greatest. During midday, these leaves received only 20% of available sunlight and were approximately 10 °C cooler than they would have been if fully exposed to the sun. Earlier and later in the day, the leaves received about 500 $\mu mol\,m^{-2}\,s^{-1}$ and leaf temperatures were 30 °C, conditions nearly optimal for photosynthesis (Ball et al. 1988). Such maintenance of leaf temperature close to air temperature by avoidance of high irradiance in leaves of *Rhizophora stylosa* was critical to maximizing the total integrated gain of carbon for a minimum expenditure of water during a day (Andrews and Muller 1985). Thus, maximizing carbon gain for a fixed total expenditure of water involves a complex balance between stomatal behavior in relation to photosynthesis and variation in leaf properties in relation to light interception and evaporative demand (Cowan and Farquhar 1977; Cowan 1986).

12.4 Implications of Conservative Water Use for Display and Properties of Leaves

Interspecific differences in the display and properties of foliage reflect the increasingly conservative water-use characteristics associated with increasing salinity tolerance. This is shown by variation in three major characteristics of leaves that contribute to maintenance of favorable leaf temperatures with minimal evaporative cooling. As discussed above, leaf angle affects the radiant heat loading on the leaf. The greater the leaf angle, the lower the proportion of leaf area projected on a horizontal surface. In the Rhizophoraceae, leaf angle in foliage fully exposed to the sun was greater, and hence the proportion of projected leaf area was smaller, the greater the salinity tolerance of the species (Table 12.1). Thus, the species which are more conservative in water use are those that tend most to avoid intense radiation (Ball et al. 1988).

A second leaf property influencing leaf temperature is that of leaf size. Heat convection between a leaf and its environment depends on resistance to transfer imposed by a boundary layer, the characteristics of which are a function of leaf geometry and wind speed. Decrease in leaf size enhances boundary layer conductance and results in the temperature of the leaf being closer to ambient air temperature without putting the leaf at a disadvantage in terms of light interception. Leaf size in the Rhizophoraceae is smallest in the most salt-tolerant (and most water-conservative) species (Table 12.1), and decreases with increasing exposure (Ball et al. 1988). Similarly, leaves of mangroves species that dominate humid low salinity wetlands (e.g., *Heritiera littoralis*, *Rhizophora mucronata*, and *Xylocarpus granatum*) are much larger than those of species that dominate hypersaline environments along the arid coasts of North Australia (e.g., *Avicennia marina*, *Ceriops australis*, *Excoecaria ovalis*, *Lumnitzera racemosa* and *Osbornia octodonta*). Apparently, mangrove leaves are smallest under conditions in which, due to intense radiation and/or limitations to evaporative cooling, they experience the greatest heat load (Ball et al. 1988).

Heat capacity per unit area, which increases with dry weight and water content per unit area, is a third leaf property influencing leaf temperature. Among the Rhizophoraceae, specific leaf weight and succulence, and thus also heat capacity, increase with salinity (Camilleri and Ribi 1983), exposure (Ball et al. 1988), and with increase in the salinity tolerance of the species (Table 12.1). The heat capacities of the leaves in Table 12.1 range from 1.1 to 2.2 $\times$ 10^3 $Jm^{-2}C^{-1}$ in *Bruguiera gymnorrhiza* and *Ceriops australis*, respectively. Leaf temperatures of both species would increase during a lull in air movement because of reduction in boundary layer conductance. However, the rate of temperature increase would be slower in the leaf with a greater heat capacity. Thus, there is a tendency for mangrove leaves to have a greater mass per unit area under conditions in which, due to intense irradiation and/or limitations to evaporative cooling, they would be most vulnerable to rapid fluctuations in leaf temperature (Ball et al. 1988).

The display and properties of foliage contribute to conservative water use, but not without costs to the plant. Increasing the angle of inclination reduces heat loading on a leaf, but is at the expense of light harvesting in that a larger leaf area index is required to intercept a given amount of light. Decrease in leaf size enhances heat transfer rates, but this requires greater investment in supportive and conductive tissue per unit of exposed leaf area than in large leaves. Increase in heat capacity of leaves buffers against rapid changes in temperatures, but at the expense of leaf carbon which might otherwise be invested in expansion of leaf area. Thus, maintenance of favorable leaf temperatures with minimal evaporative cooling is at the expense of the assimilative capacity of the plant, with the expense increasing as water use becomes more conservative (Ball et al. 1988).

12.5 Coping with Excessive Light: Another By-Product of Conservative Water Use

Photoinhibition is light-dependent loss in photosynthetic functioning of photosystem II, which is manifest in whole leaves as a decline in the quantum efficiency of photosynthesis (i.e., mol CO_2 fixed or mol O_2 evolved per mol photons absorbed) under limiting light intensities (see Chap. 10, this Vol.). Photoinhibition occurs when more light is absorbed than can be used in photosynthetic photochemistry (Osmond 1981). One form of photoinhibition is actually photo-protection in that excessive excitation energy is deflected away from PS II and dissipated harmlessly, primarily as heat. Protective dissipation occurs mainly by means of the transthylakoid pH gradient (Krause and Behrend 1986) and xanthophyll cycle pigments (Demmig-Adams 1990), enabling a so-called "down-regulation" to balance the light energy received by PS II with its capacity to use it (Chow 1993).

It follows from the low photosynthetic rates of mangroves that light requirements for maximal photosynthesis are considerably less than the amounts of light available on bright, sunny days. Rates of photosynthesis in field-grown mangrove leaves generally become light-saturated at incident quantum flux densities ranging from 25 to 50% full sunlight (Ball and Critchley 1982; Björkman et al. 1988; Cheeseman et al. 1991), consistent with the light levels received in their normal orientations under field conditions (Ball et al. 1988). Nevertheless, naturally displayed sun leaves can have lower quantum yields than shaded leaves (Björkman et al. 1988; Cheeseman et al. 1991). This depression in quantum yield of leaves naturally receiving high irradiances is correlated with the concentration per unit leaf area of zeaxanthin pigment (Lovelock and Clough 1992; Lovelock and De'ath 1993), implying that the loss of photosynthetic activity at limiting irradiances is due to protective dissipation of excessive excitation energy through the xanthophyll cycle (Demmig-Adams et al. 1989).

Clearly, there are both coarse and fine adjustments to irradiance regimes. The coarse adjustments occur primarily through the display and physical properties of the leaves to avoid prolonged exposure to excessive irradiance, whereas fine adjustments are made at the biochemical level of chloroplast functioning. Naturally, there are limits to the extent to which a leaf can provide protection from excessive irradiance, and when the light absorbed exceeds the capacities for both useful photochemistry and protective dissipation mechanisms, then the leaf becomes vulnerable to another form of photoinhibition which results from photodamage to PS II (Kyle 1984). The relative importance of photoprotection and photodamage to the carbon balance of mangroves with spatial and temporal variation in environmental factors affecting photosynthesis is an exciting area for future research.

12.6   Into the Future: Coping with Global Increase
In Atmospheric CO_2 Concentration

It is difficult to predict changes in climate and sea level due to the greenhouse effect, but change in one environmental factor is certain. The concentration of atmospheric CO_2 is increasing and is expected to double current levels during the latter half of the next century. As previously noted, enhancement of water use efficiency within the constraints of C_3 photosynthetic biochemistry can occur at the expense of the assimilation rate. This is because stomatal closure reducing the efflux of water vapor also reduces the influx of CO_2, causing the leaf to operate with a low intercellular CO_2 concentration and a correspondingly low assimilation rate (Cowan and Farquhar 1977). However, this trade-off between water loss and carbon gain can be greatly ameliorated under elevated atmospheric concentrations of CO_2. Higher external levels of CO_2 enable a leaf to maintain higher intercellular concentrations of CO_2 at lower stomatal conductances, thereby enabling a leaf to operate with a higher water use efficiency without a penalty in terms of carbon gain.

Recent studies by M.C. Ball and H.M. Rawson (unpubl.) indicate that change in the CO_2 concentration, because of its effects on both carbon gain and water use characteristics, may be sufficient to change the structure and function of mangrove vegetation. In tropical Australia, the distribution of mangrove species depends strongly on salinity and aridity (Saenger et al. 1977). Salinity affects the supply of water at the roots, whereas aridity affects the demand for water at the leaves. Growth of two mangrove species, *Rhizophora apiculata* and *R. stylosa*, in response to different combinations of salinity and humidity, was consistent with the distributions of these species along natural salinity and aridity gradients in northern Australia. However, their growth responses to salinity and humidity changed with a doubling in the atmospheric CO_2 concentration. Elevated CO_2 had little effect on growth when salinity was limiting but stimulated growth when it was limited by humidity. Both species benefited most from elevated CO_2 under the low salinity conditions in which they grow well, but under elevated CO_2, growth was enhanced more in the less salt-tolerant and more rapidly growing species, *R. apiculata*. It appears that mangrove species will not expand their distributions into areas in which the salinities exceed current tolerances, but elevated CO_2 may change competitive rankings of species which could change forest composition in relation to salinity regimes, particularly along aridity gradients.

References

Andrews TJ, Muller GJ (1985) Photosynthetic gas exchange of the mangrove, *Rhizophora stylosa* Griff, in its natural environment. Oecologia 65: 449–455

Andrews TJ, Clough BF, Muller GJ (1984) Photosynthetic gas exchange and carbon isotope ratios of some mangroves in North Queensland. In: Teas HJ (ed) Physiology and management of mangroves. Tasks for Vegetation Science, vol 9. Junk, The Hague, pp 15–23

Ball MC (1986) Photosynthesis in mangroves. Wetlands (Aust) 6: 12–22

Ball MC (1988a) Ecophysiology of mangroves. Trees 2: 129–142

Ball MC (1988b) Salinity tolerance in the mangroves, *Aegiceras corniculatum* and *Avicennia marina*. I. Water use in relation to growth, carbon partitioning and salt balance. Aust J Plant Physiol 15: 447–464

Ball MC, Critchley C (1982) Photosynthetic responses to irradiance by the grey mangrove, *Avicennia marina*, grown under different light regimes. Plant Physiol 74: 7–11

Ball MC, Farquhar GD (1984a) Photosynthetic and stomatal responses of two mangrove species, *Aegiceras corniculatum* and *Avicennia marina*, to long term salinity and humidity conditions. Plant Physiol 74: 1–6

Ball MC, Farquhar GD (1984b) Photosynthetic and stomatal responses of the grey mangrove, *Avicennia marina*, to transient salinity conditions. Plant Physiol 74: 7–11

Ball MC, Munns R (1992) Plant responses to salinity under elevated atmospheric concentrations of CO_2. Aust J Bot 40: 515–525

Ball MC, Cowan IR, Farquhar GD (1988) Maintenance of leaf temperature and the optimisation of carbon gain in relation to water loss in a tropical mangrove forest. Aust J Plant Physiol 15: 263–276

Björkman O, Demmig B, Andrews TJ (1988) Mangrove photosynthesis: response to high irradiance stress. Aust J Plant Physiol 15: 43–61

Camilleri JC, Ribi G (1983) Leaf thickness of mangroves (*Rhizophora mangle*) growing in different salinities. Biotropica 15: 139–141

Cheeseman JM, Clough BF, Carter DR, Lovelock CE, Eong OJ, Sim RG (1991) The analysis of photosynthetic performance in leaves under field conditions: a case study using *Bruguiera* mangroves. Photosynth Res 29: 11–22

Chow WS (1993) Photoprotection and photoinhibition. In: Barber J (ed) Molecular processes of photosynthesis, vol 9. Advances in molecular and cell biology. JAI Press, Greenwich, Connecticut (in press)

Cowan IR (1982) Regulation of water use in relation to carbon gain in higher plants. In: Lange OL, Nobel PS, Osmond CB, Ziegler H (eds) Physiological plant ecology II. Water relations and carbon assimilation. Springer, Berlin Heidelberg New York, pp 589–614

Cowan IR (1986) Economics of carbon fixation in higher plants. In: Givinish TJ (ed) On the economy of plant form and function. Cambridge University Press, Cambridge, pp 133–170

Cowan IR, Farquhar GD (1977) Stomatal function in relation to leaf metabolism and environment. In: Jennings DH (ed) Integration of activity in the higher plant. Cambridge University Press, Cambridge, pp 471–505

Demmig-Adams B (1990) Carotenoids and photoprotection in plants: a role for the xanthophyll zeaxanthin. Biochim Biophys Acta 1020: 1–24

Demmig-Adams B, Winter K, Krüger A, Czygan F-C (1989) Zeaxanthin and the induction and relaxation kinetics of the dissipation of excess excitation energy in leaves in 2% O_2, 0% CO_2. Plant Physiol 90: 887–893

Flowers TJ, Hajibagheri MA, Clipson NJW (1986) Halophytes. Q Rev Biol 61: 313–337

Gill AM, Tomlinson PB (1977) Studies on the growth of red mangrove (*Rhizophora mangle* L.). 4. The adult root system. Biotropica 9: 145–155

Jones HG, Sutherland RA (1991) Stomatal control of xylem embolism. Plant Cell Environ 14: 607–612

Komiyama A, Ogino K, Aksornkoae S, Sabharsi S (1987) Root biomass of a mangrove forest in southern Thailand. I. Estimation by the trench method and the zonal structure of root biomass. J Trop Ecol 3: 97–108

Krause GH, Behrend U (1986) ΔpH-dependent chlorophyll fluorescence quenching indicating a mechanism of protection against photoinhibition of chloroplasts. FEBS Lett 200: 298–302

Kyle DJ (1984) The 32 000 dalton Q_B protein of photosystem II. Photochem Photobiol 41: 107–116

Lovelock CE, Clough BF (1992) Influence of solar radiation and leaf angle on xanthophyll concentrations in mangroves. Oecologia 91: 518–525

Lovelock CE, De'ath G (1993) Influence of photosynthetic rate, solar radiation and leaf temperature on zeaxanthin concentrations in mangrove leaves. Plant Cell Environ (in press)

Miller PC (1972) Bioclimate, leaf temperature and primary production in red mangrove canopies in South Florida. Ecology 53: 22–45

Moon GJ, Clough BF, Peterson CA, Allaway WG (1986) Apolastic and symplastic pathways in *Avicennia marina* (Forsk.) Vierh. roots revealed by fluorescent tracer dyes. Aust J Plant Physiol 13: 637–648

Moore RT, Miller PC, Albright D, Tieszen LL (1972) Comparative gas exchange characteristics of three mangrove species in winter. Photosynthetica 6: 387–393

Moore RT, Miller PC, Ehleringer J, Lawrence W (1973) seasonal trends in gas exchange characteristics of three mangrove species. Photosynthetica 7: 387–394

Munns R (1985) Na^+, K^+ and Cl^- in xylem sap flowing to shoots of NaCl-treated barley. J Exp Bot 36: 1032–1042

Munns R, Termaat A (1986) Whole plant responses to salinity. Aust J Plant Physiol 13: 143–160

Osmond CB (1981) Photorespiration and photoinhibition. Some implications for the energetics of photosynthesis. Biochim Biophys Acta 639: 77–89

Passioura JB, Ball MC, Knight JH (1992) Mangroves may salinise the soil and in so doing limit their transpiration rate. Funct Ecol 6: 476–481

Pitman MG (1977) Ion transport into the xylem. Annu Rev Plant Physiol 28: 71–88

Popp M (1984a) Chemical composition of Australian mangroves. I. Inorganic ions and organic acids. Z Pflanzenphysiol 113: 395–409

Popp M (1984b) Chemical composition of Australian mangroves. II. Low molecular weight carbohydrates. Z Pflanzenphysiol 113: 411–421

Popp M, Larher F, Weigel P (1984) Chemical composition of Australian mangroves. III. Free amino acids, total methylated onium compounds and total nitrogen. Z Pflanzenphysiol 114: 15–25

Rada F, Goldstein G, Orozco A, Montilla M, Zabala O, Azocar A (1989) Osmotic and turgor relations of three mangrove ecosystem species. Aust J Plant Physiol 16: 477–486

Saenger P, Moverley J (1985) Vegetative phenology of mangroves along the Queensland coastline. Proc Ecol Soc Aust 13: 257–265

Saenger P, Specht MM, Specht RL, Chapman VJ (1977) Mangal and coastal salt-marsh communities in Australasia. In: Chapman VJ (ed) Wet coastal ecosystems. Elsevier, Amsterdam, pp 293–345

Scholander PF (1968) How mangroves desalinate seawater. Physiol Plant 21: 251–261

Scholander PF, Hammel HT, Hemmingsen EA, Garey W (1962) Salt balance in mangroves. Plant Physiol 37: 722–729

Scholander PF, Hammel HT, Hemmingsen EA, Bradstreet ED (1964) Hydrostatic pressure and osmotic potential in leaves of mangroves and some other plants. Proc Natl Acad Sci USA 52: 119–125

Scholander PF, Bradstreet ED, Hammel HT, Hemmingsen EA (1966) Sap concentrations in halophytes and some other plants. Plant Physiol 41: 529–532

Smith JAC, Popp M, Luttge U, Cram WJ, Diaz M, Griffiths H, Lee HSJ, Medina E, Schafer C, Stimmel K-H, Thonke B (1989) Ecophysiology of xerophytic and halophytic vegetation of a coastal alluvial plain in northern Venezuela. VI. Water relations and gas exchange of mangroves. New Phytol 111: 293–307

Sperry JS, Tyree MT (1988) Mechanism of water stress-induced xylem embolism. Plant Physiol 88: 581–604

Sperry JS, Tyree MT, Donnelly JR (1988) Vulnerability of xylem to embolism in a mangrove vs an inland species of Rhizophoraceae. Physiol Plant 74: 276–283

Tomlinson PB (1986) The botany of mangroves. Cambridge University Press, Cambridge, pp 62–115

Tyree MT, Sperry JS (1988) Do woody plants operate near the point of catastrophic xylem dysfunction caused by dynamic water stress? Answers from a model. Plant Physiol 88: 574–580

Wolanski E, Ridd P (1986) Tidal mixing and trapping in mangrove swamps. Estuarine Coastal Shelf Sci 25: 43–51

13 Photosynthesis as a Tool
for Indicating Temperature Stress Events

W. Larcher

13.1 Introduction

Photosynthesis is a major topic in the biophysical and biochemical approaches to plant physiology, as well as in molecular biology. Photosynthesis forms the central theme in the ecophysiology of carbon assimilation and carbon budgets of plants and plant stands, and is the basis upon which heuristic and prognostic production models are constructed. Additionally, photosynthesis provides an indicator for the quantitative characterization of states of stress and of functional limitations imposed by environmental factors.

It is with good reason that methods for studying photosynthesis are frequently employed in stress physiology. A deterioration in environmental conditions quickly leads to decreased photosynthesis, resulting either from reduced CO_2 uptake due to narrowing of the stomata or from direct inhibition of primary and secondary processes in the chloroplasts. It has long been recognized that changes in photosynthesis provide an earlier indication of stress caused by cold and heat than other criteria of cell stress and damage, such as abnormal respiration and leakiness of biomembranes (Alexandrov 1964; Kislyuk 1964).

In this chapter it is not the intent to explain the mechanisms leading to impairment of photosynthetic function, but rather to present examples demonstrating how, by means of measuring photosynthetic parameters, the impact of extreme temperatures can be identified and quantified.

13.2 Development of Temperature Stress and Characteristic Responses of Photosynthesis

Temperature is an important limiting factor for vigor and distribution of plants. Each individual process is geared to an optimal temperature, above and below which its performance drops. At temperatures too high or too low the photosynthetic yields decrease steadily until CO_2 uptake ceases. After a small and short-lived deviation from the favorable temperature range, the photosynthetic function returns to the previous level of activity. Exposure to more extreme temperatures results in perturbations that are not immediately reversible upon return to optimal conditions. After-effects of temperature stress indicate serious structural and functional impairments that can only be restored to normal by repair processes. For a comprehensive analysis of temperature stress events, the response of the plant must be recorded *not only after* its exposure to extreme temperatures (which is easier to measure), but also *during* the strain: only in this way can readily reversible deviations from normal be distinguished from irreparable pathological impairments.

The following threshold values can be employed for the characterization of progressive temperature stress:

1. The temperature at which the activity of a vital function drops below half its maximal value (A_{50}). For CO_2 uptake, this temperature is easily determined. If the CO_2 uptake sinks frequently or for longer periods of time to below 50% of its optimal value, the cumulative carbon assimilation is severely reduced; the photosynthetic function, however, is merely reduced reversibly. A biochemical criterion that correlates with a decrease in the production yield of agricultural crops is the deviation of key enzymes for

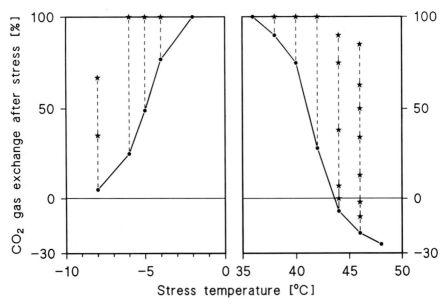

Fig. 13.1. Depression of CO_2 uptake (as percentage of the rate before treatment) of twigs of *Abies alba* following 12-h exposure to frost temperatures in October (*left diagram*) and after 30-min heating at the indicated stress temperatures in winter (*right diagram*). *Broken lines* Recovery during the subsequent days; each *asterisk* represents recovery after 1 day. The low temperature limit of apparent CO_2 uptake was at $-4\,°C$, the high temperature limit was at $38\,°C$. Necrotic damage appeared at -8 and $48\,°C$. (After Pisek and Kemnitzer 1968; Bauer 1972)

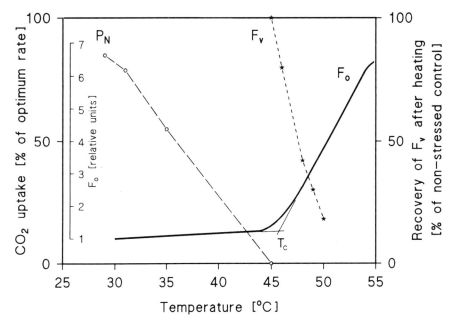

Fig. 13.2. CO_2 uptake (P_N; ○–○) and basic fluorescence (F_o; *solid line*) during progressive heating of leaves of *Persea americana*, and recovery of F_v (★–★) after preheating for 30 min at the temperatures indicated (assessed by measurement of variable fluorescence at 20 °C). The high temperature limit for CO_2 uptake was 45 °C, the breakpoint, T_c, of the F_0-curve was 46 °C. (Larcher et al. 1991a)

metabolic processes from the "thermal kinetic window", defined by Burke et al. (1988) "as the range of plant temperatures at which the apparent Michaelis constant, K_m, is below twice of the observed value at temperature optimum". In analogy to the thermal kinetic window, the temperature range between the thresholds for half-maximal activity of net photosynthesis might be defined as the "photosynthetic thermal window", and be employed for the delimitation of the favorable temperature range for dry matter productivity.

2. The temperature limits for apparent CO_2 uptake. The temperatures at which CO_2 uptake ceases, i.e., the low and the high temperature limits for net photosynthesis, indicate a state of extreme stress. As a rule, after-effects of the depression of photosynthesis are observed if the temperature sinks below or rises above these respective limits (see Fig. 13.2).

3. The temperature limits for a long-lasting impairment of photosynthesis. Subsequent to heat or cold inhibition of CO_2 assimilation, there is usually some delay in the recovery of photosynthesis (Fig. 13.1). This suggests that enzymes involved in the photosynthetic processes have been inactivated (Berry and Raison 1981) or that perturbations of thylakoid organization have occurred (Terzaghi et al. 1989). The temperature limit for

Table 13.1. Temperature thresholds (°C) for apparent CO_2 uptake and survival limits of terrestrial vascular plants during the growing season. Average ranges, based on data of numerous authors

Plant group	Cold injury (LT_i)[a]	Cold limit for CO_2 uptake	Photosynthetic thermal window		Heat limit for CO_2 uptake	Heat injury (LT_i)[a]
			Low A_{50}[b]	High A_{50}[b]		
Woody plants						
Tropical and humid-subtropical trees	−2 to +5	0 to +5	10–15	ca. 40	45–50	50–55
Sclerophyll trees and shrubs of semiarid and arid regions	−5 to −2	−5 to 0	15–20	40–45	45–50	50–60
Deciduous broadleaved trees of temperate zones	ca. −2	ca. −2	8–10	ca. 35	40–45	ca. 50
Evergreen conifers of cool temperate regions	−5 to −3	−5 to −3	ca. 5	ca. 30	35–42	45–50
Dwarf shrubs of heathlands and tundra	−5 to −3	−5 to −3	5–10	ca. 35	40–45	45–50
Herbaceous plants						
C4-Grasses	ca. 0	0 to +5	15–20	35–45	50–60	60–65
C3-Crop plants	−1 to −3	−1 to −3	5–15	35–40	45–50	45–55
CAM plants (at night)	−3 to −5	0 to −3	5–10	20–25	ca. 35	60–65
Winter annual desert plants	−5 to 0		10–15	ca. 40	45–55	50–60
High mountain plants	ca. −5	ca. −5	0–5	ca. 25	38–42	45–55

[a] Temperature at which initial necrotic damage occurs.
[b] 50% of maximum net photosynthesis.

the appearance of after-effects of stress thus indicates considerable destabilization of chloroplast structure and functions.

4. Temperature limits for irreversible damage to the photosynthetic apparatus and for lethality. Lethal injuries appear as cell and tissue necrosis; conventional thresholds are the temperatures at which visual injuries are first recognized (LT$_i$), at 50% damage (LT$_{50}$) and at complete tissue death (LT$_{100}$). Even before acute cell death due to freezing or lethal heat, the integrity of the chloroplast ultrastructure may be severely disrupted. The original photosynthetic activity can no longer be restored and this function ceases. The occurrence of irreversible inactivation of photosynthesis has been used as a convenient viability test in the determination of frost survival limits of difficult objects (Lange 1965; Nash et al. 1987; Rütten and Santarius 1992).

The limiting temperatures of the range over which photosynthesis is impaired, and the temperature limits for lethal injuries, are specific to the individual plants and depend on species and genotype, on morphological,

Table 13.2. Temperature thresholds (°C) for apparent CO_2 uptake for lichens from different environments, in the hydrated state

Species origin	Cold limit for CO_2 uptake	Photosynthetic thermal window		Heat limit for CO_2 uptake	Reference
		Low A$_{50}$	High A$_{50}$		
Dictyonema glabratum (syn. *Cora pavonia*) Paramo, Venezuelan Andes 1770–3600 m a.s.l.	0	10	35	40	Larcher and Vareschi (1988)
Ramalina maciformis Rocks, Negev desert	−6	5	32	37	Lange (1969)
Cladonia alcicornis Sandy terraces, Upper Rhine	−5	−3	18	26	Lange (1965)
Stereocaulon alpinum Moraine, Central Alps 2400 m a.s.l.	−10	−5	18	22	Lange (1965)
Usnea aurantiaco-atra Lichen tundra, maritime Antarctic					Kappen and Redon (1987)
Erect habit	−6	0	25	29	
Prostrate habit	−4	0	18	23	
Usnea sphacelata Continental Antarctic	−10	0	22	28	Kappen (1989)
Buellia sp. Endolithic in rocks Dry valleys, Antarctic	−6	−1	12	16	Kappen and Friedman (1983)

functional, and distributional type (Tables 13.1 and 13.2). These temperatures are not constant but depend on the ontogenetic stage and the state of activity (growth period, dormancy) of the plant. Furthermore, the temperature thresholds vary under the influence of changing environmental conditions within the specific range of the reaction norm.

13.3 Use of Photosynthetic Responses for Determining Heat Tolerance

Heat damage to photosynthesis arises from inactivation of the highly sensitive water-splitting reaction, disconnection of PS II centers from the bulk pigments, thermal uncoupling of photophosphorylation, and biomembrane lesions (Berry and Björkman 1980; Santarius and Weis 1988; McCain et al. 1989; Santarius et al. 1991).

The pronounced sensitivity of photosynthesis to heat can be used to detect early disturbances and injuries caused in green plant tissues. Particularly suitable for this purpose is the use of the *in vivo* chlorophyll fluorescence as an intrinsic indicator of thylakoid organization and changes in membrane fluidity. Measurements of CO_2 exchange are less reliable since they can be influenced by stomatal closure not induced primarily by heat.

When a leaf is heated slowly ($1\,K\,min^{-1}$), the heat-induced curve of the basic fluorescence F_0 shows a sharp discontinuity (Schreiber and Berry 1977). At the threshold temperature at which F_0 begins to increase, the quantum yield for CO_2 fixation suddenly drops. Seeman et al. (1984) found in desert plants, and Larcher et al. (1991a) in *Persea* species, close agreement between the temperature for the heat limit for apparent CO_2 uptake and the critical breakpoint, T_c, in the heat-induced F_0-curve (Fig. 13.2). From a comparison of the F_0-curves as a function of temperature of 26 herbaceous and woody plants of different origin, Bilger et al. (1984) derived a clear relationship between the critical breakpoint temperature and the temperature that causes necrotic damage (LT_{50}) in the leaves after 30 min heating.

Under natural conditions, few studies have been devoted to the temperature thresholds for the heat inactivation of photosynthesis and for heat damage. In the field, overheating of the leaves results from strong insolation. The midday depression of photosynthesis on clear, hot days is due, among other factors, to heat-related reversible photoinhibition (Demmig-Adams et al. 1989). On the other hand, weak light ($30-50\,\mu mol$ photons $m^{-2}\,s^{-1}$) can alleviate a heat-induced inhibition of photosynthesis (Havaux et al. 1991).

A number of different species are able to raise their heat threshold for photosynthetic functioning (Larcher 1980) and for viability by $2-4\,K$ from morning to afternoon (Alexandrov 1977; Kappen 1981). A rise in heat tolerance can be achieved within a few hours by synthesis of polypeptides and heat shock proteins, which also protect the chloroplasts (Sachs and Ho 1986; Yordanov et al. 1989).

13.4 Photosynthetic Function as a Criterion
for Screening Chilling Susceptibility

Exposure to low temperatures brings about phase transitions in the bio-membranes of chilling-susceptible plants (Lyons 1973), with consequent metabolic disorders, especially of photosynthetic function. Only in very rare cases does acute damage occur; in most cases necroses develop during the process of decay following irreversible impairment. The progress of chilling damage depends not only on the degree and duration of cooling, but also on the age of the leaves, the ontogenetic stage of the individual plant, its state of acclimatization, and numerous boundary conditions (e.g., speed of temperature change, soil temperature, air humidity, incoming light during and after chilling). Especially in agriculture and plant breeding there is great interest in an early warning and in quantitative criteria for grading plant species and varieties according to their chilling susceptibility.

A useful method for the recognition of critical threshold temperatures for stress-induced functional deviations and for the onset of pathological processes is provided by the determination of abnormalities in respiration and photosynthetic CO_2 or O_2 exchange rates at low but nonfreezing temperatures. An important criterion for chilling susceptibility is a depression of photosynthesis continuing for hours or even days. In addition to the effect of cold on chloroplast functioning, stomatal closure is also involved in the reduction of CO_2 uptake. The latter cannot only be evoked by an increased internal CO_2 partial pressure, but also by altered water relations (Bauer et al. 1985) and, probably, by hormonal signals. A direct insight into chilling disturbances in chloroplast functioning under non-destructive conditions is provided by biophysical methods, such as leaf absorbance changes, prompt and delayed chlorophyll fluorescence, and photoacoustic signals (Havaux and Lannoye 1985).

In chilling-sensitive plants the thermotropic properties of the biomembranes change at higher temperatures than in chilling-tolerant species. Lateral phase separation and segregation of chlorophyll-protein complexes occur (Berry and Raison 1981; Mäenpää et al. 1988), the water-splitting side of PS II becomes inhibited (Havaux and Lannoye 1984; Shen et al. 1990) and reoxidation of Q_A^- is drastically depressed (Havaux 1987). Such thylakoid impairments are reflected in changes of chlorophyll fluorescence during cooling by an increasing delay in peak fluorescence (Melcarek and Brown 1977; Smillie and Hetherington 1983), a decrease in the variable fluorescence decay following the peak (Havaux and Lannoye 1984; Larcher and Bodner 1987; Lichtenthaler 1988), and a lowering of the photochemical quenching coefficient at steady state (Havaux 1987; Neuner and Larcher 1990).

Employing photosynthetic parameters as stress indicators, the temperature thresholds for chilling damage were determined on related plant

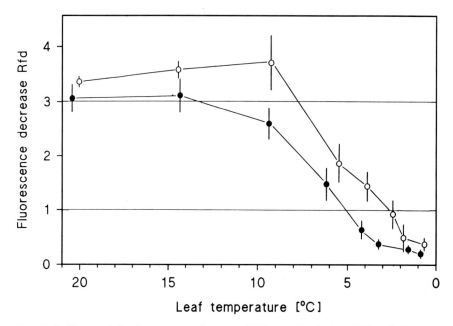

Fig. 13.3. Decay of the fluorescence decrease (*Rfd* a vitality index; Lichtenthaler et al. 1986) at steady state of the photosynthetic induction transient during stepwise cooling from 20 to 0 °C of leaves of soybean plants in the trifoliate stage. Rfd lower than 1 indicates that the leaf no longer exhibits apparent CO_2 uptake. With leaves of cv. Maple Arrow (●) Rfd drops below 1 at 4.8 °C; with those of cv. Evans (○), which is known to be less chilling-susceptible in the field, not until at 2.5 °C. (Neuner and Larcher 1990)

species and cultivars of crop plants: e.g., for wild and cultivated tomatoes (Smillie and Nott 1979; Kamps et al. 1987), wild and cultivated potatoes (Greaves and Wilson 1986), tropical forage C_4 grasses (Havaux 1989), soybean cultivars (Neuner and Larcher 1990; Fig. 13.3), wild *Saintpaulia* species from different altitudes (Bodner and Larcher 1989) and C_4 grasses along an elevational transect in Papua-New Guinea (Earnshaw et al. 1990).

For screening chilling susceptibility, stress criteria based on the assay of photosynthetic function are important since they allow recognition of susceptibility in the noninjurious temperature range. They are not always in entire agreement with the grading based on necrotic chilling injuries since the temperature at which the lipid phase transition of thylakoids occurs may be different from the temperature at which cytoplasmic biomembranes undergo destabilization (Critchley et al. 1978; MacRae et al. 1986; Bodner and Larcher 1989). Therefore in each case the applicability of criteria based on chloroplast functions should be well considered, just as the results of laboratory tests should also be verified by field measurements.

Light during and subsequent to chilling immediately after a cold night has considerable influence on the extent and duration of photosynthetic impairment and of recovery rates. The overriding effect of photoinhibition can lead to different assessments of the susceptibility of cultivars (Smillie et al. 1988; Greer and Hardacre 1989). On the other hand, weak light has a protective effect on photosynthetic function and recovery (Wise et al. 1990; Neuner and Larcher 1991). It is therefore important to differentiate between effects of chilling per se and combined effects of low temperature and light.

13.5 Assay and Analysis of Freezing Events by Monitoring Photosynthesis

Freezing represents a much more serious constraint to cell organization than the mere effect of the low thermodynamic state accompanying cold. Formation of ice crystals in plant tissues sets off contraction and concentration effects which, via toxic accumulation of ions and by membrane lesions, eventually lead to cell death (Heber and Santarius 1973; Steponkus and Webb 1992).

The CO_2 uptake by assimilatory organs is interrupted as soon as ice forms in the tissue, which is between -2 and $-10\,°C$ according to plant species and season (Pisek et al. 1967). If the leaves manage to survive freezing without suffering injury, in many plants photosynthetic activity remains low even after thawing, and this is more pronounced with greater degree and duration of the frost. These post-freezing effects are primarily due to inhibition of chloroplast functions at the ultrastructural and biochemical levels following freeze-dehydration (Senser and Beck 1977; Öquist and Martin 1986). The inhibition of photosynthesis seems to be based on diminished activation of Calvin cycle enzymes; more severe freezing stress impairs the water-oxidation system (Krause et al. 1988).

The onset of tissue freezing is indicated by characteristic changes in the fluorescence induction kinetics. Even before ice nucleation occurs, the fluorescence rise to the peak becomes delayed as the temperature is lowered, and the decrease in the variable chlorophyll fluorescence following the peak disappears. As soon as ice forms in the tissue, the chlorophyll fluorescence attains the height of maximal fluorescence (Melcarek and Brown 1979). If the mesophyll cells have survived the freezing stress, the typical induction transient reappears when the ice thaws. Partial damage to photosynthetic function results in a lower fluorescence yield, whereas in totally frozen tissues the variable fluorescence disappears after thawing (see Fig. 13.4).

By monitoring the course of freezing and thawing via chlorophyll fluorescence important cryophysiological characteristics can be determined:

1. Discrimination between freezing avoidance and freezing tolerance. Plants are able to survive subfreezing temperatures by two basically different mechanisms, i.e., by delay or prevention of ice nucleation in the tissues ("freezing avoidance" according to Levitt 1980), and by tolerance to extra-

Freezing sensitive

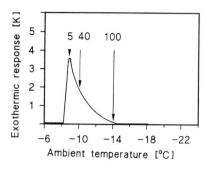

Freezing tolerant

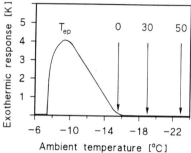

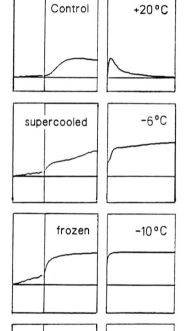

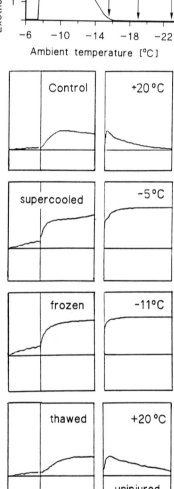

cellular ice formation and freeze-induced dehydration of the protoplasm ("freezing tolerance"). Plants that are only protected by avoidance mechanisms are subject to immediate injury as soon as ice forms in their tissues. Nevertheless, in freezing-sensitive plants of this kind ice nucleation can be prevented for a considerable time by persistent supercooling (Larcher 1982). Supercooling as a survival mechanism is not only important for bud primordia, xylem parenchyma and seeds, but also for leaves of certain plants of regions subject to episodic frosts down to about $-10\,°C$ (e.g., some palms, bamboos, *Olea, Polylepis*; Sakai and Larcher 1987). By means of differential thermal analysis combined with chlorophyll fluorescence measurements, it was shown on *Trachycarpus fortunei*, a palm species growing in warm temperate regions, that the leaves can be persistently supercooled down to $-14\,°C$ and that their frost resistance is based solely upon freezing avoidance (Larcher et al. 1991b). Leaves that survive frost temperatures by maintaining supercooling resume full photosynthetic activity upon rewarming and exhibit no inhibitory after-effects.

2. Detection of different states of acclimation. A moderate improvement in frost resistance can be achieved by freezing point depression and enhanced supercooling. This is normally effected by cold acclimation during the course of a few cooler days. However, lowering of the tissue freezing point and stabilization of the cell membranes can also be observed in connection with a general rise in resistance under the influence of stress factors (e.g., drought). *Vigna unguiculata* plants grown under salt stress (by addition of 100 to 200 mM NaCl to the substrate) were less susceptible to freezing than control plants; in the salinized plants photosynthesis was less depressed at the temperatures below $+5\,°C$ and the tissue-freezing temperature was lowered by 2 to 3 K as compared to nonsalinized plants (Larcher et al. 1990). The rise in temperature resistance under chronic saline stress points to adjustments at the level of proteins and biomembranes, resulting in a general stabilization of protoplasmic structures.

In regions with cold winters there is a regular progression to different levels of hardiness. In woody plants, there is not only a rise in the degree of

---

Fig. 13.4. Progress of freezing in leaves of *Rhododendron ferrugineum*. *Left side* Freezing-sensitive 1-year-old leaves in June. *Right side* Fully frost-hardy, freezing-tolerant leaves in January. *Top diagrams* Typical differential thermal analysis profiles and development of necrotic injuries. Experimental cooling rate was $1\,K\,min^{-1}$. T_{ep} temperature at the peak of the exotherm. *Numbers inside the diagrams* indicate the degree of injury (as %) at various stages of the freezing process assessed by leakage of electrolytes. *Lower diagrams* Original fluorescence transients before cooling ($20\,°C$), at the supercooled (-5 and $-6\,°C$) and the frozen state (-10 and $-11\,°C$), and after thawing. Time scales of the recordings: *left* 0–0.1 s, *middle* 0.1–5 s, *right* 0–60 s. Details of the method are given in Larcher et al. (1991b). A 1-mm-thick cork ring and a thin (0.15 mm) cover glass interposed between the leaf and the light-emitting diode of the SF-10 sensor head (Brancker, Ottawa) of the Pocket Computer Fluorometer (Larcher and Cernusca 1985) delayed heat exchange and prevented the transfer of humidity. (Data by E. Ralser)

frost hardening as winter approaches, but also a changeover from a freezing-sensitive to a freezing-tolerant state of the cells. An example is shown in Fig. 13.4: the course of freezing of unhardened 1-year-old leaves of *Rhododendron ferrugineum* from the alpine tree line is compared with that of frost-hardy leaves in January. The freezing curve, measured by differential thermal analysis, shows the threshold supercooling temperature before ice nucleation in the leaf, the freezing exotherm, and the extent of mesophyll damage during the progress of freezing. In early summer the leaf already suffers considerable damage during the fastest phase of ice formation (peak of exotherm) and when freezing is complete it is dead, whereas in winter the mesophyll is still alive at the end of the phase of ice spread. This indicates that leaves that were freezing-sensitive in June became freezing-tolerant in January. The same result is obtained by a noninvasive method, from the sequence of the fluorescence transients, of which Fig. 13.4 shows only those that are characteristic for optimal temperatures ("control"), for the supercooled state, for frozen leaves, and for recovered and killed leaves after thawing. Since the fluorescence measurements during the process of freezing are all carried out on the same site on the leaf, the transients of any single series are comparable.

3. Investigation of freezing processes under field conditions. To date, our knowledge of the process of freezing, of specific threshold freezing temperatures, of functional types of freezing resistance and of frost hardening rests almost solely upon artificial laboratory tests. Hence it is understandable that agronomists, forestry researchers, and ecologists repeatedly and justifiably

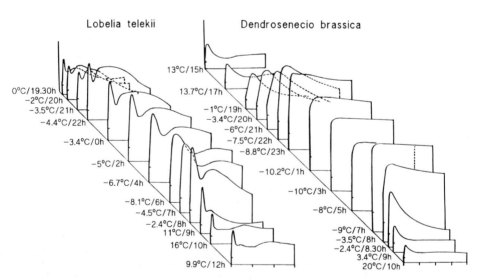

Fig. 13.5. Hourly recorded induction curves of in vivo chlorophyll fluorescence of attached leaves of *Lobelia telekii* and *Dendrosenecio brassica* during frost nights at 4200 m a.s.l. on Mt. Kenya. (After Bodner and Beck 1987)

question the relevance and applicability of such results for plants in the field. The use of portable systems for measuring photosynthesis and data acquisition now make it possible to follow the freezing process in situ (Schroeter et al. 1991). Using the noninvasive chlorophyll fluorescence technique, Bodner and Beck (1987) recorded at an altitude of 4200 m in the upper Teleki valley on Mt. Kenya, the progress of supercooling and freezing of intact leaves of giant Afroalpine rosette plants during several nights. Hourly recorded fluorescence transients during a day-night cycle are presented in Fig. 13.5. In the night of the 24th to 25th February 1985 the leaves of *Lobelia telekii* remained supercooled, and on rewarming the following morning the immediate onset of photosynthetic activity was indicated by normal induction transients. In the more severe frost in the night of the 28th February to March 1st, with temperatures below $-10\,°C$, the leaves of *Dendrosenecio brassica* remained supercooled down to $-8.8\,°C$, after which they froze stiff. The following morning photosynthesis was also completely reactivated. In an environment in which night frosts can occur at any time of the year, including the growing and flowering seasons, the plants have evolved peculiar mechanisms of protection against the effects of apoplastic freezing (Beck et al. 1987). These mechanisms could only be identified and investigated in the field, since these plants did not attain the frost-hardy state when cultivated in a greenhouse.

13.6 Conclusions

The knowledge gained by research on photosynthesis and the development of new methods employing easily used apparatus for the nondestructive measurement of photosynthetic parameters mark a great step forward not only in physiological ecology, but also in botanical stress physiology and pathology. Since photosynthesis is especially easily disturbed, it is recognized as being a sensitive, early warning indicator for many types of stress events, and especially for temperature stress.

The purpose of this chapter is to show the way in which deviations from normal photosynthesis or its component processes can provide us with information about the nature and degree of stress undergone by a plant. Nevertheless, measurement and analysis of photosynthetic perturbations, although they provide valuable insight into stress events, can only give a valid picture of the state of stress in combination with additional diagnostic data.

In the case of heat stress, procedures involving gas exchange analysis and in vivo chlorophyll fluorometry give clear and reliable information about the limits of metabolic performance and of viability. The onset of low temperature stress in chilling-susceptible plants can also be detected conveniently by measuring photosynthesis, whereas for determining severe impairment additional stress criteria such as respiratory anomalies, protein catabolism, and leakage through cellular membranes have to be employed. For monitoring the progress of freezing and thawing and for analyzing frost survival mechanisms, methods providing information about the integrity of the photosynthetic membranes, combined with methods for detecting ice formation, are particularly appropriate.

Above all, the highly developed techniques for studying photosynthesis are of great assistance in the field. Even if, particularly under natural conditions, a mechanistic analysis of all phenomena is not possible, such methods nevertheless provide relevant and important clues concerning conditions during stress as well as information about specific threshold temperatures for survival.

References

Alexandrov VY (1964) Cytophysiological and cytoecological investigations of resistance of plant cells toward the action of high and low temperature. Q Rev Bio 39: 35–77

Alexandrov VY (1977) Cells, molecules and temperature. Conformational flexibility of macromolecules and ecological adaptation. Springer, Berlin Heidelberg New York

Bauer H (1972) CO_2-Gaswechsel nach Hitzestress bei *Abies alba* Mill. und *Acer pseudoplatanus* L. Photosynthetica 6: 424–434

Bauer H, Wierer R, Hatheway WH, Larcher W (1985) Photosynthesis of *Coffea arabica* after chilling. Physiol Plant 64: 449–454

Beck E, Scheibe R, Hansen J (1987) Mechanisms of freezing avoidance and freezing tolerance in tropical alpine plants. In: Li PH (ed) Plant cold hardiness. Liss, New York, pp 155–168

Berry J, Björkman O (1980) Photosynthetic response and adaption to temperature in higher plants. Annu Rev Plant Physiol 31: 491–543

Berry JA, Raison JK (1981) Responses of macrophytes to temperature. In: Lange OL, Nobel PS, Osmond CB, Ziegler H (eds) Encyclopedia of plant physiology, vol 12A. Springer, Berlin Heidelberg New York, pp 277–338

Bilger HW, Schreiber U, Lange OL (1984) Determination of leaf heat resistance: comparative investigation of chlorophyll fluorescence changes and tissue necrosis methods. Oecologia 63: 256–262

Bodner M, Beck E (1987) Effect of supercooling and freezing on photosynthesis in freezing tolerant leaves of Afroalpine "giant rosette" plants. Oecologia 72: 366–371

Bodner M, Larcher W (1989) Chilling susceptibility of wild *Saintpaulia* species of different altitudinal origin. Angew Bot 63: 501–512

Burke JJ, Mahan JR, Hatfield JL (1988) Crop-specific thermal kinetics windows in relation to wheat and cotton biomass production. Agron J 80: 553–556

Critchley Ch, Smillie RM, Patterson BD (1978) Effect of temperature on photoreductive activity of chloroplasts from passionfruit species of different chilling sensitivity. Aust J Plant Physiol 5: 443–448

Demmig-Adams B, Adams III WW, Winter K, Meyer A, Schreiber U, Pereira JS, Krüger A, Czygan FC, Lange OL (1989) Photochemical efficiency of photosystem II, photon yield of O_2 evolution, photosynthetic capacity, and carotenoid composition during the midday depression of net CO_2 uptake in *Arbutus unedo* growing in Portugal. Planta 177: 377–387

Earnshaw MJ, Carver KA, Gunn TC, Kerenga K, Harvey V, Griffiths H, Broadmeadow MSJ (1990) Photosynthetic pathway, chilling tolerance and cell sap osmotic potential values of grasses along an altitudinal gradient in Papua New Guinea. Oecologia 84: 280–288

Greaves JA, Wilson JM (1986) Assessment of the non-freezing cold sensitivity of wild and cultivated potato genotypes by chlorophyll fluorescence analysis. Potato Res 29: 509–520

Greer DH, Hardacre AK (1989) Photoinhibition of photosynthesis and its recovery in two maize hybrids varying in low temperature tolerance. Aust J Plant Physiol 16: 189–198

Havaux M (1987) Effects of chilling on the redox state of the primary electron acceptor Q_A of photosystem II in chilling-sensitive and resistant plant species. Plant Physiol Biochem 25: 735–743

Havaux M (1989) Fluorimetric determination of the genetic variability existing for chilling tolerance in sweet sorghum and Sudan grass. Plant Breed 102: 327–332

Havaux M, Lannoye R (1984) Effects of chilling temperatures on prompt and delayed chlorophyll fluorescence in maize and barley leaves. Photosynthetica 18: 117–127

Havaux M, Lannoye R (1985) In vivo chlorophyll fluorescence and delayed light emission as rapid screening techniques for stress tolerance in crop plants. Z Pflanzenzücht 95: 1–13

Havaux M, Greppin H, Strasser RJ (1991) Functioning of photosystems I and II in pea leaves exposed to heat stress in the presence or absence of light. Planta 186: 88–98

Heber U, Santarius KA (1973) Cell death by cold and heat, and resistance to extreme temperatures. Mechanisms of hardening and dehardening. In: Precht H, Christophersen J, Hensel H, Larcher W (eds) Temperature and life. Springer, Berlin Heidelberg New York, pp 232–292

Kamps TL, Isleib TG, Herner RC, Sink KC (1987) Evaluation of techniques to measure chilling injury in tomato. Hortic Sci 22: 1309–1312

Kappen L (1981) Ecological significance of resistance to high temperature. In: Lange OL, Nobel PS, Osmond CB, Ziegler H (eds) Encyclopedia of plant physiology, vol 12A. Springer, Berlin Heidelberg New York, pp 439–474

Kappen L (1989) Field measurements of carbon dioxide exchange of the Antarctic lichen Usnea sphacelata in the frozen state. Antarct Sci 1: 31–34

Kappen L, Friedmann EI (1983) Ecophysiology of lichens in the dry valleys of southern Victoria Land, Antarctica. II. CO_2 gas exchange in cryptoendolithic lichens. Polar Biol 1: 227–232

Kappen L, Redon J (1987) Photosynthesis and water relations of three maritime antarctic lichen species. Flora 179: 215–229

Kislyuk IM (1964) Issledovanie povrezhdayushchego deistviya okhlazhdeniya na kletki listev rastenii, chuvstvitelnykh k kholodu. Nauka, Moskau

Krause GH, Grafflage S, Rumich-Bayer S, Somersalo S (1988) Effects of freezing on plant mesophyll cells. In: Long SF, Woodward FI (eds) Plants and temperature. Comp Biol, Cambridge, pp 311–327

Lange OL (1965) Der CO_2-Gaswechsel von Flechten bei tiefen Temperaturen. Planta 64: 1–19

Lange OL (1969) Experimentell-ökologische Untersuchungen an Flechten der Negev-Wüste. I. CO_2-Gaswechsel von Ramalina maciformis (Del.) Bory unter kontrollierten Bedingungen im Laboratorium. Flora 158 B: 324–359

Larcher W (1980) Klimastress im Gebirge – Adaptationstraining und Selektionsfilter für Pflanzen. Rheinisch-Westfälische Akad Wiss, Vortr N 291. Westdeutscher Vlg, Leverkusen, pp 49–88

Larcher W (1982) Typology of freezing phenomena among vascular plants and evolutionary trends in frost acclimation. In: Li PH, Sakai A (eds) Plant cold hardiness and freezing stress, vol 2. Academic Press, New York, pp 417–426

Larcher W, Bodner M (1987) Criteria for chilling stress in Saintpaulia ionantha. Angew Bot 61: 309–323

Larcher W, Cernusca A (1985) Mikrocomputergesteuerte mobile Anlage zum fluorometrischen Nachweis von Photosynthesestörungen. Sitzungsber Österr Akad Wiss Math-Naturwiss Kl I 194: 45–64

Larcher W, Vareschi V (1988) Variation in morphology and functional traits of *Dictyonema glabratum* from contrasting habitats in the Venezuelan Andes. Lichenologist 20: 269–277

Larcher W, Wagner J, Thammathaworn A (1990) Effects of superimposed temperature stress on in vivo chlorophyll fluorescence of *Vigna unguiculata* under saline stress. J Plant Physiol 136: 92–102

Larcher W, Wagner J, Neuner G, Mendez M, Jimenez MS, Morales D (1991a) Thermal limits of photosynthetic function and viability of leaves of *Persea indica* and *Persea americana*. Acta Oecol 12: 529–541

Larcher W, Meindl U, Ralser E, Ishikawa M (1991b) Persistent supercooling and silica deposition in cell walls of palm leaves. J Plant Physiol 139: 146–154

Levitt J (1980) Responses of plants to environmental stresses, vol I. Chilling, freezing, and high temperature stresses, 2nd edn. Academic Press, New York

Lichtenthaler HK (1988) In vivo chlorophyll fluorescence as a tool for stress detection in plants. In: Lichtenthaler HK (ed) Applications of chlorophyll fluorescence. Kluwer, Dordrecht, pp 129–142

Lichtenthaler HK, Buschmann C, Rinderle U, Schmuck G (1986) Application of fluorescence in ecophysiology. Radiat Environ Biophys 25: 297–308

Lyons JM (1973) Chilling injury in plants. Annu Rev Plant Physiol 24: 445–466

MacRae EA, Hardacre AK, Ferguson IB (1986) Comparison of chlorophyll fluorescence with several other techniques used to assess chilling sensitivity in plants. Physiol Plant 67: 659–665

Mäenpää P, Aro EM, Somersalo S, Tyystjärvi E (1988) Rearrangement of the chloroplast thylakoid at chilling temperature in the light. Plant Physiol 87: 762–766

McCain DC, Croxdale J, Markley JL (1989) Thermal damage to chloroplast envelope membranes. Plant Physiol 90: 606–609

Melcarek PK, Brown GN (1977) Effects of chill stress on prompt and delayed chlorophyll fluorescence from leaves. Plant Physiol 60: 822–825

Melcarek PK, Brown GN (1979) Chlorophyll fluorescence monitoring of freezing point exotherms in leaves. Cryobiology 16: 69–73

Nash TH, Kappen L, Lösch R, Larson DW, Matthes-Sears U (1987) Cold resistance of lichens with *Trentepohlia*- or *Trebouxia* photobionts from the North American West coast. Flora 179: 241–251

Neuner G, Larcher W (1990) Determination of differences in chilling susceptibility of two soybean varieties by means of in vivo chlorophyll fluorescence measurements. J Agron Crop Sci 164: 73–80

Neuner G, Larcher W (1991) The effect of light, during and subsequent to chilling, on the photosynthetic activity of two soybean cultivars, measured by in vivo chlorophyll fluorescence. Photosynthetica 25: 257–266

Öquist G, Martin B (1986) Cold climates. In: Baker NR, Long SP (eds) Photosynthesis in contrasting environments. Elsevier, Amsterdam, pp 237–293

Pisek A, Kemnitzer R (1968) Der Einfluß von Frost auf die Photosynthese der Weißtanne (*Abies alba* Mill.). Flora 157 B: 314–326

Pisek A, Larcher W, Unterholzner R (1967) Kardinale Temperaturbereiche der Photosynthese und Grenztemperaturen des Lebens der Blätter verschiedener Spermatophyten. I. Temperaturminimum der Netto-Assimilation, Gefrier- und Frostschadensbereiche der Blätter. Flora 157 B: 239–264

Rütten D, Santarius KA (1992) Age-related differences in frost sensitivity of the photosynthetic apparatus of two *Plagiomnium* species. Planta 187: 224–229

Sachs MM, Ho THD (1986) Alteration of gene expression during environmental stress in plants. Annu Rev Plant Physiol 37: 363–376

Sakai A, Larcher W (1987) Frost survival of plants. Responses and adaptation to freezing stress. Springer, Berlin Heidelberg New York

Santarius KA, Weis E (1988) Heat stress and membranes. In: Harwood JL, Walton TJ (eds) Plant membranes – structure, assembly and function. Biochem Soc, London, pp 97–112

Santarius KA, Exner M, Thebud-Lassak R (1991) Effects of high temperature on the photosynthetic apparatus in isolated mesophyll protoplasts of *Valerianella locusta* (L.) Betcke. Photosynthetica 25: 17–26

Schreiber U, Berry JA (1977) Heat-induced changes of chlorophyll fluorescence in intact leaves correlated with damage of the photosynthetic apparatus. Planta 136: 233–238

Schroeter B, Kappen L, Moldaenke C (1991) Continuous in situ recording of the photosynthetic activity of Antarctis lichens – established methods and a new approach. Lichenologist 23: 253–265

Seemann JR, Berry JA, Downton WJS (1984) Photosynthetic response and adaptation to high temperature in desert plants. A comparison of gas exchange and fluorescence methods for studies of thermal tolerance. Plant Physiol 75: 364–368

Senser M, Beck E (1977) On the mechanisms of frost injury and frost hardening of spruce chloroplasts. Planta 137: 195–201

Shen JR, Terashima I, Katoh S (1990) Cause for dark, chilling-induced inactivation of photosynthetic oxygen-evolving system in cucumber leaves. Plant Physiol 93: 1354–1357

Smillie RM, Hetherington SE (1983) Stress tolerance and stress-induced injury in crop plants measured by chlorophyll fluorescence in vivo. Plant Physiol 72: 1043–1050

Smillie RM, Nott R (1979) Assay of chilling injury in wild and domestic tomatoes based on photosystem activity of the chilled leaves. Plant Physiol 63: 796–801

Smillie RM, Hetherington SE, He J, Nott R (1988) Photoinhibition at chilling temperatures. Aust J Plant Physiol 15: 207–222

Steponkus PL, Webb MS (1992) Freeze-induced dehydration and membrane destabilization in plants. In: Somero GN, Osmond CB, Bolis CL (eds) Water and life. Springer, Berlin Heidelberg New York, pp 338–362

Terzaghi WB, Fork DC, Berry JA, Field CB (1989) Low and high temperature limits to PS II. A survey using trans-parinaric acid, delayed light emission, and F_0 chlorophyll fluorescence. Plant Physiol 91: 1494–1500

Wise RR, Terashima I, Ort DR (1990) The effect of chilling in the light on photophosphorylation. Analysis of discrepancies between in vitro and in vivo results. Photosynth Res 25: 137–139

Yordanov IT, Goltsev V, Doltchinkova V, Kruleva L (1989) Effect of some polyamines on the functional activity of thylakoid membranes. Photosynthetica 23: 314–323

14 Air pollution, Photosynthesis and Forest Decline: Interactions and Consequences

U. Heber, W. Kaiser, M. Luwe, G. Kindermann,
S. Veljovic-Javonovic, Z. Yin, H. Pfanz, and S. Slovik

14.1 Introduction

Industrialization has led, and is still leading, to the emission of large quantities of sulfur dioxide and nitrogen oxides in many countries. Atmospheric photochemistry has added ozone, which is formed by the interaction between light, NO_2, and oxygen, and is one of the strongest oxidants known. There are also other air pollutants with the potential to damage plants, but since, according to Paracelsus, "the dose defines a poison", only those pollutants can cause plant injury which occur at concentrations high enough to produce appreciable diffusional fluxes into the plants, fluxes that cannot be mastered by the chemical fluxes engaged in detoxifying entering pollutants.

Forest decline has become a matter of public concern in many industrialized countries. Nevertheless, it has been, and still is, a matter of dispute whether or not air pollution and forest decline are causally related. Understanding possible interactions is complicated by the fact that agricultural productivity has remained high in countries where forest decline is extensive. The multiplicity of factors acting in any ecosystem is another reason why causal analysis of the phenomenon of forest decline, which requires separation of individual factors, is so difficult to achieve. In this situation, analysis must be based on a combination of fieldwork, laboratory experiments under controlled conditions, which necessarily deviate from those in the field, and, finally, on data integration and deduction.

In the following, we attempt to justify our conviction that, depending on climatic conditions and exposure, ozone and SO_2 are important both in triggering and causing, in part, the forest decline observed in many mountainous areas of central Europe and Northern America. The main emphasis will be placed here on a consideration of problems caused by SO_2, because analysis of ozone effects has not yet progressed as far as that of SO_2 effects. Nevertheless, owing to the chemical reactivity of ozone, problems caused by ozone are well defined in principle, although much remains to be done before quantitative conclusions can be drawn. For SO_2, a quantitative approach appears to be possible already at present (Slovik et al. 1992a,b). Although both SO_2 and ozone are known to affect photosynthesis of leaves at high concentrations, there is good reason to doubt that photosynthesis of leaves at high concentrations, there is good reason to doubt that photosynthesis is a primary target at ambient concentrations. For this reason, our discussion does not focus on the direct effects of air pollutants on photosynthesis which can often be observed only at concentrations far above ambient concentrations. Rather, it considers the role of chloroplasts in detoxification of air pollutants within a broader context.

14.2   Sites of Interaction of Air Pollutants with Plants

Leaves are highly effective gas exchange systems when stomata are open. When they are closed, gas exchange is severely curtailed. Epidermal tissues are protected, to different degrees, by thick cell walls incrusted with cutin-like substances or suberin, or by a cuticle which may be covered by waxes. These layers act to slow diffusion and to intercept reactive agents, preventing them from reaching living tissues at concentrations which are toxic (Lendzian 1984; Lendzian and Kerstiens 1988, 1991). In consequence, all gaseous air pollutants act mainly on internal tissues in direct contact with the intercellular air space of leaves and needles.

Since the leaf area index in forests, i.e., the ratio of total projected leaf area to the underlying ground area, is usually high, interaction of gaseous air pollutants with the soil is negligible compared with interactions in the canopy. With rain or fog, air pollutants dissolve in droplets according to Henry's distribution law, thereby effectively cleaning the air. This alters the site of impact on plants. SO_2 and NO_2 are hydrated in water to form sulfurous, nitrous, and nitric acid. Ozone decomposes in water, finally yielding H_2O_2 and H_2O. Sulfurous and nitrous acid are oxidized, producing sulfuric and nitric acid. Since the surface tension of water revents aqueous solutions from entering leaves through stomata (Ziegler 1988), rain water, after wetting those plant surfaces which are not water-repellent (thereby exchanging protons of the acids with available cations; cf. Mitterhuber et al. 1989) reaches the soil, where acids interact with soil particles which may be considered as complex cation and anion exchangers. According to the Hofmeister lyotropic power series, which defines relative exchange strength and groups cations in this order

$$Al^{3+} > H^+ > Ca^{2+} > Mg^{2+} > K^+ = NH_4^+ > Na^+,$$

protons of the acids exchange with cations of the soil particles. Liberated cations are, together with their counter anions, primarily nitrate, chloride, and sulfate, subject to leaching, which depletes the soil of essential nutrients (Ulrich 1980). It is doubtful whether direct effects of acidification on root growth are generally important (Kreutzer and Göttlein 1991), but, as will be shown, decreased cation availability is an important factor in the toxicity of gaseous air pollutants which enter plants via leaves or needles.

14.3 The Magnitude of Fluxes into Leaves

Henry's distribution law not only determines to what extent the air is cleaned by precipitation, it also determines the rate of uptake of air pollutants by leaves. High solubility of a reactive gas in aqueous or lipophilic cellular phases, where it undergoes hydration or other reactions, leads to fast diffusional flux from the atmosphere to the site of solubilization. For SO_2 and ozone, the stomata are the main diffusion barrier (Pfanz et al. 1987; Laisk et al. 1989). This makes an approximate calculation of fluxes into leaves easy, because concentrations in the intercellular gas phase are lowered by fast solubilization to a negligible level. This assumption is valid only for SO_2, and for ozone because of its high reactivity, but not for NO_2. In this situation, a simplified version of Fick's law can be used to calculate fluxes of SO_2 or ozone into leaves. It is sufficient to know the pollutant concentrations in the atmosphere outside the leaves and to determine the sum of the boundary layer and stomatal flux resistances. They can be calculated from transpiration measurements (Nobel 1983). Measured resistances to the flux of H_2O must then be divided by 0.53 for SO_2 and by 0.61 for ozone. The flux F is directly proportional to the concentration c and inversely proportional to the sum of the flux resistances R_{Sum}:

$$F = c/R_{Sum}.$$

NO_2 fluxes are slower than indicated by the equation because NO_2 concentrations in the intercellular gas phase of the leaf interior cannot be neglected. When stomata are reasonably open, boundary layer and stomatal flux resistances are, on a slightly windy and sunny day, often between 2 and $10 \, s \, cm^{-1}$ in herbaceous plants and between 10 and $20 \, s \, cm^{-1}$ in woody plants. At a pollutant concentration in air of 100 ppb, which is rare for SO_2 but not uncommon for ozone on a sunny summer day, fluxes are then between 0.4 and $2 \, pmol \, s^{-1} \, cm^{-2}$ leaf area or, on a chlorophyll basis, roughly between 40 and 200 nmol in herbaceous plants $(mg \, chlorophyll)^{-1} \, h^{-1}$ ($25 \, cm^2$ leaf area often contain 1 mg chlorophyll). In woody plants, the fluxes are between 20 and 40 nmol $(mg \, chlorophyll)^{-1} \, h^{-1}$.

Reactions capable of detoxifying air pollutants entering the leaf must not only be fast enough to handle these fluxes, but they must in addition be able to outcompete damaging reactions. The detoxifying reactions must therefore possess a reaction capacity far in excess of the actual reaction rate required by the influx rate of pollutants.

14.4 Toxicity

The toxicity of air pollutants is a consequence of their reactivity which enables them to undergo fast general or specific reactions with cellular constituents (Elstner 1990). For instance, SO_2 reacts with water to form sulfurous acid. Since acidification must be avoided, bases need to be produced for neutralization. Anions formed slightly above neutrality are HSO_3^- (bisulfite) and SO_3^{2-} (sulfite). In the presence of oxygen radicals, which are normal products of photosynthetic oxygen reduction in the Mehler reaction, an electron may be stripped of sulfite. This results in the highly reactive sulfur trioxide anion radical which, together with oxygen radicals, initiates a radical chain reaction which is capable of oxidizing sulfite to sulfate at extremely high rates (Asada and Takahashi 1987). Within the chain reaction, and actually terminating those reaction chains, cellular materials are oxidized. Other reactions of sulfite, for instance the reaction with carbonyl and disulfide groups, lead to the formation of addition compounds. Sulfitolysis of oxidized thioredoxin interferes with regulation and activity of Calvin cycle enzymes (Würfel et al. 1990). There is abundant literature on other SO_2 effects (e.g., Ziegler 1975; Hällgren 1978; Rennenberg 1984).

NO_2 has by itself a radical structure. Together with the nitrite which is formed during its hydration, it is an oxidant (Elstner 1984). However, much stronger oxidants, actually some of the strongest oxidants known, are ozone and the hydroxyl radical which is formed when ozone reacts with water. They attack oxidizable cell constituents. In a specific reaction, ozone splits the double bonds of fatty acid residues in biomembranes forming ozonides which are then subject to degradation. All these reactions are potentially damaging. There is growing evidence that ambient levels of ozone decrease productivity in tree and crop species (Reich and Amundson 1985; Matyssek et al. 1992).

14.5 Detoxification

To be effective, detoxifying reactions must intercept reactive air pollutants or their toxic products before cell damage can occur. Cells possess mechanisms to scavenge radicals and to cope with oxidative stress. Ascorbate and glutathione are not only important radical scavengers, they are also easily oxidized by less aggressive oxidants. In contrast to glutathione, ascorbate is not only localized in the cytoplasm and the vacuole of cells, but, at lower concentrations than in the cytoplasm, also outside the plasmalemma in the apoplasm (Polle et al. 1991; Takahama et al. 1993; Luwe et al. 1993). Apoplasmic phenolics also act to intercept ozone or hydroxyl radicals, or to reduce hydrogen peroxide, which can be formed from ozone (Takahama et al. 1993). Particularly well protected against oxidative stress are chloroplasts (Halliwell and Foyer 1978; Halliwell 1978; Asada and Takahashi 1987). They contain superoxide dismutase (SOD) for the effective detoxification of oxygen radicals which are formed by isolated thylakoid membranes in the

Mehler reaction at rates exceeding $30\,\mu mol\,(mg\ chlorophyll)^{-1}h^{-1}$, i.e., far more than 150 times the rate of SO_2 or ozone influx into leaves, when the external pollutant concentration is 100 ppb. The H_2O_2 produced by disproportionation of the oxygen radicals is destroyed by chloroplastic ascorbate peroxidase. Ascorbate is oxidized when H_2O_2 is reduced. It is regenerated from the resultant monodehydroascorbate and dehydroascorbate by specific reductases. Electron donors for the regeneration of ascorbate are glutathione and NADPH. Oxidized glutathione is itself reduced by NADPH.

In spite of the potentially high rate of oxidant formation in photosynthesis, the activities of enzymes and the concentrations of substrates involved in detoxification of the reactive oxygen species are so high that oxidant production in the chloroplasts in the light does not interfere with photosynthesis although enzymes of the chloroplast stroma are highly sensitive to oxidants (Kaiser 1976). In this situation, oxidants derived from incoming air pollutants will be effectively reduced together with endogenously produced oxidants as long as pollutant concentrations are not excessive.

As in the chloroplasts, ascorbate and glutathione act as antioxidants and radical scavengers also in other cytoplasmic compartments. Catalase, which destroys H_2O_2, may be concentrated in the peroxisomes to an extent that crystallization occurs. In the vacuoles, peroxidases use H_2O_2 as an electron acceptor to oxidize phenols and other substrates (Takahama 1993). Ascorbate is also a vacuolar solute. In biomembranes, tocopherol is an effective radical scavenger and antioxidant (Elstner 1990).

It appears that the primary function of these and other antioxidative systems consists in the detoxification of radicals and oxidants produced during the normal life span of a cell by respiration, photosynthesis, and by other cellular activities. Nevertheless, they also help to cope with the oxidative stresses produced by anthropogenic air pollutants.

14.5.1 The Path of Air Pollutants

The first cellular barrier that air pollutants have to overcome once they have passed through the stomata is the aqueous phase of the apoplasm, which includes the cell wall. The apoplasm contains ascorbate at a concentration approximately one-tenth or less of the cytoplasmic concentration (Luwe et al. 1993). Antioxidants of the apoplasm include phenolic substances. Apoplasmic peroxidases oxidize phenolics using hydrogen peroxide as electron acceptor (Pfanz et al. 1990; Takahama et al. 1992). Phenolic radicals produced by the peroxidase reaction are either intermediates of lignin formation or are reduced by ascorbate. Air pollutants not intercepted in the apoplasm have to cross the plasmalemma before they can interact with cytoplasmic constituents. While it is very unlikely that ozone will ever reach the central vacuole of fully differentiated leaf cells owing to its extremely high reactivity (Urbach et al. 1989; Lange et al. 1989a), the vacuole is a site of deposition of products of detoxification such as nitrate, in the case of NO_2, or sulfate, in the case of SO_2 (Kaiser et al. 1989).

14.5.2   The Fate of Nitrogen Oxides

Owing to its lipid solubility, NO_2 can penetrate biomembranes. However, before it reaches the plasmalemma of leaf cells, it reacts with water in the apoplasm. The resultant nitrate and nitrite anions enter the cells. In the cytosol, nitrate is either reduced to nitrite or transported into the vacuole, where it may be stored. Nitrite is taken up into chloroplasts by a specific nitrite carrier of the chloroplast envelope (Krämer et al. 1988).

In vitro, illuminated spinach chloroplasts may evolve oxygen during the reduction of substrate concentrations of nitrite at rates close to $100\,\mu mol$ (mg chlorophyll)$^{-1}$h^{-1} (Heber and Purczeld 1978). In contrast, fluxes of NO_2 into leaves are distinctly less than $40\,nmol$ (mg chlorophyll)$^{-1}$h^{-1} at commonly observed NO_2 concentrations in air (10 or 20 ppb). In view of the very high detoxification capacity of nitrite reductase, it is perhaps not surprising that short exposure of leaves from herbaceous plants to 10 000 or even 20 000 ppb NO_2 failed to produce persisting leaf damage (Yin 1990). Photosynthesis was not even appreciably suppressed by such high NO_2 levels. The slight inhibition occasionally observed was reversed rapidly after termination of fumigation. Instead of the expected acidification of the cytosol by nitrous and nitric acid (cf. Wellburn 1990), transient alkalization was observed in the presence of high concentrations of NO_2 in the air (Heber et al. 1989a). This is attributed to the production of hydroxyl ions during nitrite and nitrate reduction. The oxidative capabilities of NO_2 and nitrite were efficiently neutralized by the antioxidative defenses of herbaceous leaf cells. We conclude that oxidative interference of NO_2 or nitrite with leaf metabolism is unlikely to cause acute leaf damage at commonly observed NO_2 (or NO) concentrations. In herbaceous plants, NO_2 detoxification is fast and highly effective. There, nitrite reduction proceeds even in the dark, although at slower rates than in the light (Kaiser et al. 1992). Even in spruce, which was fumigated for a period of 85 days, 10 h per day with 300 ppb NO_2, neither nitrite nor nitrate accumulation was observed in the needles (Kaiser et al. 1991). Apparently, fast reductive detoxification of NO_2 is not restricted to herbaceous plants. Nevertheless, there are plants, which accumulate nitrite in leaves after fumigation with NO_2 (Zeevaart 1976; Yoneyama et al. 1979). Wellburn (1990) proposes that in leaves of plants which assimilate nitrogen mainly in roots, damage by NO_2 immisssion may be possible because of insufficient nitrite detoxification.

The permanent NO radical is potentially more phytotoxic than NO_2 (cf. Wellburn 1990), but its water solubility is low. For this reason, its diffusional flux into leaves which is controlled by stomatal opening, is slow. In addition, the NO concentration is usually about one order of magnitude smaller than the NO_2 concentration in the field. There is still confusion as to whether and how nitrogen oxides affect plants directly at ambient concentrations in air (cf. Wellburn 1990). We conclude that the indirect toxic action of NO_2 via ozone formation is much more important as a damaging factor than the direct action.

14.5.3   The Fate of Ozone

Ozone and its highly reactive degradation products react with all plant surfaces. However, it is doubtful whether reactions with epidermal structures can cause damage comparable with that which ensues when ozone, after penetrating open stomata, fails to be completely intercepted in the apoplasmic space of internal leaf tissues. Apoplasmic detoxification is the main defense against ozone. On sunny days, peak ozone concentrations in air may be 200 ppb or more in Germany (Umweltbundesamt, Jahresbericht 1990). In North America, concentrations up to 800 ppb have been observed (Smith 1991). In spinach leaves, the redox state of apoplasmic ascorbate changed from 90% reduced to 90% oxidized within 6 h of fumigation with 300 ppb ozone (Luwe et al. 1993). Ascorbate regeneration could not keep pace with oxidation which, however, accounted for only 10% or less of calculated ozone fluxes into the leaves. Oxidized ascorbate could not be reduced in the apoplasm. Rather, it was transferred into the cytosol and reduced there. Cytosolic ascorbate was exported into the apoplasm, but transport was slow. There is little doubt that apoplasmic antioxidants other than ascorbate participated in apoplasmic ozone detoxification. Nevertheless, after about 12 h of fumigation with 300 ppb ozone, intracellular glutathione (but not intracellular ascorbate) started to become oxidized. After 48 h of fumigation, oxidation extended to intracellular ascorbate. Total ascorbate levels declined. Simultaneously, necrotic leaf damage became visible. This shows that, at 300 ppb ozone, apoplasmic antioxidative defenses were overwhelmed. Presumably, ozone reached the plasmalemma. This made ozonization of unsaturated fatty acid residues and subsequent reactions inevitable. Urbach et al. (1989) have calculated that only a smally percentage of the ozone which reaches the plasmalemma actually manages to diffuse into the cytosol. Most is intercepted within this biomembrane, damaging it by ozonization of double bonds and subsequent reactions. This alters permeability properties. At increased levels of ozone, intracellular ascorbate leaks into the apoplasm. Whereas this may contribute to protection, apoplasmic levels of cations are also increased, indicating a general loss of intracellular solutes. Even though, by itself, this would not necessarily decrease the viability of affected cells, tissues already suffering from mineral deficiency due to the loss of essential minerals may be severely damaged.

It should be noted that, under field conditions, ozone is most unlikely to reach concentrations in air which would affect intracellular organelles such as chloroplasts directly. It damages cells before it actually reaches the chloroplasts. Photosynthesis of intact chloroplasts which are suspended in an isotonic sorbitol medium remains uninhibited when the suspension is exposed to ozone at concentrations which kill leaves (Urbach, pers. comm.). The simple explanation is that ozone reacts with constitutents of the medium before it can damage the chloroplasts, even though these are highly sensitive to oxidants.

Nevertheless, even low concentrations of ozone in air have been observed to decrease the productivity of birch trees (Matyssek et al. 1992). Apparently, effects of ozone on photosynthesis are of indirect nature. If forest trees, where leaves or needles are exposed to air polluted with ozone during their life span, do not possess antioxidative defenses which are superior to those of birch trees (or of spinach), ozone will contribute to forest decline. Sandermann et al. (1989) summarized subacute effects of ozone on the metabolism of conifers. He observed that latent damage after fumigation with ozone found expression only in the following vegetation period ("memory effect").

14.5.4 The Fate of SO_2

Like NO_2, SO_2 may be hydrated already in the aqueous phase of the apoplasm. Sulfite anions of the resultant sulfurous acid can be oxidized to sulfite radicals by the oxidized phenolics which are formed in the apoplasm by the reaction of peroxidase with hydrogen peroxide and apoplasmic phenolics (Pfanz et al. 1990; Takahama et al. 1993). The radicals are readily reduced by apoplasmic ascorbate. However, since little or no oxidation of apoplasmic ascorbate has been observed during fumigation of leaves with high concentrations of SO_2 (Takahama et al. 1993), a radical chain oxidation of sulfite to sulfate seems to be prevented in the apoplasm by ascorbate (but see Pfanz and Oppmann 1991). Irrespective of what happens in the apoplasm, the inhibition of photosynthesis observed 1 or 2 min after onset of fumigation of leaves with high concentrations of SO_2 (2 to 6 ppm, Veljovic-Jovanovic et al. 1993) shows that SO_2 and its hydration products rapidly enter the cytoplasm. Sulfurous acid decreases the cytoplasmic pH, stimulating the cytoplasmic pH-stat mechanisms which counter acidification (Pfanz et al. 1987; Pfanz and Heber 1989; Pfanz and Heber et al. 1989b; Yin 1990; Veljovic-Jovanovic et al. 1993). The phosphate translocator of the chloroplast envelope catalyzes not only exchange of phosphate and phosphate esters between cytosol and chloroplast stroma, but also permits transfer of sulfite and sulfate (Hampp and Ziegler 1977; Hampp et al. 1980). In the chloroplasts, sulfite is detoxified either by reduction to the sulfide level or by oxidation to sulfate (Dittrich et al. 1992). Effective reduction which leads to the amino acid cysteine depends on the presence of O-acetylserine. Chloroplastic radical scavengers prevent the fast radical chain oxidation of sulfite to sulfate which is observed in the presence of isolated thylakoids. Actually, reductive and oxidative detoxification of SO_2 have a comparable maximum capacity in spinach chloroplasts (Dittrich et al. 1992). At 20 °C, it is about 2 μmol sulfite (mg chlorophyll)$^{-1}$h^{-1} which is either reduced or oxidized. This must be compared to a SO_2 flux into leaves between 20 and 100 nmol (mg chlorophyll)$^{-1}$h^{-1}, when the SO_2 concentration in air is about 50 ppb and the sum of boundary layer and stomatal resistances is between 2 and 10 s cm^{-1}. Apparently, detoxification capacities of leaves for SO_2 are high, although they do not approach the detoxification capacity for NO_2. However, reductive detoxification of SO_2 appears to be effective only when sinks are available for reduced sulfur. Proteins constitute a main sink. In trees, net protein synthesis is a seasonal affair. During most of the year, protein synthesis and protein degradation match each other and there is, in the absence of growth, no additional demand for reduced sulfur. It is therefore not surprising that in Norway spruce sulfate accumulation accounts for practically all SO_2 which has been taken up (Kaiser et al. 1991). Herbaceous plants have a much higher capacity for the reductive detoxification of SO_2 than woody plants. Grasses thrive where spruce has been killed by over-exposure to SO_2 (Heber et al. 1987).

Very high detoxification capacities are actually required for a successful competition with toxic reactions. Even very brief exposure of leaves to 5 ppm SO_2 produces considerable inhibition of photosynthesis. This concentration is more than about 25 times the peak concentrations of SO_2 observed in the field during inversion periods in the winter. If fumigation with 5 ppm SO_2 is discontinued after little more than 5 min, recovery of photosynthesis may be observed within 30 min or more, indicating the presence of repair mechanisms (Veljovic-Jovanovic et al. 1993). Effective repair requires light, ambient air levels of oxygen, and summer temperatures. Very little is known about acute toxicity and detoxification of SO_2 at winter temperatures.

In contrast to reductive detoxification, oxidative detoxification, which leads to sulfuric acid, burdens the organism with protons. Short-term pH regulation is highly effective at 20 °C, but slower at 16 °C (Heber et al. 1989b). Normally, its capacity is far higher than needed to deal with the acidification caused by SO_2 influx. However, problems arise in the long term. In needles of spruce from the Ore Mountains (Erzgebirge), average sulfate concentrations have been measured which approach or even exceed 100 mM (Kaiser et al. 1991; Kaiser et al. 1993b; Pfanz and Beyschlag 1991). Accordingly, 200 mEq l^{-1} H$^+$ would have burdened the cells, if all of this sulfate had been airborne. In fact, comparisons with sulfate contents in needles from unpolluted areas show that most of the sulfate must have been derived from SO_2. Nevertheless, the pH of homogenates of needles from spruce grown in polluted and unpolluted areas proved to be almost identical. Apparently, needles usually manage to neutralize the sulfuric acid which slowly accumulates during long-term exposure to SO_2-polluted air. Persisting acidification of needles was observed only in exposure chambers during fumigation of young spruce trees for several weeks with concentrations of sulfur dioxide which were much higher than those that spruce has to endure in areas exposed to polluted air (Kaiser et al. 1991, 1993a).

14.5.5 Acid-Dependent Cation Requirements

There are several mechanisms by which a cell can cope with imported acid which cannot be degraded (for a detailed discussion, see Slovik et al. 1992a,b):

1. A high proton buffering capacity might prevent a deleterious decrease in pH (Pfanz and Heber 1986). However, buffering does not prevent acidification. It only decreases it. For cells to remain viable during prolonged influx of SO_2, cytoplasmic pH values must be maintained under strict control.
2. Degradation of endogenous organic anions and reduction of nitrate yields hydroxyl ions which can neutralize airborne sulfuric acid (Heber et al. 1987). Such reactions are the chemical basis of cellular pH-stat mechanisms (Raven 1986). However, it appears that the chemical pH-stat is overtaxed in the long term, when sulfate accumulates to high levels in cells.
3. Airborne acid may be sequestered intracellularly, e.g., into vacuoles.
4. Airborne acid is exported from the cells or bases are imported.

All of these possibilities occur. Of particular importance is sequestration and/or export when acid stress is chronic. The tonoplast membrane which separates the cytosol from the vacuole contains two proton-translocating enzymes, an ATPase, and a pyrophosphatase in addition to anion transporters (Walker and Leigh 1981; Sze 1985). Sulfate is transported into the vacuoles of leaf cells, where it is sequestered (Kaiser et al. 1989). This transport is energy-dependent. In principle, not only sulfate but also the protons of sulfuric acid could be stored inside the vacuoles. However, it seems that proton storage in the vacuoles as an effective means of cytoplasmic pH regulation can be observed only temporarily under acute acid stress (Heber et al. 1987). If it were effective in the long term, homogenates of needles from polluted areas with high sulfate contents should be more acidic than homogenates of needles from unpolluted areas. This is not the case (Pfanz and Beyschlag 1993; cf. Slovik et al. 1992a,b). Also, the vacuolar acidification observed when leaves are fumigated with SO_2 is reversed after termination of fumigation (Heber et al. 1989a). There is the question of how vacuolar sulfuric acid is neutralized. Obviously, bases can be mobilized, but there is a limit to the extent of base mobilization in leaves. However, not only the tonoplast, but also the plasmalemma possesses a proton-translocating ATPase. Its function is to power energy-dependent ion uptake into the cells. After SO_2 has entered cells and has been detoxified by oxidation to sulfuric acid, protons need to be exported. The ATPase functions in this export. On the other hand, cation import is required to satisfy the need for counterions of sulfate. Proton export solves the problem of cellular acidification at the level of the leaf, but constitutes a burden for other tissues. It functions on the basis of proton/cation exchange. A final solution to the problem of cellular or tissue acidification is proton export into the soil at the root/soil interface (Kaupenjohann et al. 1988; Thomas and Runge 1992; Kaiser et al. 1993a). This requires proton transport from the leaves to the roots. In exchange for exported protons, cations must be imported. They are taken up from the soil and transported from the roots into leaves via the transpiration stream. In this way, protons produced as a consequence of hydration of SO_2 in the leaves finally reach the soil in exchange for cations which are needed as counterions of sulfate (Slovik et al. 1992a,b; Kaiser et al. 1993a,b). Roots have long been known to excrete protons. ATPase-mediated proton export is required for energy-dependent ion uptake by roots. This import is necessary to satisfy the requirements of growth for nutrient ions.

In the presence of SO_2 in the atmosphere, cations are needed not only for growth, but also for the neutralization of the sulfuric acid which is formed in the leaves. The mechanisms of cellular pH stabilization compete successfully with growth for cations. In consequence, growth is retarded. It has long been known that growth is reduced in areas where the atmosphere is polluted by SO_2. Moreover, so-called novel forest decline ("neuartige Waldschäden") is often characterized by symptoms of cation deficiency. Different nutrient cations (e.g., K^+, Ca^{2+}, Mg^{2+}) may be immobilized in vacuoles as counterions of airborne sulfate at different ratios depending on cation availability in the soil. The nutrient cation which is closest to the minimum supply in the soil will be most sensitive to vacuolar immobilization. Particularly well known is the expression of magnesium deficiency in Norway spruce by the extensive yellowing of needles (Lange et al. 1989a; Schulze et al. 1989). On soils poor in magnesium, SO_2 will enhance or promote magnesium deficiency. This situation is particularly likely to arise on soils exposed to acid leaching. It has been mentioned above that potentially acidic air pollutants are washed out of the atmosphere when it rains. In acid rain, protons exchange for nutrient cations of the soil according to the Hofmeister lyotropic power series. This decreases cation availability for roots. It is often not recognized that cation deficiency may be a consequence of acid stress which is exerted both above and below the soil surface, i.e., in leaves and at the level of the root system (Slovik et al. 1992a,b).

14.5.6 Interactions Between Different Air Pollutants

Perhaps it is not surprising that in fumigation experiments with elevated concentrations of different air pollutants simultaneously, more damage is produced than is caused by the individual components alone. However, the combination of NO_2 and SO_2 actually appears to be more toxic than would be expected on the basis of individual toxicities (Heath 1980; Wellburn 1984; Yin 1990). It is not known whether this experimental observation can be extrapolated to the lower levels of different pollutants experienced simultaneously by plants in the field. Oren and Schulze (1989) introduced the idea of nutritional disharmony to explain effects of increased N availability on forest decline. By stimulating growth, nitrogen increases cation demand, adding to the problems caused by the necessity to supply cations as counterions of airborne sulfate.

For the combination of ozone and SO_2, rational explanation of enhanced toxicity as observed by Heggestad and Bennet (1981) and Yin (1990) has been proposed by Slovik et al. (1992b). If ozone cannot be completely detoxified in the apoplasm before reaching the plasmalemma, increased plasmalemma permeability caused by ozonization of unsaturated fatty acid residues could lead to cation losses exacerbating already existing $SO_2$2-dependent cation deficiency.

14.5.7 Interactions with Climatic Conditions

The primary sites of forest damage are mountainous areas where plants are exposed to harsh climatic conditions. Drought can in principle be protective by limiting the direct impact of air pollutants through stomatal closure. However, drought also represents a stress. Added stress is unlikely to relieve a stress produced by air pollutants during prolonged exposure. Rather, it is likely to result in damage which neither stress alone would cause. This may be rather general. SO_2 is known to increase the frost sensitivity of plants (Davison and Bailey 1982). Frost periods decrease the tolerance of spruce trees to SO_2 (Slovik et al. 1992a). Low temperatures will also decrease the rate of metabolic detoxification and repair reactions.

14.6 Tolerance Limits

At elevated concentrations, air pollutants are toxic not only for plants, but also for man. Legislative bodies and concerned organizations have defined limits of tolerance for both (for Germany: Bundesimmissionsschutzgesetz; Jäger 1989). It appears that legal limits are a result of both damage risk assessment and compromise. Control of industrial and household emissions is costly. Main problems in defining (or redefining) limits of tolerance are that different organisms possess different capacities for detoxification, and that tolerance of a particular organism may vary with its physiological state and the season. For SO_2 pollution in exposed forest sites with severe climatic conditions, a general tolerance limit recommended by IUFRO (International

Union of Forest Research Organisations, Vienna, 1978) is $25\,\mu g\,m^{-3}$ or 8.7 ppb (annual means). This limit is doubled for forest sites with moderate or good growth conditions. The WHO (World Health Organization) and the UNECE (United Nations Economic Commission for Europe, Paris) recommend, in 1987 and 1988, a tolerance limit of $30\,\mu g\,m^{-3}$ (10.5 ppb, annual means) SO_2 for sensitive forest plants. The German TA-Luft (TA = Technische Anleitung) of 1986 is less restrictive permitting emissions up to $50\,\mu g\,m^{-3}$ (17.4 ppb) on an annual basis. The same limit is proposed by the German VDI-Richtlinie 2310 (VDI = Verein Deutscher Ingenieure) of 1978 for "very sensitive" plants and a higher limit of $80\,\mu g\,m^{-3}$ (28 ppb) for "sensitive" plants within vegetation periods of 7 months.

In 1987, the VDI considered 8-h averages of 40 ppb, 85 ppb, and 150 ppb ozone tolerable for very sensitive, sensitive, and less sensitive plants, respectively. The corresponding 1/2-h averages were 150, 250, and 500 ppb. In 1988, the UNECE recommended 30 ppb (8-h-average) and 150 ppb (1/2-h average) ozone as limits for sensitive plants and ecosystems. During the summer of 1992, ozone values measured in various locations in Bavaria exceeded lower limits in many cases.

For NO_2, tolerance limits recommended in 1978 by the VDI for sensitive plants were 180 ppb averages during a vegetation period of 7 months. The UNECE distinguished between the vegetation period with a tolerance limit of 30 ppb average and the winter period with a lower tolerance limit of 20 ppb average. In combination with SO_2 and ozone, a less than 15 ppb average was considered tolerable.

Such recommended limits are usually based on fumigation experiments with elevated concentrations of the pollutants performed in climatic chambers or open top chambers. The duration of such experiments is necessarily restricted. They are performed under climatic or growth conditions which rarely correspond to the climatic or growth conditions of exposed forest sites.

For SO_2, Slovik et al. (1992a,b) have recently attempted to derive limits of tolerance of healthy spruce on the basis of measured physiological data. Form measurements of cation contents in xylem sap during the course of a year, and considering the necessity of cation circulation in the tree, they calculated a limit of long-term acidification tolerance which was below 10 ppb SO_2 during part of the winter, but much higher during the summer. At any time of the year, tolerance levels are lowered when cation availability is restricted. In this situation, cation deficiency symptoms will develop even when cation determinations in leaf extracts fail to indicate deficiency because of vacuolar cation sequestration. In comparing the calculations with accepted tolerance limits which permit averaging of pollution levels on an annual basis, it must be emphasized that the highest levels actually occur during the winter, when sensitivity of SO_2 is particularly high because metabolic activity is low and cation transport in the xylem is slow. High winter levels of SO_2 are presently legally tolerable because low summer levels produce an acceptable average. The obvious conclusion is that legal (or recommended) tolerance limits for SO_2 must be changed in accordance with the changing sensitivity of plants. They should be lower during the winter when evergreen plants are more sensitive than in the summer. If this is difficult to implement, permissible annual averages must be reduced. Depending on growth rates of spruce which provide information on the extent of possible reductive detoxification of SO_2 and on cation availability, Slovik et al. (1992b) calculated SO_2 immissions between 2 and 5 ppb SO_2 (annual means) to be tolerable. This is below international and German tolerance limits. Depending on growth rates, the annual cation demand has been calculated to be doubled in the presence of SO_2 levels between 16 and 40 ppb (annual average). Doubling the cation demand on soils where growth is nutrient-limited must, in the long term, have disastrous consequences on forests. However, it would appear that increasing the nutrient supply by proper application of fertilizers or by liming could reduce or even stop the decline of spruce forests (cf. Lange et al. 1989b).

14.7 Conclusions

In the middle and eastern part of Europe, forest decline is particularly apparent in alpine regions, the Black Forest, the Bavarian Forest, the Thuringian Forest, the Harz, the Fichtelgebirge, the Ore Mountains (Erzgebirge), the northern and northeastern part of Czechoslovakia, and in Upper and Lower Silesia (Riesengebirge). Except for the Alps and the Black Forest, where ozone appears to predominate, SO_2 pollution is frequent in these locations. Although the evidence for direct involvement of air pollutants in forest decline is overwhelming, the different distribution of individual air pollutants, differing soil qualities, and climatic conditions make it impossible to attribute forest decline to a particular offender. Rather it appears that forest decline is caused by a complex combination of different factors, including not only air pollutants but also natural stress factors such as drought and frost. Ozone and SO_2 decrease cellular viability, thereby increasing the susceptibility to natural stress factors. Decreased forest health cannot be explained on the basis of direct effects of air pollutants on photosynthesis. Rather, the photosynthetic apparatus appears to be well protected against ozone by its location in cells and against NO_2 and SO_2 by antioxidative defenses. These are particularly efficient in scavenging radicals. The cellular pH-stat is effective in regulating cellular pH values during influx or production of acid as long as bases are available or protons of the acid can be exchanged for cations. Decline is inevitable and leaves or needles are shed when increased cation demands cannot be met. Of particular importance in this respect is the decreased availability of nutrient cations from poor soils or soils exposed to acid leaching in situations where cations are either sequestered as counterions of sulfate or viability is decreased under the influence of ozone. Agricultural productivity may remain high owing to high detoxification capacities of crop plants and high nutrient availability in agricultural used soils where nearby forests decline under the influence of air pollution in soils which are poor or depleted of essential nutrients. For SO_2, calculations based on known physiological parameters of spruce (Slovik et al. 1992a,b) suggest that legal tolerance limits are too high and must be reduced. Depending on growth conditions, tolerance limits between 2 and 5 ppb are recommended for spruce (annual average). No similar calculations of tolerance limits which are based on physiological data are available for ozone. Nevertheless, direct observations show that ambient levels of ozone decrease plant productivity (Matyssek et al. 1992). Concentrations only 50% higher than frequently observed peak concentrations of ozone cause dramatic oxidative damage in spinach within 2 days (Luwe et al. 1993). Only in the case of NO_2 do measured concentrations in the atmosphere appear to be too low for plant damage since plants possess considerable detoxification capacity for this oxidant. However, emissions of NO_x and NH_3 may stimulate growth, thereby increasing the cation demand. This may feed back on the sensitivity to other pollutants (Schulze 1989).

Acknowledgments. Experiments leading to the conclusions presented in this chapter have been performed within the research efforts of the Sonderforschungsbereich 251 of the University of Würzburg. We are grateful to the Bavarian Ministry for Development and Environmental Problems for support within the Projektgruppe Bayern zur Erforschung der Wirkung von Umweltschadstoffen (PBWU).

References

Asada K, Takahashi M (1987) Production and scavenging of active oxygen in photosynthesis. In: Kyle DJ, Osmond CB, Arntzen CJ (eds) Photoinhibition. Elsevier, Amsterdam, pp 227–288

Davison AW, Bailey IF (1982) SO₂ pollution reduces the freezing resistance of ryegrass. Nature 297: 400–402

Dittrich A, Pfanz H, Heber U (1992) Oxidation and reduction of SO₂ by chloroplasts and formation of sulfite addition compounds. Plant Physiol 98: 738–744

Elstner EF (1984) Schadstoffe die über die Luft zugeführt werden. In: Hock B, Elstner EF (eds) Pflanzentoxikologie. B.I. Wissenschaftsverlag, Mannheim, pp 67–94

Elstner EF (1990) Der Sauerstoff. B.I. Wissenschaftsverlag, Mannheim

Hällgren JE (1978) Physiological and biochemical effects of sulfur dioxide on plants. In: Nriagu IO (ed) Sulfur in the environment. Part II. Ecological impacts. Wiley, New York, pp 163–209

Halliwell B (1978) The Chloroplast at work. A review of modern developments in our understanding of chloroplast metabolism. Prog Biophys Mol Biol 33: 1–54

Halliwell B, Foyer CH (1978) Properties and physiological function of a glutathione reductase purified from spinach leaves by affinity chromatography. Planta 139: 9–17

Hampp R, Ziegler I (1977) Sulfate and sulfite translocation via the phosphate translocator of the inner envelope membrane of chloroplasts. Planta 137: 309–312

Hampp R, Spedding DJ, Zieger I, Ziegler H (1980) The efflux of inorganic sulfur from spinach chloroplasts. Z Pflanzenphysiol 99: 113–119

Heath RL (1980) Initial events in injury to plants by air pollutants. Annu Rev Plant Physiol 31: 395–431

Heber U, Purczeld P (1978) Substrate and product fluxes across the chloroplast envelope during bicarbonate and nitrite reduction. In: Hall DO, Coombs J, Goodwin TW (eds) Photosynthesis 77, Proc 4th Int Congr Photosynth. Biochemical Society, London, pp 299–310

Heber U, Laisk A, Pfanz H, Lange OL (1987) Wann ist SO_2 Nähr- und wann Schadstoff? Ein Beitrag zum Waldschadensproblem. Allg Forstztg 27/28/29: 700–705

Heber U, Yin Z-H, Dittrich A, Pfanz H, Lange O-L (1989a) The response of leaves to potentially acidic gases. In: Ulrich B (ed) International congress on forest decline research: state of knowledge and perspectives, Friedrichshafen, October 2–6, 1989, BMFT, pp 499–516

Heber U, Yin Z-H, Dittrich A, Ghisi R, Wagner U (1989b) Response of mesophyll cells and its organelles to the stresses produced by the atmospheric pollutant SO_2. In: Tazawa M, Katsumi M, Masuda Y, Okamoto H (eds) Pant water relations and growth under stress. Yamada Science Foundation and Myu KK, Tokyo, pp 93–100

Heggestad HE, Bennet HJ (1981) Photochemical oxidants potentiate yield losses in snap beans attributable to sulfur dioxide. Science 213: 1008–1010

Jäger HJ (1989) Stand der Diskussion über Richtwerte für Schadstoffkonzentrationen in der Luft. In: Ulrich B (ed) International congress on forest decline research: state of knowledge and perspectives, Friedrichshafen, Oct. 2–6, 1989, Lecture Volume II, BMFT, pp 717–731

Kaiser G, Martinoia E, Schröppel-Meier G, Heber U (1989) Active transport of sulfate into the vacuole of plant cells provides halotolerance and can detoxify SO_2. J Plant Physiol 133: 756–763

Kaiser WM (1976) The effect of hydrogen peroxide on CO_2 fixation of isolated intact chloroplasts. Biochim Biophys Acta 440: 476–482

Kaiser WM, Dittrich APM, Heber U (1991) Sulfatakkumulation in Fichtennadeln als Folge von SO_2-Belastung. In: PBWU (ed) Proc 2. Statusseminar der PBWU zum Forschungsschwerpunkt "Waldschäden", Projektgruppe Bayern zur Erforschung der Wirkung von Umweltschadstoffen. GSF-Ber 26/91: 425–437

Kaiser WM, Spill D, Brendle-Behnisch E (1992) Adenine nucleotides are apparently involved in the light-dark modulation of spinach-leaf nitrate reductase. Planta 186: 236–240

Kaiser WM, Höfler M, Heber U (1993a) Can plants exposed to SO_2 excrete sulfuric acid through the roots. Physiol Plant 87: 61–67

Kaiser WM, Dittrich A, Heber U (1993b) Sulfate concentrations in Norway spruce needles in relation to atmospheric SO_2: a comparison of trees from various forests in Germany with trees fumigated with SO_2 in growth chambers. Tree Physiol 12: 1–13

Kaupenjohann M, Schneider BU, Hantschel R, Zech W, Horn R (1988) Sulfuric acid rain treatment of *Picea abies* (L.) Karst.: effects on nutrient solution, throughfall chemistry, and tree nutrition. Z Pflanzenernähr Bodenkd 151: 123–126

Krämer E, Tischner R, Schmidt A (1988) Regulation of assimilatory nitrate reduction at the level of nitrite in *Chlorella fusca*. Planta 176: 28–35

Kreutzer K, Göttlein A (1991) Ökosystemforschung Höglwald. Paul Parey, Hamburg

Laisk A, Kull O, Moldau H (1989) Ozone concentration in leaf intercellular air spaces is close to zero. Plant Physiol 90: 1163–1167

Lange O-L, Heber U, Schulze ED, Ziegler H (1989a) Atmospheric pollutants and plant metabolism. In: Schulze ED, Lange OL, Oren R (eds) Forest decline and air pollution. Ecological Studies 77. Springer, Berlin Heidelberg New York, pp 237–273

Lange O-L, Weikert R, Wedler M, Gebel I, Heber U (1989b) Photosynthese und Nährstoffversorgung von Fichten aus einem Waldschadensgebiet auf basenarmem Untergrund. Allg Forstztg 3/1989: 55–64

Lendzian KJ (1984) Permeability of plant cuticles to gaseous air pollutants. In: Koziol MJ, Whatley FR (eds) Gaseous air pollutants and plant metabolism. Butterworths, London, pp 77–81

Lendzian KJ, Kerstiens G (1988) Interactions between plant cuticles an gaseous air pollutants. Aspects App Biol 17: 97–104

Lendzian KJ, Kerstiens G (1991) Sorption and transport of gases and vapors in plant cuticles. Rev Environ Contam Toxicol 121: 65–128

Luwe MWF, Takahama U, Heber U (1993) Role of ascorbate in detoxifying ozone in the apoplast of spinach leaves. Plant Physiol 101: 969–976

Matyssek R, Günthardt-Goerg MS, Saurer M, Keller T (1992) Seasonal growth, $\delta^{13}C$ in leaves and stem, and phloem structure of birch (*Betula pendula*) under low ozone concentrations. Trees 6: 69–76

Mitterhuber E, Pfanz H, Kaiser WM (1989) Leaching of solutes by the action of acidic rain: a comparison of efflux from twigs and single needles of *Picea abies* (L.) Karst. Plant Cell Environ 12: 93–100

Nobel PS (1983) Biophysical plant physiology and ecology. Freeman, New York

Oren R, Schulze ED (1989) Nutritional disharmony and forest decline: a conceptual model. In: Schulze ED, Lange OL, Oren R (eds) Forest decline and air pollution. Ecological Studies 77. Springer, Berlin Heidelberg New York, pp 425–443

Pfanz H, Beyschlag W (1991) Photosynthetic performance of Norway spruce (*Picea abies* (L.) Karst.) in relation to the nutrient status of the needles. A study in the forests of the Ore Mountains. In: PBWU (ed) Expertentagung Waldschäden im östlichen Mitteleuropa und in Bayern. GSF-Ber 24/91: 523–527

Pfanz H, Beyschlag W (1993) Photosynthetic performance and nutrient status of Norway spruce [*Picea abies* (L.) Karst.] in the forests of the Ore Mountains (Erzgebirge). Trees 7: 115–122

Pfanz H, Heber U (1986) Buffer capacities of leaf cells and leaf cell organelles in relation to fluxes of potentially acidic air pollutants. Plant Physiol 81: 597–602

Pfanz H, Heber U (1989) Determination of extra- and intracellular pH values in relation to the action of acidic gases on cells. In: Linshens HF, Jackson IF (eds) Gases in plant and microbial cells. Mod Meth Plant Anal NS, vol 9. Springer, Berlin Heidelberg New York, pp 322–343

Pfanz H, Oppmann B (1991) The possible role of apoplastic peroxidases in detoxifying the air pollutant sulfur dioxide. In: Lobarzewski J, Greppin H, Peuel C, Gaspar Th (eds) Biochemical, molecular, and physiological aspects of plant peroxidases. University of Geneva, pp 401–417

Pfanz H, Martinoia E, Lange OL, Heber U (1987) Mesophyll resistances to SO_2 fluxes into leaves. Plant Physiol 85: 922–927

Pfanz H, Dietz K-J, Weinerth I, Oppmann B (1990) Detoxification of sulfur dioxide by apoplastic peroxidases. In: Rennenberg H, Brunold Ch, De Kok IJ, Stulen I (eds) Sulfur nutrition and assimilation in higher plants; fundamental, environmental and agricultural aspects. SPB Acad Publ, The Hague, pp 229–233

Polle A, Chakrabarti K, Rennenberg H (1991) Entgiftung von Peroxyden in Fichtennadeln (*Picea abies* L.) am Schwerpunktsstandort Kalkalpen (Wank). In: PBWU (ed) Proc 2. Statusseminar der PBWU zum Forschungsschwerpunkt "Waldschäden", Projektgruppe Bayern zur Erforschung der Wirkung von Umweltschadstoffen. GSF-Ber 26/91: 151–160

Raven JA (1986) Biochemical disposal of excess H^+ in growing plants. New Phytol 104: 175–206

Reich PB, Amundson RG (1985) Ambient levels of ozone reduce net photosynthesis in tree and crop species. Science 230: 566–570

Rennenberg H (1984) The fate of excess sulfur in higher plants. Annu Rev Plant Physiol 35: 121–153

Sandermann H, Schmitt R, Heller W, Rosemann D, Langebartels C (1989) Ozone-induced early biochemical reactions in conifers. In: Longhurst JWS (ed) Acid deposition. Sources, effects and controls. The British Library, London, pp 243–254

Schulze E-D (1989) Die Wirkung von Immissionen auf Fichtenökosysteme – Ergebnisse der Waldschadensforschung im Fichtelgebirge. In: PBWU (ed) Proc 1. Statusseminar der PBWU zum Forschungsschwerpunkt "Waldschäden", Projektgruppe Bayern zur Erforschung der Wirkung von Umweltschadstoffen. GSF-Ber 6/89: 95–106

Schulze E-D, Lange O-L, Oren R (1989) Forest decline and air pollution. A study of spruce (*Picea abies*) on acid soils. Springer, Berlin Heidelberg New York, 475 pp

Slovik S, Kaiser WM, Körner Ch, Kindermann G, Heber U (1992a) Quantifizierung der physiologischen Kausalkette von SO_2-Immissionsschäden für Rotfichten [*Picea abies* (L.) Karst]. I. Ableitung von SO_2-Immissionsgrenzwerten für akute Schäden. Allg Forstztg 15/1992: 800–805

Slovik S, Heber U, Kaiser WM, Kindermann G, Körner Ch (1992b) Quantifizierung der physiologischen Kausalkette von SO_2-Immissionsschäden für Rotfichten [*Picea abies* (L.) Karst]. II. Ableitung von SO_2-Immissionsgrenzwerten für chronische Schäden. Allg Forstztg 17/1992, 17: 913–920

Smith WH (1991) Air pollution and forest damage. Chem Eng News 11: 30–43

Sze H (1985) H^+ translocating ATPase S: advances using membrane vesicles. Annu Rev Plant Physiol 36: 175–208

Takahama U (1993) Hydrogen peroxide scavenging systems in vacuoles of mesophyll cells of *Vicia faba* L., Phytochemistry (in press)

Takahama U, Veljovic-Jovanovic S, Heber U (1993) Effects of the air pollutant SO_2 on leaves: inhibition of sulfite oxidation in the apoplast by ascorbate and of apoplastic peroxidase by sulfite. Plant Physiol (in press)

Thomas FM, Runge M (1992) Proton neutralization in the leaves of English oak (*Quercus robur* L.) exposed to sulfur dioxide. J Exp Bot 43: 803–809

Ulrich B (1980) Die Wälder in Mitteleuropa. Meßergebnisse ihrer Umweltbelastung, Theorie ihrer Gefährdung, Prognose ihrer Entwicklung. Allg Forstztg 35: 1198–1202

Umweltbundesamt (ed) (1991) Jahresbericht 1990. Berlin

Urbach W, Schmidt W, Kolbowski J, Rümmele S, Reisberg E, Steigner W, Schreiber U (1989) Wirkungen von Umweltschadstoffen auf Photosynthese und Zellmembranen von Pflanzen. In: PBWU (ed) Proc 1. Statusseminar der PBWU zum Forschungsschwerpunkt "Waldschäden", Projektgruppe Bayern zur Erforschung der Wirkung von Umweltschadstoffen. GSF-Ber 6/89: 195–206

Veljovic-Jovanovic S, Bilger W, Heber U (1993) Inhibition of photosynthesis, stimulation of zeaxanthin formation and acidification in leaves by SO_2 and reversal of these effects. Planta (in press)

Walker RR, Leigh RA (1981) Mg^{2+}-dependent, cation-stimulated inorganic pyrophosphatase associated with vacuoles isolated from storage roots of red beet (*Beta vulgaris* L.). Planta 153: 150–155

Wellburn AR (1984) The influence of atmospheric pollutants and their cellular products upon photophosphorylation and related events. In: Koziol MJ, Whatley FR (eds) Gaseous air pollutants and plant metabolism. Butterworths, London, pp 203–221

Wellburn AR (1990) Why are atmospheric oxides of nitrogen usually phyototoxic and not alternative fertilizers? Tansley Rev 24, New Phytol 115: 395–429

Würfel M, Häberlein I, Follmann H (1990) Inactivation of thioredoxin by sulfite ions. FEBS Lett 268: 146–148.

Yin Z-H (1990) Durch Licht oder Luftschadstoffe induzierte pH-Änderungen in verschiedenen Kompartimenten der Blätter höherer Pflanzen. PhD Thesis, University of Würzburg

Yoneyama T, Saskawa H, Ishizuka S, Totsuka T (1979) Absorption of atmospheric NO_2 by plants and soils. II. Nitrite accumulation, nitrite reductase activity and diurnal change of NO_2 absorption in leaves. Soil Sci Plant Nutr 25: 267–276

Zeevaart AJ (1976) Some effects of fumigating plants for short periods with NO_2. Environ Pollut 11: 97–108

Ziegler I (1975) The effect of SO_2-pollution on plant metabolism. Res Rev 56: 79–105

Ziegler H (1988) Weg der Schadstoffe in die Pflanze. In: Hock B, Elstner EF (eds) Schadwirkungen auf Pflanzen, 2nd edn. Bibliographisches Institut, Mannheim, pp 35–46

**Part C
Plant Performance
in the Field**

15 Photosynthesis in Aquatic Plants

J.A. Raven

15.1 Introduction

To address the topic of the ecophysiology of photosynthesis in aquatic plants in the space allotted is a daunting task, and the coverage must of necessity be very selective. Since this Volume is honoring Professor Dr. Lange, I shall emphasize those aspects which interface with his work, and since the rest of the Volume deals with terrestrial plants, I shall emphasize more generally the comparison with terrestrial plants. My third objective is to address particularly those aspects which most interest me and which are, or could be, growth-points in research.

15.2 Definition of the Aquatic Habitat

Nature's distinction between aquatic and terrestrial environments is less rigid than that of many cartographers or civil engineers (see Sand-Jensen et al. 1992). I shall define an aquatic habitat as one which has a permanent, or temporally determinate but impermanent, occurrence of water over the solid substratum. I thus include not only permanent water bodies, but also their fringes subject to tidal or seasonal submersion of the substrata and thus of any plants growing on them. I also (arbitrarily) consider mainly the plants of such habitats whose photosynthetic organs are below the water surface during periods of submersion. Another arbitrary exclusion is those terrestrial poikilohydric plants which have a regular hydration period (e.g., desert lichens supplied with dew at dawn each day: Lange et al. 1970).

15.3 The Diversity of Aquatic Plants

Primarily, aquatic plants (Fig. 15.1; cf. the usage of Sand-Jensen et al. 1992) comprise essentially all aquatic O_2-evolvers at the prokaryotic level of organization [cyanobacteria; chloroxybacteria (= prochlorophytes)] as well

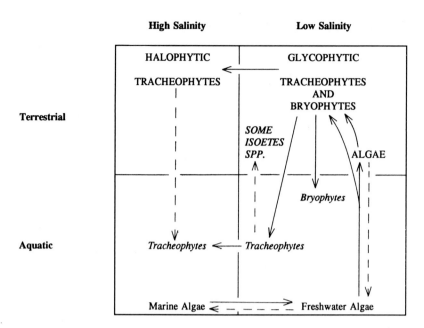

Fig. 15.1. Likely evolutionary scheme for the evolution of primarily (*roman*) and secondarily (*italic*) aquatic plants, and of primarily (*ROMAN CAPITALS*) and secondarily (*ITALIC CAPITALS*) terrestrial plants. *Full arrows* indicate major pathways; *dashed arrows* indicate minor pathways . (After Raven et al. 1980, 1988; Raven and Johnston 1991a)

as essentially all aquatic algae. This group also includes O_2-evolving pro- and eukaryotes in symbiotic relationships with aquatic invertebrates. These primaryily aquatic O_2-evolvers have never lived on land in their evolutionary history (see Raven 1984a; Raven et al. 1980). It is important to note that these primarily aquatic plants are much more diverse than the higher plants, since they contain organisms of many Divisions, while the higher plants, comprising however many Divisions the taxonomists presently see fit to recognize, were all derived from a single class (Charophyceae) of one Division (Chlorophyta) of the primarily aquatic O_2-evolvers (see Raven 1987; Raven and Johnston 1991a). The greatest diversity of primary aquatics (in terms of species number) is in marine environments, although for some classes there are more freshwater species (Fig. 15.1).

Secondarily, aquatic plants are those which have clearly spent some of their evolutionary history on land as primarily terrestrial plants and which have returned to aquatic life. Most of these organisms are "higher plants", but this category also includes aquatic lichens and, perhaps, some nonlichenized (free-living) algae (Raven et al. 1980, 1990a,b, Stebbins and Hill 1980; Raven 1984a; Raven and Johnston 1991a); see Fig. 15.1. It is of interest that some secondarily aquatic plants are apparently in the process of becoming secondarily terrestrial (e.g., *Isoetes* (= *Stylites*) *andicola*, *Isoetes andina*, *Isoetes triquetra*): Raven et al. (1988); see Fig. 15.1. All (secondarily) aquatic bryophytes and pteridophytes, and many (secondarily) aquatic flowering plants are from freshwaters.

This taxonomically and evolutionarily based scheme (Fig. 15.1) can be complemented by one based on life forms. These life forms (sensu Raunkaier 1934) are, for aquatic plants, those of Luther (1947; cf. Den Hartog and Segal 1964; Raven 1981, 1984a); see Table 15.1. It should be noted that the planophytes (represented mainly by the microscopic planktophytes) are predominantly primarily aquatic plants, as are the haptophytes, while many rhizophytes are secondarily aquatic (Table 15.1). By analogy with the primarily-secondarily aquatic distinction (Fig. 15.1), we can distinguish

Table 15.1. Life-forms of aquatic plants: definitions, and occurrence among primarily and secondarily aquatic O_2-evolvers. (After Luther 1947; Den Hartog and Segal 1964; Raven 1981, 1984a, 1992a)

Life-form and definition	Occurrence in primarily aquatic plants	Occurrence in secondarily aquatic plants
Planophyte (unattached) a) Planktophyte (small, planktonic)	Many cyanobacteria, chloroxybacteria, Bacillariophyceae, Chlorophyceae, Dinophyceae, Micromonadophyceae; radiolarians, ciliates symbiotic with microalgae	None
b) Pleustophyte (macrophytic)	Unattached macroalgae (Chlorophyceae, Phaeophyceae, Rhodophyceae, Ulvophyceae), foramenifera symbiotic with microalgae	*Ceratophyllum*, *Utricularia*
Haptophyte (attached to substratum of a particle size which is large relative to the size of the organisms)	Some cyanobacteria, Bacillariohyceae, Chlorophyceae; many Ulvophyceae; essentially all Phaeophyceae, Rhodophyceae; some ciliates, coelenterates, poriferans and bivalves symbiotic with microalgae; ascidians symbiotic with *Prochloron*	Some freshwater (Podostemaceae) and marine (*Phyllospadix*) flowering plants; many bryophytes
Rhizophytes (roots or rhizoids embedded in a substratum of particle size which is small relative to the size of the organism)	Some Chlorophyceae, Charophyceae, Tribophyceae, Ulvophyceae	Most secondarily aquatic vascular plants (freshwater and marine)

benthic plants which are primarily or secondarily haptophytic, and planophytes which are primarily or secondarily planophytic (Table 15.1). Raven
(1981, 1984a) suggests that the planophyte/haptophyte/rhizophyte distinction
is very significant in terms of the extent to which that part of the surface
area of the plant which is involved in photosynthesis is also involved in
the acquisition of resources other than carbon and energy, and thus in the
extent to which the area involved in photosynthesis is optimized for the
photosynthetic process. Raven (1981, 1984a) argues that rhizophytes, with
their ability to obtain much of their resources other than inorganic carbon
and photons via their rhizoids and roots (they rarely obtain most of their
inorganic carbon via their roots: Raven et al. 1988), have a photosynthetic
surface which is more optimized for photosynthesis per se than are planophytes and haptophytes, in which the large fraction of the total surface used
for photosynthesis is also the only surface area available for acquisition of
dissolved inorganic nitrogen, phosphorus, iron, etc. Raven (1981, 1984a,b,
1992a, 1993a) also notes that *Utricularia*, certain flagellates and alga-
invertebrate symbioses (planophytic and haptophytic) which are phagotrophic as well as photosynthetic, and thus obtain nitrogen, phosphorus, and
iron in particulate form, share with rhizophytes the possibility of optimizing
their photosynthetic surface for this function alone. The necessary respiratory loss of carbon as CO_2 in converting particulate organic carbon into
phagotroph carbon, and the frequently lower carbon:nitrogen, carbon:
phosphorus, and carbon:iron ratios in food particles than in the phagotroph,
make particle intake a less quantitatively significant source of carbon than of
nitrogen, phosphorus or iron for phagotrophic algal-invertebrate symbioses,
the phagotrophic flagellates, and *Utricularia* (Raven 1981, 1984a, 1992a,
1993a).

15.4   Contribution of Aquatic Plants to Global Net Primary Productivity

Table 15.2 shows that aquatic photolithotrophs contribute less to global net
primary productivity than would be expected on a pro rata basis from the
fraction of the earth's surface which they occupy. Marine phytoplankton,
which as planophytes are the only photolithotrophic life-form able to grow
where water depth exceeds a few hundred meters (Littler et al. 1985, 1986),
and thus are the sole photolithotrophs growing on more than half of the
earth's surface, and the main life-form accounting for the low global aquatic
productivity. Raven (1981, 1984a, 1986a,b) argues that the small size of the
open-ocean planophytes (i.e., planktophytes) relates to the large surface
area per unit volume of smaller cells and its influence on nutrient and
photon absorption in the frequently nutrient-limited and deep-mixed, and
hence photon-limited, environment as well as to its influence in reducing

Table 15.2. Net primary productivity of various aquatic habitats and for comparison, the values for terrestrial habitats. (After Raven 1991c)

Habitat, organisms	Total area/Mm^2	Net productivity	
		gC m^{-2}y^{-1}	10^{15} gC world^{-1}y^{-1}
Marine planktophytes	370	81	30
Marine benthic (primarily and secondarily aquatic organisms)	7.15[a]	572	4.09
Inland waters (planktophytes, pleustophytes, benthic organisms; primarily and secondarily aquatic)	2	290	0.58
Terrestrial, all phototrophs	150	400	60

[a] Area overlaps with that of marine planktophytes.

sinking rate. The large fraction of marine phytoplankton net primary productivity which is consumed by grazers may *increase* net primary production by recycling nutrients in a nutrient-limited environment.

Benthic marine primary productivity is almost ten times that of marine phytoplankton on a habitat area basis (Table 15.2) "New" nutrient input from land runoff and upwellings, water movement over the attached plants minimizing diffusion boundary layer constraints on nutrient uptake by bulky plants, and longevity permitting optimal use of seasonally available resources, all contribute to maintaining a high biomass and thus high fractional absorption of incident photons by plants as opposed to the water and other inanimate components as occurs in the open ocean (Raven and Richardson 1986).

Freshwater aquatic plants only have a small global area of habitat (Table 15.2), and achieve a lower area-based net productivity than at least the benthic marine photolithotrophs. As far as global CO_2 fluxes are concerned, many freshwater habitats are greatly CO_2-enriched due to the activity of terrestrial biota, and the photolithotrophs only achieve sufficient net primary productivity to decrease the net evasion of CO_2 to the atmosphere, not to cause net CO_2 invasion (see Sect. 15.6 below).

15.5 Photon Absorption and Use by Aquatic Plants

Kirk (1985) has elegantly discussed the optics of natural waters. By contrast with terrestrial environments, the aquatic environment frequently features more attenuation by inanimate material relative to plants than occurs on land. Thus, instead of shade habitats depleted mainly in the blue and red regions of the spectrum due to attenuation by chlorophylls and carotenoids as occurs on land, aquatic shade habitats can be predominantly depleted in the red region (low plant biomass; low content of blue-absorbing organic solutes) or even in the blue (low plant biomass; very high content of blue-absorbing solutes).

This diversity of spectral quality of shade light in aquatic habitats is paralleled by a diversity of photosynthetic pigments in at least the algal component of the aquatic flora relative to that of most terrestrial vegetation (Rowan 1989). It must be emphasized that this diversity of pigments increases as one moves outward (counter to the direction of movement of the excitation energy derived from absorbed photons) from the reaction centers of the two photosystems. It is clear that the two reaction centers are homologous throughout the O_2-evolvers; some pigment substitutions occur in the dedicated antenna pigment-protein complexes of the two photosystems, while major differences occur in both chromophores and polypeptides among the analogous light-harvesting pigment-protein complexes (Raven 1984a,b; Rowan 1989). The three major pigment groups are chlorophyll a plus phycobilins, chlorophyll a plus chlorophylls c and chlorophyll a plus chlorophyll b. In addition to their photoprotective role, the carotenoids of at least the two latter groups can have a major photon-harvesting role, serving to extend the blue peak of chlorophyll(s) absorption into the green, although not to the extent which is possible with phycobilins (Rowan 1989). Before considering the ecological role of these pigments, it is worth noting that recent evidence suggests a polyphyletic origin of chlorophyll b- but not chlorophyll c-based light-harvesting systems (Palenik and Haselkorn 1992; Urbach et al. 1992; Fujiwara et al. 1993). Furthermore, several chlorophyll b-containing primarily aquatic O_2-evolvers have a chlorophyll c-like pigment (Mg 2,4 divinyl, 2,4 desethyl, pheoporphyrin a_5 monomethyl ester or MgDVP) which functions in light harvesting (Rowan 1989). Spanning of two of the three pigment groups in the Cryptophyceae relates to monophylly of the chlorophyll a plus c and phycobilin-containing plastids. Finally, the carotenoids of primarily aquatic photolithotrophs show a great diversity which is difficult to accommodate in phyletic schemes without assuming multiple origins of some carotenoids (Rowan 1989; Fawley 1991).

Consideration of the package effect, i.e., the decreased achieved specific absorption coefficient of pigments when they are present at high areal densities (mol m^{-2}) in terms of the projection of individual cells or extended thalli at right angles to a vector photon field, or (via a more complicated analysis) to a scalar photon field, is necessary if the potential ecological implications of the variations in pigmentation are to be explored (e.g., Ramus 1978; Dring 1981). The decrease in the achieved in vivo specific absorption coefficient of the pigment at high areal chromophore densities means that, in addition to the decreased catalytic utility of individual pigment molecules, there is also a much smaller ratio of "peaks" to "troughs" in the

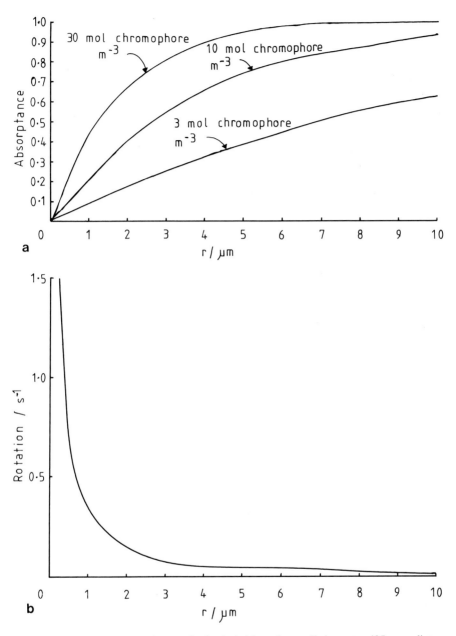

Fig. 15.2. **a** Computed absorbance of spherical chlorophyte cells in vector 435 nm radiaton as a function of cell radius (r) for three chromophore concentrations (mol chromophore per m^3 cell volume). (After Raven 1984b). Three mol chromophore m^{-3} corresponds to 40 g thylakoid per kg dry weight; 10 mol chromophore m^{-3} corresponds to 133 g thylakoid per kg dry weight; 30 mol chromophore m^{-3} corresponds to 400 g thylakoid per kg dry weight (Raven 1984b). **b** Computed rate of rotation of spherical cells at 20 °C as a function of cell radius (r). (Berg 1983; Mitchell 1991)

absorption sepctrum as a result of a greater package effect. The end-point of this process is a black organism with no features in its absorption spectrum: absorptance is 1.0 at all wavelengths between 400–700 nm. Accordingly, the nature of the pigments has a much greater influence on the effectiveness of absorption of particular wavelengths in small cells with minimal package effects. Thus, while fully concurring with the conclusions of Dring (1981) as to the minimal contribution of "chromatic adaptation" to photosynthetic performance of marine benthic macroalgae in situ, the situation could well be very different for smaller cells with much lower absorptances (Raven 1986a). This tendency is, of course, most marked for the very smallest O_2-evolving photolithotrophic cells, the picoplankton (Raven 1986a) where, even with the highest plausible chromophore concentration per unit volume, a 0.5 μm radius chlorophyte cell only absorbs 0.35 of incident vector radiation at 435 nm (the blue absorption peak): Fig. 15.2a. This argument, based on photophysical first principles, is consistent with the finding that the greatest diversity of tetrapyrrol-based chromophores within a single organism, and of novel carotenoids, occurs in phytoplankton organisms. Of course, the phylogenetic influence must not be ignored; the organisms in which this great diversity of chromophores occurs are found only in the plankton. However, I would argue that this is in itself significant; the selective advantage of diverce pigmentation is greatest in small organisms with a minimal package effect. Coming down to specifics, the eukaryotic organisms which have the greatest diversity of tetrapyrrols and carotenoids are in the Cryptophyceae and the Micromonadophyceae. The Cryptophyceae are the organisms which have chlorophylls a and c_2 as well as phycoerythrobilin and phycocyanobilin, and thus combine the attributes of the chlorophyll a plus chlorophyll c organisms and the chlorophyll a plus phycobilin organisms; all of them are planktonic, and a few approach the picoplankton size range (Thomsen 1986). The Micromonadophyceae include many picoplankters (Thomsen 1986; Guillard et al. 1991); all of them have the chlorophyll c-like MgDVP as well as light-harvesting carotenoids (e.g., prasinoxanthin; siphonein/siphonoxanthin) with significant absorption in vivo out to 540 nm. It is also worthy of note that the greatest diversity of c-type chlorophylls (c_3 and other "new" molecular species as well as c_1 and c_2) occurs in the smaller chromophytes (see Wilhelm and Wiedemann 1991, and references therein). Among the prokaryotic O_2-evolvers the picoplanktonic *Prochlorococcus marinus* is distinguished by having not only divinyl analogs of chlorophylls a and b, but also probably the chlorophyll c-like Mg 2,8 divinyl, 2,8 desethyl, pheoporphyrin a_5 (Goericke and Repeta 1992).

I would not, of course, attempt to contest data showing that modification of the basal chromophore composition of members of the three main pigment groups such as to extend the range of wavelengths at which a high absorptance is attained is also found in organisms with a very substantial package effect (high chromophore content per m^2 surface, and/or very large internal scattering). Examples are some macroscopic marine Ulvophyceae, especially deep water representatives (see Raven 1987), which have siphonein and/or siphonoxanthin, extending absorptance into the green region from the blue peak. The high *overall* absorptance of at least some of these algae makes this extension of limited value (see Ramus 1978; Dring 1981). In any case, the occurrence of these light-harvesting carotenoids is limited to primarily aquatic plants; secondarily aquatic plants with chlorophyll a plus chlorophyll b appear to lack them. Although deep water (secondarily) aquatic bryophytes have a high content, relative to chlorophyll a, of a number of the usual higher plant carotenoids, these do not seem to have any additional light-harvesting role (Boston et al. 1991).

The role of "anomalous" light-harvesting chromophores in large and small photolithotrophs clearly needs further investigation to test the suggestion that such additional chromophore species are of greater selective advantage in small cells, granted the radiation climate in which the organisms normally occur. Such studies should be integrated with energetic cost-benefit analyses of photon harvesting as a function of cell size and the nature and cost of synthesis of chromophores and the polypeptides which bind them (Raven 1984b, 1986a,b; Alberte 1989; Andrews 1991).

A further aspect of light effects on aquatic plants is that of photo-inhibition. The homologous nature of the photoreaction 2 reaction center in all O_2-evolvers is consistent with the potential for photoinhibition in all O_2-evolvers (Raven 1984a,b; Kyle et al. 1987). The only point I wish to make here is that the differences in pigmentation and thylakoid organization among the three main pigment groups of aquatic plants means that some of the means of *avoiding* photoinhibition damage are different in, for example, the chromophytes as compared to the green plants in the strict sense. Thus, the option of detaching light-harvesting complexes *specifically* from photo-reaction 2 reaction centers ("state transition") does not seem to be of general occurrence in chromophytes (Raven et al. 1989; Lichtlé et al. 1992; Pyszniak and Gibbs 1992; Raven 1993a). Furthermore, the option of a "carotenoid cycle" as a means of quenching excess excitation of the photoreaction 2 centers is present in chromophytes as well as green plants, but with different carotenoids (Willemoes and Monas 1991). As with the nature of the light-harvesting complexes discussed above, cost-benefit analyses of the various approaches which photolithotrophs have to photo-inhibition are possible (Raven and Samuelsson 1986; Raven 1989). It is possible (Raven 1989) to compare the energetic costs and benefits of *avoidance* of photoinhibition (taking into account any decrease in the potential for light-saturated or light-limited photosynthesis as well as the costs of avoidance of damage per se) with those of permitting photoinhibition to occur, with subsequent *repair* (again taking into account the decrease in photosynthetic rate in high or low light as well as the direct energy costs of repair): Raven (1989). The rather specialized avoidance mechanism modeled by Raven (1989) should be extended to a wider range of avoidance mechanisms, such as the state transitions and carotenoid cycles discussed above.

The cost-benefit analysis of photon absorption for photosynthesis as a function of the nature of the light-harvesting pigment-protein complex and of the size of the organism (Raven 1984b, 1986a) was extended by Raven (1991a) to screening sensitive sites (DNA; plastoquinone (PQ)) from UV-B as a function of cell size. Such analyses could profitably be extended (cf. the photoinhibition analysis above) with those of repair after damage (cf. Karentz et al. 1991) for organisms of different size; just as smaller organisms make more effective use of each molecule of chromophore of light-harvesting machinery, they have restrictions on the effectiveness of UV-B screening (Raven 1984b, 1989, 1991a).

A final aspect of size effects on photosynthesis by aquatic plants as a function of variations in photon flux density relates to movement of organisms. This has been addressed previously in terms of, for example, wave-frequency variations in shading of understorey marine macroalgae (Greene and Gerard 1990) and the optimization of resource acquisition from inverse gradients of photon flux density (only available in daytime) and nutrients by diel vertical migration of flagellates in stratified water bodies (Raven and Richardson 1984). I now wish to address rotation of planktophytes in a vector radiation field. Pienaar (1980) shows that swimming flagellates (including photolithotrophic flagellates) rotate at about 1 Hz; Raven (1993a) points out that this could, in cells with a significant fractional photon absorption and thus a significant photon flux density gradient across their diameter, serve to expose peripheral parts of the photosynthetic apparatus to significant variations (twofold or more for a densely pigmented 5 μm radius cell: Fig. 15.2a) when the cell is swimming at right angles to the vector radiation. This variation is within the range of frequencies at which constructive interactions of variations in photon flux density on photosynthetic rate, at least at high photon flux densities, occur (see Greene and Gerard 1990, for data on a benthic macrophyte in which the light field is in a constant spatial relationship to individual parts of the photosynthetic apparatus).

In addition to the rotation around the axis along which flagellar motion is occurring, all planktonic cells are subject to rotation under the influence of very small-scale turbulence (grading into thermal vibration) which occurs even in macroscopically still water bodies (Berg 1980; Mitchell 1991). This rotation is inversely related to cell size; the frequency as a function of cell radius at 20 °C is indicated in Fig. 15.2b. The size range in which rotation occurs at frequencies (0.1–1.0 Hz) at which constructive interactions of variations in photon flux density are found (Greene and Gerard 1990) is ~0.5–1.2 μm radius, i.e., picoplankton-size cells. The extent to which this rotation can yield increased photosynthesis and growth in a vector light field is, of course, a function of the difference in illumination between the "lit" and "shaded" side of the cell; even with a high chromophore content per unit volume, the difference for a 0.5 μm radius cell is not greater than 0.35 of incident photon flux density taken as 1.0, i.e., the photon flux density at the "shaded" side is 0.65 that at the "lit" side (Fig. 15.2a). A larger difference (up to 0.5) could occur for 1 μm radius cells, again with chromophore content per unit volume at the upper end of the observed range (Fig. 15.2a). Such differences could be significant for photosynthetic rates in vector radiation fields close to light saturation for photosynthesis. We note that an analogous analysis for rather larger cells by Babin et al. (1992) concludes that there is unlikely to be a constructive effect of rotation on photosynthetic rate. While concurring with this *conclusion* for these larger, slowly rotating cells (cf. Greene and Gerard 1990), Babin et al. (1992) used the specific reaction rate of redox reactions of photosynthesis ($\sim$20–200 mol electron transferred [mol photoreaction 2]$^{-1}$ s^{-1}) to estimate the rotation frequency (20–200 Hz) at which constructive effects on photosynthesis could be expected, rather than the observed lower frequency at which such stimulations do occur; the extension of stimulatory effects to lower frequencies relates to the finite pool sizes of intermediates of photosynthetic reactions. Since the rotation rate is a linear function of the viscosity of the medium, the occurrence of intermittency effects on photosynthesis as a function of rotation rate could be examined by altering the viscosity of the medium with solutes (e.g., polymers) which do not impose an osmotic (or other) metabolic challenge to the cells.

15.6 Inorganic Carbon Acquisition by Aquatic Plants: When Does It Limit Net Productivity?

Much work has been devoted over the past two decades to the study of inorganic carbon acquisition by aquatic plants (see Raven 1991b). It has been established that some aquatic plants acquire their inorganic carbon in situ in a manner analogous to that of terrestrial C_3 plants, i.e., by CO_2 diffusion from the bulk phase to ribulose bisphosphate carboxylase-oxygenase (Rubisco), although in the great majority of aquatics all of the diffusion pathway is in the aqueous phase. However, this mechanism may yield a relatively low potential rate of photosynthesis if CO_2 concentration in solution is at or below the air-equilibrium concentration, especially at high temperatures, in view of the low diffusion coefficient of CO_2 in water (about 10^{-4} of the value in air) which is not offset by the thinner diffusion boundary layers in aquatic environments except in the case of the smallest (picoplanktonic) cells (Raven 1991b,c). Accordingly, a majority of aquatic plants rely on more complex mechanisms of inorganic carbon acquisition. Some of these are listed in Table 15.3. A common feature of all of these mechanisms is that they give a steady-state CO_2 concentration at the site of Rubisco activity which exceeds that in air-equilibrium solution. This has theoretical implications (in some cases tested and verified) for the *rate* of CO_2 fixation

310 J.A. Raven

Table 15.3. Mechanisms of inorganic carbon acquisition by aquatic plants, other than by diffusion of CO_2 from a bulk phase (at or below air equilibrium) to Rubisco, using a totally aqueous phase pathway. All of these mechanisms result in a higher CO_2 concentration at the site of Rubisco activity than could be achieved by CO_2 diffusion from air-equilibrium solution

Mechanism	Example	Reference
1 Inhabit environment enriched in CO_2 by chemoorganotrophy acting on products of photosynthesis at the expense of atmospheric CO_2; diffusive CO_2 entry in aqueous phase; fixation by Rubisco	Many phototrophs in small bodies of freshwater; supply maybe enhanced by rapid flow reducing thickness of diffusion boundary layer (primarily and secondarily aquatic plants)	Raven (1991b); Raven (1992b)
2 Rhizophyte rooted in sediment enriched in CO_2 by chemoorganotrophy acting on photosynthetic material sedimented from bulk phase; CO_2 transport from roots to shoots in intercellular gas spaces; fixation by Rubisco	All vascular plants of the isoetid life-form (secondarily aquatic); all freshwater	Raven (1991b); Raven et al. (1988)
3 Use of HCO_3 by extracelluar catalysis of HCO_3^- to CO_2 followed by CO_2 entry, *or* HCO_3^- active influx followed by catalysis of HCO_3^- conversion to CO_2; includes (freshwater only) HCO_3^- use based on acidic and alkaline zones on the plant surface; fixation by Rubisco	Many primarily and secondarily aquatic plants in marine and freshwater environments	Raven (1991b)
4 Active influx of CO_2; fixation by Rubisco	Many primarily aquatic plants	Raven (1991b)
5 Primary fixation by phosphoenolpyruvate carboxylase or phosphoenolpyruvate carboxykinase, followed by (C_4-C_1) decarboxylation and refixation by Rubisco	Spatial separation: C_4-like metabolism by the marine (primarily aquatic) *Udotea flabellum*. Temporal separation: CAM-like behaviour of some isoetids, aquatic Crassulaceae	Raven (1991b); Reiskind and Bowes (1991); Raven et al. (1988); Newman (1991)

on a biomass basis, and the *cost* of CO_2 fixation in terms of photons, nitrogen, iron, manganese, and molybdenum (Raven 1991b,c).

The plethora of mechanisms of inorganic carbon acquisition whereby aquatic plants increase the CO_2 supply to Rubisco in excess of the air-equilibrium value suggests that the fitness of aquatic plants is significantly enhanced by such mechanisms. Does this mean that productivity of aquatic plants would routinely be limited by diffusive CO_2 supply from air-equilibrium solution under natural conditions of growth? The answer to this question is undoubtedly "yes" for larger organisms in otherwise resource-rich environments; rapid growth by such organisms requires the operation of a CO_2-concentrating mechanism (entries 3–5 in Table 15.3) which is, indeed, present in these organisms (Raven 1991b). The occurrence of such mechanisms in smaller aquatic photolithotrophs, and especially picoplankton cells, is less obviously essential if CO_2 diffusion from an air-equilibrium solution to Rubisco is considered if the kinetics of Rubisco are taken as those of terrestrial C_3 plants. However, most aquatic plants have Rubisco kinetics which are markedly less favorable to net CO_2 fixation in air-equilibrium solutions than those of the terrestrial C_s plant Rubisco (Raven 1984a; Raven et al. 1990a), thus requiring a CO_2-concentrating mechanism if CO_2 fixation per unit Rubisco is to be more than a small fraction of the potential rate. Furthermore, the possibility, (proved in some cases), that the use of CO_2 concentrating mechanisms economizes in the use of other resources could evolutionarily favor their occurrence in many natural environments, including HCO^-_3-rich seawater (Raven 1991b,c).

A final, vital, question on the ecophysiology of inorganic carbon acquisition by aquatic plants is the extent to which carbon supply limits net productivity in aquatic environments. My tentative answer is "relatively infrequently". CO_2-concentrating mechanisms, and/or CO_2 enrichment from terrestrial inputs to freshwater, mean that aquatic plants are more frequently limited by other nutrients, by light or by temperature (Maberly 1985a,b; Raven and Richardson 1986; Raven and Johnston 1991b; Raven 1993b). This is not to say that the inorganic carbon supply might not have a significant influence on the relative performance of different species, e.g., organisms with CO_2 concentrating mechanisms (e.g., diatoms) relative to those with diffusive CO_2 entry (e.g., coccolithophorids): Raven (1991c, 1993b). Limitation of the growth of aquatic plants by factors other than inorganic carbon supply is exemplified by many flowing bodies of freshwater, supplied with CO_2 (at above air-equilibrium concentrations) and other nutrients as "leaks" from terrestrial communities in groundwater. Here there is typically a net CO_2 loss from the water body to the atmosphere; growth of the aquatic plants is limited by some factor other than CO_2 supply to Rubisco (e.g., light if riparian vegetation shades the stream, or a nutrient other than CO_2), so these organisms merely serve to decrease the net evasion of CO_2 to the atmosphere which could occur from a similar but sterile stream (Raven 1992b).

15.7   Water Relations of Intertidal Aquatic Plants
in Relation to Photosynthesis

Primarily aquatic plants are all poikilohydric. Those which live in habitats in which exposure to air is predictable (the marine intertidal, seasonal freshwater pools, streams subject to seasonal drawdown) and is sufficiently prolonged to substantially desiccate the populations, have a desiccation-tolerant stage in the life cycle. For the marine algae the desiccation tolerance involves the vegetative phases(s) (Rugg and Norton 1987). For freshwater algae desiccation tolerance may involve the vegetative body (e.g., the diploid perennial "*Chantransia*" phase, but not the annual semi-erect haploid gametophyte phase, of the batrachospermalean red algae) or phases associated with sexual reproduction and/or dispersal (Raven 1992b).

Among secondarily aquatic plants, the bryophytes are (like their terrestrial ancestors) poikilohydric; the vascular plants which are most thoroughly aquatic have lost the homoiohydric capacities of their terrestrial ancestors. Perennial haptophytes of predictably annually desiccated environments are desiccation-tolerant (Raven 1986b).

Before considering the implications of these attributes for photosynthesis, it is worth pointing out that at least one primarily aquatic marine macroalga, the high intertidal brown *Pelvetia canaliculata*, requires periodic emersion for its survival (Rugg and Norton 1987). The habitat of such (secondarily) aquatic marine lichens as *Lichina pygmaea*, i.e., high intertidal of exposed shores, is consistent with a requirement for periodic emersion (Raven et al. 1990a,b) such as might be expected from the known properties of other lichen symbioses.

The acquisition of inorganic carbon by periodically emersed aquatics as a function of wetting-drying cycles is analogous to the terrestrial lichen situation which Professor Dr. Lange has so successfully addressed (Lange et al. 1970; Lange 1989). Much of the work on intertidal algae has suffered from the use of O_2 evolution as a measure of submersed photosynthesis and CO_2 fixation to measure emersed photosynthesis (e.g., Surif and Raven 1990). Recent work has overcome this by measuring inorganic C uptake (Maberly and Madsen 1990) or O_2 evolution (Britting 1992) for both emersed and submersed thalli, in each case using natural inorganic carbon levels in both air and seawater. A further problem concerns putting short-term measurement into the context of environmental variability. Integration of the observed dependence of emersed CO_2 fixation on water content during a drying cycle, and the delay in resumption of submersed photosynthesis upon re-immersion while any damage incurred during desiccation is repaired, with the in situ variables (tides, light-dark cycles, weather) is complex. The best-investigated example is that of the mid-high intertidal brown macroalga *Fucus spiralis*; Maberly and Madsen (1990) and Madsen and Maberly (1990) showed that this alga in situ fixes about a quarter of its total carbon during emersion. This finding is of interest in the context of the extent of limitation of productivity by inorganic carbon supply considered under Section 15.6 above. If light is limiting productivity, then emersed CO_2 acquisition in the light might be advantageous for plants growing in the turbid coastal waters. However, if nutrients other than carbon are limiting productivity, carbon acquisition during emersion (when the supply of combined nitrogen or phosphorus is negligible) serves to further exacerbate the potential imbalance in supply of carbon relative to the more rate-limiting nutrient. Thomas et al (1987) have shown a substantial post-emersion stimulation of uptake of (noncarbon) nutrients when submersed which could occur if nitrogen or phosphorus supply were limited by diffusion through boundary layers unless very large changes in the capacity for transport across the plasmalemma at lower concentrations at the transporter site. Further work is clearly needed.

Two further aspects are of current interest. One relates to the relationship between carbon gain during dehydration during emersion and the mechanism of inorganic carbon assimilation. Surif and Raven (1990) have suggested that intertidal macroalgae with CO_2-concentrating mechanisms could, by maintaining a larger CO_2 concentration gradient through gaseous diffusion boundary layers than could plants with a purely diffusive CO_2 supply to Rubisco, fix more CO_2 per unit water evaporated. This means that, with a fixed supply of thallus water (the dowry left by the retreating tide), more inorganic carbon could be fixed before the extent of water loss curtailed photosynthesis in each desiccation event. This suggestion should be tested. Analogous work on terrestrial lichens (Cowan et al. 1992) is more complex since the sites of CO_2 uptake and water vapor loss are more likely to be different in these lichens than in the intertidal macroalgae.

The other aspect of water relations of intertidal macroalgae which can be related to those of certain terrestrial lichens and related free-living terrestrial algae is that of the possibility of uptake of water vapor from an unsaturated atmosphere by desiccated algae. Terrestrial lichens with green photobionts, and related free-living terrestrial green algae, have this capacity to rehydrate without contact with liquid water (Bertsch 1966; Lange 1989). Such a water uptake mechanism has not been reported for intertidal algae, and may be unlikely (although it should be tested for in such possibly secondarily aquatic littoral fringe algae as *Prasiola*, which furthermore forms lichen-like associations with fungi: Raven and Johnston 1991a). However, results of Cooper and De Niro (1989) on the $^2H/^1H$ and $^{18}O/^{16}O$ of water in intertidal macroalgae and seagrasses may give a pointer here. Net loss of water from emersed plants by evaporation preferentially removes water depleted in the heavy isotopes ^2H and ^{18}O; however, the water in emersed specimens of the intertidal plants is itself depleted in ^2H and ^{18}O, instead of being enriched in the heavy isotopes as would be expected from net loss of ^2H- and ^{18}O-depleted water vapour (Cooper and De Niro 1989). The suggested explantation is exchange of water in the plant with ^2H- and ^{18}O-depleted water vapor overcoming the effect of net evaporative loss (Cooper and De Niro 1989). The extent to which this constitutes evidence for rehydration from water vapor is debatable, but evidence for such a mechanism in intertidal plants should be sought. A start has been made by Britting (1992) in her work on the high intertidal red alga *Endocladia muricata*.

15.8 Conclusions

Photosynthesis in aquatic plants must be viewed in the context of the great diversity of these plants. The primarily aquatic plants have a much greater diversity of light-harvesting pigments than do secondarily aquatic plants (or their terrestrial ancestors). The ecophysical significance of the variety of pigments is likely to be greater in the smallest (picoplanktonic) aquatic plants than in larger plants as result of differences in the package effect.

The majority of aquatic plants acquire inorganic carbon by a mechanism more complex than CO_2 diffusion in solution from the bulk phase to Rubisco. The occurrence of these CO_2-concentrating mechanisms is related to low CO_2 diffusivity in water (especially significant for larger plants) and the properties of Rubisco in aquatic plants. The diversity of mechanisms of inorganic carbon acquisition is, by contrast to light-harvesting machinery, greater in secondarily than primarily aquatic plants. The diversity of CO_2 acquisition mechanisms may have significance for the extent of carbon assimilation during emersion in intertidal plants.

The availability of light, and of nutrients other than carbon, is probably a more significant abiotic factor limiting net primary productivity of aquatic plants than is the supply of inorganic carbon. However, the mechanism by which inorganic carbon is acquired by aquatic plants may affect the need for photons or other nutrients for photosynthesis.

The most important conclusion is, however, that we are still very ignorant of the ecophysiology of photosynthesis by aquatic plants relative to what is known about their terrestrial counterparts; it would be very helpful if some Lange clones tackled some of these problems.

Acknowledgments. Work on resource acquisition by aquatic plants in the author's laboratory has been supported by S.E.R.C., N.E.R.C., the Nuffield Foundation and the British Council; it is currently supported by N.E.R.C. Interaction with past and present colleagues has been very important in arriving at the views expressed here.

References

Alberte RS (1989) Physiological and cellular features of *Prochloron*. In: Lewin RA, Cheng L (eds) *Prochloron*: a microbial enigma. Chapman and Hall, New York, pp 31–52

Andrews JH (1991) Comparative ecology of microorganisms and macroorganisms. Springer, Berein Heidelberg New York

Babin M, Levasseur M, Michaud D, Legendre L (1992) Effect of angular distribution of light on photosynthesis at the scale of a phytoplankton cell. In: Maestrini S (ed) Poster abstracts of symposium on measurement of primary production from the molecular to the global scale. ICES, Copenhagen, p 1

Berg M (1983) Random walks in biology. Princeton University Press, Princeton

Bertsch A (1966) CO_2-Gaswechsel und Wasserhaushalt der aerophilen Grünalge *Apactococcus lobatus*. Planta 70: 46–62

Boston HL, Farmer AM, Madsen TD, Adams MS, Hurley JP (1991) Light-harvesting caratenoids in two deep-water bryophytes. Photosynthetica 25: 61–66

Britting SA (1992) Effect of emergence on the physiology and biochemistry of a high intertidal alga, *Endocladia muricata* (Post & Rupt) J Ag (Cryptonemiates, Rhodophyta). PhD dissertation, University of California, Los Angeles

Cooper LW, De Niro MJ (1989) Detection of heavy isotopes of oxygen and hydrogen in tissue water of intertidal plants: implications for water economy. Mar Biol 101: 397–400

Cowan IR, Lange OL, Green TGA (1992) Carbon-dioxide exchange in lichens: determination of transport and carboxylation reactions. Planta 187: 282–284

Den Hartog C, Segal C (1964) A new classification of water plant communities. Acta Bot Neerl 13: 367–393

Dring MJ (1981) Chromatic adaptation of photosynthesis in benthic marine algae: an examination of its ecological significance using a theoretical model. Limnol Oceanogr 26: 271–284

Fawley MW (1991) Disjunct distribution of the xanthophyll loroxanthin in the green algae. J Phycol 27: 544–548

Fujiward S, Iwahashi H, Someya J, Nishikawd S (1993) Structure and cotranscription of the plastid-encoded rbcL and rbcS genes of *Pleuroehrysis carterae* (Prymnesiophyta). J Phycol 29: 347–355

Goericke R, Repeta DJ (1992) The pigments of *Prochlorococcus marinus*: the presence of divinyl chlorophyll *a* and *b* in a marine procaryote. Limnol Oceanogr 37: 425–433

Greene RM, Gerard VA (1990) Effects of high-frequency light fluctuation on growth and photoacclimation of the red alga *Chondus crispus*. Mar Biol 105: 337–344

Guillard RRL, Keller MD, O'Kelly CJ, Floyd GL (1991) *Pycnococcus provasolii* gen. et sp. nov., a coccoid prasinoxanthin-containing phytoplankton from the western North Atlantic and Gulf of Mexico. J Phycol 27: 39–47

Karentz D, Cleaver JE, Mitchell DL (1991) Cell survival characteristics and molecular responses of phytoplankton to ultraviolet B radiation. J phycol 27: 326–341

Kirk JTO (1985) Light and photosynthesis in aquatic ecosystems. Cambridge University Press, Cambridge

Kyle DJ, Osmond CB, Arntzen CJ (eds) (1987) Photoinhibition. Elsevier, Amsterdam

Lange OL (1989) Ecophysiology and photosynthesis: performance of poikilohydric and homoiohydric plants. In: Greuter W, Zimmer B (eds) Proceedings of the XIV International Botanical Congress. Koeltz, Königstein/Taunus, pp 357–383

Lange OL, Schulze ED, Koch W (1970) Experimental-ökologische Untersuchungen an Flechter der Negev-Wüste II CO_2-Gaswechsel und Wasserhaushalt von *Ramalina maciformis* (De) Bory am natürlichen Standort während der sommerlichen Trockenperiode. Flora (Jena) 159: 38–62

Lichtlé C, Spilar A, Dural JC (1992) Immunogold localization of light-harvesting and photosystem I complexes in the thylakoids of *Fucus serratus* (Phaeophyceae). Protoplasma 166: 99–106

Littler MM, Littler DS, Blair SM, Norris JN (1985) Deepest known plant life discovered on an unchanted seamount. Science 227: 57–59

Littler MM, Littler DS, Blair SM, Norris JN (1986) Deep water plant communities from an uncharted seamount off San Salvador Island, Bahamas: distribution, abundance and primary productivity. Deep-Sea Res 33: 881–892

Luther H (1947) Vorschlag zu einer ökologischen Grundeinteilung der Hydrophyten. Acta Bot Fenn 44: 1–15

Maberly SC (1985a) Photosynthesis by *Fontinalis antipyretica*. I. Interaction between photon irradiance, concentration of carbon dioxide and temperature. New Phytol 100: 127–140

Maberly SC (1985b) Photosynthesis by *Fontinalis antipyretica*. II. Assessment of environmental factors limiting photosynthesis and production. New Phytol 100: 141–155

Maberly SC, Madsen TV (1990) Contribution of air and water in the carbon balance of *Fucus spiralis*. Mar Ecol Prog Ser 62: 175–183

Madsen TV, Maberly SC (1990) A comparison of air and water as environments for photosynthesis by the intertidal alga *Fucus vesiculosus* (Phaeophyta). J Phycol 26: 24–30

Mitchell JG (1991) The influence of cell size on marine bacterial mobility and energetics. Microb Ecol 22: 227–238

Newman JR (1991) Carbon assimilation by freshwater aquatic macrophytes. PhD Thesis, University of Dundee

Palenik B, Haselkorn R (1992) Multiple evolutionary origins of prochlorophytes, the chlorophyll *b*-containing prokaryotes. Nature 355: 265–267

Pienaar RN (1980) Chrysophytes. In: Cox ER (ed) Phytoflagellates. Elsevier, New York, pp 213–242

Pyszniak AM, Gibbs SP (1992) Immunocytochemical localization of photosystem I and the fucoxanthin-chorophyll a/c light-harvesting complex in the diatom *Phaeodactylum tricornutum*. Protoplasma 166: 208–217

Ramus J (1978) Seaweed anatomy and photosynthetic performance: the ecological significance of light guides, heterogenous absorption and multiple scatter. J Phycol 14: 352–362

Raunkaier C (1934) The life forms of plants and statistical plant geography. Clarendon Press, Oxford

Raven JA (1981) Nutritional strategies of submerged benthic plants: the acquisition of C, N and P by rhizophytes and haptophytes. New Phytol 88: 1–30

Raven JA (1984a) Energetics and transport in aquatic plants. Liss, New York

Raven JA (1984b) A cost-benefit analysis of photon absorption by photosynthetic unicells. New Phytol 98: 593–625

Raven JA (1986a) Physiological consequences of extremely small size for autotrophic organisms in the sea. In: Platt T, Li WKW (eds) Photosynthetic picoplankton. Can Bull Fish Aquat Sci 214: 1–70

Raven JA (1986b) Evolution of life forms. In: Givnish TV (ed) On the economy of plant form and function. Cambridge University Press, Cambridge, pp 421–492

Raven JA (1987) Biochemistry, biophysics and physiology of chlorophyll b-containing algae: implications for taxonomy and phylogeny. Prog Phycol Res 5: 1–122

Raven JA (1989) Fight or flight: the economics of repair and avoidance of photoinhibition of photosynthesis. Funct Ecol 3: 5–19

Raven JA (1991a) Responses of aquatic photosynthetic organisms to increased solar UV-B. J Photochem Photobiol B: Biology 9: 239–244

Raven JA (1991b) Implications of inorganic C utilization: ecology, evolution and geochemistry. Can J Bot 68: 905–924

Raven JA (1991c) Physiology of inorganic C acquisition and implications for resource use efficiency by marine phytoplankton: relation to increased CO_2 and temperature. Plant Cell Environ 14: 779–794

Raven JA (1992a) Energy and nutrient acquisition by autotrophic symbioses. Symbiosis 14: 33–60

Raven JA (1992b) How benthic macroalgae cope with flowing freshwater: resource acquisition and retention. J Phycol 28: 133–146

Raven JA (1993a) Comparative aspects of chrysophyte nutrition with emphasis on carbon, phosphorus and nitrogen. In: Sandgren CD, Smol JP, Kristiansen J (eds) Chrysophyte algae: ecology, phylogeny and development. Cambridge University Press, Cambridge (in press)

Raven JA (1993b) Carbon: a phycocentric view. In: Evans GT, Fasham MJR (eds) Towards & Model of Ocean Biogeochemical Cycles. Springer, Berlin Heidelberg New York, pp 123–152

Raven JA, Johnston AM (1991a) Photosynthetic inorganic carbon assimilation by *Prasiola stipitata* (Prasiolales, Chlorophyta) under emersed and submersed conditions: relationship to the taxonomy of *Prasiola*. Br Phycol J 26: 247–247

Raven JA, Johnston AM (1991b) Carbon assimilation mechanisms. Implications for intensive culture of seaweeds. In: Garcia-Reina G, Pedersen M (eds) Seaweed cellular biotechnology, physiology and intensive cultivation. Las Palmas de Gran Canaria, Espana, pp 151–166

Raven JA, Johnston AM (1991c) Mechanisms of inorganic carbon acquisition in marine phytoplankton and their implications for the use of other resources. Limnol Oceanogr 36: 1701–1714

Raven JA, Richardson K (1984) Dinophyte flagella: a cost-benefit analysis. New Phytol 98: 259–276

Raven JA, Richardson K (1986) Marine environments. In: Baker NR, Long SP (eds) Photosynthesis in contrasting envioronments. Elsevier, Amsterdam, pp 337–398

Raven JA, Samuelsson G (1986) Repair of photoinhibity damage in *Anacystis nidulans* 625 (*Synechococcus* 6301): relation to catalytic capacity for, and energy supply to, protein synthesis, and implications for μ_{max} and the efficiency of light-limited growth. New Phytol 103: 625–643

Raven JA, Smith FA, Smith SE (1980) Ions and osmoregulation. In: Rains DW, Valentine RC, Hollaender A (eds) Genetic engineering of osmoregulation: impact on plant productivity for food, chemicals and energy. Plenum Press, New York, pp 101–118

Raven JA, Handley LL, MacFarlane JJ, McInroy S, McKenzie L, Richards JH, Samuelsson G (1988) The role of root CO_2 uptake and CAM in inorganic C acquisition by plants of the isoetid life form. A review, with new data on *Eriocaulon decangulare*. New Phytol 108: 125–148

Raven JA, Johnston AM, Surif MB (1989) The photosynthetic apparatus as a phyletic character. In: Green JC, Leadbeater BSC, Diver WL (eds) The chromophyte algae. Problems and perspectives. Oxford Science Publications, Oxford, pp 63–84

Raven JA, Johnston AM, MacFarlane JJ (1990a) Carbon metabolism. In: Sheath RG, Cole KM (eds) The biology of the red algae. Cambridge University Press, New York, pp 171–202

Raven JA, Johnston AM, Handley LL, McInroy SG (1990b) Transport and assimilation of inorganic carbon by *Lichina pygmaea* under emersed and submersed conditions. New Phytol 114: 407–417

Reiskind JB, Bowes G (1991) The role of phosphoenolpyruvate carboxykinase in a marine macroalga with C_4-like photosynthetic characteristics. Proc Natl Acad Sci USA 88: 2993–2887

Rowan KS (1989) Photosynthetic pigments of algae. Cambridge University Press, Cambridge

Rugg DA, Norton TA (1987) *Pelvetia canaliculata*, a high-shore seaweed that shuns the sea. In: Crawford RMM (ed) Plant life in aquatic and amphibious habitats. Blackwell, Oxford, pp 347–358

Sand-Jensen K, Pedersen MF, Nielsen SL (1992) Photosynthetic use of inorganic carbon among primary and secondary water plants in streams. Freshwater Biol 27: 283–293

Stebbins GL, Hill GJC (1980) Did multicellular plants invade the land? Am Nat 115: 342–353

Surif MB, Raven JA (1990) Photosynthetic gas exchange under emersed conditions in eulittoral and normally submersed members of the Fucales and the Laminariales: interpretation in relation to C isotope ratio and N and water use efficiency. Oecologia 82: 68–80

Thomas TE, Turpin DH, Harrison PJ (1987) Desiccation enhanced nitrogen uptake rates in intertidal seaweeds. Mar Biol 94: 293–298

Thomsen HA (1986) A survey of the smallest eucaryotic organisms of the marine phytoplankton. In: Platt TL, Li WKW (eds) Photosynthetic picoplankton. Can Bull Fish Aquat Sci 214: 121–158

Urbach E, Robertson DL, Chisholm SW (1992) Multiple evolutionary origins of prochlorophytes within the cyanobacterial radiation. Nature 355: 267–270

Wilhelm C, Wiedemann I (1991) Evidence of protein-bound chlorophyll c_3 in a light-harvesting protein isolated from the flagellate alga *Prymnesium parvum* (Prymnesiophyceae). Photosynthetica 25: 249–255

Willemoes M, Monas E (1991) Relationship between growth irradiance and xanthophyll cycle pool in the diatom *Nitzschia palea*. Physiol Plant 83: 459–456

16 Photosynthesis in Poikilohydric Plants: A Comparison of Lichens and Bryophytes

T.G.A. Green and O.L. Lange

16.1 Introduction

Poikilohydric plants are those in which water status is completely dependent on their environment (Walter 1931) so that, in terrestrial habitats, the water vapor partial pressure of the plant body comes into equilibrium with the humidity of the atmosphere. These plants must be able to tolerate some desiccation although the minimum water potentials reached will depend on the environment and some poikilohydric plants may not be desiccation-tolerant (Green et al. 1991). In contrast, homoiohydric, terrestrial, vascular plants maintain their water status within fairly tight limits. Poikilohydry is exhibited by some ferns, club mosses, and even several phanerogams, but the major groups are algae, cyanobacteria (e.g., cyanobacterial mats), fungi and, of course, lichens, and bryophytes. Taxonomically, lichens and bryophytes are widely separated since, under presently accepted systems of classification (Raven et al. 1986), they fall into two separate Kingdoms; lichens in the Mycota and bryophytes in the Plantae. The situation is further complicated by the phycobionts of lichens which may be in the Monera (cyanobacteria) or Protista (chlorophyta) or in both, when multiple photobionts are present.

Despite the taxonomic divergence, lichens and bryophytes are, to a greater or lesser extent, present in almost every terrestrial ecosystem and may dominate some plant communities. The single factor that both lichens and bryophytes are poikilohydric may be sufficient to explain their co-occurrence in so many habitats. Although the two groups often coexist, there is a general tendency for bryophytes to dominate in wetter environments, such as in rainforests, and lichens to predominate in dry areas, such as deserts.

Poikilohydry also affects the ecophysiology of photosynthesis for both lichens and bryophytes. There is a general perception that both groups have low photosynthetic rates compared to phanerogams. In this chapter an attempt will be made to analyze bryophytes and lichens in terms of their photosynthetic features: the mechanisms and how morphology interacts with CO_2 exchange. The analysis is not a complete review but builds on past reviews including those of Proctor (1979, 1980, 1981, 1982, 1984, 1990),

Kappen (1973, 1988) and Longton (1988). No previous work seems to have compared and analyzed lichens and bryophytes on the basis of their photosynthetic features. It is hoped that such an analysis could provide insights into several aspects of the groups, for instance, similarities that might explain coexistence, or characteristics that favor one group over the other in certain habitats.

16.2 CO$_2$ Exchange of Lichens and Bryophytes

16.2.1 Net Photosynthetic Rates

Maximum net photosynthetic rates (A) of lichens and bryophytes have been widely portrayed as being lower than those of other terrestrial plants (Table 16.1, see also Bannister 1976; Larcher 1984). Indeed, rates on a dry weight or area basis are much lower than those of C$_3$ and C$_4$ phanerogams. Only CAM plants are lower on a dry weight basis but even they perform better on an area basis. There has been considerable debate on how best to present A

Table 16.1. Maximum values for net photosynthesis (A)

	Net CO$_2$ uptake	
	$mgCO_2\,dm^{-2}\,h^{-1}$	$mgCO_2\,mg^{-1}\,h^{-1}$
Herbaceous plants		
C$_4$	30–80	60–140
Crop plants (C$_3$)	20–45	30–60
Deciduous trees		
Sun leaves	10–20	15–25
Shade leaves	5–10	
Evergreen trees	5–18	4–18
CAM		
Light	3–20	0.3–2.0
Dark	10–15	1–1.5
Ferns	3–5	
Lichens	0.5–2.0	0.3–5.0
Mosses	<3.0	0.6–3.5

Data mainly from Kallio and Kärenlampi (1975), Bannister (1976), Larcher (1984), Longton (1988)

of bryophytes and lichens for purposes of comparison. Dry weight is the most common basis but only because of ease of use. The choice of the basis for A can affect interpretation of results: large differences in A on a dry weight basis for different sizes of *Umbilicaria* thalli disappear when expressed on an area basis (Larson 1979, 1984).

Chlorophyll content has only rarely been used as the basis to calculate A. The limited data available show that A, for lichens and bryophytes, fall in the same range as those for higher plants (Table 16.2; see also Cowan et al. 1992). The agreement is even more remarkable when it is considered that the plants listed range from the deep shade lichen, *Pseudocyphellaria dissimilis* to the full sun tree, *Alnus glutinosa*. Chlorophyll contents in lichens and bryophytes have often been suggested as the limiting factor for A because, on a dry weight basis, they are so low compared to higher, vascular plants (for summary see Tretiach and Carpanelli 1992). Although A and chlorophyll content are often correlated (see Nash et al. 1980, for fruticose lichens) it seems premature to conclude that a causal relationship exists. If gross photosynthesis is calculated as A + R (R = dark respiration), then bryophytes and lichens may even perform better than higher plants, because R is so large compared to A.

In conclusion, the general concensus that lichens and bryophytes have low A is supported but only when A is on a dry weight or area basis. On a chlorophyll basis both groups perform similarly to phanerogams, indicating that the photosynthetic processes are fundamentally similar.

Table 16.2. Maximum net photosynthesis on a chlorophyll basis

Plant	Net photosynthesis $mg\,CO_2\,mg^{-1}Chl\,h^{-1}$	Refs.
Vascular		
Alnus glutinosa	4.0	Green (unpubl.)
Picea abies	4.8	Lange et al. (1989)
Picea abies	5.3	Oren and Zimmerman (1989)
(10) Desert plants	1.5–7.0	Schulze et al. (1975)
Lichens		
Pseudocyphellaria dissimilis	5.6	Snelgar (1971)
Parmelia caperata	4.0	Tretiach and Carpanelli (1992)
Cladonia rangiformis	1.5–3.0	Hawksworth and Hill (1984)
Peltigera praetextata	18.5 (Chl. a)	Hawksworth and Hill (1984)
Umbilicariaceae	0.26–1.01	Sancho and Kappen (1989)
Bryophytes		
Grimmia laevigata	1.4–2.0	Alpert (1988)
Leucobryum glaucum	2.5	McCall and Martin (1991)
Thuidium delicatulum	4.0	McCall and Martin (1991)
Dicranum scoparium	3.5	McCall and Martin (1991)

16.2.2 Compensation Points and Photorespiration

Bryophyte photosynthesis appears similar in several respects to that of C_3 higher plants. Carbon isotope ratios ($\partial^{13}C$ from -24 to $-35‰$, Rundel et al. 1979; Teeri 1981) fall within the range expected for C_3 plants and bicar-

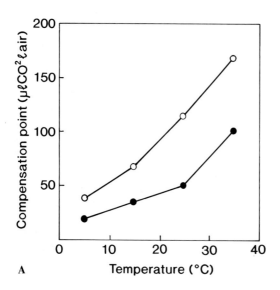

A

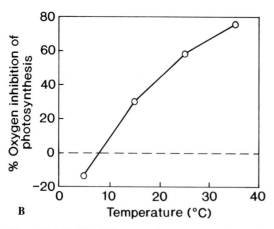

B

Fig. 16.1A,B. The influence of oxygen concentration (2 and 21%) and temperature on: *top panel* CO_2 compensation point, 2% oxygen *solid symbols*; 21% oxygen *open symbols*; and *lower panel* oxygen inhibition of A, for *Marchantia berteroana*. Carbon dioxide exchange was determined with a closed-loop gas-exchange system as described in Green and Snelgar (1981a)

bonate seems never to be used as an inorganic carbon source (Bain and Proctor 1980). CO_2 compensation points range from $45-160\,\mu l CO_2 l^{-1}$ air (Dilks 1976; Proctor 1981) and, for *Marchantia berteroana*, are much lower in 2% oxygen and show the expected increase with rise in temperature (Fig. 16.1, top). Photosynthesis is also depressed at 21% oxygen compared to 2% oxygen and the depression increases at higher temperatures (Fig. 16.1, bottom).

Lichens, on the other hand, appear atypical when compared to C_3 phanerogams. Carbon isotope ratios show a wide range from $-14\permil$, for homoiomerous, cyanobacterial species to below $-30\permil$ for some green species (Lange and Ziegler 1986; Lange et al. 1988). This is a very broad range compared to values typical for C_3 or C_4 phanerogams and similar to that found for CAM plants. The values have been explained in terms of different thallus diffusion resistances (Lange et al. 1988) but this may need re-evaluation in the light of the discovery of carbon-concentrating mechanisms (CCM) in lichens (Badger et al. 1993, and later this section). CO_2 compensation points range from typical C_3 values to very low values ($<10\,\mu l CO_2 l^{-1}$ air), more in the range of C_4 or C_3-C_4 intermediate higher plants (Snelgar and Green 1980; Green and Snelgar 1981a; Bauer 1984). The sensitivity of A to oxygen level also varies and is low, as expected, when CO_2 compensation is low, but the situation is not simple (Green et al. 1985). Green and Snelgar (1981a) attempted to explain low CO_2 compensation values as a result of CO_2 refixation within the lichen thallus. CO_2 refixation may well occur but cannot produce a lower compensation value than that of the photobiont, the CO_2 compensation concentration at the photobiont must always be the same, or lower than the entire lichen value (Cowan et al. 1992). Although it is not yet totally clear how it is achieved, it is obvious that lichens may be able to photosynthesize at much lower CO_2 levels than bryophytes. The existence of CO_2-concentrating mechanisms is well known for free-living green algae and cyanobacteria (Badger and Price 1992) and has also been suggested for the photobionts of lichens (Raven et al. 1990; Máguas and Griffiths 1992). The existence of CCM in lichens has recently been demonstrated by Badger et al. (1993), and seems to be particularly active in cyanobacterial lichens and, to a smaller extent (perhaps only 10% of the cyanobacterial rate), in green lichens. The occurrence of CCM would certainly explain the low CO_2 compensation points and low CO_2 saturation in lichens.

16.2.3   Dark Respiration Rates

Dark respiration rates, on a dry weight basis (R), of lichens and bryophytes show similar response patterns to thallus water content and temperature. Typically, R increases with the degree of hydration until a maximum value is reached at a thallus water content that varies between species but is often in the range of 100 to 200%, related to plant dry weight (Dilks and Proctor 1979; Kershaw 1985). Dilks and Proctor (1979) noted for bryophytes that, where CO_2 exchange is linearly related to thallus water content at low hydration, both A and R respond similarly. The same situation seems to apply in lichens. At higher thallus water content, R is normally fairly constant although there has been a controversy over this point in lichens that still needs to be adequately resolved (Lange and Matthes 1981; Kershaw 1985; Green et al. 1985). Response of R to increase in temperature has, as expected, a Q_{10} close to 2, at least for mosses (Longton 1988).

Lichen studies have often noted the high value of R relative to A. The Photosynthetic Efficiency Coefficient [$K_F = (A + R)/R$, Larcher 1984] for lichens, at optimal thallus water content, is 2–4 (Snelgar et al. 1980) and is low compared to the 10–20 of higher plant sun leaves (Larcher 1984). This high R (relative to A) has been explained as being a consequence of the high, nonphotosynthetic, fungal component of lichens (Lange and Kappen 1972; Richardson 1974). Yet lichen R does not appear conspicuously different from that of other groups of plants (Table 16.3). Although they are almost 100% photosynthetic tissue, bryophytes have similar R to lichens (dry weight basis, Table 16.3), and a similar K_F of around 3 (from publications in Table 16.3).

The similarity in R for lichens and bryophytes (and with many higher plants) suggests that some other factor common to all groups tends to

Table 16.3. Respiration rates at 20 °C for different plant groups, (Data mostly from Larcher 1984)

Plant group	Respiration rate ($mg\,CO_2\,g^{-1}\,dw\,h^{-1}$)
Crop plants	5–8
Winter deciduous trees	
Sun leaves	3–4
Shade leaves	1–2
Evergreen conifers	
Sun leaves	ca. 1
Shade leaves	ca. 0.2
Desert shrubs	1–2
Lichens (several)	1–3 (Snelgar et al. 1980)
Bryophytes	
Weymouthia mollis	1.0 (Snelgar et al. 1980)
Polytrichum alpinum	1.2 (Longton 1988)
Polytrichum alpinum	2.0 (Billings et al. 1973)
Hylacomium splendens	1.0 (Proctor 1982)
Porella platyphylla	1.5 (Proctor 1982)

determine R. In higher plants, maintenance respiration is often modeled entirely from crude protein content or tissue nitrogen content (see Merino 1987; Buwalda 1991). Using these relationships, rates of $70-100$ mg CO_2 $g^{-1} N h^{-1}$ would give a value of around $0.7-1.0$ mg CO_2 g^{-1} dw h^{-1} (calculated at 1% tissue nitrogen and 15 °C). Crude protein contents of bryophytes are only slightly below those of other plant types in a tundra biome although lichens may be lower (Table 7.9 in Longton 1988). From this viewpoint, a similarity in R between lichens and bryophytes does not seem so surprising since it would simply indicate similar, underlying metabolic processes. There is a need for more respiration data for lichens and bryophytes that have been calculated on a nitrogen or protein basis.

In conclusion, the respiration rates of both lichens and bryophytes are very high relative to A, but this is because of the low A maxima, not because of unusually high respiration. Because respiration increases rapidly with temperature, it becomes a major load at moderate to high temperatures; it has been suggested that this load is severe enough to exclude bryophytes from lowland tropical areas (Frahm 1990a,b). From the previous discussion, such an exclusion would also be expected to apply to lichens.

16.2.4   Lichens and Bryophytes as Shade Plants

Valanne (1984) has suggested that the characteristics of shade plants seem to be obligate within the photosynthetic apparatus of mosses. All mosses investigated had large amounts of the light-harvesting chlorophyll a/b protein complex, a low chlorophyll a/b ratio and rapid in vivo fluorescence induction kinetics. Thylakoid structure resembled that of higher, shade plants. The cyanobacterial lichen, *Pseudocyphellaria dissimilis*, also shows a suite of characters typical of shade plants (Green et al. 1991). Certainly the light compensation and light saturation values for bryophytes and lichens are low when compared to phanerogam sun plants (on an area and dw basis). Extreme low values can be found in some cases such as saturation of A at $20 \mu mol\, m^{-2} s^{-1}$ PFD for *P. dissimilis* (Green et al. 1991) and very low values for compensation and saturation in submerged mosses (Priddle 1980).

It follows that, if shade adaptation is entrenched, species growing in higher light must utilize mechanisms to protect against damage. The most obvious mechanism is the poikilohydric nature of lichens and mosses which allows them to dry out and thus be metabolically inert. Dry lichens are protected against photoinhibition (Demmig-Adams et al. 1990) and *Ramalina maciformis*, a lichen which occurs in open, desert habitats, seems to be adapted to low light and temperatures, these being the conditions to which it is exposed in the mornings when it is active and before it dries out (Lange 1969a). Antarctic lichens have also been shown to have peak photosynthesis in dim light and to show photoinhibition if exposed to high light (Kappen et al. 1991). Green lichens can apparently withstand excessive light by the use of the xanthophyll cycle (Demmig-Adams et al. 1990) and others may use some form of filter. Increased parietin production may act as a filter in *Xanthoria parietina* (Hawksworth and Hill 1984) and a dark pigment has been shown to lower light levels at the photobiont within the thallus of *Peltula* species to less than 5% of ambient (Büdel 1987). The black upper cortex of the Antarctic lichen, *Buellia frigida*, strongly reduced fluorescence signals, which could be much increased by scraping away the cortex (Schroeter et al. 1992). Also, the Antarctic moss *Bryum argenteum* shows a lower quantum efficiency for CO_2 uptake in sun forms, but still has fully active photosystems, strongly suggesting that excess light is lost through refraction by the hyaline leaf tips (Green, unpubl. results).

16.2.5   Thallus Water Content and Photosynthesis

Lichens and bryophytes show a similar relationship between A and thallus water content (Kershaw 1985; Dilks and Proctor 1979). Initially, A increases rapidly with thallus water content until a maximum value is reached. Over this range of hydration it is accepted that A is biochemically limited through low water potential effects. At higher thallus water content, A may remain relatively constant or decline; the decline can be explained by increased diffusion resistances as water constricts or blocks gas diffusion pathways (Lange and Tenhunen 1981). The situation has been thoroughly analyzed in the lichen *Ramalina maciformis* and the increased diffusion resistances quantified (Cowan et al. 1992). No similar analysis has been done for mosses or hepatica but the blocking effect of excess water can be shown for those with complex, ventilated tissues (Table 16.4).

The total range of thallus water content reported varies considerably both within and between lichen and bryophyte species. Bryophytes generally have much higher maximum thallus water content; values up to 2000% occur and 1000–1500% are common (Dilks and Proctor 1979). In general, the depression of A found at high thallus water content for lichens (Kershaw 1985) is not so obvious in bryophytes (Dilks and Proctor 1979; Rundel and Lange 1980) and, for example, in *Weymouthia mollis*, no depression was

Table 16.4. Effect of supraoptimal water on photosynthesis of *Polytrichum commune*

Treatment	Photosynthesis (% of untreated control)	
	Shoot tips	Individual leaves
Normal-untreated	100	100
Flooded[a]	40	19
Flooded + 5 min drying	105	69

[a] Flooded: shoots or leaves vacuum infiltrated with water. Photosynthesis measured with $^{14}CO_2$ incubation (Green unpubl.).

found, even at 1500% water content (Snelgar et al. 1980). There seems little doubt that bryophytes can deal with high thallus water content much better than lichens. It must be noted that the complex Polytrichales, with ventilated tissues, have a low maximum thallus water content at around 250%, much more similar to thallus water content of lichens.

Although A declines at low thallus water content, it can continue at very low water potentials. When lichens, bryophytes, and higher plants were equilibrated on solutions of sucrose, mannitol, or sorbitol, the higher plants (*Nerium oleander* and *Spinacea oleracea*) reached zero A at water potentials of about −5 to −6 MPa. Bryophytes were similar or reached slightly lower potentials (Kaiser 1987; Dilks and Proctor 1979). Lichens with green (Chlorophycean) photobionts continued to photosynthesize below −20 MPa and, in equilibrium with water vapor, could reach −38 MPa, in the case of *Dendrographa minor* (Lange 1988). Moreover, these lichens showed similar photosynthetic rates at low potentials whether equilibrated with an osmotic solute or with atmospheric water vapor (Nash et al. 1990). This ability to utilize atmospheric water vapor is confined to green algal lichens (Lange et al. 1988). Interestingly, Kappen (1993) has suggested that the behavior of lichens when frozen may be identical to that when they are in equilibrium with water vapor. In both cases the photobiont cells are subjected to low water potentials following concentration, by water loss, of the cell contents. Cyanobacterial lichens cannot achieve positive net photosynthesis in equilibrium with water vapor but, in equilibrium with a solute potential, behave similarly to higher plants. It seems possible, based on limited evidence, that some bryophytes may also show identical photosynthetic behavior when in equilibrium with water vapor or liquid water (Lange 1969b). Their ability to use water vapor for reactivation of net photosynthesis after desiccation seems to be much more restricted than with lichens (see Rundel and Lange 1980). The green lichens stand out quite distinctly and can exhibit considerable discrimination against $^{13}CO_2$, suggesting that some may carry out most of their photosynthesis in the vapor-equilibrated state since liquid water would lead to less discrimination (Lange et al. 1988).

16.2.6 Environmental CO_2 Concentration

Lichens and bryophytes typically grow in close contact with their substrate. When this substrate has an organic component, it is likely that there will be CO_2 evolution from respiratory processes in the substrate, such as in tundra, peat mire, and forest ecosystems (Sveinbjörnsson and Oechel 1992). Because of this CO_2 efflux, local CO_2 levels around bryophytes and lichens may be considerably higher than normal ambient levels ($340-350 \mu l CO_2 l^{-1}$ air). This topic has recently been reviewed and good evidence for substantial CO_2 effluxes exists but there are only a few data on local CO_2 elevation (Sveinbjörnsson and Oechel 1992). Measurements from a New Zealand rainforest show atmosphere CO_2 inside the forest to be 11% above ambient outside the forest, while, amongst the lichen and bryophyte layer, it was 23 and 55% higher, respectively (Tarnawski et al. 1994). Also, initial studies of Antarctic mosses have shown even higher CO_2 levels, up to $1500\,\mu lCO_2 l^{-1}$ air in the upper layer of moss turfs (Tarnawski et al. 1992). It is clear, although from a limited literature, that bryophytes appear to be strongly CO_2-limited at ambient CO_2 levels of around $350\,\mu lCO_2 l^{-1}$ air and that saturation may not occur until around $2000\,\mu lCO_2 l^{-1}$ air (Silvola 1985; Coxson and Mackey 1990; Adamson et al. 1990; pers. unpubl. data). If natural CO_2 levels are higher than normal amongst bryophytes then this will have three major effects. First, actual A in the field will be greater than the values reported in the literature measured at $350\,\mu lCO_2 l^{-1}$ air. A good example would be the Antarctic bryophytes *Bryum argenteum* and *B. antarcticum* (correct name *Pottia heimii*), that have much lower A at $350\,\mu lCO_2 l^{-1}$ air than at higher CO_2 levels (Rastorfer 1970; Green 1981). Second, optimal temperatures for A will also increase with increased CO_2 concentration in the same manner that it would increase with PFD at levels below saturation (Bannister 1976). In general, reported temperature optima for A in bryophytes appear to be lower than for other C_3 plants, although it must be remembered that the relatively high dark respiration rates, especially at higher temperatures, will also act to lower these optima in both bryophytes and lichens (Larcher 1984). Third, increased CO_2 will lead to higher PFD being required for A saturation (Green et al. 1991). The situation for lichens is more complex since the response of A to CO_2 concentration depends strongly on thallus water content. Some lichens are close to CO_2 saturation at the normal ambient CO_2 levels of $350\,\mu lCO_2 l^{-1}$ air and optimal thallus water content. At supraoptimal thallus water content, lichens are also substantially CO_2-limited.

16.3 Plant Morphology and Photosynthesis

Both lichens and bryophytes show a wide range of growth forms. Bryophytes are particularly diverse with ten defined life forms described from the solid thalli of the Metzgeriales to the thin leafy Bryales and the complex tissues of the Polytrichales and Marchantiales (Mägdefrau 1982). Lichens have traditionally been divided into a smaller number of forms; foliose, fruticose, and crustose, which may overlap, and also there is the complication of heteromerous and homoiomerous photobiont arrangement (Hawksworth and Hill 1984). An attempt is made here to define lichens and bryophytes in terms of the presentation of their photosynthetic tissue. This approach results in two basic photosynthetic forms for bryophytes and one for lichens (Fig. 16.2). The approach is simplistic but takes into account

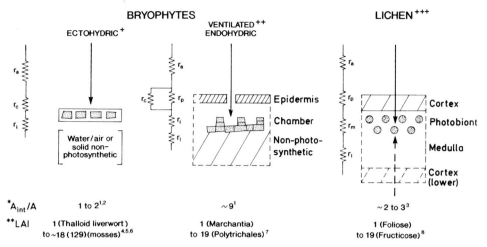

Fig. 16.2. Representation of the photosynthetic structures of bryophytes and lichens as defined in the text. The abbreviations are: + bryophytes obtaining water through their surfaces; ++ bryophytes obtaining their water mainly by an internal pathway, with variable water-proofed surfaces and a pore or slit leading to a gas-exchange "chamber"; +++ lichen with upper and/or lower cortex; * ratio of internal gas-exchange surface area to external projected area; ** Leaf Area Index: projected surface area of all "leaves" above unit ground area. Resistances: r_a boundary layer; r_p pore or slit; r_c cuticle; r_i internal gas pathways; r_l liquid and carboxylation within the photosynthetic cells. References: *1* Green and Snelgar (1982a); *2* Nobel (1977); *3* Collins and Farrar (1978); *4* Proctor (1979); *5* Vitt (1990); *6* Simon (1987); *7* Proctor (1980); Coxson and Lancaster (1988)

whether the cell surface, through which the carbon dioxide finally enters the liquid phase of its diffusion to the site of carboxylation, is exposed or protected. Protection, such as a cell layer interrupted by pores, or similar structure, involves greater gas diffusion resistance between the CO_2 exchange surface and the atmosphere. It is suggested that these forms cover the vast majority of the groups.

16.3.1 Bryophytes

Bryophytes can be divided into two photosynthetic forms (Fig. 16.2):
   a) Solid structures, photosynthetic tissue at surface:

1. thalloid liverworts (with solid thalli),
2. leafy liverworts and mosses (with solid "leaves").

This grouping contains bryophytes that would be classified as ectohydric from their water relations. They take up water over all, or most, of their surface and have no internal water transport system. The gas-phase CO_2 diffusion pathway in the solid bryophytes is not complex, being composed only of the boundary layer resistance. The lack of any other gas-phase resistance is clearly shown by helox experiments on the thalloid liverwort, *Monoclea forsteri* (Cowan et al. 1992). In this work photosynthetic rate is measured both in air and in helox (21% oxygen, 79% helium). Since CO_2 diffuses 2.3 times faster in helox, an increased A is obtained at any chosen CO_2 concentration, in helox compared to in air if a gas-phase diffusion resistance is present and CO_2 is limiting photosynthesis. In *M. forsteri*, A is identical in both air and helox indicating the absence of any significant gas-phase resistance. Some form of cuticle (resistance r_c in Fig. 16.2) is now known to exist, even on leafy mosses, and any increase in cuticular resistance to water loss will also affect CO_2 uptake (Proctor 1981). Most of the resistance to CO_2 diffusion is between the exchange surface and the photosynthetic centers (Nobel 1977). The presence of these large, non-gas-phase resistances is suggested by the response of A in bryophytes to CO_2 concentration. Typically A does not saturate until well above $1000\,\mu l\,CO_2\,l^{-1}$ air (see Sect. 16.2.6). One value cited for the CO_2 diffusion resistance is $95\,s\,cm^{-1}$ (*Mnium ciliare*, Nobel 1977).

The boundary layer resistance can be increased in leafy liverworts and mosses by arrangement of shoots into clumps or turfs, or by terminal, nonphotosynthetic structures such as hair points (Proctor 1981). Moreover, the photosynthetic structures, "leaves", can be arranged in a form of microcanopy that can achieve the following high leaf area indices: 1, *Monoclea forsteri*, thalloid liverwort (Green and Snelgar 1982a); 6, *Tortula intermedia*; 18, *Mnium hornum*; 20–25, *Pseudoscleropodium purum* (Proctor 1979); 15, *Drummordia prorepens* (Vitt 1990); 44, *Tortula ruralis*; 129, *Ceratodon purpureus* (Simon 1987). Entire microcanopies would be expected to achieve higher photosynthetic rates than individual stems through better light utilization.

b) Ventilated structures, photosynthetic tissue not at the surface (Fig. 16.2).

1. behind a pore: thalloid liverworts (Marchantiales),
2. behind a slit: members of the Polytrichales (parallel photosynthetic lamellae on the upper leaf surface have enlarged, wax-covered terminal cells that confine gas-exchange to a fine slit between the lamellae).

This grouping contains the endohydric mosses and liverworts, plants that have variously water-proofed surfaces, often particularly well developed near gas-exchange pores (Schönherr and Ziegler 1976; Clayton-Greene et al. 1985), and a significant internal water-transport pathway. Although poikilohydric, they are, in many ways, equivalent to homoiohydric higher plants with "leaves" having an increased internal area to surface area ratio, for instance 9 for *Marchantia foliacea* (Green and Snelgar 1982a). According to Nobel (1977), a higher internal area to surface area should allow higher A and certainly some of the higher values for A in mosses are those of *Polytrichum* species.

Ventilated bryophytes have the additional diffusion resistance of the pore structure between the photosynthetic cell surface and the atmosphere. Values are available only for the thalloid liverworts, being $1.6\,s\,cm^{-1}$ (*Marchantia foliacea*, Green and Snelgar 1982a) and between 0.6 and $5.0\,s\,cm^{-1}$ (various thalloid liverworts, Proctor 1981). The only estimate reported for cellular CO_2 diffusion resistance in these species is $68\,s\,cm^{-1}$ (Green and Snelgar 1982a), again much larger than that of the gas pathway resistance. Ventilated tissues are susceptible to blockage by water (Table 16.4) and it has been calculated that solid thalli are advantageous in habitats with abundant liquid water (Green and Snelgar 1982a).

16.3.2 Lichens

The photosynthetic gas exchange surface of lichens is the surface of the photobiont cells. Lichens, unlike bryophytes, cannot construct two-dimensional surfaces entirely of photosynthetic tissue. This is a consequence of the nature of the photobiont, often unicellular or, at best, filamentous or in small groups. Any photosynthetic cells will always be surrounded by fungal tissue to maintain structural integrity. In the majority of lichens (heteromerous) the cells form a layer beneath the upper, fungal cortex (Fig. 16.2). The cortex is a compact structure in comparison to the lichen medulla, which may have up to 18% air space (Collins and Farrar 1978). Although water can be lost from the entire cortex surface, in many cases CO_2 exchange has been found to be confined to surfaces with special structures such as the cyphellae or pseudo-cyphellae of the genera *Sticta* and *Pseudocyphellaria* (Green and Snelgar 1982b; Green et al. 1982), or small pores (Peveling 1970; Hale 1981) estimated to be 0.1–0.2% of the cortex surface for *R. maciformis* (Cowan et al. 1992), or to air spaces between fungal tissue. The values of the diffusion resistances at the various points along the CO_2 diffusion pathway have been calculated for *Pseudocyphellaria* and *Sticta* species (Snelgar et al. 1981) and measured for *Ramalina maciformis* (Cowan et al. 1992). In both cases, at optimal water content, the resistance of the gas diffusion pathways was found to be similar to non-gas-phase diffusion resistances in the photobiont. In a comparison with a herbaceous higher plant (*Phaseolus vulgaris*), the photobiont resistance to CO_2 uptake was lower than the internal resistance of a mesophyll cell (Cowan et al. 1992). Both Green and Snelgar (1981b) and Cowan et al. (1992) found that the gas diffusion resistances had little effect on A (except at high thallus water content). This is a consequence of the A response to external CO_2 concentration, which was nonlinear at $350\,\mu l\,CO_2\,l^{-1}$ air, with A often at about 90% of maximum rate. Thus, the majority of optimally wetted lichens contrast with the situation for bryophytes where the CO_2 response is still almost linear at normal ambient CO_2 levels (Silvola 1985). The hypothesis still seems to hold that the lichen thallus structure creates a moist, low VPD environment around the photobiont cells, which are not surrounded by a cuticle, in a manner similar to the mesophyll environment of higher plant leaves (Green et al. 1985).

The actual photobiont gas exchange surface area in a lichen does not appear to be high. A ratio of internal surface area to projected surface area of $4.6\,m^2\,m^{-2}$ has been calculated for *Xanthoria parietina* (Collins and Farrar 1978) but, since only about half the photobiont surface is actually in contact with air space, this reduces to about $2\,m^2\,m^{-2}$. Measurements of leaf area index in lichens are rare but must be around 1 for foliose and crustose lichens, and for fruticose species known values are 6, *Stereocaulon virgatum*, and 19, *S. tomentosum* (Coxson and Lancaster 1988).

16.4 Water Location and Transport

16.4.1 Bryophytes

The mechanisms by which bryophytes can store and move water have been elegantly described by Proctor (1979, 1981, 1982, 1984, 1990). Bryophytes can form sheets of photosynthetic cells and these can be arranged so that photosynthetic exchange surfaces and stored water can be kept separate. An excellent example would be *Pleurozium schreberi* where the shoot is effectively a column of water with a shell of photosynthetic tissue around it (Proctor 1990). Each photosynthetic cell has a water store on one side and an open CO_2 exchange surface on the other. Overlapping of leaves and the presence of papillae or folds on surfaces are also systems for the movement and storage of water (Proctor 1980). Areas can be kept free of water by slight cuticle development (water repellancy), convex surfaces, and papillae. Perhaps the most impressive example of the separation of water storage and photosynthetic tissue is in the genus *Sphagnum*, where the water is held in large, dead, hyaline cells between the living, photosynthetic cells. Proctor (1980, 1984) has pointed out that the capillary systems developed by "leaf" arrangements and structure can transport water from a high potential source as long as evaporative demand is not great. Effective use of photosynthetic tissue structures allows bryophytes to have high thallus water content with only slight depression of A. The above comments apply to the vast majority of bryophytes that are ectohydric. Endohydric bryophytes, with their capacity for some internal water supply, behave more like homoiohydric higher plants and lichens, and show depression in A when excess water is present (Table 16.4).

16.4.2 Lichens

Water storage is still quite an enigma in lichens. Various tissues, such as the cortex, photobiont layer, and medulla, have been suggested as storage sites

Table 16.5. Water-holding capacity of lichens

Lichen[a]	Dry weight $(g\,m^{-2})$	Water held $(g\,m^{-2})$	Maximum thallus water content $(gH_2O\,g^{-1}DW)$	Habitat
1 *Pseudocyphellaria dissimilis*	59	110	2	Deep shade-forest floor
1 *P. dissimilis*	74	150	2	Medium shade-forest interior
1 *P. dissimilis*	91	190	2	Open shade-forest interior
2 *P. homoeophylla*	200	440	1.5–2.5	Forest floor or trunks
2 *P. lividofusca*	150	460	3.1	Forest floor or trunks
2 *P. rufovirescens*	160	250	1.7	Forest floor or trunks
2 *Sticta latifrons*	150	330	2.2	Forest floor or trunks
2 *P. colensoi*	200	280	1.4	Forest canopy
3 *Umbilicaria antarctica*	500	850–1300	1.5	Open rocks
4 *Stereocaulon tomentosum*	540	1080	2.1	Open woodland ground
4 *S. virgatum*	1800	5400	3.3	Open mountain

[a] 1, Snelgar et al. (1980); 2, Snelgar (1981); 3, Harrison et al. (1989); 4, Coxson and Lancaster (1988).

(Rundel 1988). There is evidence that the medulla is maintained as an open air-filled tissue (Snelgar et al. 1981) and that flooding of the medulla can severely depress A (Cowan et al. 1992). External water storage has been demonstrated but there are obvious problems through interference with CO_2 exchange (Snelgar et al. 1981; Green et al. 1985; Kappen and Breuer 1991). Because thallus water content (maximum value) is fairly constant for hetero-merous, green algal lichens it follows that increase in stored water requires an increase in biomass (on an area basis) and this is clearly seen in the rather crude data in Table 16.5. No lichen seems to possess structures equivalent to those of bryophytes with a two-dimensional, continuous layer of photosynthetic cells exposed to the air on one side and water on the other. It seems that either the maintenance of a low VPD environment around the photobiont cells is paramount or, alternatively, the unicellular algae and fungal hyphae cannot produce such a structure.

In contrast to bryophytes there is little evidence of a water transport system in lichens. Water placed on the thallus may be taken up rapidly but with little lateral movement (Rundel 1988). Lichens may well behave like soils, which are also composed of fine capillaries, where water movement only occurs if a continual supply of high potential water is available to allow movement of the wetting front.

16.5 An Upper Limit for Photosynthetic Rate?

In previous sections, various factors tending to limit A, such as resistances and respiration rate, have been discussed. Maximum net photosynthetic rates (on a dry weight or area basis) for lichens and bryophytes are not only similarly low but, for species that have not dried out in open sites, there is little increase in A with the higher light but, instead, evidence of decreased light utilization efficiency (Sect. 16.2.4). The A are similar to phanerogams on a chlorophyll basis, indicating that any limitation is not in the metabolic processes of photosynthesis. Neither is R a suitable explanation (Sect. 16.2.3). Alternative explanations for the consistently low, and similar, A for both lichens and bryophytes are now required.

One possible explanation for low A is one based on the morphology of the plants. Raven (1977) demonstrated theoretically that solid, photosynthetic tissues could never achieve the photosynthetic rates of ventilated tissues; a factor of 10–20% was suggested. This work has been amplified by Nobel (1977, 1983), who has shown a clear positive relationship between mesophyll area to leaf area ratio and maximum photosynthetic rate (Fig. 16.3). Lichens and bryophytes have low internal (mesophyll) to external (leaf) area ratios (2 and 1–2, respectively) and, it seems, can only have low maximum photosynthetic rates. A detailed investigation of these ratios with respect to A in the Marchantiales and Polytrichales would certainly provide some important evidence on this concept. The proposal is that lichens and bryophytes are confined to low A by their structure, in particular through their low internal to external ratios.

16.6 Lichens and Bryophytes as Early Land Plants?

Although simple in form and with low A, the bryophytes and lichens are obviously very successful ecologically. It is interesting to consider their possible evolutionary history, particularly with respect to colonization of the

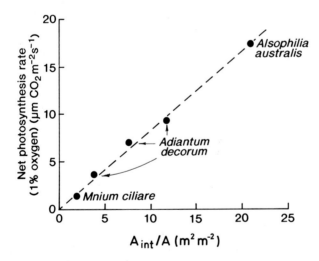

Fig. 16.3. Relationship between maximum photosynthetic rates and the ratio A_{int}/A. (Nobel 1977)

land. The fossil record is exceptionally poor for both groups and is little help in determining their origins. Standard texts consider the colonization of the land during the Silurian period, over 400 million years ago, as being by multicellular, telomic (cylindrical) plants with a suite of adaptations, namely cuticle, stomata, vascular strand, and dispersal spores. These are all homoiohydric characters and it is interesting that the first fossil records are of spores (Raven 1977). It is not unreasonable to assume that the first successful land colonists were the cyanobacteria, which form, today, the dominant plant communities in the phanerogam-free Antarctic, and desiccation-resistant chlorophycean algae. Both these groups are involved in lichen symbioses, and the fungi, which are known as fossils in the Silurian, can also be desiccation-resistant, and grow using only atmospheric humidity. Lichens could also be early colonizers, the prime adaptation being poikilohydry but the major advantage being extension of habitat for both symbionts. The earliest lichen fossils are from the 220-million-year-old Keuper formation in Germany (Ziegler 1991) and it seems quite possible that problems of fossilization rather than their absence, are the reason for their late occurrence in the fossil record.

What of the bryophytes, since fossil bryophyte-like plants are now known from the Devonian (Remy 1982). Despite the emphasis in textbooks on the requirement for liquid water during fertilization, the bryophyte life cycle is highly suited for colonization of the early land areas. Bryophyte spores are an excellent dispersal system and continual vegetative reproduction is possible until the sporophyte can be produced. Combine the spores and vegetative reproduction with poikilohydry and a successful colonizing lifeform exists. A small sporophyte would be advantageous rather than being regarded as a reduced structure. Bryophytes are common in the Antarctic but no mature sporophytes have been reported, so vegetative dispersal has clearly been sufficient to effect colonization and maintain populations on that continent.

Perhaps the land was first colonized by algae and cyanobacteria, then by the lichens and bryophytes. The low A enforced by their simple structure would not be a disadvantage at that time. If this is correct, then the major bryophyte and lichen communities in the shade of forests and other higher plant-dominated ecosystems, would be secondary. Bryophytes and lichens are successful in these communities because they are poikilohydric and, in particular, because of their structurally imposed low A which preadapts them to shade. Perhaps future research will shed more light on this idea.

16.7 Conclusions

Lichen and bryophytes provide fascinating examples of the interaction between plant structure and photosynthetic metabolism. Both groups have remarkably similar and low, maximal A compared with phanerogams on an area or weight basis. Bryophytes and lichens also have very similar dark respiration rates which only appear high because A is low. The relatively high respiration rate does have potentially major effects on net photosynthesis at high temperatures and may even limit distributions in areas of consistently high temperature, especially where the plants are wet at night. Respiration-induced depressions in photosynthesis are found in lichens from Antarctica at temperatures as low as 12°C (Kappen et al. 1989). It is suggested that the low A is a consequence of the simple photosynthetic structure of lichens and bryophytes, in particular the low ratio of the internal photosynthetic tissues to external surface area. Photosynthetic rates on a chlorophyll basis are similar to those found in phanerogams. Structurally imposed, low A was not a disadvantage in early colonization of the land, which is suggested to have been by cyanobacteria, lichens, and bryophytes, but had the potential to become a problem when other higher plants evolved. Phanerogams opened a new series of shade environments that have been exploited by both lichens and bryophytes.

The two groups are quite distinct in their abilities to deal with liquid water. Bryophytes have the capacity to produce complex, two-dimensional, self-supporting photosynthetic tissue, where the CO_2 exchange surface is effectively a "skin" over the surface of stored water. These tissues can be arranged, not only to generate water storage volumes separated from gas exchange areas, but also to encourage capillary water movement over, and between, surfaces. In contrast, lichens have an internal photosynthetic surface which has relatively compact tissues between it and the atmosphere. Any water storage will tend to block either the compact tissues or the outer lichen surface. Lichens tend to have lower maximal water content on a dry weight basis with high risk of A depression at high thallus water content. This difference is probably the explanation for the dominance of bryophytes in very wet habitats. Green algal lichens, in contrast, have a major advantage in dry habitats because of their ability to attain positive net photosynthesis using only air humidity.

The data base for detailed comparison of photosynthetic processes in lichens and bryophytes is small. Therefore, the comparisons and conclusions in this chapter must be regarded as tentative, but they pose areas for future research.

Acknowledgments. Much of the work reported here was supported by the Deutsche Forschungsgemeinschaft (Bonn) as a program of the Sonder-forschungsbereich 251 der Universität Würzburg. The Alexander von Humboldt Stiftung is also thanked for financial support to TGAG. Julie Cooke is thanked for preparing the text and Frank Bailey for drawing the figures. The New Zealand Department of Conservation is thanked for their permission, over many years, to collect and study bryophytes and lichens in reserved and park areas.

References

Adamson E, Post A, Adamson H (1990) Photosynthesis in *Grimmia antarctici*, an endemic, antarctic bryophyte, is limited by carbon dioxide. Curr Res Photosynth 4: 639–642

Alpert P (1988) Survival of a desiccation-tolerant moss, *Grimmia laevigata*, beyond its observed microdistributional limits. J Bryol 15: 219–227

Badger MR, Price GD (1992) The CO_2 concentrating mechanism in cyanobacteria and microalgae. Physiol Plant 84: 606–615

Badger MR, Pfanz H, Büdel B, Heber U, Lange OL (1993) Evidence for the functioning of photosynthetic CO_2 concentrating mechanisms in lichens containing green algal and cyanobacterial photobionts. Planta (in press)

Bain J, Proctor MCF (1980) The requirements of aquatic bryophytes for free CO_2 as an inorganic carbon source: some experimental evidence. New Phytol 87: 269–283

Bannister P (1976) Introduction to physiological plant ecology. Blackwell, Oxford

Bauer H (1984) Net photosynthetic CO_2 compensation concentrations of some lichens. Z Pflanzenphysiol 114: 45–50

Billings WD, Shauer GR, Trent AN (1973) Temperature effects on growth and respiration of roots and rhizomes in tundra graminoids. In: Bliss LC, Wielgolaski FE (eds) Primary production and production processes, tundra biome. Proceedings of the conference, Dublin, Ireland, April 1973. Tundra Biome Steering Committee, Dept Botany, Univ Alberta, Edmonton, Alberta, pp 57–63

Büdel B (1987) Zur Biologie und Systematik der Flechtengattungen *Heppia* and *Peltula* im südlichen Afrika. Bibl Lichenol 23: 1–150

Buwalda JG (1991) A mathematical model of carbon acquisition and utilisation by kiwifruit vines. Ecol Model 57: 43–64

Clayton-Greene KA, Collins NJ, Green TGA, Proctor MCF (1985) Surface wax, structure and function in leaves of Polytrichaceae. J Bryol 13: 549–562

Collins CR, Farrar JF (1978) Structural resistances to mass transfer in the lichen *Xanthoria parietina*. New Phytol 81: 71–83

Cowan IR, Lange OL, Green TGA (1992) Carbon dioxide exchange in lichens: determination of transport and carboxylation characteristics. Planta 187: 282–294

Coxson DS, Lancaster J (1988) The morphological basis of variation in net photosynthetic and respiratory responses of the mat-forming lichen species *Stereocaulon virgatum* and *S. tomentosum*. Can J Bot 67: 167–176

Coxson DS, Mackey RL (1990) Diel periodicity of photosynthetic response in the subalpine moss *Pohlia wahlenbergii*. Bryologist 93: 417–422

Crawford RMM (1989) Studies in plant survival. Blackwell, Oxford

Demmig-Adams B, Máguas C, Adams WA, Meyer A, Kilian E, Lange OL (1990) effect of high light on the efficiency of photochemical energy conversion in a variety of lichen species with green and blue-green phycobionts. Planta 180: 400–409

Dilks TJK (1976) Measurement of the carbon dioxide compensation point and the rate of loss of $^{14}CO_2$ in the light and dark in some bryophytes. J Exp Bot 27: 98–104

Dilks TJK, Proctor MCF (1979) Photosynthesis, respiration and water content in bryophytes. New Phytol 82: 97–114

Frahm JP (1990a) The ecology of epiphytic bryophytes on Mt Kinabalu, Sabah (Malaysia). Nova Hedwigia 51: 121–132

Frahm JP (1990b) The effect of light and temperature on the growth of bryophytes of tropical rain forests. Nova Hedwigia 51: 151–164

Green TGA (1981) Photosynthesis of antarctic bryophytes: field studies in the McMurdo Dry Valley region. In: Carr DJ (ed) XIIIth International Botanical Congress, Sydney Abstracts, p 290

Green TGA, Snelgar WP (1981a) Carbon dioxide exchange in lichens: apparent photorespiration and possible role of CO_2 refixation in some members of the Stictaceae (Lichenes). J Exp Bot 32: 661–668

Green TGA, Snelgar WP (1981b) Carbon dioxide exchange in lichens: partition of total CO_2 resistances at different thallus water contents into transport and carboxylation components. Physiol Plant 52: 411–416

Green TGA, Snelgar WP (1982a) A comparison of photosynthesis in two thalloid liverworts. Oecologia 54: 275–280

Green TGA, Snelgar WP (1982b) Carbon dioxide exchange in lichens: relationship between the diffusive resistance of carbon dioxide and water vapour. Lichenologist 14: 255–260

Green TGA, Snelgar WP, Brown DH (1982) Carbon dioxide exchange through the cyphellate lower cortex of *Sticta latifrons* Rich. New Phytol 88: 421–426

Green TGA, Snelgar WP, Wilkins AL (1985) Photosynthesis, water relations and thallus structure of Stictaceae lichens. In: Brown DH (ed) Lichen physiology and cell biology. Plenum Press, New York, pp 57–76

Green TGA, Kilian E, Lange OL (1991) *Pseudocyphellaria dissimilis*: a desiccation-sensitive, highly shade-adapted lichen from New Zealand. Oecologia 85: 498–503

Hale ME (1981) Pseudocyphellae and pored epicortex in the Parmeliaceae: their delimitation and evolutionary significance. Lichenologist 13: 1–10

Harrison PM, Walton DH, Rothery P (1989) The effects of temperature and moisture on CO_2 uptake and total resistance to water loss in the antarctic foliose lichen *Umbilicaria antarctica*. New Phytol 111: 673–682

Hawksworth DL, Hill DJ (1984) The lichen-forming fungi. Blackie, Glasgow

Kaiser WM (1987) Methods for studying the mechanism of water stress effects on photosynthesis. In: Tenhunen JD, Catarino FM, Lange OL, Oechel WC (eds) Plant response to stress. Springer, Berlin Heidelberg New York, pp 77–94

Kallio P, Kärenlampi L (1975) Photosynthesis in mosses and lichens. In: Cooper JP (ed) Photosynthesis and productivity in different environments. Cambridge University Press, Cambridge, pp 393–424

Kappen L (1973) Responses to extreme environments. In: Ahmadjian V, Hale ME (eds) The lichens. Academic Press, New York, pp 310–380

Kappen L (1988) Ecological and physiological relationships in different climatic regions. In: Galun M (ed) CRC Handbook of lichenology, vol II, VII B 2. CRC Press, Boca Raton, pp 37–100

Kappen L (1993) Plant activity under snow and ice, with particular reference to lichens. Arctic 10 (in press)

Kappen L, Breuer M (1991) Ecological and physiological investigations in continental Antarctic cryptogams II. Moisture relations and photosynthesis of lichens near Casey Station, Wilkes Land. Antarct Sci 3: 273–278

Kappen L, Lewis Smith RI, Meyer M (1989) Carbon dioxide exchange of two ecodemes of *Schistidium antarctici* in continental Antarctica. Polar Biol 9: 415–422

Kappen L, Breuer M, Bölter M (1991) Ecological and physiological investigations in continental Antarctic cryptogams 3. Photosynthetic production of *Usnea sphacelata*: diurnal courses, models, and the effect of photoinhibition. Polar Biol 11: 393–401

Kershaw KA (1985) Physiological ecology of lichens. Cambridge University Press, London

Lange OL (1969a) Die funktionellen Anpassungen der Flechten an die ökologischen Bedingungen arider Gebiete. Ber Dtsch Bot Ges 82: 3–22

Lange OL (1969b) CO_2-Gaswechsel von Moosen nach Wasserdamnfaufnahme aus dem Luftraum. Planta 89: 90–94

Lange OL (1988) Ecophysiology of photosynthesis: performance of poikilohydric lichens and homoiohydric Mediterranean sclerophylls. The seventh Tansley Lecture. J Ecol 76: 915–937

Lange OL, Ziegler H (1986) Different limiting processes of photosynthesis in lichens. In: Marcelle R, Clijsters H, van Poucke M (eds) Biological control of photosynthesis. Nijhoff, Dordrecht, pp 147–161

Lange OL, Kappen L (1972) Photosynthesis of lichens from Antarctica. In: Llano GA (ed) Antarctic terrestrial biology. Geophysical Union, Washington, pp 83–95

Lange OL, Matthes U (1981) Moisture-dependent CO_2 exchange of lichens. Photosynthetica 15: 555–574

Lange OL, Tenhunen J (1981) Moisture content and CO_2 exchange of lichens. II. Depression of net photosynthesis in *Ramalina maciformis* at high water content is caused by increased thallus carbon diffusion resistance. Oecologia 51: 426–429

Lange OL, Green TGA, Ziegler H (1988) Water status related photosynthesis and carbon isotope discrimination in species of the lichen genus *Pseudocyphellaria* with green or blue-green photobionts and in photosymbiodemes. Oecologia 75: 494–501

Lange OL, Heber V, Schulze ED, Ziegler H (1989) Atmospheric pollutants and plant metabolism. In: Schulze ED, Lange OL, Oren R (eds) Forest decline and air pollution. Ecological Studies 77. Springer, Berlin Heidelberg New York, pp 238–276

Larcher W (1984) Physiological plant ecology. Springer, Berlin Heidelberg New York

Larson DW (1979) Lichen water relations under drying conditions. New Phytol 82: 713–731

Larson DW (1984) Thallus size as a complicating factor in the physiological ecology of lichens. New Phytol 97: 87–97

Longton RE (1988) Biology of polar bryophytes and lichens. Cambridge University Press, Cambridge

McCall KK, Martin CE (1991) Chlorophyll concentrations and photosynthesis in three forest understory mosses in northeastern Kansas. Bryologist 94: 25–29

Mägdefrau K (1982) Life-forms of bryophytes. In: Smith AJE (ed) Bryophyte ecology. Chapman and Hall, London, pp 45–58

Máguas C, Griffiths H (1992) Different carbon isotope discrimination characteristics in green algal and cyanobacterial lichens. In: Kärnefelt I (ed) Abstracts, Second International Lichenological Symposium. Lund, 1992

Merino J (1987) The cost of growing and maintaining leaves of mediterranean plants. In: Tenhunen JD, Catarino FM, Lange OL, Oechel WC (eds) Plant response to stress. Springer, Berlin Heidelberg New York, pp 553–564

Nash TH, Egan RS (1988) The biology of lichens and bryophytes. In: Nash TH, Wirth V (eds) Lichens, bryophytes and air quality. Bibl Lichenol 30: 11–22

Nash TH, Moser TJ, Link SO (1980) Non-random variation of gas exchange within arctic lichens. Can J Bot 58: 1181–1186

Nash TH, Reiner A, Demmig-Adams B, Kilian E, Kaiser WM, Lange OL (1990) The effect of atmospheric desiccation and osmotic water stress on photosynthesis and dark respiration of lichens. New Phytol 116: 269–276

Nobel PS (1977) Internal leaf area and cellular CO_2 resistance: photosynthetic implications of variations with growth conditions and plant species. Physiol Plant 40: 137–144

Nobel PS (1983) Biophysical plant physiology and ecology. Freeman, San Francisco

Oren R, Zimmerman (1989) CO_2 assimilation and carbon balance of healthy and declining Norway spruce stands. In: Schulze ED, Lange OL, Oren R (eds) Forest decline and air pollution – a study of spruce (*Picea abies*) on acid soils. Ecological Studies 77. Springer, Berlin Heidelberg New York, pp 352–369

Peveling E (1970) Die Darstellung der Oberflächenstrukturen von Flechten mit dem Raster-Elektronmikroskop. Vortr Gesamtgeb Bot NF 4: 89–101

Priddle J (1980) The production ecology of benthic plants in some Antarctic lakes. I. In situ production studies. J Ecol 68: 141–153

Proctor MCF (1979) Structure and ecophysiological adaptation in bryophytes. In: Clarke CGS, Duckett JG (eds) Bryophyte systematics. Academic Press, London, pp 479–509

Proctor MCF (1980) Diffusion resistances in bryophytes. In: Grace J, Ford ED, Jarvis PG (eds) Plants and their atmospheric environment. Blackwell, British Ecological Society Symposia. Oxford, pp 219–230

Proctor MCF (1981) Physiological ecology of bryophytes. Adv Bryol 1: 79–166

Proctor MCF (1982) Physiological ecology: water relations, light and temperature responses, carbon balance. In: Smith AJE (ed) Bryophyte ecology. Chapman and Hall, London, pp 333–381

Proctor MCF (1984) Structure and ecological adaptation. In: Dyer AF, Duckett JG (eds) The experimental biology of bryophytes. Academic Press, London, pp 9–37

Proctor MCF (1990) The physiological basis of bryophyte production. Bot J Linn Soc 104: 61–77

Rastorfer JR (1970) Effects of light intensity and temperature on photosynthesis and respiration of two East Antarctic mosses *Bryum argenteum* and *Bryum antarcticum*. Bryologist 73: 544–556

Raven JA (1977) The evolution of vascular land plants in relation to supracellular transport processes. Adv Bot Res 5: 153–219

Raven JA, Johnston AM, Handley LL, McInroy SG (1990) Transport and assimmilation of inorganic carbon by *Lichina pygmaea* under emersed and submersed conditions. New Phytol 114: 407–417

Raven PH, Evert RF, Eichhorn SE (1986) Biology of plants. Worth, New York

Remy WR (1982) Die Vorfahren der Landpflanzen. Forschung, Mitt DFG 82: 12–13

Richardson DHS (1974) Photosynthesis and carbohydrate movement. In: Ahmadjian V, Hale ME (eds) The lichens. Academic Press, New York, pp 250–288

Rundel PW (1988) Water relations. In: Galun M (ed) CRC Handbook of lichenology, vol II. CRC Press, Boca Raton, pp 17–36

Rundel PW, Lange OL (1980) Water relations and photosynthetic response of a desert moss. Flora 169: 329–335

Rundel PW, Stichler W, Zander RH, Ziegler H (1979) Carbon and hydrogen isotopes of bryophytes from arid and humid regions. Oecologia 44: 91–94

Sancho LG, Kappen L (1989) Photosynthesis and water relations and the role of anatomy in Umbilicariaceae (lichens) from central Spain. Oecologia 81: 473–480

Schönherr J, Ziegler H (1975) Hydrophobic cuticular wedges prevent water entering the air pores of liverwort thalli. Planta 124: 51–60

Schroeter B, Green TGA, Seppelt RD, Kappen L (1992) Monitoring photosynthetic activity of crustose lichens using a PAM-2000 fluorescence system. Oecologia 92: 457–462

Schulze ED, Lange OL, Kappen L, Evenari M, Buschbom V (1975) Physiological basis of primary production of perennial higher plants in the Negev Desert. In: Cooper JP (ed) Photosynthesis and productivity in different environments. Cambridge University Press, Cambridge, pp 107–120

Silvola J (1985) CO_2 dependence of photosynthesis in certain forest and peat mosses and simulated photosynthesis at various actual and hypothetical CO_2 concentrations. Lindbergia 11: 86–93

Simon T (1987) The leaf-area index of three moss species (*Tortula ruralis, Ceratodon purpureus* and *Hypnum cupressiforme*) In: Pocs T, Simon T, Tuba Z, Podani J (eds) Proceedings of the IAB Conference of Bryoecology Budapest – Vacratot, Hungary, 5–10 August 1985, part B. Akademiai Kiado, Budapest, pp 699–706

Snelgar WP (1981) The ecophysiology of New Zealand forest lichens with special reference to carbon dioxide exchange. PhD Thesis, Waikato University, Hamilton

Snelgar WP, Green TGA (1980) Carbon dioxide exchange in lichens: low carbon dioxide compensation levels and lack of apparent photorespiratory activity in some lichens. Bryologist 83: 505–507

Snelgar WP, Brown DH, Green TGA (1980) A provisional survey of the interaction between net photosynthetic rate, respiratory rate, and thallus water content in some New Zealand cryptogams. NZ J Bot 18: 247–256

Snelgar WP, Green TGA, Beltz CK (1981) Carbon dioxide exchange in lichens: estimation of internal thallus CO_2 transport resistance. Physiol Plant 52: 417–422

Sveinbjörnsson B, Oechel WC (1992) Controls on growth and productivity of bryophytes: environmental limitations under current and anticipated conditions. In: Bates JW, Farmer AM (eds) Bryophytes and lichens in a changing environment. Oxford Scientific Publications, Oxford, pp 77–102

Tarnawski M, Melik D, Roser D, Adamson E, Adamson H, Seppelt R (1992) In situ CO_2 levels in cushion and turf forms of *Grimmia antarctici* at Casey Station, East Antarctica. J Bryol 17: 241–249

Tarnawski MG, Green TGA, Budel B, Meyer A, Zellner H, Lange OL (1994) Diel changes of atmospheric CO_2 concentration within, and above, cryptogam stands in a New Zealand temperate rainforest. NZ J Bot 32(1) (in press)

Teeri JA (1981) Stable carbon isotope analysis of mosses and lichens growing in xeric and moist habitats. Bryologist 84: 82–84

Tretiach M, Carpanelli A (1992) Chlorophyll content and morphology as factors influencing the photosynthetic rate of *Parmelia caperata*. Lichenologist 24(1): 81–90

Valanne N (1984) Photosynthetic products in mosses. In: Dyer AF, Duckett JG (eds) The experimental biology of bryophytes. Academic Press, London, pp 257–273

Vitt DH (1990) Growth and production dynamics of boreal mosses over climatic, chemical and topographic gradients. Bot J Linn Soc 104: 35–59

Walter H (1931) Die Hydratur der Pflanze und ihre physiologisch-ökologische Bedeutung. Fischer, Jena

Ziegler R (1991) Komplex-thallöse, fossile Organismen mit blattflechtenartigem Bau aus dem mittleren Keuper (Trias, Karn) Unterfrankens. In: Kovar-Eder J (ed) Palaeovegetational Developments in Europe and regions relevant to its palaeofloristic evolution. Proc Pan-European Palaeobotanical Conf, Museum of Natural History Vienna, pp 341–349

17 The Consequences of Sunflecks for Photosynthesis and Growth of Forest Understory Plants

R.W. Pearcy and W.A. Pfitsch

17.1 Introduction

Plants in forest understories are subjected to light environments consisting of a very low background of diffuse light that is punctuated by often much brighter sunflecks lasting from a few seconds to several minutes. These sunflecks, although usually present for less than 10% of the time, typically contribute 10 to 80% of the photosynthetically active photon flux density (PFD) (Chazdon 1988). Therefore much of the photosynthesis of understory plants may occur under transiently changing light conditions that characterize sunflecks. The environmental and physiological controls on photosynthesis during transient light changes are not necessarily the same as those that determine photosynthetic performance under steady-state conditions. The shade-plant syndrome of understory plants has been widely studied but mostly in terms of the controls on steady-state photosynthetic characteristics. Until recently, relatively little attention has been given to the mechanisms regulating the use of sunflecks.

Essentially all of the components of the photosynthetic apparatus of a leaf respond either directly or indirectly to a change in light, but on very different time scales. Steps in charge separation and electron transport occur virtually instantaneously. However, increases in metabolite concentrations necessary for maximal photosynthetic rates require longer times. Finally, slow changes in activation state of enzymes and stomatal opening result in a slow increase in photosynthesis towards a final steady-state level. To further complicate matters, metabolite concentrations also exhibit slow changes due to the light activation of the enzymes that regulate the flux through the pools. Since light fluctuations can occur on a time scale much faster than stomatal opening or enzyme activation, the response to any given light change will in part be a function of the previous light environment.

In this chapter, we will discuss the spatial and temporal variation in photosynthesis in relation to sunflecks in a forest understory. We will also present results of studies designed to understand the role of this variation in the carbon balance and growth in the forest understory. Most of work was completed utilizing *Adenocaulon bicolor*, (Asteraceae) a common plant in moist redwood forst understories of northern California, as the experimental plant. This perennial species forms a basal rosette of typically 6 to 20 leaves in March and may remain vegetatively active until September. Although sexual reproduction is common there is no apparent vegetative reproduction.

17.2 Sunflecks in Forest Understories

The light regimes of forest understories exhibit tremendous spatial and temporal variation largely because of the occurence of sunflecks resulting from canopy gaps of varying size. When these gaps are aligned with the solar path, sunflecks occur that may reach PFDs up to 20-fold greater than the shade light. In total, they typically are present at a given location <10% of the day but typically contribute 35 to 80% of the daily PFD on clear days (Björkman and Ludlow 1972; Pearcy 1983, 1987; Chazdon and Fetcher 1984). The redwood forest understory where *Adenocaulon* was abundant received on average on clear days 214 sunflecks that contributed 69% of the PFD and were present for 12% of the time (Pfitsch and Pearcy 1989a). The earth's rotation and canopy movement cause the duration of sunflecks to be quite brief. Penumbral effects caused by the finite (1/2°) angular diameter of the solar disc result in a reduction of the maximum PFD in small sunflecks in tall canopies and a spreading of the sunfleck so that the edges are indistinct (Miller and Norman 1971). While the consequences of penumbral spreading of sunflecks has been investigated in tree canopies where the effect is small (Oker-Blom 1984), no work has been done in forest understories where a larger effect could be expected.

Canopy structure is the dominant factor influencing the temporal characteristics of sunflecks. Overstory canopies generally are clumped so that when viewed from the understory there are areas with more gaps while others have few gaps. As a result, sunflecks tend to occur in clusters of several to 100 or more, separated by periods when there few or none. In total, there may be >300 sunflecks recorded at a single site during a day. Therefore most are brief, resulting from canopy movement and lasting less than 10 s. However, longer sunflecks have been shown to contribute a disproportionate share of the PFD in redwood forest understories (Pfitsch and Pearcy 1989a). Open, flexible canopies such as those of quaking aspen stands have many more sunflecks and proportionately more of the PFD is contributed by brief sunflecks (Roden and Pearcy 1993). Cloudiness is also important in determining the frequency of sunflecks. In climates where clouds build in the afternoon, for example, sunflecks will be restricted to the morning.

17.3   Mechanisms Regulating the Utilization of Sunflecks

The lack of repeatability and predictability of natural light regimes means that most studies of the physiological regulation of sunfleck use have utilized lamp and shutter systems to simulate the temporal nature of sunflecks in a spatially uniform and repeatable manner. These simulated sunflecks have been termed lightflecks (Pearcy et al. 1985). The requirement for fast-response gas-exchange systems and computer-based data acquistion systems have also tended to restrict most studies to laboratories. However, newer systems and portable computers mean that fast-response systems suitable for field use are now more readily constructed. Pfitsch and Pearcy (1989b), for example, examined the response to lightflecks of leaves on *Adenocaulon* plants in situ and found responses to be generally similar to those measured for other species in the laboratory.

Using systems of the type briefly discussed above, considerable progress has been made in the last 5–10 years in understanding the controls on sunfleck utilization (see review by Pearcy 1990). These controls are now known to be of two basic types: the slow responses to light changes that comprise the induction requirement of photosynthesis and the much faster changes in assimilation that relate to increases and decreases in metabolite levels within the chloroplast. The induction requirement largely sets the capacity of the system to respond to a light increase. The changes in metabolite pools determine the rate of increase and decrease of photosynthesis occurring in response to a sunfleck itself. Metabolites built up during a lightfleck can be used to support continued CO_2 fixation for a short period after the lightfleck (post-illumination CO_2 fixation), substantially enhancing the total carbon gain in some circumstances (Pearcy et al. 1985; Chazdon and Pearcy 1986b).

The induction requirement photosynthesis in understory plants is now known to be a function of light-dependent stomatal opening and of light regulation of the primary CO_2-fixing enzyme, ribulose-1,5-bisphosphate carboxylase/oxygenase (Rubisco) and the light-activated enzymes involved with RuBP regeneration (Seemann et al. 1988; Woodrow and Mott 1989). During induction, activation of RuBP regenerating system may limit for the first minute or so (Sassenrath-Cole and Pearcy 1992) while later, the increase in Rubisco activity and stomatal conductance become the primary factors controlling the rate of increase in photosynthetic capacity (Kirschbaum and Pearcy 1988). The increase in Rubisco activity is generally complete within 5–10 min, whereas stomatal conductance may continue to increase for 30 to 60 min. The relative role of these two limitations depends on the initial conductance established in the low light period prior to the beginning of induction. However, the high humidities characteristic of understories often cause stomatal conductance to be high, shifting more of the limitation to Rubisco. For example, there appeared to be very little limitation imposed by stomatal conductance in *Adenocaulon* leaves in the field because of the high humidities.

An important feature of both Rubisco and stomatal conductance responses to lightflecks is their hysteretic behavior. The decreases in both stomatal conductance and Rubisco are slower than the increases. Moreover, for short (<10 min) sunflecks, stomatal opening continues for up to 20 min afterwards. Thus, the occurrence of a sunfleck primes the leaf so that it is better able to use subsequent sunflecks. Tinoco-Ojanguren and Pearcy (1992) found that use of series of short (1–16 s) lightflecks was significantly enhanced by the occurrence 20 min earlier of a 4-min lightfleck. This enhancement was related to the higher conductances following the 4-min lightfleck. Since sunflecks often occur in clusters, this priming effect helps to minimize the induction limitations during subsequent sunflecks in the clusters.

Under natural sunfleck regimes, the induction requirement of photosynthesis, post-illumination CO_2 fixation and the steady-state photosynthetic characteristics all interact to determine the carbon gain. Throughout much of the day, leaves are unlikely to be fully induced, which will limit the utilization of sunflecks. Daily courses of stomatal conductance show a continuous modulation, increasing during sunfleck periods and decreasing in low light periods (Björkman et al. 1972; Pearcy 1987), but rarely, if ever, reaching steady-state values. Similar studies of Rubisco activity have not been conducted with understory plants, but simulations of its behavior (Pearcy and Gross, unpubl.) suggest that it, too, would be continuously modulated at intermediate activity levels. Post-illumination CO_2 fixation may offset these induction limitations to a certain extent, but its effect will depend on the frequency of short (<10 s) sunflecks where its contribution can be a large fraction of the total CO_2 assimilation due to a sunfleck.

17.4 Photosynthesis in Natural Sunfleck Pegimes

Despite their potentially large contribution to the carbon gain of understory plants, few studies have been carried out in the field to assess the importance of the spatial and temporal variation of sunflecks or of the mechanisms that regulate their use. Pfitsch and Pearcy (1989a) measured the daily course of gas exchange of *Adenocaulon* leaves at six microsites on different days. These days were all similar, with the fog and overcast that is common in the morning dissipating by 0830 to 0930 h and the remainder of the day being clear. Thus differences among sites probably reflect more the spatial variation in sunflecks than any seasonal or day-to-day variation. The microsites were selected to be typical of those where *Adenocaulon* occurred and to give a range of sunfleck activity.

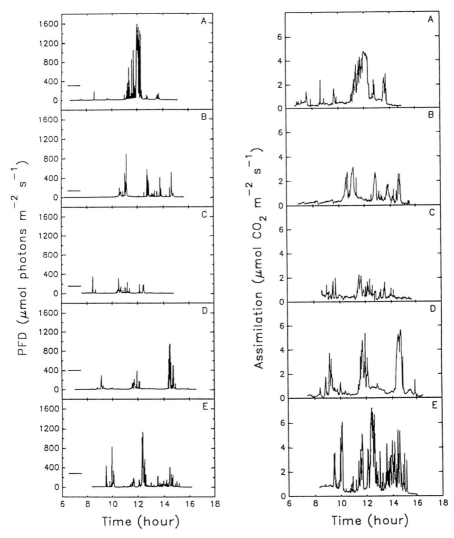

Fig. 17.1. Daily courses of photon flux density (*left*) and CO_2 assimilation of *Adenocaulon bicolor* leaves (*right*) at five different microsites in a redwood forest understory. Each was measured on a different day (day *A* to *E*). (Pfitsch and Pearcy 1989a)

Figure 17.1 shows the daily course of PFD and assimilation on the 5 days on which sunflecks occurred. By comparing figures for PFD and assimilation on the left and right side, respectively, it can be seen that there is a close correspondence between the sunflecks and peaks of photosynthetic activity. The total daily PFD on these 5 days ranged from 0.45 (day C) to 1.93 (day A) $mol\,m^{-2}\,day^{-1}$. Sunflecks contributed from 47 (day C) to 85% (day A) of the daily PFD but were present only 7 (day C) to 20% of the time (day E). Most (75 to 85%) sunflecks did not exceed the PFDs required to light saturate these particular *Adenocaulon* leaves ($200-400\,\mu\text{mol}\,\text{m}^{-2}\,\text{s}^{-1}$; see the horizontal lines on the left side of Fig. 17.1).

Integration of the curves in Fig. 17.1 gives the total daily carbon gain. Since the diffuse light levels in the understory were low and fairly constant, and the response to sunflecks was distinctly resolved, it was possible to separate the assimilation due to sunfleck utilization from that due to utilization of diffuse light alone. Total daily carbon gain ranged from 15 to 43 mmol $\text{m}^{-2}\,\text{d}^{-1}$. Measurements on another day not shown in Fig. 17.1 were done with the chamber shaded by a circular disc mounted away from

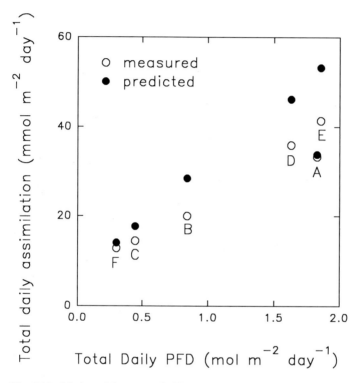

Fig. 17.2. Measured (*open symbols*) and modeled (*closed symbols*) daily total assimilation of *Adenocaulon bicolor* leaves for the 5 days shown in Fig. 17.1. The *letter* under each symbol pair corresponds to the daily course of assimilation and PFD in Fig. 17.1. *F* is for a day in which the leaf received only diffuse light. The model was of the form:

$$A = R_d + \frac{qI + A_m - [(qI + A_m)^2 - 4qtIA_m]^{0.5}}{2t},$$

where A is the assimilation rate at PFD = I, R_d is the day respiration rate, A_m is the maximum assimilation rate plus Rd, q is the apparent quantum yield, and t is a curving factor for the transition from light limitation to light saturation. The model was fit by a least-squares procedure to steady-state light response curves for each leaf and then used to calculate the predicted carbon gain in the absence of any dynamic responses (induction and post-illumination CO_2 fixation) for the daily courses of PFD shown in Fig. 17.1. (Pfitsch and Pearcy 1989a)

the chamber so that sunflecks were blocked but diffuse light was received. Daily assimilation on this day (day F) was $14\,\text{mmol}\,\text{m}^{-2}\text{d}^{-1}$. The carbon gain resulting from utilization of sunflecks ranged from about $3\,\text{mmol}\,\text{m}^{-2}\text{d}^{-1}$ on the day with the least sunflecks (day C) to over $30\,\text{mmol}\,\text{m}^{-2}\text{day}^{-1}$ (day E). These values represent 20 to 65% of the total carbon gain.

In addition to the daily course shown in Fig. 17.1, steady-state light response curves for assimilation were determined for each leaf on the next day using an artificial light source. These curves made it possible to accurately fit a simple model of the steady-state photosynthetic light response for each leaf, which would take as an input the light data shown in Fig. 17.1 and calculate photosynthesis at 1-s intervals. The model assumes an instantaneous response to the new light input and thus ignores induction limitations or enhancements due to post-illumination CO_2 fixation. Therefore, comparisons between the measured and simulated daily carbon gains reveal the effects of the dynamic responses that are important to sunfleck utilization.

Figure 17.2 shows the measured and modeled daily assimilation totals as a function of the daily PFD. Both gave nearly identical totals on day F when the leaf received no sunflecks and on day A when most of the assimilation occurred during a single period of sunfleck activity dominated by one long sunfleck. On the other days, the measured assimilation totals were 15 to 30% lower than the predicted values. On these days, there was no apparent relationship between the nature of the sunfleck activity and the difference between the measured and predicted values. Both the measured and predicted values on day A were low relative to the trend of daily assimilation and daily PFD for other measurements, probably because the sunflecks had much higher PFD on this day as compared to the others.

The lower measured than predicted carbon gain is evidence that induction is a significant limitation to sunfleck utilization by *Adenocaulon* in its natural environment. Post-illumination CO_2 fixation may partially offset some of the induction limitation, but it is clear that it is much less than needed to overcome it. Although the vast majority of sunflecks are brief (Fig. 17.3, top), most of the PFD is supplied in longer sunflecks (Fig. 17.3, bottom) for which post-illumination CO_2 fixation makes only a small contribution to the total. Moreover, many of the brief sunflecks have too low a PFD to cause significant postillumination CO_2 fixation. The situation may, however, be quite different in aspen canopies where post-illumination CO_2 fixation resulting from the many brief but high PFD sunflecks may be sufficient to overcome any induction limitations (Roden and Pearcy 1992).

17.5 The Significance of Sunflecks to annual Carbon Gain

Cloudiness and other factors make it difficult to extrapolate directly from the contribution of sunflecks on clear days to a contribution on an annual

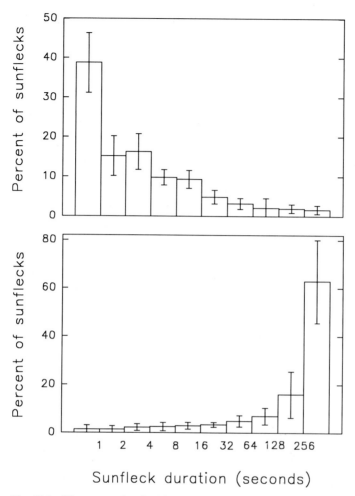

Fig. 17.3. Histograms showing the percent of sunflecks (*top*) and percent of sunfleck PFD received in sunflecks (*bottom*) of different duration in the redwood forest understory

basis. Extrapolation from gas exchange measurements such as those shown above is difficult because of the required sampling on days with different cloud covers and climatic conditions, and because of the extreme spatial variation. Consequently, a more integrative approach is needed. Since the carbon isotope ratio ($\delta^{13}C$) of plant tissue is a function of the ratio of intercellular to ambient CO_2 pressure (p_i/p_a) at which the carbon was fixed, it is a measure of the behavior of stomatal conductance and assimilation relative to one another. It is an integrative measure over the period that the carbon making up the sample is assimilated. Although applied primarily to water use efficiency problems (e.g., Farquhar and Richards 1984), it is also applicable to any response that causes a change in p_i/p_a, such as occurs during sunflecks (Pearcy 1987).

We examined the relationship between $\delta^{13}C$ and the direct PFD received by a plant with the objective of using the differences in isotopic discrimination as an integrative measure of the importance of sunflecks to carbon gain in different microsites. Farquhar et al. (1982) showed that the $\delta^{13}C$ of the plant tissue is related to p_i/p_a following the relationship:

$$\delta^{13}C_{leaf} = \delta^{13}C_{air} - 44.4 - 22^*p_i/p_a. \tag{1}$$

Since p_i/p_a approaches 1 in the shade because of low assimilation rates but decreases markedly during sunflecks (Weber et al. 1985; Pearcy 1987), variation in $\delta^{13}C$ of plants in different microsites should be an indicator of the fraction of carbon that was fixed during sunflecks. The expected $\delta^{13}C$ of the plant ($\delta^{13}C_p$) is given by:

$$\delta^{13}C_p = f\delta^{13}C_{sf} + (1 - f)\delta^{13}C_d, \tag{2}$$

where the subscripts sf and d refer to the carbon fixed during sunflecks and diffuse light, respectively, and f is the fraction of carbon fixed during sunflecks. Rearranging Eq. (2) yields

$$f = \frac{\delta^{13}C_d - \delta^{13}C_p}{\delta^{13}C_d - \delta^{13}C_{sf}}. \tag{3}$$

Use of Eq. (3) requires that the expected $\delta^{13}C$ values in sunflecks and diffuse light be known. Measured values of p_i/p_a in diffuse light and during sunflecks were 0.95 and 0.76, respectively (Pfitsch and Pearcy 1992) and from these an expected $\delta^{13}C$ for each could be calculated. However, the $\delta^{13}C$ value is also dependent on the source air, which in the understory is influenced by respired carbon and can be 1 to 2‰ less than in the bulk atmosphere (Schleser and Jayasekera 1985; Sternberg et al. 1989). The source air value could be determined by sampling the air, but this might require many samples over the season and under different weather conditions. We adopted an alternative approach of sampling plants that had been exposed to only diffuse light and therefore should have $\delta^{13}C$ values that depend only on p_i/p_a during diffuse light and on the source air. These plants were obtained by erecting shadow bands over plants in different microsites, which remained in place for 2 years before harvesting. Nearby plants that received both direct and diffuse light in varying proportions, depending on the particular microsite, were sampled at the same time. The relative amounts of direct and diffuse PFD received by each plant over the growing season were estimated from hemispherical photographs. The shadow bands were semi-circular 10-cm-wide aluminum strips supported over the plant in a 0.7 m diameter semi-circle corresponding to the solar path and thus blocked all sunflecks but still allowed receipt of most of the diffuse light. Adjustments were made on a 7- to 14-day basis to account for changes in solar zenith angle with season.

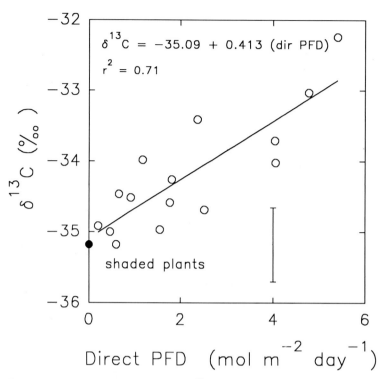

Fig. 17.4. Relationship between the $\delta^{13}C$ of *Adenocaulon bicolor* plants and the mean daily total of direct (sunfleck) PFD received by these plants. The *closed symbol* and the *error bar* are the mean (n = 13) and 95% confidence interval of the mean for the plants shaded by the shadow bands so that they received no sunflecks. (Pearcy and Pfitsch 1991)

Figure 17.4 shows the relationship between $\delta^{13}C$ values of plants and the estimated mean daily exposure to direct PFD (sunflecks). The fisheye photographic technique used to estimate direct PFD does not take into account cloudiness or penumbral effects and therefore overestimates the amount actually received. However, it is well correlated with the actual amount received and therefore provides a good relative estimate of differences among microsites (Pfitsch and Pearcy 1992). The $\delta^{13}C$ were highly correlated with direct PFD estimates, increasing on average 0.41‰ for each extra mol photons $m^{-2} d^{-1}$ received. The $\delta^{13}C$ values of plants under the shadow bands was not correlated with either the diffuse light they received ($r^2 = 0.08$) or with the estimates of the direct PFD they would have received had the shadow bands not been in place ($r^2 = 0.16$). The mean $\delta^{13}C$ value for these banded plants was −35.1‰ which was very close to the intercept the regression of $\delta^{13}C$ with direct PFD ($\delta = -35.09‰$).

The $\delta^{13}C$ of carbon fixed during diffuse light can also be calculated from the p_i/p_a using Eq. (1). In this case, using a value of −7.8 for the bulk atmosphere, a value of −33.6 results. The 1.2‰ lower measured than calculated value is the effect of the source air. The value of the source air was thus -9.0%, which is slightly higher than values reported for tropical forest understories (Sternberg et al., 1989). The respiration rates may be lower because of cooler temperatures and slower decomposition, or the mixing rates of the understory air with the bulk atmosphere greater in the redwood forest than in the tropical forests. Using the corrected value for the source air the $\delta^{13}C$ expected during sunflecks can be obtained from the p_i/p_a and is equal to -29.1. Thus the carbon gained during sunflecks should be 4.5% lighter than the carbon gained during diffuse light.

The values for carbon fixed during sunflecks and diffuse light can be used with Eq. (3) to calculate the fractional contribution of sunflecks. For 5 mol photons $m^2 day^{-1}$, which is representative for the brightest microsites sampled, Eq. (3) predicts that 46% of the carbon gain that went to production of the biomass of these plants was due to sunflecks. By contrast, at 1 mol photons $m^{-2} day^{-1}$, which is characteristic of the microsites with the smallest amount of direct PFD, only about 9% of the carbon gain was due to sunflecks. Thus utilization of sunflecks was responsible for a substantial fraction of the total annual carbon gain in the growing season but, as already shown with the daily gas exchange measurements (Fig. 17.1), it varied greatly from one microsite to another as a function of the direct PFD.

Comparison of the long-term fraction of assimilation due to sunflecks as estimated from carbon isotope ratios with the short-term ratios from gas exchange reveals that the latter are substantially greater. This is to be expected since the diurnal measurements were done on mostly clear days when sunflecks would obviously be most abundant whereas the carbon isotope measurements integrate over both clear and cloudy days. The brightest microsites used for the carbon isotope studies were brighter than the brightest of those for the diurnal run, so that in fact the difference may be even larger than indicated by the maximum values.

17.6 Consequences of Sunflecks for Growth and Reproduction

The studies with gas exchange and carbon isotopes both indicate that utilization of sunflecks contributes a significant fraction of the carbon gain in the brightest microsites but much less in the most shaded sites. The obvious question is how does this translate into performance in the understory in terms of growth and reproduction and to the variablity of these parameters? To examine this question, the shadow band system was used and growth and allocation were compared between plants that received sunflecks and those from which sunflecks were excluded. Aboveground growth was measured nondestructively via allometric techniques for 2 years, followed by harvest of both the above- and belowground parts. Since the plants were in micro-

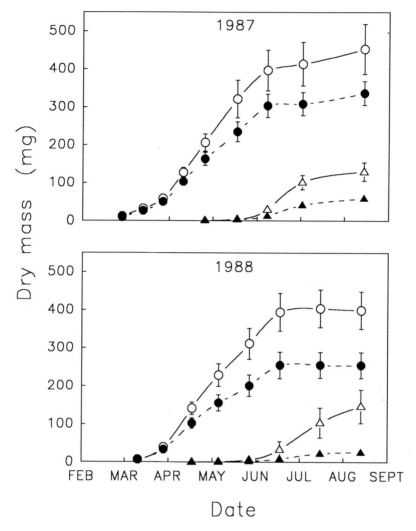

Fig. 17.5. Seasonal courses of estimated aboveground vegetative (*circles*) and reproductive (*triangles*) biomass for plants under the shadow bands that blocked all sunflecks (*dashed line, closed symbols*) and nearby plants that received sunflecks (*solid line, open symbols*). Data for 1987 (*top*) and 1988 (*bottom*) are shown. (Pfitsch and Pearcy 1992)

sites with different average daily direct and diffuse PFD totals, the relationship between growth and light environments in the field could also be examined.

Figure 17.5 shows the results of 2 years of growth under the shadow bands. After an initially slow phase in the spring of each year, vegetative expansion proceeded at a relatively constant rate until flowering commenced. Further increases in biomass were then largely confined to the inflorescences. There were no differences in the rate of growth during the rapid vegetative phase, which may have depended on reserves in the roots. However, by the end of the second year, vegetative biomass was significantly greater for plants receiving sunflecks than for those from which sunflecks had been excluded. Large differences in reproductive allocation were evident with the infloresence mass reduced by 83% in the plants from which sunflecks had been excluded. Overall, plants from which sunflecks had been excluded were 51% smaller than those receiving sunflecks. These results show that the fraction of carbon gain contributed by utilization of sunflecks takes on an even greater importance when growth is considered. This is because the initial growth and maintenance costs must be borne before resources can be allocated to reproduction. When the light is reduced by removal of sunflecks, the effect is largest for reproductive output. Some caution must be expressed, however, because it was not possible to rule out a role for variation in diffuse light. Performance of the banded plants was significantly correlated with estimates from fisheye photographs of the diffuse light they received, as would be expected. As will be discussed below, there was also evidence for an important role for diffuse light for the plants receiving sunflecks.

Examination of the growth and reproductive performance in microsites differing in the direct PFD received revealed some at first surprising patterns. First, there was, with one exception, no correlation between direct PFD and any measure of growth or reproductive performance in the different microsites. The one exception was that leaf area per plant was negatively correlated with direct PFD. Plant size and reproductive effort were, however, positively related to estimates of diffuse PFD, both for plants under the shadow bands and for plants also receiving sunflecks. A greater sensitivity to a given amount of diffuse PFD could be expected since it is in the quantum yield-dependent region of the light response where the efficiency of use is greatest. Many sunflecks, particularly those from canopy gaps that contribute the most to estimates from fisheye photographs, are, on the other hand, more than saturating.

It might be expected on the basis of the positive relationship between carbon gain and direct PFD and the effects observed when sunflecks were "removed", that more sunflecks would cause an increase in vegetative growth or reproductive output. However, plants under the shadow bands shrank as a result of a reduced carbon balance rather than the plants receiving sunflecks growing more. Other studies with tree seedlings and saplings (Pearcy 1983; Oberbauer et al. 1988) have shown a positive correlation between direct PFD and growth. However, tree seedlings have considerable potential for additional growth, whereas most *Adenocaulon* plants that were receiving sunflecks were not increasing in size. Possibly *Adenocaulon* plants had already reached a size that reflected some other limitations. There may be additional competition for nutrients or water in areas receiving more sunflecks. In addition, sunflecks themselves may cause stresses that limit growth at the same time as additional carbon is gained.

An example the stress created by a sunfleck is shown in Fig. 17.6. Sunflecks increase the leaf temperature and hence the transpiration rate. Consequently, water potentials decrease rapidly, reaching the turgor loss point, before recovering more slowly after the sunfleck passes. Visible leaf wilting as evidenced by a change in reflectance and curling of the leaf margins occurred during prolonged (>10 min) sunflecks. Both stomatal conductance and light-saturated assimilation rate decrease by 30 to 50% during sunflecks, presumably because of the wilting, but recovery as measured by g and light-saturated A was 80–90% complete within 1.5 h after the sunfleck. The only exceptions were during long, intense sunflecks on a warm day when leaf temperatures exceeded 37 °C. Recovery of photosynthetic rates was less than 50% 2 h after the sunfleck and only 75% complete on the next day in this case.

17.7 Conclusions

It is clear from both the daily gas exchange measurements and the seasonal estimates from carbon isotope discrimination that sunflecks are of considerable importance to the carbon gain of *Adenocaulon* plants. The 30 to 65% of daily carbon gain that was due to utilization of sunflecks is similar to values reported for understory tree seedlings in Australian and Hawaiian tropical forests (Pearcy and Calkin 1983; Pearcy 1987) but is considerably higher than the 6 to 19% previously estimated for tree seedlings and understory plants in deciduous forests (Schulze 1972; Weber et al. 1985). Two factors may account for the difference. First, diffuse light levels were considerably higher in the deciduous forest, decreasing the relative importance of assimilation due to sunflecks. Second, deciduous forest species had lower photosynthetic capacities per unit leaf area than *Adenocaulon* or the tropical forest species studied, which also reduces the contribution by sunflecks. The exception reported by Schulze was *Deschampsia flexuosa*, which had higher photosynthetic capacities and depended on sunflecks for 27% of its assimilation. The reasons for the generally lower photosynthetic capacities despite a higher overall light level in the deciduous forest are unclear. It may possibly indicate an additional limitation by nutrients or some other factor. Alternatively, it may be that selection or acclimation has favored increased efficiency of utilization of diffuse light in the deciduous forest plants, where it is relatively more important as compared to tropical and redwood forest plants, where sunflecks are a more important component of the available light. In the latter, higher photosynthetic capacities might be advantageous in allowing better utilization of sunflecks. Better use of diffuse light could

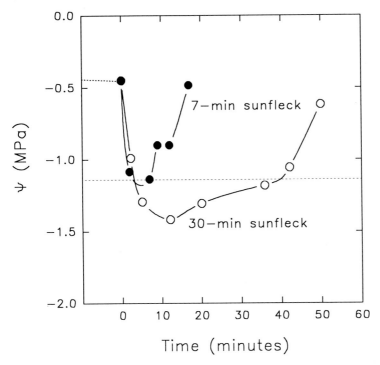

Fig. 17.6. Changes in leaf water potential of *Adenocaulon bicolor* leaves during a sunfleck of 7 min and a 30 min duration. The zero turgor point shown by the *dashed horizontal line* was determined from pressure-volume curves

have occurred through decreases in leaf mass per unit area, allowing more leaf area per plant but also resulting in lower photosynthetic capacities (Pearcy and Sims 1993). In support of this proposition is the observation that removal of sunflecks by the shadow bands resulted in lower photosynthetic capacities and lower leaf mass per unit areas of *Adenocaulon* plants as compared to plants receiving sunflecks, even at the same daily total PFD (Pearcy and Pfitsch 1991). More comparative work is needed to evaluate the relative roles of diffuse light and sunflecks in different habitats in both determining the nature of the photosynthetic responses and the daily carbon balance in understory environments.

Working against the increased photosynthetic capacity as a mechanism for enhancing the utilization of sunflecks is the induction requirement of photosynthesis. The induction requirement means that for most sunflecks the maximum photosynthetic capacity will not be realized and consequently the carbon gain during these sunflecks will be reduced. Comparison to the steady-state model shows that this induction requirement limits carbon gain of *Adenocaulon* leaves by 20 to 30%. Because these plants are extremely light-limited and have fixed costs that must be met before any growth and reproduction is possible, the 20 to 30% reduction in assimilation because of the induction requirement is probably very significant to the ecology of this species. Thus, minimization of induction could be advantageous. This appears to have occurred in *Adenocaulon* through minimization of stomatal limitations that allows for a relatively rapid induction response (Pfitsch and Pearcy 1989b). However, the biochemical limitations are still present and under some circumstances may be even greater in *Adenocaulon* than observed in other species.

Both the daily and seasonal estimates of the contribution of sunflecks show the extreme heterogeniety in this resource in the understory. While a good picture of the effects of the temporal heterogeniety of sunflecks on photosynthesis has been developed, much less is known about the spatial heterogeniety and its role in the ecology of these plants. In particular, there needs to be more work on the constraints on sunfleck utilization and also on the interactions between diffuse and direct light in terms of resource allocation.

Acknowledgments. This research was supported by NSF Grants BSR 86-00065 and BSR 88-17820

References

Björkman O, Ludlow M (1972) Characterization of the light climate of a Queensland rainforest. Carnegies Inst Wash Yearb 71: 85–94

Björkman O, Ludlow M, Morrow P (1972) Photosynthetic performance of two rainforest species in their native habitat and analysis of their gas exchange. Carnegie Inst Wash Yearb 71: 94–102

Chazdon RL (1988) Sunflecks and their importance to forest understory plants. Adv Ecol Res 18: 1–63

Chazdon RL, Fetcher N (1984) Photosynthetic light environments in a lowland tropical forest in Costa Rica. J Ecol 72: 553–564

Chazdon RL, Pearcy RW (1986a) Photosynthetic responses to light variation in rain forest species. I. Induction under constant and fluctuating light conditions. Oecologia 69: 517–523

Chazdon RL, Pearcy RW (1986b) Photosynthetic responses to light variation in rain forest species. II. Carbon gain and light utilization during lightflecks. Oecologia 69: 524–531

Farquhar GD, Richards RA (1984) Isotopic composition of plant carbon correlates with water use efficiency of wheat genotypes. Aust J Plant Physiol 11: 539–552

Farquhar GD, O'Leary MH, Berry JA (1982) On the relationship between carbon isotope discrimination and intercellular carbon dioxide concentration in leaves. Aust J Plant Physiol 9: 121–137

Kirschbaum MUF, Pearcy RW (1988) Gas exchange analysis of the relative importance of stomatal and biochemical factors in photosynthetic induction in *Alocasia macrorrhiza*. Plant Physiol 86: 782–785

Miller E, Norman JM (1971) A sunfleck theory for plant canopies. II. Penumbra effect: intensity distributions along sunfleck segments. Agron J 63: 739–748

Oberbauer SF, Clark DB, Clark DA, Quesada MA (1988) Crown light environments of saplings of two species of rain forest emergent trees. Oecologia 75: 207–212

Oker-Blom P (1984) Penumbral effects of within-plant shading on radiation distribution and leaf photosynthesis: a Monte-Carlo simulation. Photosynthetica 18: 522–528

Pearcy RW (1983) The light environment and growth of C_3 and C_4 tree species in the understory of a Hawaiian forest. Oecologia 58: 19–25

Pearcy RW (1987) Photosynthetic gas exchange responses of Australian tropical forest trees in canopy, gap and understory micro-environments. Funct Ecol 1: 169–178

Pearcy RW (1990) Sunflecks and photosynthesis in plant canopies. Annu Rev Plant Physiol Plant Mol Biol 41: 421–453

Pearcy RW, Calkin H (1983) Carbon dioxide exchange of C_3 and C_4 tree species in the understory of a Hawaiian forest. Oecologia 58: 26–32

Pearcy RW, Pfitsch WA (1991) Influence of sunflecks on the $\delta^{13}C$ of *Adenocaulon bicolor* plants occurring in contrasting forest understory microsites. Oecologia 86: 457–462

Pearcy RW, Osteryoung K, Calkin HW (1985) Photosynthetic responses to dynamic light environments by Hawaiian trees. The time course of CO_2 uptake and carbon gain during sunflecks. Plant Physiol 79: 896–902

Pearcy RW, Sims DA (1993) Photosynthetic acclimation to changing light environments: scaling from the leaf to the whole plant. In: Caldwell MM, Pearcy RW (eds) Exploitation of environmental heterogeneity by plants: ecophysiological processes above and below ground. Academic Press, San Diego (in press)

Pfitsch WA, Pearcy RW (1989a) Daily carbon gain by *Adenocaulon bicolor*, a redwood forest understory herb, in relation to its light environment. Oecologia 80: 465–470

Pfitsch WA, Pearcy RW (1989b) Steady-state and dynamic photosynthetic response of *Adenocaulon bicolor* in its redwood forest habitat. Oecologia 80: 471–476

Pfitsch WA, Pearcy RW (1992) Growth and reproductive allocation of *Adenocaulon bicolor* following experimental removal of sunflecks. Ecology 73: 2109–2117

Roden JS, Pearcy RW (1993) Effect of leaf flutter on the light environment of poplars. Oecologia 93: 201–207

Sassenrath-Cole GF, Pearcy RW (1992) The role of ribulose-1,5-bisphosphate regeneration in the induction requirement of photosynthetic CO_2 exchange under transient light conditions. Plant Physiol 99: 227–234

Schleser GH, Jayasekera R (1985) $k^{13}C$ variation of leaves in forest as an indicator of reassimilated CO_2 from soil. Oecologia 65: 536–542

Schulze E-D (1972) Die Wirkung von Licht und Temperatur auf den CO_2-Gaswechel verschiedener Lebensformen aus der Krautschicht eines montanen Buchenwaldes. Oecologia 9: 223–234

Seemann JR, Kirschbaum MUF, Sharkey TD, Pearcy RW (1988) Regulation of ribulose 1,5-bisphosphate carboxylase activity in *Alocasia macrorrhiza* in response to step changes in irradiance. Plant Physiol 88: 148–152

Sternberg LSL, Mulkey SS, Wright SJ (1989) Ecological interpretation of leaf carbon isotope ratios: influence of respired carbon dioxide. Ecology 70: 1317–1324

Tinoco-Ojanguren C, Pearcy RW (1992) Dynamic stomatal behavior and its role in carbon gain during lightflecks of a gap phase and an understory *Piper* species acclimated to high and low light. Oecologia 92: 222–228

Weber JA, Jurik TW, Tenhumen JD, Gates DM (1985) Analysis of gas exchange in seedlings of *Acer saccharum*: integration of field and laboratory studies. Oecologia 65: 338–347

Woodrow IE, Mott KA (1989) Rate limitation of non-steady-state photosynthesis by ribulose-1,5-bisphosphate carboxylase in spinach. Aust J Plant Physiol 16: 487–500

18 Variation in Gas Exchange Characteristics Among Desert Plants

J.R. Ehleringer

18.1 Introduction

Professor Otto Lange and colleagues have made significant contributions to our understanding of ecophysiological aspects of plant performance under arid land conditions. It was their pioneering field research in the Negev Desert during the 1960s and 1970s that firmly established our understanding of the impacts of drought and high temperature on photosynthetic gas exchange, respiration, and transpiration in extreme environments. These field studies described the mechanistic bases for photosynthetic adjustment to desert habitats and evaluated the carbon gain significance of different acclimation patterns. During the past decade, Lange and colleagues have extended this research by examination of plant performance in sclerophyllous shrub and tree species in the arid mediterranean climate zones of Portugal, and these studies have provided further insights into the cellular adjustments of plants to arid conditions.

In this chapter, I would like to build on the foundation laid down by Lange and colleagues and to examine a question related to the physiological and evolutionary ecology of plants in arid zones. Namely, given the spatial and temporal diversity of deserts: what kinds of photosynthetic gas exchange characteristics are expected under different selective regimes? In particular, what patterns are expected within different life forms and how do life-form-related physiological characters relate to species composition in different desert locations? Are the known gas exchange patterns from field studies consistent with these relationships? The primary focus of this chapter will be the deserts of North America with special emphasis on the Sonoran Desert, since it is the most diverse of the North American deserts (Shreve and Wiggins 1964). However, the patterns that emerge should be applicable to other desert regions of the world.

18.2 Species Distribution Gradients in the Desert

The deserts of western North America (Chihuahuan, Colorado Plateau, Great Basin, Mojave, and Sonoran) are defined by a limited-precipitation regime, but differ quite substantially in biological, climatic, and geomorphic aspects (Brown 1968; Bender 1982; MacMahon 1985; Osmond et al. 1990). Plant growth and activity are, of course, constrained by limited soil moisture availability in each of these deserts. In the northerly and higher elevation Colorado Plateau and Great Basin Deserts, cold winter temperatures further restrict activity to the spring and early summer months (Caldwell 1985; Comstock and Ehleringer 1992). Seasonal precipitation patterns further differentiate these deserts, with the Mojave Desert being primarily a winter-precipitation desert and the Chihuahuan Desert a summer-precipitation desert. The Sonoran Desert, geographically at the center of the deserts,

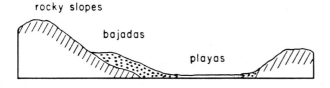

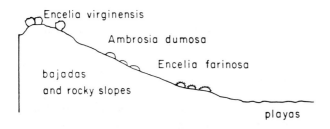

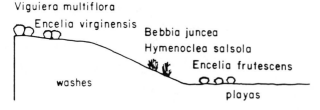

Fig. 18.1. *Top* Cross section of elevational transect in the Sonoran Desert showing the three dominant macrohabitats – rocky slopes, bajadas (coarse alluvium), and playas (fine alluvium). With the bajadas and rocky slopes are washes of varying size and dimension. *Middle* Distribution of common drought-deciduous shrubs along elevational transect on bajadas and rocky slopes near Needles, California. *Bottom* Distribution of common drought-deciduous shrubs along elevational transect in washes near Needles, California

experiences both winter and summer precipitation regimes; since the Sonoran Desert is a low-elevation desert, plants may be active at any time of the year when precipitation is received.

Species and life-form abundance vary substantially along elevation clines (Hastings and Turner 1965b), reflecting both changes in precipitation and temperature. Whittaker and Niering (1965) described a progression from predominance of trees to increased abundance of suffrutescent shrubs as precipitation decreases. In particular, they noted a higher species diversity and greater life-form diversity in the kinds of leaf and stem morphologies of plants in drier desert locations. Similar trends in the Sonoran Desert vegetation have also been described by Phillips and MacMahon (1978) and Bowers and Lowe (1985). As first discussed by Schulze (1982), these patterns relate to variation in gas exchange characteristics in terms of both photosynthetic pathways and seasonality of carbon gain activities.

Within the Sonoran Desert, elevation-related transects are usually also related to substrate variation (Fig. 18.1). Upper elevations of a range typically have shallow, coarse soils; the gradient from upper rocky slopes along a bajada and ending up at a playa is associated with an increase in the fine particle structure of soils. At various places along a bajada, washes erode through the alluvium, creating microhabitats that are coarser in structure and typically have greater soil moisture availability.

Of interest to this discussion is that a dominant life form will often persist over this entire transect, including wash and slope microhabitats. One species is typically dominant at a site and there is usually a single-species replacement pattern along these elevation gradients. For instance, drought-deciduous shrubs are a key component of virtually all Sonoran Desert plant communities. In most situations, there is just one dominant species at a location (Table 18.1), but in both wash and bajada slope microhabitats, there is a continual replacement of species with elevation. *Encelia virginensis* may dominate slope bajada sites at upper elevations, but it is replaced first by *Ambrosia dumosa*, and then by *E. farinosa* at lower elevation sites (Fig. 18.1). Similarly, while *Viguiera multiflora* predominates at upper elevation wash sites, *E. frutescens* dominates the lower elevation sites.

18.3   Variation in Moisture and Temperature as Selective Forces for Photosynthetic Variation

Availability of moisture is the primary feature influencing plant productivity and overall performance. To better understand the possible benefits and disadvantages of different gas exchange characteristics, it is useful to know the absolute moisture inputs and the variability of those inputs. The predictability of moisture inputs on both a seasonal and an interannual basis is

Table 18.1. Percentage abundance of shrub species on slope and in wash microhabitats along an elevational gradient in the Sonoran Desert near Needles, California

	Elevation (m)				
	180	365	615	715	975
Slope					
Encelia farinosa	–	89	93	0	0
Ambrosia dumosa	–	6	4	88	0
Krameria parviflora	–	6	2	12	55
Encelia virginensis	–	0	0	0	29
Viguiera multiflora	–	0	0	0	16
Wash					
Encelia frutescens	80	0	0	0	0
Hymenoclea salsola	20	96	78	37	3
Bebbia juncea	0	4	14	58	14
Chrysothamnus paniculatus	0	0	6	5	0
Encelia virginensis	0	0	0	0	41
Viguiera multiflora	0	0	0	0	24
Krameria parviflora	0	0	0	0	14

key to understanding aspects of plant carbon gain such as whether or not annual versus perennial life forms are favored, whether or not acclimation to drought or drought avoidance will lead to higher rates of carbon gain, and whether or not conditions are sufficiently predictable to favor the emergence of one particular set of photosynthetic characteristics over another in a particular habitat.

18.3.1 Predictability of Precipitation

Rainfall is the life's blood of deserts, yet precipitation amounts are usually sufficiently low that primary productivity and precipitation are linearly related (Noy-Meir 1973; Evanari et al. 1976; Ehleringer and Mooney 1983; Le Houérou 1984). Plants of arid regions not only face low annual amounts of precipitation, but the interannual variation in that precipitation is high, resulting in a variability of primary production that exceeds the variability in precipitation input (Le Houérou et al. 1988). The distribution of mean annual precipitation in deserts can be approximated by a negative binomial or gamma distribution (Thom 1958; Hershfield 1962; Hastings 1965; Nicholson 1980). What this indicates is that the coefficient of variation increases as the mean precipitation decreases (Fig. 18.2). As a consequence, the interannual variation in precipitation becomes progressively greater at drier sites. Total precipitation in most years then becomes progressively lower than the arithmetic mean precipitation values.

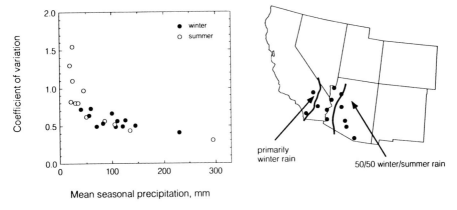

Fig. 18.2. *Left* The relationship between the coefficient of variation of precipitation and mean seasonal precipitation for selected sites in the arid regions of the southwestern United States. Sites are those indicated on *right*. Precipitation data are presented for winter (November–May) and summer (June–October) seasons. Calculations are based on data from the US Weather Bureau records (NOAA) during the period 1929–1979. *Right* Approximate geographic boundaries of predominant winter and summer rain environments in the arid regions of the southwestern United States. Shown also are the sites used for meteorological analyses in Figs. 18.3–18.7

Of particular interest is that front-related winter storms do not generate a significantly different relationship between mean and coefficient of variation from those of convectional summer storms (Fig. 18.2). Contrary to the popular belief that summer rainfall is more variable, that applies only to short-term spatial variability (McDonald 1956). In the long term, the year-to-year variability for summer and winter rains is the same for a given mean; it is only the long-term precipitation values that are needed to generate information on the interannual predictability of that precipitation.

This leads to two patterns. First, drier sites will be characterized by drought of increased length. Second, single-storm events will have a greater impact on plants growing in the driest regions. Goude and Wilkinson (1980) have shown that the maximum daily rainfalls in South African deserts constitute a much larger percentage of the annual precipitation than in wetter regions. In other words, episodic large-storm events would be expected to have a much greater impact on the dynamics of arid zones than in semi-arid or mesic regions.

When evaluating the impact of storms on the annual precipitation, it appears that the frequency distribution of storm sizes is the same for sites differing widely in total annual precipitation. Phoenix, Arizona, receives 185 mm precipitation annually, whereas Indio, California, receives only 80 mm. However, the frequency distributions of storms sizes have similar shapes (Fig. 18.3). Needles, California, is intermediate at 110 mm, but still has a distribution curve similar to those at Phoenix and Indio. The same conclusion is reached if the data are evaluated on a seasonal basis instead of

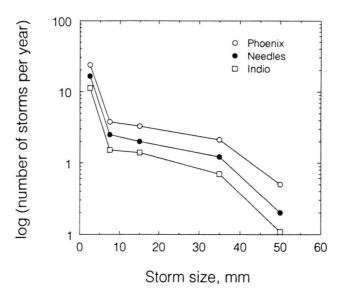

Fig. 18.3. The frequency distribution of storm sizes for sites receiving different average amounts of precipitation in either winter (November–May) or summer (June–October) seasons. Calculations are based on data from the US Weather Bureau records (NOAA) during the period 1929–1979. A storm is defined as the total cumulative precipitation at a site in a contiguous time period (typically 1–3 days)

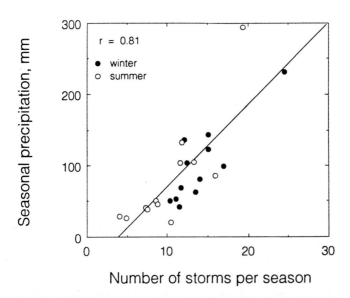

Fig. 18.4. The correlation between mean precipitation in either winter (November–May) or summer (June–October) seasons at a location and the number of storms. Calculations are based on data from the US Weather Bureau records (NOAA) during the period 1929–1979. The combined correlation is significant at the $P < 0.01$ level

an annual basis (data not presented). The vast majority of storm events result in less than 10 mm precipitation, and it may be questionable as to just how useful this input is to increasing soil moisture at soil depths where plants can effectively use the precipitation input. Sala and Lauenroth (1982) examined the impact of small precipitation events (~5 mm) on plant water relations in a semi-arid glassland. They concluded that for shallow-rooted grasses these small precipitation events could significantly improve gas exchange for 1–2 days following the storm.

Given the frequency distributions in Fig. 18.13, total annual precipitation and the number of storm events should be tightly correlated. Figure 18.4 shows that not only is there indeed a significant linear relationship between

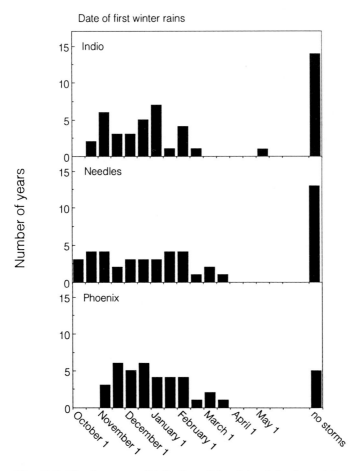

Fig. 18.5. The frequency distribution of the date of the first winter rains for three sites differing in the predictability of that winter precipitation. The first winter storm is defined as the first storm after October 1 that was ≥10 mm. Calculations are based on data from the US Weather Bureau records (NOAA) during the period 1929–1979. The three figures are arranged in a west-to-east transect from *top to bottom*

these two parameters, but that the relationships for both summer and winter storms are the same.

18.3.2 Drought Duration

Of importance in trying to understand the constraints imposed on plant performance by drought is length or duration of the drought period. One way of examining this character is to examine how long it takes, after the "winter season" or "summer season" begins, before significant precipitation occurs. A 10-mm event is often considered a minimal trigger for plant growth activities in the desert (Beatley, 1974a,b). Winter precipitation is driven by frontal storms coming off the Pacific Ocean. If we examine the date of occurrence of the first 10-mm precipitation for the winter season at sites differing in the mean annual precipitation, we see that this date is highly variable (Fig. 18.5). The "beginning" of the winter season may be essentially any date between October 1 and May 15 in the Sonoran Desert; the frequency distribution is flat enough that there is little tendency for the winter growing period to begin during any specific window. Moreover, at the two driest sites (Indio and Needles), there was not a single storm large enough to trigger growth according to the criteria of Beatley (1974b) in almost one-third of the years. As such, we would obviously expect that perennial plants at the driest sites would require features to insure persistence throughout an entire year without precipitation.

The entry of moist tropical air into the Sonoran Desert is more reliable. In the more southerly sites, the dates of the first summer monsoonal rains are predictable (Fig. 18.6). The date of the first summer rains is, however, very much dependent on the average amount of summer rains received. In the regions receiving less summer precipitation, the date of these first rains is less predictable, such that at the drier sites this summer moisture (if any) could come at virtually any time during the summer.

Once the first rains have arrived, just how good is early season moisture as a predictor of the overall quality of the growing season? If the first rains events portend a generally wet and favorable season, then plants should respond (break dormancy, germinate) shortly after these rains events to capitalize on the available soil moisture and to be positioned to effectively use later moisture inputs. However, if the first rains of the season do not provide qualitative information about the remainder of the season, then perhaps responding to those early rains is a chance event. For 10 of the 12 stations listed in Fig. 18.2, there is a highly significant, positive correlation ($P < 0.01$) between the amount of winter precipitation falling in the first 30 days of the growing season and the total amount for that growing season. For 7 of 12 stations, there was an equivalently positive and significant correlation for summer rains as well. Thus, early season precipitation is statistically a reliable indicator of future moisture inputs. Pianka (1967) ex-

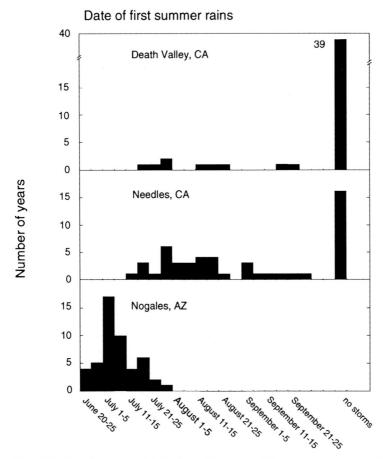

Fig. 18.6. The frequency distribution of the date of the first summer rains for three sites differing in the predictability of that summer precipitation. The first summer storm is defined as the first storm after June 1 that was ≥10 mm. Calculations are based on data from the US Weather Bureau records (NOAA) during the period 1941–1989. The three figures are arranged in a north-to-south transect from *top to bottom*

amined autocorrelations of monthly precipitation for different sites throughout the Great Basin, Mojave, and Sonoran Deserts. As in the previous analysis, he noted a strong autocorrelation for up to 2–3-month periods in the Mojave and Sonoran Deserts. However, for the Great Basin, Pianka (1967) found no autocorrelation between months and concluded that precipitation had little predictability from month to month.

18.3.3   Predictability of Temperature

Air temperatures exhibit less variance than precipitation (Fig. 18.7). However, there is a strong tendency for temperatures in the cooler months of the

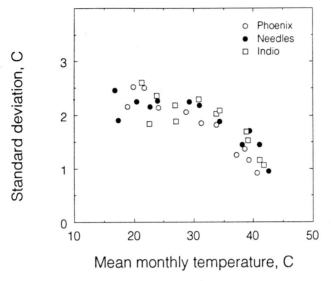

Fig. 18.7. The relationship between the mean monthly temperature and the standard deviation of values about the mean for different locations in arid regions of the western United States. Calculations are based on data from the US Weather Bureau records (NOAA) during the period 1929–1979

year to be more variable than those in the hotter months (Hinds and Rotenberry 1979). Variability of air temperatures during the winter months in the Sonoran Desert is almost twice that in summer months. One possible interpretation of these data is that drier winters are warmer and have a greater impact on absolute temperatures than do differences in temperature arising from changes in the number of convection storms that occur during summer months. An alternative explanation is that infrequently during the winter months Arctic air moves into the southern desert regions for a brief period of time. These episodic events would influence calculations of temperature variability, and would undoubtedly have a significant impact on freezing and plant distribution. In probably the best known example, Shreve (1911) showed that exposure to freezing temperatures for only a day was sufficient to kill the giant saguaro cactus. Episodic low-temperature events are known to play a major role in defining the distribution limits of saguaro and other cacti (Nobel 1980).

18.4 Gas Exchange Patterns Among Life-Forms

18.4.1 Photosynthetic Pathway Distribution Among Life-Forms

Deserts contain a much larger range of life-forms than are found in more mesic regions, particularly in those deserts that receive bimodal precipitation

(Shreve 1942; Werger 1986; Shmida and Burgess 1988). Shreve and Wiggins (1964) described 25 categories of life forms within the Sonoran Desert. While many of the different life-forms relate to overall physiognomy (tree, subtree, shrub, annual, etc.), 19 relate to succulence and to the high diversity associated with shrubs as a life-form. For photosynthetic purposes, we will examine the diversity of life-forms related to leaf types and succulence differences, because these are of particular importance to understanding the distribution patterns of photosynthetic pathways.

Within the C_3 photosynthetic pathway, there is wide variation in photosynthetic tissue types in that leaves, twigs, stems, or any combinations of these serve as the photosynthetic surface. Green-stem or green-photosynthetic nonleaf tissues are perhaps the most frequent variant (Ehleringer et al. 1987; Osmond et al. 1987; Smith and Osmond 1987; Comstock and Ehleringer 1988, 1990). Species exhibiting photosynthesis by nonleaf tissues increase both as precipitation decreases and as the fraction of summer precipitation increases (Shmida and Burgess 1988; Cody 1989). Twig and stem photosynthesis contribute significantly to the overall carbon balance of these species, and in some cases, leaves no longer serve a significant role in the plant's carbon balance (Comstock et al. 1988). In all instances, the intercellular CO_2 concentrations of twigs are lower than those of leaves and the twig tissues are more drought-tolerant than leaf tissues. As elaborated in Section 18.5.4, leaves of C_3 species in the desert exhibit over a $100\,\mu l\,l^{-1}$ variation in intercellular CO_2 concentrations, which will have significant bearing on both the extent to which stomata limit photosynthesis and on water-use efficiency.

In terms of carbon gain, advantages and disadvantages can be associated with each of the three photosynthetic pathways depending on environmental conditions. Under today's CO_2 environment, the C_4 photosynthetic pathway offers little or no intrinsic advantage over C_3 under cool temperatures ($\leqslant 25\,^\circ C$). In fact, the increases in quantum yield and reductions in photorespiration associated with C_3 photosynthesis under cooler temperatures suggest an advantage to plants possessing that pathway in cooler environments (Ehleringer 1978; Osmond et al. 1982). While C_4 photosynthesis may not have originally evolved under hot, arid conditions (Ehleringer et al. 1991b), C_4 photosynthesis does result in a reduced photorespiration rate and increased quantum efficiency at high temperatures ($\geqslant 35\,^\circ C$). Thus, we should expect C_4 photosynthesis to predominate in hot temperature environments and C_3 photosynthesis in cool temperature environments, provided that adequate moisture for growth is available. There is no evidence that intrinsic aspects of each photosynthetic pathway should confer any competitive advantage with respect to drought tolerance. However, because of the higher water-use efficiency of C_4 photosynthesis, these plants are expected to be at a competitive advantage in saline environments (Pearcy and Ehleringer 1984).

CAM photosynthesis provides a mechanism for a potential advantage over C_3 and C_4 plants under limited moisture conditions (Lange et al. 1974,

1975, 1978; Smith and Nobel 1986). The extremely high water-use efficiency of the CAM pathway may be of advantage during drought periods, when carbon gain by other pathways approaches zero. However, it is unclear whether or not the low photosynthetic-capacity constraints of CAM plants during wetter periods of the year offsets its advantage during drought. It may be that CAM plants exhibit an advantage only if competition for light is insufficient to exclude these plants during the wetter periods. Consistent with this is the observation of an increase in the frequency of CAM plants along gradients of decreasing precipitation in coastal regions of southern California and northern Chile (Mooney et al. 1974).

Following the initial observations by Bender (1968) that photosynthetic pathways could be distinguished on the basis of their carbon isotope ratio ($\delta^{13}C$), there was an extensive attempt to survey the world's flora to determine the ecological and taxonomic distribution of photosynthetic pathway types. Over the past 20 years, these efforts resulted in the development of a relatively complete picture of the distribution of photosynthetic pathway types in arid zones (summarized in Ehleringer 1989, and Ehleringer and Monson 1993). Life-form related aspects of these patterns are summarized

Table 18.2. Taxonomic distribution of photosynthetic pathway types according to life form. Abundance estimates are for floristic abundance (not ecological abundance). Abundance estimates are $+ =$ infrequent occurrences known, $++ =$ occasionally, $+++ =$ common, and $++++ =$ essentially exclusive

	C_3	C_4	C_3-CAM	CAM
Annuals				
Winter anual	++++			
Summer annual	++	+++		
Perennial succulents				
Leaf deciduous	++		+++	
Leaf succulent	++		+++	
Stem succulent				++++
Perennial arborescents				
Subtree	++++	+		
Tree	++++			
Perennial shrubs				
Evergreen-leaved	+++	++		
Drought-deciduous	+++			
Winter-deciduous	++++			
Photosynthetic twigs				
Twigs only	+++			+++
Leaves and twigs	++++			
Perennial herbs				
Graminoid	+++	+++		
Geophytes	++++			

in Table 18.2. From this table it is evident that C_4 plants are not so taxonomically common in deserts. The C_4 pathway is essentially found in only three life-forms: graminoids, evergreen shrubs, and summer annuals. The distribution of C_4 photosynthesis among evergreen perennials is essentially restricted to halophytic plants (e.g., *Atriplex*). When the distribution of photosynthetic pathway types is analyzed with respect to the different North American deserts, C_4 photosynthesis is rare in deserts that do not receive significant amounts of summer rain (Teeri and Stowe 1976; Stowe and Teeri 1978), except in the case of halophytes (Caldwell et al. 1977; Pearcy and Ehleringer 1984). The distribution of CAM plants closely parallels that of C_4 species (Teeri et al. 1978). CAM plants are most frequent in habitats receiving summer rains and those regions without cold winter temperatures.

Predictability and patterning of precipitation provide a basis for understanding the distribution of C_3 and C_4 photosynthesis in perennials. Throughout much of the Sonoran Desert, the most effective precipitation comes during the winter periods when temperatures are cooler and evaporative gradients are lower (see Sect. 18.5). If winter and spring are the primary periods of active growth, then C_3 photosynthesis may have an advantage over C_4 photosynthesis in perennial species. Under the moderate temperatures prevailing at that time, photorespiration rates would be expected to be reduced in C_3 plants and quantum yields higher in C_3 relative to C_4

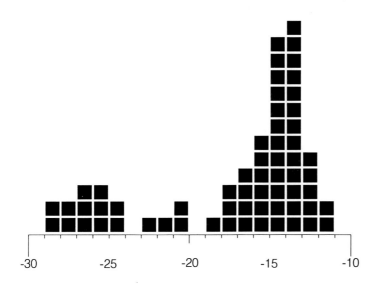

Carbon isotope ratio, ‰

Fig. 18.8. Variation in the carbon isotope ratio ($\delta^{13}C$) of *Aloe* species from South Africa. The $\delta^{13}C$ values approaching -10% indicate a species that uses only CAM photosynthesis; values approaching -30% indicate a species using only C_3 photosynthesis. Intermediate $\delta^{13}C$ values indicate a species using both C_3 and CAM photosynthesis (After Vogel 1980)

plants (Ehleringer and Björkman 1977; Ehleringer and Pearcy 1983; Pearcy and Ehleringer 1984). While such an explanation may be satisfactory for the northern and central portions of the Sonoran Desert, it falls short on explaining the lack of C_4 photosynthesis in perennials from the southern desert regions where precipitation comes primarily during the summer months and where temperatures are high. At present, there is no satisfactory explanation (other than historical) as to why C_4 photosynthesis is not so taxonomically common in the southern Sonoran Desert.

Two interesting patterns emerge with respect to shifts in photosynthetic pathway within a plant in response to increased soil moisture deficit. The first is tissue-dependent differentiation of photosynthetic pathway. Lange and Zuber (1977) were the first to note this with *Frerea indica*, a south African perennial having a succulent stem with CAM photosynthesis and drought-deciduous C_3 leaves. Presumably, by such a mechanism, the plants is able to gain carbon longer into the drought period than were it to have only C_3 metabolism. More common, though, is a second pattern in which the same leaf tissue switches from C_3 to CAM photosynthesis depending on the external soil moisture stress (Troughton et al. 1977; Winter et al. 1978; Bloom and Troughton 1979). Perhaps nowhere is this better developed than in the south African genus *Aloe*, whose members span the entire range from 100% C_3 to 100% CAM photosynthesis (Fig. 18.8).

18.4.2 Environment and Life-Form Distribution

It is difficult to evaluate or predict leaf-level gas exchange characteristics of different life-forms in the absence of information about other possible constraints within the plant that influence overall carbon gain. For photosynthetic gas exchange, two of the most critical aspects will be factors related to mineral nutrition and to water acquisition and transport. Mineral nutrition (particularly nitrogen) limits those aspects of gas exchange most closely associated with capacity (Field and Mooney 1986; Evans 1989). Models exist to predict the optimal allocation of nitrogen to maximize carbon gain (Field 1983). Photosynthetic capacity in desert species is linearly related to leaf nitrogen contents (Mooney et al. 1981; Field and Mooney 1986). A priori, there is no reason to expect that life form should impose constraints on the maximum photosynthetic capacity of leaves in desert plants, although field observations indicate distinct trends (Mooney and Gulmon 1982; Smith and Nobel 1986). Stomatal and biochemical aspects of gas exchange are known to be closely integrated (Wong et al. 1979; Woodrow and Berry 1988), suggesting that water-related aspects of gas exchange may account for the life-form-dependent patterns recognized by Mooney and Gulmon (1982).

There are reasons to expect that life-form characters should impose constraints on actual photosynthetic rates, particularly in perennial plants

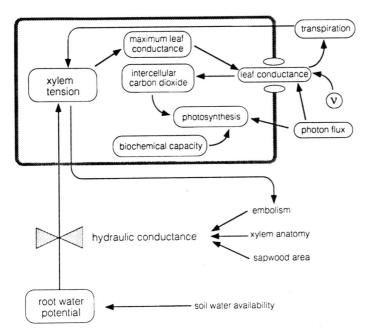

Fig. 18.9. A conceptual model of how higher intercellular CO_2 concentrations driven by increased stomatal conductances may contribute to enhanced xylem embolism rates, particularly under conditions of low root water potentials as would be expected during periods of extreme soil water deficit. (After Ehleringer 1993b)

which must persist through repeated drought periods. Tyree and Sperry (1989) have shown that under low water potentials cavitation events within the xylem increased, leading to a reduced capacity to conduct water. Structural adaptations exist which permit tolerance of reduced water potentials, but appear to come at the expense of a reduced capacity to conduct water. Given equal stem hydraulic conductances between two species, a relatively higher stomatal conductance by one species should result in an increased transpiration rate and a decreased leaf water potential (Fig. 18.9). As soil moisture availability decreases during the season, the water potential gradient between leaf and root should increase and ultimately under extreme or prolonged drought stress, water potentials may reach the point at which cavitation events occur with high frequency. In theory, progressive accumulation of these cavitation events could restrict water flow sufficiently to reduce transpiration and result in stem death if drought persisted over an extended period.

If drought-induced cavitation events are irreversible or mostly irreversible, then persistence through time and stem hydraulic conductivity should be inversely related. That is, perennials would be expected to have lower hydraulic conductances than annuals. Long-lived perennials, which must persist through repeated droughts of varying duration, would be expected to have lower hydraulic conductances than short-lived perennials in order to avoid extensive xylem cavitation. With respect to perennial life forms accessing the same limited soil moisture, trees would be at a significant disadvantage to shrubs (an overall shorter conducting system). On this point Balding and Cunningham (1974) showed that plant height became progressively shorter in *Acacia constricta* (tree) and *Larrea tridentata* (shrub) as drought severity increased along an aridity gradient. Stem xylem cavitation events would be expected to proceed from the stem tip where transpiration is occurring toward basal regions, creating progressive stem dieback under conditions of increasing drought duration. This may in part be the explanation for why suffrutescent growth (many stems emerging from a common root base) is so common among aridland shrubs.

Life-form diversity of perennial species is positively related to climatic diversity. Cody (1989) showed that life-form diversity among Sonoran Desert sites increased under extreme conditions – conditions likely to reflect both extremes in drought and a high year-to-year variability in precipitation. Shmida and Burgess (1988) found that biseasonality of precipitation was the major driving factor accounting for the increased growth-form diversity in the Sonoran Desert. Biseasonal precipitation was also correlated with these same life-form patterns within the Great Basin and Colorado Plateau (Comstock and Ehleringer 1992). Together, these data suggest that in North American deserts it is the diversity of incoming moisture sources that drives the evolution of differences in life-form.

The coefficient of variation is a measure of unpredictability (Sect. 18.2.1) and an unpredictable environment will preclude the evolution of perennial plants dependent on frequent rainfall episodes or not capable of persisting through extended drought periods. The loss of constancy (the degree to which conditions are similar from year to year) in drier sites should have a direct impact on the distribution of life-forms along an aridity gradient. Hastings and Turner (1965a) examined the amounts and variability of winter and summer precipitation in Baja California. They noted that the variability in precipitation corresponded to the distribution of vegetation types described by Shreve (1934) and Shreve and Wiggins (1964). After Colwell (1974) showed that predictability and constancy were mathematically related, Schaffer and Gadgil (1975) then followed with a prediction of the distribution abundances of annual versus perennial life-forms as predictability of rainfall decreased. The data of Harper (unpubl. data cited in Schaffer and Gadgil 1975) supported the prediction that annuals become an increasing fraction of the total flora as predictability of that precipitation decreases. What these data sets did not reveal was any information on the distribution of photosynthetic pathways among these annuals. Such information can be extracted from Shreve and Wiggins (1964), who noted that there were distinct winter- and summer-annual floras, even within a particular region. It turns out that winter annuals have C_3 photosynthesis exclusively, while summer annuals are primarily C_4 plants (Mulroy and Rundel 1977). Such a distribution pattern is consistent with the observation that C_4 photosynthesis, through its effects on reduced photorespiration and increased quantum yields at high temperature, provides a significant potential carbon gain advantage over the C_3 pathway under hot summer conditions if soil moisture is adequate.

18.5 Longevity and Gas Exchange

While absolute photosynthesis and transpiration rates among species may exhibit substantial variation, these flux rates decrease as water stress increases (Smith and Nobel 1986; Smith and Nowak 1990). Variation in maximum flux capacities is loosely associated with life-form, with annuals typically having higher rates than perennials (Mooney and Gulmon 1982; Smith and Nobel 1986). However, there are enough counter-examples of annuals having low photosynthetic capacities (Seeman et al. 1980; Werk et al. 1983) and perennials having high photosynthetic capacities (Ehleringer and Björkman 1977) that generalities of this type cannot be drawn with a high degree of certainty. An alternative approach to understanding life-history variation in gas exchange parameters is to evaluate ratios of activity (such as water-use efficiency) or possible set points in photosynthetic gas exchange.

18.5.1   Water Use in Relation To Carbon Gain

Gas-exchange responses at the leaf level can be viewed from two perspectives: what causes changes in absolute flux rates and what controls changes in flux rates. Changes in maximum photosynthesis (A) and transpiration (E) rates of desert plants in response to soil moisture availability (or any other measures of plant water status for that matter) have been described in numerous studies (Lange et al. 1976; Ehleringer and Mooney 1983; Smith and Nobel 1986; Smith and Nowak 1990). Changes in flux rates and canopy photosynthetic area almost always show a linear response to plant stress (e.g., water potential), and species vary widely in their capacity to maintain photosynthetic activity under water stress. As a consequence, instantaneous measures of gas exchange activity at a single point in time may provide limited insight into primary productivity and plant fitness, although the parameters are ultimately linked with each other (Fig. 18.10).

18.5.2   Gas Exchange Flux Versus Set Point

An alternative approach to examining absolute flux rates and their impact on gas exchange performance is to examine control points or set points in

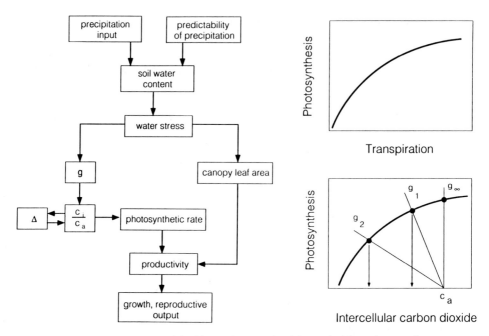

Fig. 18.10. A conceptual model of how plant productivity and ultimately growth, reproductive output, and plant fitness are influenced by water stress and several of the gas
exchange characters that influence photosynthetic rate. Δ = carbon isotope discrimination,
c_i and c_a are the intercellular and ambient CO_2 concentrations, respectively, and g is the
leaf conductance to water vapor. (After Ehleringer 1993b)

gas exchange activity. Set points may be more stable than absolute flux
rates, thereby providing a better indicator of whole-plant constraints. That
is, whereas absolute flux rates will vary greatly in response to resource levels
over the short term or to stress levels over the long term, changes in the set
point may be substantially less. The intercellular CO_2 concentration will be
considered as a set point for photosynthetic activity, providing integrated
information about photosynthesis without providing information on the
absolute fluxes. In this manner, the intercellular CO_2 concentration is analogous to body temperature in warm-blooded animals, providing a system-
level measure of metabolic activity.

Photosynthesis requires the simultaneous inward diffusion of carbon
dioxide from outside the leaf and its fixation into organic compounds by
light and dark reactions within the chloroplast. One set point illustrated in
Fig. 18.10 is the intercellular CO_2 concentration (c_i), which represents a
balance between rates of inward CO_2 diffusion (controlled by stomatal
conductance, g) and CO_2 assimilation (controlled by photosynthetic light/
dark reactions). In principle, there is no expected relationship between flux
rate and set point. A primary advantage of set point analysis over flux rate
would arise if set points remained relatively fixed among plants under

nonstressed conditions, and, if in response to abiotic stresses, any changes in the absolute value of a set point resulted in the relative rankings of plants remaining constant.

Using long-term estimates of the c_i value as a measure of the set point for gas exchange metabolism provides a comparative estimate of the extent to which stomatal conductance and water-related process limit photosynthesis instead of photosynthetic capacity and mineral-nutrition-related processes under a specific set of climatic conditions. This is exactly analogous to previous indications of tradeoffs between water-use efficiency and nitrogen-use efficiency (Field et al. 1983). Both parameters cannot be simultaneously increased and the operational point represents a relative difference in the extent to which gas exchange is controlled by mineral-related components versus water-related components. Annuals and perennials would be expected to represent two ends of this spectrum, with limitations in the gas exchange of perennial plants largely controlled by water-related components (e.g., leaf conductance, hydraulic conductance).

18.5.3 Carbon Isotope Discrimination as a Measure of Intercellular Carbon Dioxide Concentration

18.5.3.1 C_3 Plants

Over extended time periods, the intercellular CO_2 concentration can be estimated through measurement of the carbon isotopic composition of plant material (Farquhar et al. 1989). Carbon isotope discrimination (Δ) in C_3 plants is related to photosynthetic gas exchange; because Δ is in part determined by c_i/c_a, the ratio of CO_2 concentrations in the leaf intercellular spaces to that in the atmosphere. This ratio, c_i/c_a, differs among plants because of variation in stomatal opening (affecting the supply rate of CO_2), and because of variation in the chloroplast demand for CO_2. Of the models linking C_3 photosynthesis and $^{13}C/^{12}C$ composition, the one developed by Farquhar et al. (1982) has been the most extensively tested. In its simplest form, their expression for discrimination in leaves of C_3 plants is

$$\Delta = a + (b - a)\frac{c_i}{c_a},$$

where a is the fractionation occurring due to diffusion in air (4.4‰), and b is the net fractionation caused by carboxylation (mainly discrimination by RuBP carboxylase, approximately 27‰). The result of these constant fractionation processes during photosynthesis is that the leaf carbon isotopic composition represents the assimilation-weighted intercellular CO_2 concentration during the lifetime of that tissue. Farquhar et al. (1989) and Ehleringer et al. (1992) summarize the data showing that Δ values of leaf material are a reliable estimate of c_i/c_a during the lifetime of that leaf for C_3 species.

18.5.3.2 C_4 Plants

Farquhar (1983) developed an expression for carbon isotope discrimination in C_4 plants,

$$\Delta = a + (b_4 + b_3\phi - a)\frac{c_i}{c_a},$$

where b_3 and b_4 are the fractionation of gaseous CO_2 by Rubisco and the fixation of bicarbonate by PEP carboxylase, respectively, and ϕ is the leakage of CO_2 from the bundle sheath cells. Depending on the value of ϕ, the relationship with c_i/c_a could be positive, negative, or zero. The value of ϕ cannot be measured directly, but only indirectly estimated by biochemical or gas-exchange techniques. Early gas-exchange studies suggested that the value of ϕ was close to 0.34, resulting in a zero slope between Δ and c_i/c_a for C_4 plants (Evans et al. 1986). Alternative biochemical calculations suggested that ϕ should be in the range of 0.1–0.13 (Hatch and Osmond 1976; Jenkins et al. 1989). The most recent estimates of ϕ for a broad range of C_4 species are 0.21 (Henderson et al. 1992), resulting in a negative relation between Δ and c_i/c_a.

There is limited information on variation in Δ of C_4 species. O'Leary (1988) and Hubick et al. (1989) report variation on the order of 1–1.5‰ for two crop species. While this may appear to be a small amount of variation, at $\phi = 0.21$, it does represent significant variation in c_i/c_a (Henderson et al. 1992). Bowman et al. (1989) reported that salt stress would induce an equivalent change in online Δ values and attributed increased Δ values to an increased bundle sheath leakage during stress. More recently, Walker and Sinclair (1992) have reported changes in Δ of *Atriplex* along salinity gradients in Australian deserts. The extent to which the changes in Δ values in the field represent genetic differences and/or environmental acclimation is unclear at this time. Should the differences in Δ of native C_4 species represent genetic differences, then it is possible that c_i values and life-history patterns may vary in a manner analogous to that of C_3 plants, but opposite in sign since Δ and c_i are negatively related at $\phi = 0.21$.

18.5.4 Intercellular CO_2 and Life History in C_3 Plants

The Δ values vary among C_3 species in the Sonoran Desert (Fig. 18.11). Carbon isotope discrimination was positively correlated with expected soil moisture availability, suggesting that species in wetter habitats had higher c_i values. Of ecological interest, Δ values of species within a microbabitat were inversely related to the life expectancy of the shrub. Longer-lived species has lower c_i values than shorter-lived species, indicating a more conservative efficiency of water use than in shorter-lived species. Similar negative correlations between life expectancy and Δ have also been observed among

Oatman, Arizona

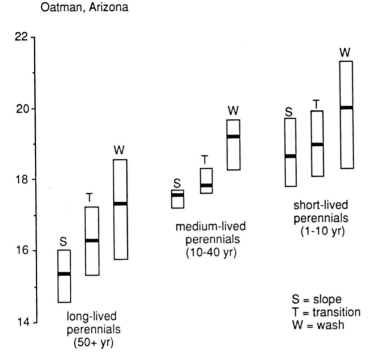

Fig. 18.11. Ranges of carbon isotope discrimination (Δ) values of short-lived (1–10 years), medium-lived (10–40 years), and long-lived (>50 years) species growing in different microhabitats at a Sonoran Desert site. (After Ehleringer and Cooper 1988)

species in grassland communities (Smedley et al. 1991) and between bean cultivars (Ehleringer et al. 1990) and wheat cultivars (Richards and Condon 1993) with respect to date of flowering. This pattern of Δ values among desert species is maintained through time, indicating a high degree of stability in this parameter (Ehleringer and Cook 1991). The slope of the correlation of isotope observations at the two time intervals was less than 1, indicating plasticity or acclimation potential in Δ in response to seasonal environmental fluctuations. From other data sets on agricultural and range species, there is strong evidence that Δ is a stable character and that differences in Δ among genotypes are maintained through time and across sites (see references in Farquhar et al. 1989) so that time-series analyses in individual plants (such as for tree rings) does provide meaningful valuable information of the long-term responses of the genotype to environmental fluctuation.

In the driest desert environments, such as the Atacama Desert of northern Chile, Δ values are quite low with effective c_i values as low as $125\,\mu l\,l^{-1}$ (Ehleringer et al. 1993). Such low c_i values might once have been thought to occur only in C_4 species, and they indicate that stomata are very nearly closed during the main periods of carbon gain. Such low Δ values have been

found in a large number of species from deserts throughout the world
(Winter and Troughton 1978; Winter 1981; Rundel and Sharifi 1993), suggesting that long-lived species are very conservative in their set point.

Within populations, there can be variation in Δ values corresponding to c_i
value differences of $30\,\mu l\,l^{-1}$ (Schuster et al. 1992), suggesting life-history
variation not only at the species level but also variation within a species.
Ehleringer (1993a) examined variation in Δ values among adjacent *Encelia
farinosa* shrubs and showed that high-Δ genotypes grew faster than low-Δ
genotypes, but were also more sensitive to drought. In response to the
extremes in precipitation patterns that characterize the desert, there appeared
to be tradeoffs, with one end of the temporal water-availability spectrum
favoring high-Δ genotypes and the other favoring low-Δ genotypes. Thus,
variation at the population level in this case mirrored patterns also seen at
the community level in terms of variation in c_i.

18.6   Integrating Gas Exchange Across Complex Environmental Gradients

18.6.1   Evaporative Gradients

A common observation is that in response to a decreased humidity level,
stomata partially or completely close (Lange et al. 1971), resulting in a
reduced c_i value. When plants are grown under reduced humidity levels, c_i
values are reduced, as indicated by heavier $\delta^{13}C$ values (Winter et al. 1982).
If plants show this environmental plasticity, it seems reasonable to expect
that populations adapted to different climatic regimes should show corresponding differences. On an instantaneous basis, transpiration (E) is the
product of leaf conductance (g) and the leaf-to-air water vapor gradient
divided by total atmospheric pressure (v). Temporal variations in the growing
season among sites can be incorporated without bias by averaging the
saturation vapor pressure expressed as a mole fraction ($e_{a,sat}/P_{total}$) over
each month of the year, using the monthly ratio of precipitation (P) to
potential evapotranspiration (E_p) as a weighting factor (Comstock and
Ehleringer 1992). The effective seasonal leaf-to-air water vapor gradient (ω)
is then calculated as

$$\omega = \frac{\dfrac{1}{P_{total}} \sum_{Jan}^{Dec} \left(e_{a,sat}\dfrac{P}{E_p} \right)}{\sum_{Jan}^{Dec} \dfrac{P}{E_p}}.$$

ω is an index which can be used to rank sites according to the mean evaporative demand expected during the most likely growing seasons throughout
the year and has the same units as v ($mbar\,bar^{-1}$). In the low humidity

environments that characterize aridlands, the value of ω converges on the actual mean growing-season value of v (Comstock and Ehleringer 1992). ω is but one way of characterizing the evaporative gradient in leaves, but provides a means for predicting plant gas exchange parameters for plants that would be active at different seasons through the year. Habitats in which the evaporative gradients are higher during the growing season are characterized by higher ω values. As such, ω appears to be a useful means for differentiating habitats and seasonal variation in the evaporative gradient across sites in a way that encompassed both the driving potential for transpiration as well as moisture input into the soil.

At the ecotypic level, Comstock and Ehleringer (1992) have shown that variation in Δ values reflected shifts in habitat quality in *Hymenoclea salsola*, a common shrub in the Mojave and Sonoran Deserts. The carbon isotope discrimination values in *H. salsola* can vary by >2‰ in the field, suggesting that c_i values among populations differ by more than $30\,\mu l\,l^{-1}$. Under common garden conditions, the isotopic variation was greater than 2‰ and was negatively related to the ω (the average leaf-to-air water vapor gradient weighted for periods when soil moisture was available) values for the habitats from which the plants originated (Fig. 18.12). *Hymenoclea salsola* has both photosynthetic twigs and leaves, with twigs always having lower c_i and Δ values. Since leaves and twigs both have small diameters, resulting in strong convective exchange and equivalent tissue temperatures, twigs also always have a greater water-use efficiency (Comstock and Ehleringer 1988). The fraction of leaf to twig photosynthetic areas is also negatively related to ω, resulting in plants from drier habitats (atmospheric drought) having both lower Δ values (higher water-use efficiencies) at the leaf level as well as a greater allocation to the more water-use efficient twig tissues in these environments. Overall, this results in a combined morphological-physiological progression towards canopies of greater water-use efficiency in climates with drier atmospheric conditions. This pattern of decreasing Δ values in plants from drier environments and an increased allocation to photosynthetic twigs is consistent with possible tradeoffs between Δ, as a set point for gas exchange, and drought stress.

The implication of the Comstock and Ehleringer (1992) study is that the seasonality of soil moisture inputs is important in affecting absolute Δ values; in desert habitats where precipitation occurred during the hotter summer months, plants had lower Δ values than from sites receiving equivalent amounts of precipitation during cooler winter-spring periods of the year. Implicit in this interpretation is that those ecotypes growing in summer-wet habitats have the capacity to utilize summer precipitation. For *H. salsola*, this is the case (Ehleringer and Cook 1991). In broader terms, it is expected that along a north-to-south transect through the Sonoran Desert, which is a gradient of increasing summer precipitation, intraspecific c_i values of leaves of plants should decrease even though there might be more summer moisture inputs at those locations.

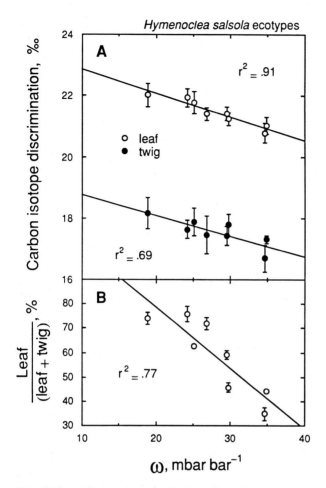

Fig. 18.12. A Carbon isotope discrimination of leaves and twigs from different ecotypes of *Hymenoclea salsola* grown under common garden conditions as a function of ω the leaf-to-air evaporative gradient weighted for seasonal precipitation input. **B** The ratio of leaf area as a proportion of total photosynthetic area plotted as a function of ω. (Comstock and Ehleringer 1992)

18.6.2 Utilization of Summer Moisture Inputs

The evolution of greater life-form diversity in desert regions with bimodal precipitation patterns may relate niche differentiation to the use of moisture resources. Cannon (1911) and Cody (1986) both noted striking differences in the root distributions of desert plants, suggesting that plants were using different moisture resources. From illustrations provided in those studies, it is evident that there can be substantial root overlap and thus it is unclear that plants have nonoverlapping soil moisture sources. The water resource used by plants can be distinguished by examining the hydrogen isotope ratio (δD) of water in xylem sap of stems and quantitative estimates of the uptake of different moisture sources can be determined (White et al. 1985).

In desert ecosystems, Ehleringer and Cook (1991) used δD observations of xylem sap to analyze the extent to which Sonoran Desert perennials near Needles, California, utilized summer precipitation. They observed that in moderately dry summers, long-lived species (includes both trees such as *Acacia greggii*, *Carcidium floridum*, and *Chilopsis linearis* as well as shrubs such as *Ephedra viridis*) did not utilize summer rain, but instead relied on moisture from deeper soil layers. In contrast, shorter-lived perennials (such as *Ambrosia dumosa* and *Encelia farinosa*) used moisture from those summer rains whenever it was available. It was only during years of above-average summer rains, when moisture penetrated to greater depths, that long-lived species increasingly utilized summer rains.

Winter rains come as slow-moving frontal systems, and if they are of sufficient magnitude, saturate the soil profile. In contrast, summer rains occur as shorter, more intense convectional storms, often saturating only the upper soil layers. As shown earlier, the frequency distributions of summer and winter storms are not different at sites receiving the same average amount of precipitation. Thus, it would appear that because of the short intense duration of summer rains, the depth of penetration for storms of equivalent rainfall would be less. For plant water use in the desert, the situation is further compounded by the greater ω values for plants growing in summer-wet habitats. Thus, we might expect that Δ values to be lower for plants using summer moisture.

Parts of southern Utah receive 30% or more of the annual precipitation during the summer months. Ehleringer et al. (1991a) studied desert species near Wahweap (Utah) over a 2-year period and observed that following the onset of summer rains (δD value of approximately $-25\permil$), annuals, herbaceous perennials and CAM perennials used water from the upper soil layers wetted by summer rains (Fig. 18.13). A fraction of the woody perennials had δD values intermediate between the summer rains and the deeper soil layers (which were approximately $-80\permil$), implying that both water sources were being utilized in equal proportion. On the other hand, the δD values in a second group of woody perennials did not use any moisture from the summer rains. Similar apparent niche separations for summer what use by perennial shrubs have been described by Donovan and Ehleringer (1992) in the Great Basin and Valentini et al. (1992) in the macchia of Italy.

Taken together, these data suggest that some perennial species may have rooting patterns allowing them to utilize both summer and winter rains, whereas other perennial species have effective rooting patterns restricted to deeper soil layers and thus the plants are unable to utilize summer rains. Functionally dimorphic root systems for water uptake are a possibility, but since the carbon cost of root turnover is high, this should only occur if the

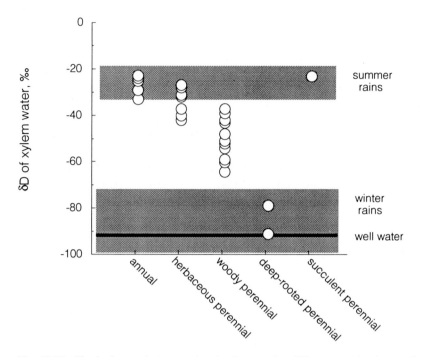

Fig. 18.13. The hydrogen isotope ratio of xylem sap for different species (categorized by life form) at a desert site in southern Utah following summer rains. The *gray areas* represent the range of hydrogen isotope ratios for both summer and winter rain events. The *solid line* represents the hydrogen isotope ratio of goundwater at this site. (Based on a figure and data in Ehleringer et al. 1991a)

summer surface moisture input is reliable (Ehleringer 1993b). In regions characterized by both heavy winter and summer rains, plants would be expected to have summer-active roots in the upper layers to capture summer rains and also a deeper root system to capture water from the deeper perco-lating winter storms. In regions characterized by less predictable summer rains, some species may be adapted to use that summer surface moisture, whereas other species do not, opening the possibility of increased life-form diversity in less predictable environments. The gain to be derived from being able to capitalize on surface moisture in unpredictable environments is that a greater fraction of the total annual precipitation comes from these infrequent storm events. In terms of understanding water dynamics within an ecosystem or possible competitive interactions between plants in an ecosystem, it is critical to know to what extent plants are capable of utilizing different moisture sources.

18.7 Conclusions

Desert environments are characterized by both low precipitation inputs and high year-to-year variability in soil moisture inputs. In response to this environmental variation, there is significant variation in plant gas exchange characteristics. While C_4 and CAM photosynthesis are associated with deserts, they do not constitute the dominant photosynthetic pathways and most of the gas-exchange variation occurs in C_3 photosynthesis. These patterns include substantial variations in the intercellular CO_2 concentrations of leaves and in the photosynthetic tissue type (leaves, twigs, and stems). Variation in the fractions of winter versus summer precipitation and of a species' ability to use summer moisture are also associated with variations in the intercellular CO_2 concentrations of C_3 plants. These environmental constraints and unpredictability of year-to-year precipitation may be the major selective force for the high life-form diversity that characterizes desert vegetation.

Acknowledgments. The support of the Ecological Research Division at the Office of Health and Environmental Research at the US Department of Energy is very much appreciated.

References

Balding FR, Cunningham GL (1974) The influence of soil water potential on the perennial vegetation of a desert arroyo. Southwest Nat 19: 241–248

Beatley JC (1974a) Effects of rainfall and temperature on the distribution and behavior of *Larrea tridentata* (creosote-bush) in the Mojave Desert of Nevada. Ecology 52: 245–261

Beatley JC (1974b) Phenological events and their environmental triggers in Mojave Desert ecosystems. Ecology 55: 856–863

Bender GL (ed) (1982) Reference handbook on the deserts of North America. Greenwood Press, Westport, Connecticut

Bender MM (1968) Mass spectrometric studies of carbon 13 variations in corn and other grasses. Radiocarbon 10: 468–472

Bloom AJ, Troughton JH (1979) High productivity and photosynthetic flexibility in a CAM plant. Oecologia 38: 35–43

Bowers MA, Lowe CH (1985) Plant-form gradients on Sonoran Desert bajadas. Oikos 46: 284–291

Bowman WD, Hubick KT, von Caemmerer S, Farquhar GD (1989) Short-term changes in leaf carbon isotope discrimination in salt- and water-stressed C_4 grasses. Plant Physiol 90: 162–166

Brown GW Jr (ed) (1968) Desert biology, vol 1. Academic Press, New York

Caldwell MM (1985) Cold desert. In: Chabot BF, Mooney HA (eds) Physiological ecology of north American plant communities. Chapman and Hall, New York, pp 198–212

Caldwell MM, White, RS, Moore RT, Camp LB (1977) Carbon balance, productivity, and water use of cold-winter desert shrub communities dominated by C_3 and C_4 species. Oecologia 29: 275–300

Cannon WA (1911) The root habits of desert plants. Carnegie Inst Wash Year Book 131: 7–96

Cody ML (1986) Structural niches in plant communities. In: Diamond J, Case TJ (eds) Community Ecology. Harper and Row, New York, pp 381–405

Cody ML (1989) Growth-form diversity and community structure in desert plants. J Arid Environ 17: 199–209

Colwell RK (1974) Predictability, constancy and contingency of periodic phenomena. Ecology 55: 1148–1153

Comstock JP, Ehleringer JR (1984) Photosynthetic responses to slowly decreasing leaf water potentials in *Encelia frutescens*. Oecologia 61: 241–248

Comstock JP, Ehleringer JR (1986) Canopy dynamics and carbon gain in response to soil water availability in *Encelia frutescens* Gray, a drought-deciduous shrub. Oecologia 68: 271–278

Comstock JP, Ehleringer JR (1988) Contrasting photosynthetic behavior in leaves and twigs of *Hymenoclea salsola*, a green-twigged, warm desert shrub. Am J Bot 75: 1360–1370

Comstock JP, Ehleringer JR (1990) Effect of variations in leaf size on morphology and photosynthetic rate of twigs. Funct Ecol 4: 209–221

Comstock JP, Ehleringer JR (1992) Plant adaptation in the Great Basin and Colorado Plateau. Great Basin Nat 52: 195–215

Comstock JP, Cooper TA, Ehleringer JR (1988) Seasonal patterns of canopy development and carbon gain in nineteen warm desert shrub species. Oecologia 75: 327–335

Donovan LA, Ehleringer JR (1993) Water sources and use of summer precipitation in a Great Basin shrub community. Funct Ecol (in press)

Ehleringer JR (1978) Implications of quantum yield differences on the distribution of C_3 and C_4 grasses. Oecologia 31: 255–267

Ehleringer JR (1989) Carbon isotope ratios and physiological processes in aridland plants. In: Rundel PW, Ehleringer JR, Nagy KA (eds) Stable isotopes in ecological research. Ecological Studies Series. Springer, Berlin Heidelberg New York, pp 41–54

Ehleringer JR (1993a) Variation in leaf carbon isotope discrimination in *Encelia farinosa*: implications for growth, competition, and drought survival. Oecologia (in press)

Ehleringer JR (1993b) Gas exchange implications of isotopic variation in aridland plants. In: Griffiths H, Smith J (eds) Plant responses to water deficit. Environmental Plant Biology Series. BIOS Scientific Publ, London (in press)

Ehleringer J, Björkman O (1977) Quantum yields for CO_2 uptake in C_3 and C_4 plants: dependence on temperature, carbon dioxide, and oxygen concentration. Plant Physiol 59: 86–90

Ehleringer JR, Cook CS (1991) Carbon isotope discrimination and xylem D/H ratios in desert plants. Stable isotopes in plant nutrition, soil fertility, and environmental studies. IAEA, Vienna, pp 489–497

Ehleringer JR, Cooper T (1988) Correlations between carbon isotope ratio and micro-habitat in desert plants. Oecologia 76: 562–566

Ehleringer JR, Monson RK (1993) Ecology and evolution of photosynthetic pathway types. Annu Rev Ecol Syst 24: 411–439

Ehleringer JR, Mooney HA (1983) Photosynthesis and productivity of desert and Mediterranean climate plants, p. 205–231. In: Encyclopedia of plant physiology, New Series, vol 12D. Springer, Berlin Heidelberg New York

Ehleringer J, Pearcy RW (1983) Variation in quantum yields for CO_2 uptake in C_3 and C_4 plants. Plant Physiol 73: 555–559

Ehleringer JR, Comstock JP, Cooper TA (1987) Leaf-twig carbon isotope ratio differences in photosynthetic-twig desert shrubs. Oecologia 71: 318–320

Ehleringer JR, White JW, Johnson DA, Brick M (1990) Carbon isotope discrimination, photosynthetic gas exchange, and water-use efficiency in common bean and range grasses. Acta Oecol 11: 611–625

Ehleringer JR, Phillips SL, Schuster WFS, Sandquist DR (1991a) Differential utilization of summer rains by desert plants: implications for competition and climate change. Oecologia 88: 430–434

Ehleringer JR, Sage RF, Flanagan LB, Pearcy RW (1991b) Climate change and the evolution of C_4 photosynthesis. Trends Ecol Evol 6: 95–99

Ehleringer JR, Phillips SL, Comstock JP (1992) Seasonal variation in the carbon isotopic composition of desert plants. Funct Ecol 6: 396–404

Ehleringer JR, Hall AE, Farquhar GD (eds) (1993) Stable isotopes and plant carbon/water relations. Academic Press, San Diego

Evanari M, Schulze E-D, Lange OL, Kappen L, Buschbom U (1976) Plant production in arid and semi-arid areas. In: Lange OL, Kappen L, Schulze E-D (eds) Water and plant life. Ecological Studies Series. Springer Berling Heidelberg New York, pp 439–451

Evans JR (1989) Photosynthesis and nitrogen relationships in leaves of C_3 plants. Oecologia 78: 9–19

Evans JR, Sharkey TD, Berry JA, Farquhar GD (1986) Carbon isotope discrimination measured concurrently with gas-exchange to investigate CO_2 diffusion in leaves of higher plants. Aust J Plant Physiol 13: 281–292

Farquhar GD (1983) On the nature of carbon isotope discrimination in C_4 species. Aust J Plant Physiol 10: 205–226

Farquhar GD, O'Leary MH, Berry JA (1982) On the relationship between carbon isotope discrimination and intercellular carbon dioxide concentration in leaves. Aust J Plant Physiol 9: 121–137

Farquhar GD, Ehleringer JR, Hubick KT (1989) Carbon isotope discrimination and photosynthesis. Annu Rev Plant Physiol 40: 503–537

Field CB (1983) Allocating leaf nitrogen for the maximization of carbon gain: leaf age as a control on the allocation program. Oecologia 56: 341–347

Field CB, Mooney HA (1986) The photosynthesis-nitrogen relationship in wild plants. In: Givnish TJ (ed) On the economy of plant form and function. Cambridge University Press, Cambridge, pp 25–55

Field CB, Merino J, Mooney HA (1983) Compromises between water-use efficiency and nitrogen-use efficiency in five species of California evergreens. Oecologia 60: 384–389

Flanagan LB, Ehleringer, JR, Marshall JD (1992) Differential uptake of summer precipitation and groundwater among co-occurring trees and shrubs in the southwestern United States. Plant Cell Environ 15: 831–836

Goudie A, Wilkinson J (1980) The warm desert environment. Cambridge University Press, Cambridge

Hastings JR (1965) On some uses of non-normal coefficients of variation. J Appl Meteorol 4: 475–478

Hastings JR, Turner RM (1965a) Seasonal precipitation regimes in Baja California, Mexico. Geog Ann 47: 204–223

Hastings JR, Turner RM (1965b) the changing mile: an ecological study of vegetation change with time in the lower mile of an arid and semi-arid region. University of Arizona Press, Tucson

Hatch MD, Osmond CB (1976) Compartmentation and transport in C_4 photosynthesis. In: Stocking CR, Heber U (eds) Encyclopedia of plant physiology New Series, vol 3. Springer, Berlin Heidelberg New York, pp 144–184

Henderson SA, von Caemmerer S, Farquhar GD (1992) Short-term measurements of carbon isotope discrimination in several C_4 species. Aust J Plant physiol 19: 263–285

Hershfield DM (1962) A note on the variability of annual precipitation. J Appl Meteorol 1: 575–578

Hinds WT, Rotenberry J (1979) Relationship between mean and extreme temperatures in diverse microclimates. Ecology 60: 1073–1075

Hubick KT, Hammer GL, Farquhar GD, Wade LJ, von Caemmerer S, Henderson SA (1989) Carbon isotope discrimination varies genetically in C_4 species. Plant Physiol 91: 534–537

Jenkins CLD, Furbank RT, Hatch MD (1989) Mechanism of C_4 photosynthesis – a model describing the inorganic carbon pool in bundle-sheath cells. Plant Physiol 91: 1372–1381

Lange OL, Zuber M (1977) *Frerea indica*, a stem succulent CAM plant with deciduous C_3 leaves. Oecologia 31: 67–72

Lange OL, Lösch R, Schulze E-D, Kappen L (1971) Responses of stomata to changes in humidity. Planta 100: 76–86

Lange OL, Schulze ED, Evenari M, Kappen L, Buschbom U (1974) The temperature-related photosynthetic capacity of plants under desert conditions I. Seasonal changes of the photosynthetic response to temperature. Oecologia 17: 97–110

Lange OL, Schulze ED, Kappen L, Evenari M, Buschbom U (1975) CO_2 exchange pattern under natural conditions of *Caralluma negevensis*, a CAM plant of the Negev Desert. Photosynthetica 9: 318–326

Lange OL, Kappen L, Schulze E-D (1976) Water and plant life: problems and modern approaches. Ecological Studies 19. Springer Berlin Heidelberg New York, 536 pp

Lange OL, Schulze ED, Evenari M, Kappen L, Buschbom U (1978) The temperature-related photosynthetic capacity of plants under desert conditions. Oecologia 34: 89–100

Le Houérou HN (1984) Rain-use efficiency: a unifying concept in arid land ecology. J Arid Environ 7: 1–35

Le Houérou HN, Bingham RL, Skerbek W (1988) Relationship between the variability of primary production and the variability of annual precipitation in world arid lands. J Arid Environ 15: 1–18

MacMahon JA (1985) Deserts. Chanticleer Press, New York

McDonald JE (1956) Variability of precipitation in an arid region: a survey of characteristics for Arizona. Technical reports on the meteorology and climatology of arid regions, University of Arizona, Tucson

Mooney HA, Gulmon SL (1982) Constraints on leaf structure and function in reference to herbivory. BioScience 32: 198–206

Mooney HA, Troughton JH, Berry JA (1974) Arid climates and photosynthetic systems. Carnegie Inst Wash Yearb 73: 793–805

Mooney HA, Ehleringer J, Berry JA (1976) High photosynthetic capacity of a winter annual in Death Valley. Science 194: 322–324

Mooney HA, Field C, Gulmon SL, Bazzaz FA (1981) Photosynthetic capacity in relation to leaf positions in desert versus old-field annuals. Oecologia 50: 109–112

Mulroy TW, Rundel PW (1977) Annual plants: adaptations to desert environments. BioScience 27: 109–114

Nicholson SE (1980) The nature of rainfall fluctuations in subtropical West-Africa. Monthly Weather Rev 108: 473–487

Nobel PS (1980) Morphology, surface temperature, and the northern limits of columnar cacti in the Sonoran Desert. Ecology 61: 1–7

Noy-Meir I (1973) Desert ecosystems: environment and producers. Annu Rev Ecol Syst 4: 25–51

O'Leary MH (1988) Carbon isotopes in photosynthesis: fractionation techniques may reveal new aspects of carbon dynamics in plants. BioScience 38: 328–336

Osmond CB, Winter K, Ziegler H (1982) Functional significance of different pathways of CO_2 fixation in photosynthesis. In: Lange OL, Nobel PS, Osmond CB, Ziegler H (eds) Plant physiological ecology II: Encyclopaedia of plant physiology New Series, vol 12B. Springer, Berlin Heidelberg New York, pp 479–547

Osmond CB, Smith SD, Gui-Ying B, Sharkey TD (1987) Stem photosynthesis in a desert ephemeral, *Eriogonum inflatum*: characterization of leaf and stem CO_2 fixation and H_2O vapor exchange under controlled conditions. Oecologia 72: 542–549

Osmond CB, Pitelka LF, Hidy GM (eds) (1990) Plant biology of the basin and range. Springer, Berlin Heidelberg New York

Pearcy RW, Ehleringer J (1984) Ecophysiology of C_3 and C_4 plants. Plant Cell Environ 7: 1–13

Phillips DL, MacMahon JA (1978) Gradient analysis of a Sonoran Desert bajada. Southwest Nat 23: 669–680

Pianka ER (1967) On lizard species diversity: North American flatland deserts. Ecology 48: 333–351

Richards RA, Condon AG (1993) Challenges ahead using carbon isotope discrimination in plant breeding programs. In: Ehleringer JR, Hall AE, Farquhar GD (eds) Stable isotopes and plant carbon water relations. Academic Press, San Diego (in press)

Rundel PW, Sharifi MR (1993) Carbon isotope discrimination and resource availability in the desert shrub, *Larrea tridentata*. In: Ehleringer JR, Hall AE, Farquhar GD (eds) Stable isotopes and plant carbon/water relations. Academic Press, San Diego (in press)

Sala OE, Lauenroth WK (1982) Small rainfall events: an ecological role in semiarid regions. Oecologia 53: 301–304

Schaffer WM, Gadgil MD (1975) Selection for optimal life histories in plants. In: Cody ML, Diamond JM (eds) Ecology and evolution of communities. Harvard University Press, Cambridge, pp 142–157

Schulze E-D (1982) Plant life forms and their carbon, water, and nutirent relations. In: Lange OL, Nobel PS, Osmond CB, Ziegler H (eds) Plant physiological ecology II: Encyclopaedia of plant physiology New Series, vol 12B. Springer, Berlin Heidelberg New York, pp 615–676

Schuster WSF, Sandquist DR, Phillips SL, Ehleringer JR (1992) Comparisons of carbon isotope discrimination in populations of aridland plant species differing in lifespan. Oecologia 91: 332–337

Seemann JR, Field CB, Berry JA (1980) Photosynthetic capacity of desert winter annuals measured in situ. Carnegie Inst Wash Year Book: 146–147

Shmida A, Burgess TL (1988) Plant growth-form strategies and vegetation types in arid environments. In: Werger MJA (ed) Vegetation structure. SPB Academic Publ, The Hague, pp 1–31

Shreve F (1911) The influence of low temperature on the distribution of giant cactus. Plant World 14: 136–146

Shreve FS (1934) Vegetation of the northwestern coast of Mexico. Bull Torrey Bot Club 61: 373–380

Shreve F (1942) The desert vegetation of North America: the Bot Rev 8: 195–246

Shreve F, Wiggins I (1964) Vegetation and flora of the Sonoran Desert. Stanford University Press, Stanford

Smedley MP, Dawson TE, Comstock JP, Donovan LA, Sherrill DE, Cook CS, Ehleringer JR (1991) Seasonal carbon isotopic discrimination in a grassland community. Oecologia 85: 314–320

Smith SD, Nobel PS (1986) Deserts. In: Baker NR, Long SP (eds) Photosynthesis in contrasting environments. Elsevier, Amsterdam, pp 13–62

Smith SD, Nowak RS (1990) Ecophysiology of plants in the intermountain lowlands. In: Osmond CB, Pitelka LF, Hidy GM (eds) Plant biology of the basin and range. Ecological studies series 80. Springer, Berlin Heidelberg New York, pp 179–241

Smith SD, Osmond CB (1987) Stem photosynthesis in a desert ephemeral, Eriogonum inflatum: morphology, stomatal conductance and water-use efficiency in field populations. Oecologia 72: 533–541

Stowe LG, Teeri JA (1978) The geographic distribution of C_4 species of the Dicotyledonae in relation to climate. Am Nat 112: 609–623

Teeri JA, Stowe LG (1976) Climatic patterns and the distribution of C_4 grasses in North America. Oecologia 23: 1–12

Teeri JA, Stowe LG, Murawski DA (1978) The climatology of two succulent plant families: Cactaceae and Crassulaceae. Can J Bot 56: 1750–1758

Thom HCS (1958) A note on the gamma distribution. Monthly Weather Rev 86: 117–122

Troughton JH, Mooney HA, Berry JA, Verity D (1977) Variable carbon isotope ratios of *Dudleya* species growing in natural environments. Oecologia 30: 307–311

Tyree MT, Sperry JS (1989) Vulnerability of xylem to cavitation and embolism. Annu Rev Plant Physiol Mol Biol 40: 19–38

Valentini R, Scarascia Mugnozza GE, Ehleringer JR (1992) Hydrogen and carbon isotope ratios of selected species of a mediterranean macchia ecosystem. Funct Ecol 6: 627–631

Vogel JC (1980) Fractionation of the carbon isotopes during photosynthesis. In: Sitzungsberichte der Heidelberger Akademie der Wissenschaften Mathematisch-naturwissenschaftliche Klass Jahrang 1980, 3. Abhandlung. Springer, Berlin Heidelberg New York, pp 111–135

Walker CD, Sinclair R (1992) Soil salinity is correlated with a decline in ^{13}C discrimination in leaves of *Atriplex* species. Aust J Ecol 17: 83–88

Werger MJA (1986) The Karoo and southern Kalahari. In: Evanari M, Noy-Meir I, Goodall DW (eds) Hot deserts and arid shrublands. Ecosystems of the World vol 12B. Elsevier, Amsterdam, pp 283–359

Werk KS, Ehleringer J, Forseth IN, Cook CS (1983) Photosynthetic characteristics of Sonoran Desert winter annuals. Oecologia 59: 101–105

White JWC, Cook ER, Lawrence JR, Broecker WS (1985) The D/H ratios of sap in trees: implications for water sources and tree ring D/H ratios. Geochim Cosmochim Acta 49: 237–246

Whittaker RH, Niering WA (1965) Vegetation of the Santa Catalina Mountains, Arizona: a gradient analysis of the south slope. Ecology 46: 429–452

Winter K (1981) C_4 plants of high biomass in arid regions of Asia – occurrence of C_4 photosynthesis in Chenopodiaceae and Polygonaceae from the Middle East and USSR. Oecologia 48: 100–106

Winter K, Troughton JH (1978) Photosynthetic pathways in plants of coastal and inland habitats of Israel and the Sinai. Flora 167: 1–34

Winter K, Luttge U, Winter ETJH (1978) Seasonal shift from C_3 photosynthesis in Crassulacean acid metabolism in *Mesembryanthemum crystallinum* growing in its natural environment. Oecologia 34: 225–237

Winter K, Holtum JAM, Edwards GE, O'Leary MH (1982) Effect of low relative humidity on $\delta^{13}C$ value in two C_3 grasses and in *Panicum milioides*, a C_3–C_4 intermediate species. J Exp Bot 33: 88–91

Woodrow IE, Berry JA (1988) Enzymatic regulation of photosynthetic CO_2 fixation in C_3 plants. Annu Rev Plant Physiol 39: 533–594

Wong SC, Cowan IR, Farquhar GD (1979) Stomatal conductance correlates with photosynthetic capacity. Nature 282: 424–426

19 Deuterium Content in Organic Material of Hosts and Their Parasites

H. Ziegler

19.1 Introduction

The basic source for the deuterium in the organic material of an autotrophic plant (using water as a hydrogen source) is the medium water, and in terrestrial plants this means the soil water. The deuterium content of this soil water again depends normally mainly on its concentration in precipitation. Condensate from the water vapor is enriched in deuterium relative to the vapor with the result that the remaining vapor will be increasingly depleted in deuterium as moisture is removed from the air. Since the absolute water content of the atmosphere is temperature-dependent, an increasing depletion of deuterium in precipitation will result as a consequence of decreasing temperature. This is termed the "climatic effect" (cf. Schiegl 1970). The variations in deuterium content of natural waters caused by condensation and evaporation lie between $+100‰$ and $-400‰$ with respect to the deuterium content of the ocean [standard mean ocean water (SMOW)].

As early as 1934, 2 years after Urey (1932) discovered deuterium, Washburn and Smith found that the leaf tissue water of terrestrial plants was heavier than the soil water. It proved later that the leaf tissue water is enriched in deuterium, as well as in oxygen-18.

This enrichment of deuterium in leaf tissue water is not due to a discrimination of the H_2O (against HDO) during uptake into the roots or during the transport of water in the xylem, but results during the transpiration in the leaves (cf. Ziegler 1988).

Bokhoven and Theeuwen (1956) were the first to describe a strong decrease in deuterium content in organic plant material, when compared with that of tissue water. As was shown by Estep and Hoering (1981) with microalgae, the deuterium content of the organic material of a plant depends on the deuterium content of the tissue (or medium) water only when plants are growing photoautotrophically. If the cells grow heterotrophically, the δD value[1] of the organic material depends largely on the deuterium

---

[1] The relative deuterium content is expressed as δD in the following way:

$$\delta D\ [‰] = \left[\frac{D/H\ \text{in sample}}{D/H\ \text{in standard}} - 1 \right] \times 1000.$$

As a standard, the D/H ratio of Standard Mean Ocean Water (SMOW) is used, which is well established (Craig 1961).

content of the food material. This is also true for heterotrophic tissues or organs in an autotrophic higher plant, e.g., for roots and seeds.

An additional fractionation occurs in the metabolism, when products of pyruvate dehydrogenase (acetyl-CoA) are involved: Lipids, for example, are always depleted in deuterium compared with the source carbohydrates (Smith and Epstein 1970; Smith and Ziegler 1990). This has to be taken into account, when lipid-rich plant material (e.g., some seeds, fruits, membrane-rich cells) is analyzed.

As mentioned shortly earlier (Ziegler 1988), there were some indications that the organic material of higher plant holoparasites, galls, and phyto-phagous fungi has a higher deuterium content (δD less negative) than that of the host or gall-producing plant. This phenomenon is described in more detail in the following.

19.2   The Relative Deuterium Content in the Host and Parasitic Organic Material in Different Kinds of Parasite Performance

The data in the following tables were obtained according to the methods described by Osmond et al. (1975). The $\delta^{13}C$ value is defined in an analogous manner as the δD:

$$\delta^{13}C \ \permil = \left[\frac{^{13}C/^{12}C \text{ in sample}}{^{13}C/^{12}C \text{ in standard}} - 1 \right] \times 1000.$$

Standard is here a belemnite from the Pee Dee formation in South Carolina, USA (PDB).

Since the data groups are not normally distributed, the Wilcoxon Matched-Pairs Signed-Ranks Test, SPSS/PC + was used. The significance of differences is expressed by:

$p < 0.05$ (<5% probability of error);

$p < 0.01$ (<1% p.e.);

$p < 0.001$ (<0.1% p.e.).

A control with a T-test (for pairs with normally distributed data) gave identical results. The correlation analysis was calculated as Pearson's correlation coefficient: K.

As far as the relative ^{13}C content (the $\delta^{13}C$ value) is concerned, it is assumed a priori that the parasites, saprophytes, or galls should be similar to their food source, in that they do not perform metabolic conversations with significant discrimination, e.g., transformation of carbohydrates to lipids which depletes the material not only in deuterium but also in ^{13}C.

19.2.1   Isotope Contents of Galls

Galls are structures of abnormal growth produced on normal plants by gall-inducing organisms, mainly insects and in some cases fungi. The gall itself is not a parasite sensu stricto, but it obtains its organic material from the host plant in a way similar to a true parasite. Table 19.1 gives the $\delta^{13}C$- and δD-values of a number of European galls.

There is a relatively weak statistical difference in the $\delta^{13}C$ values between the host plants and their galls (all together, insect-induced as well as fungi-induced ones), with less negative (^{13}C-richer) values for the galls.

The higher deuterium content in the organic material of the calls, compared with the host plants, is statistically relevant at the $p < 0.01$ level.

There is no apparent correlation between the $\delta^{13}C$ and δD values of the hosts (K = 0.057) or between the $\delta^{13}C$ and δD values of the galls (K = 0.245).

19.2.2 Isotope Contents of Holoparasites and Their Host Plants

Holoparasites are organisms which are not (or to a negligible degree only) able to produce organic material by their own photosynthesis. They obtain all necessary compounds from their host. It is understandable, therefore,

Table 19.1. Galls induced by different gall insects and by fungi and their host plants: $\delta^{13}C$- (‰, PDB) and δD- (‰, SMOW) values of the biomass

	$\delta^{13}C$	δD
Harmandia cavernosa (Cecidomyidae)	−27.44	−94.7
Populus tremula L. (leaves)	−28.21	−102.3
Grafenaschau, Bavaria (6/75)		
Mikiola fagi Htg. (Cecidomyidae)	−26.68	−99.7
Fagus sylvatica L. (leaves)	−30.17	−108.6
Eurasburg, Bavaria (10/75)		
Rhabdophaga salicis Schrk.	−27.99	−65.2
(Cecidomyidae)		
Salix waldsteiniana Willd. (leaves)	−27.85	−90.1
Laber, O'Ammergau, Bavaria (9/75)		
Stictodiplosis scrophulariae	−25.95	−84.3
(Cecidomyidae)		
Scrophularia hoppei Koch	−26.44	−113.8
S-Tyrolia, Italy (5/58)		
Diplolepis quercus-folii L. (Cynipidae)	−24.69	−66.4
Quercus pubescens Willd. (leaves)	−29.20	−75.7
Malinska, Krk, Croatia (8/75)		
Neuroterus quercus baccarum L.	−23.79	−53.0
(Cynipidae)		
Quercus robur L. (leaves)	−28.29	−65.3
Frieding, Bavaria (9/75)		

Table 19.1. *Continued*

	$\delta^{13}C$	δD
Neuroterus quercus-baccarum L. (Cynipidae)	−26.96	−24.1
Quercus pubescens Willd. (leaves) Malinska, Krk, Croatia (5/78)	−26.63	−53.3
Rhodites rosae L. (Cynipidae)	−24.19	−74.0
Rosa arvensis Huds. München (7/91)	−25.22	−88.2
Eriophyes laevis inangulis Nal. (Gall mite)	−28.97	−95.6
Alnus glutinosa Gaertn. (leaves) Murnauer Moor, Bavaria (6/75)	−28.14	−114.9
Cnaphalodes strobilobius Kalt. (Aphid)	−26.78	−76.8
Picea abies (L.) Karst. Ebersberger Forst, Bavaria (7/78)	−27.53	−98.0
Pemphigus spirothecae Pass. (Aphid)	−31.96	−70.3
Populus nigra L. (leaves) München (7/91)	−31.32	−105.5
Exobasidium rhododendri Cram. (Basidiomycetes, Exobasidiales)	−24.58	−69.9
Rhododendron ferrugineum L. Planei, Nied. Tauern, Austria (9/74)	−27.39	−133.6
Exobasidium rhododendri	−25.14	−37.6
Rhododendron ferrugineum Planei, Nied. Tauern, Austria (9/74)	−26.44	−118.6
Average $\delta^{13}C$ of host plants	−27.9‰	(±1.64)
Average $\delta^{13}C$ of galls	−26.5‰	(±2.26)
Average δD of host plants	−97.5‰	(±22.61)
Average δD of galls	−70.1‰	(±22.17)
Difference $\delta^{13}C$ values host/gall $p < 0.05$ (galls less negative)		
Difference δD values host/gall $p < 0.01$ (galls less negative)		

that for the most part they have direct contact with the assimilate conducting tissues, the phloem, of their hosts.

In Table 19.2 all parasites are higher plants (angiosperms) with phloem contact (cf. Kollmann and Dörr 1987); only *Lathraea* is reported to obtain its organic material from the xylem of its host (Ziegler 1955).

It is apparent that the $\delta^{13}C$ values of the parasites mirror those of their hosts: *Hydnora africana*, e.g., has a similar high ^{13}C content as its CAM host *Euphorbia damarana* in its natural habitat in Namibia. Also within a single species, the polyphagous *Cuscuta hyalina*, has $\delta^{13}C$ values clearly resembling those of the different hosts.

Table 19.2. δ^{13}C (‰, PDB) and δD (‰, SMOW) values of holoparasites and their host plants (leaves) (dry matter)

	δ^{13}C	δD
Orobanche alba Steph.	−28.14	−81.5
Thymus serpyllum L. (leaves)	−29.53	−134.4
München-Gräfelfing, Bavaria (7/20)		
Orobanche sp.	−29.38	−61.3
Quercus ilex L.	−26.46	−86.3
Monte Elias, Italy (4/52)		
Lathraea squamaria L.	−23.65	−42.7
Carpinus betulus L. (leaves)	−27.58	−77.6
Grafrath, Bavaria (6/56)		
Monotropa hypopitys L.	−24.01	+9.0
Pinus halepensis Mill (needles)	−30.16	−43.4
Krk, Croatia (8/75)		
Cynomorium coccineum L.	−26.95	−31.6
Halimione portulacoides (L.) Aellen	−26.68	−94.5
Trapani, Sicily (5/71)		
Cytinus rubra (Fourreau) Pavillard	−28.17	−88.6
Cistus sp.	−27.03	−106.4
S France (5/62)		
Hydnora africana Thunb.	−11.43	−17.0
Euphorbia damarana Leach. (CAM)	−11.95	−49.2
Namib, Namibia (3/86)		
Cuscuta hyalina Heyne ex Roth[a]	−12.53	−37.7
Tribulus terrestris L. (C_4)	−13.58	−53.5
Cuscuta hyalina	−15.52	−9.8
Pennisetum typhoides (Burm.f.)	−14.44	−58.7
Stapf & C.E. Hubbard (C_4)		
Cuscuta hyalina	−24.75	−31.6
Zea mays L. (C_4)	−14.63	−84.9
Cuscuta hyalina	−30.31	−22.6
Coleus sp. (C_3)	−30.52	−74.3
Cuscuta hyalina	−28.06	−39.1
Beloperone guttata (C_3)	−27.55	−72.6
Cuscuta hyalina	−22.78	−84.3
Aesculus hippocastanum L. (C_3)	−25.76	−110.5
Cuscuta hyalina	−17.92	−9.2
Sedum dendroideum Moc.et Sesse (CAM)	−18.62	−24.2
ssp. *praealtum* (DC.) R.T. Claus.		

Table 19.2. *Continued*

	$\delta^{13}C$	δD
Cuscuta hyalina	−17.63	−11.3
Sedum dendroideum (CAM)	−20.10	−28.3
ssp. *praealtum*		
Cuscuta hyalina	−13.47	−9.3
Kalanchoe tubiflora (Harvey) Hamet (CAM)	−14.18	−20.9
Cuscuta hyalina	−16.48	−11.1
Kalanchoe daigremontiana Hamet et Perr. (CAM)	−17.46	−25.6
Cuscuta hyalina	−13.04	+9.3
Crassula portulacea Lam. (CAM)	−14.34	−2.3
Cuscuta hyalina	−12.80	+22.8
Crassula portulacea (CAM)	−15.15	−0.8
Cuscuta hyalina	−12.79	−24.7
Trianthema pentandra L. (CAM)	−12.51	−52.1
Cuscuta lehmaniana	−28.09	−67.5
Sambucus nigra L.	−28.99	−103.7
Taschkent		
Cuscuta campestris Yuncker	−26.89	−116.5
Xanthium strumarium	−26.52	−131.1
Taschkent		
Cassytha filiformis L.	−28.55	−38.4
Eucalyptus sp.	−30.90	−108.8
Northern Territory, Alu-Mine (9/81)		
Cassytha filiformis	−30.22	−80.6
Eucalyptus tetrodonta	−30.49	−93.3
Australia (9/81)		
Cassytha glabella	−27.82	−58.8
Banksia marginata Cav.	−28.07	−87.4
Grampians National Park, NSW (11/91)		
Cassytha glabella	−25.38	−41.0
Casuarina equisetifolia J.R. & C. Forster	−28.54	−93.0
Grampians National Park, NSW (11/91)		
Cassytha melantha	−27.57	−31.5
Acacia melanoxylon R. Br.	−29.13	−58.4
Grampians National Park, NSW (11/91)		
Cassytha pubescens	−27.40	−48.3
Leptospermum myrsinoides	−28.15	−93.8
Grampians National Park, NSW (11/91)		

[a] The *Cuscuta* species and their host plants without data of origin were cultivated in the greenhouse at Dept. of Botany, Munich. I thank Prof. O.H. Volk, Würzburg, for supply and identification.

Due to this close dependence of the holoparasite with its host as far as organic material is concerned, it is not surprising that there is no statistical difference in the $\delta^{13}C$ values between host and parasite.

As expected, there is a clear difference in the $\delta^{13}C$ value of the hosts with C_3 and C_4 photosynthesis, as well as with their corresponding parasites (in both cases $p < 0.001$). Surprisingly, there is a statistically highly significant ($p < 0.001$) difference in the deuterium content of the organic material of the host and its holoparasite: the parasite is always considerably richer in deuterium. We discuss this phenomenon later in more detail.

In these pairs of host plants and holoparasites there is a strong correlation between the $\delta^{13}C$ and δD values both within the host plants (K = 0.672) and within the parasites (K = 0.635): those richer in ^{13}C are also richer in deuterium. It is remarkable that the correlation coefficients are so similar in both cases.

19.2.3 Isotope Contents of Mistletoes (Hemiparasites) and Their Hosts

Mistletoes are considered to be hemiparasites, since they are able to perform considerable photosynthesis. As far as is known, they have C_3 photosynthesis. Recently, several experimental data indicate that mistletoes (at least in special cases) also obtain a considerable amount of organic material from their hosts (cf. Ehleringer et al. 1985; Marshall and Ehleringer 1990; Schulze et al. 1991). Since mistletoes feed on the xylem of their hosts, this organic material must be in the transpiration stream.

As a consequence, mistletoes on C_4- or CAM hosts can have $\delta^{13}C$ values different from normal C_3 values (cf. Schulze et al. 1991).

In Table 19.3 the $\delta^{13}C$ and δD values of a large number of mistletoes and their hosts from two mistletoe-rich regions (Australia and Africa + Madagascar) are given. It can be seen that – as expected – the $\delta^{13}C$ values of mistletoes growing on CAM plants in Africa are significantly less negative (^{13}C-richer) (average $-17.4 \pm 4.46\permil$) than the ones growing on C_3 hosts (average -25.4 ± 2.07).

As was mentioned and explained earlier (Ehleringer et al. 1985; Ullmann et al. 1985; Ziegler 1986; Schulze et al. 1991), the mistletoes have in practically every case more negative $\delta^{13}C$ values than their hosts ($p < 0.001$ in any case). This mirrors the higher C_i (intercellular CO_2 concentration) over the whole vegetation period of the parasites with their higher degree of stomatal opening and their consequently higher transpiration. The average of the $\delta^{13}C$ in the C_3 hosts in Australia is $-26.5\permil$ (± 1.65), while that of the mistletoes is -28.5 (± 1.3). The corresponding values for the African/-Madagascar species are -23.8 (± 4.22) and -26.7 (± 2.9), respectively.

The $\delta^{13}C$ values of the African CAM species show an average of $-17.4\permil$ (± 4.46), and those of associated mistletoes average $-24.1\permil$ (± 3.74).

Table 19.3. δ^{13}C (‰, PDB) and δD (‰, SMOW) values of mistletoes and their hosts. All species belong to the C_3 type of photosynthetic CO_2 fixation, beside the indicated CAM species. Dry matter, mostly from leaves

Australia	δ^{13}C	δD
Amyema gibberulum (Tata) Dans.	−27.96	−57.9
Hakea eyreana (S. Moore) McGillvray	−24.80	−52.5
Ayers Rock (9/81)		
Amyema gibberulum	−29.02	−44.3
Hakea eyreana	−26.05	−49.7
Ayers Rock (9/81)		
Amyema gibberulum	−29.48	−62.6
Grevillea wickhamii Meisn.	−24.84	−100.3
Devils Marbles (9/81)		
Amyema linophyllum (Fenzl.) Tiegh.	−31.00	−38.3
Casuarina cristata Mig.	−26.72	−38.9
Hattah (9/81)		
Amyema mackayense (Blakely) Dans.	−29.53	−73.4
ssp. *cycnei-sinus* (Blakely) Barlow		
Avicennia marina (Forsk.) Vierh.	−28.03	−93.0
South Alligator River (8/81)		
Amyema maidenii (Blakely) Barlow	−30.34	−58.4
Acacia kempeana F. Muell.	−28.05	−63.6
Ayers Rock (9/81)		
Amyema maidenii	−31.19	−61.4
Acacia kempeana	−26.39	−91.9
Ayers Rock (9/81)		
Amyema maidenii	−28.53	−36.0
Acacia cowleana Tate	−26.53	−76.4
Devils Marbles (9/81)		
Amyema maidenii	−30.13	−63.3
Acacia coreacea DC.	−27.85	−85.5
Devils Marbles (9/81)		
Amyema miquelii (Lehm. ex Mig.) Tiegh.	−25.89	−61.6
Eucalyptus crebra F. Muell.	−27.37	−94.4
Norderner Range, NSW (8/81)		
Amyema miquelii	−29.54	−74.7
Eucalyptus sideroxylon A. Cunn.	−27.40	−84.7
Narderwarna (8/81)		
Amyema miquelii	−31.41	−76.4
Angophora intermedia DC.	−27.82	−96.6
Kingston (8/81)		
Amyema preissii (Mig.) Tiegh.	−26.70	−33.7
Acacia brachystachya Benth.	−25.76	−49.6
Pimba Mine (9/81)		

Amyema quandang (Lindl.) Tiegh.	−28.13	−55.0
Acacia brachystachya	−26.15	−52.6
Broken Hill (9/81)		
Amyema quandang	−28.36	−57.4
Acacia aneura F. Muell. ex Benth.	−26.08	−45.6
Ayers Rock (9/81)		
Amyema quandang	−28.53	−50.1
Acacia papyrocarpa Benth.	−27.66	−48.9
Wyalla (9/81)		
Amyema sanguineum (F. Muell.) Dans.	−28.51	−73.9
Eucalyptus sp.	−27.83	−80.4
Ayers Rock (9/81)		
Amylotheca dictyophleba (F. Muell.) Tiegh.	−29.82	−78.2
Cinnamomum camphora (L.) T. Nees & Eberm.	−28.79	−83.8
Atherton (8/81)		
Dendrophthoe vitallina (F. Muell.) Tiegh.	−27.80	−53.1
Casuarina glauca Sieb.	−25.78	−72.5
Bateman's Bay (19/81)		
Diplatia grandibracteata Tiegh.	−29.39	−75.6
Eucalyptus leucophloia Brooher	−25.53	−92.5
Camoweal (9/81)		
Diplatia grandibracteata	−28.38	−74.5
Eucalyptus leucophloia	−26.49	−116.5
Mount Isa (9/81)		
Lysiana casuarinae (Mig.) Tiegh.	−28.60	−64.0
Gossypium robinsonii F. Muell.	−27.86	−86.7
Wittenoom Gorge (8/81)		
Lysiana casuarinae	−27.04	−55.6
Acacia acradenia	−25.41	−80.2
Wittenoom Gorge (8/81)		
Lysiana exocarpi (Behr.) Tiegh.	−26.68	−36.7
Acacia brachystachya	−27.43	−58.8
Broken Hill (9/81)		
Lysiana exocarpi	−28.22	−47.5
Acacia victoriae Benth.	−28.62	−84.3
Hattah (9/81)		
Lysiana exocarpi	−28.18	−41.6
Acacia tetragonophylla F. Muell.	−28.57	−74.6
Pimba (9/81)		
Lysiana exocarpi	−27.97	−37.4
Acacia aneura F. Muell. ex Benth.	−27.62	−63.2
Ayers Rock (9/81)		
Lysiana exocarpi	−29.07	−40.0
Myoporum platycarpum R. Br.	−24.92	−56.4
Pt. Augusta (9/81)		

Table 19.3. *Continued*

Australia	$\delta^{13}C$	δD
Lysiana exocarpi	−27.71	−36.1
Templetonia egena (F. Muell.) Benth.	−25.33	—
Pt. Augusta (9/81)		
Lysiana exocarpi	−27.91	−50.1
Pittosporum phylliraeoides DC.	−28.48	−91.2
Woomera (9/81)		
Lysiana exocarpi	−28.99	−36.5
Heterodendron oleaefolium Desf.	−27.46	−53.9
Yanta (9/81)		
Lysiana exocarpi	−27.80	−28.0
Santalum acuminatum (R. Br.) DC.	−27.0	−27.6
Pimba Mine (9/81)		
Lysiana murrayi (Tate) Tiegh.	−26.63	−14.5
Acacia aneura F. Muell. ex Benth.	−26.44	−33.4
Pimba Mine (9/81)		
Lysiana spathulata (Blakely) Barlow	−30.72	−27.8
Acacia aneura	−26.56	−44.8
Pimba Mine (9/81)		
Lysiana spathulata	−28.25	−48.8
Acacia farnesiana (L.) Willd.	−26.75	−80.8
Mount Isa (9/81)		
Lysiana spathulata	−24.54	−40.7
Acacia cowleana Tate	−26.53	−76.4
Devils Marbles (9/81)		
Lysiana spathulata	−27.83	−90.1
Acacia monticola J.M. Black	−27.52	−103.3
Devils Marbles (9/81)		
Lysiana spathulata	−27.58	−35.3
Callitris columellaris F. Muell.	−25.71	−83.1
Palm Valley (9/81)		
Lysiana subfalcata (Hook.) Barlow	−28.35	−39.5
Cassia oligophylla F. Muell.	−27.01	−98.3
Mount Isa (9/81)		
Lysiana subfalcata ssp. *subfalcata* Barlow	−27.49	−60.9
Atalaya hemiglauca (F. Muell.) F. Muell. ex	−25.80	−74.8
Benth.		
Mount Isa (9/81)		
Lysiana subfalcata ssp. *maritima* Barlow	−29.44	−24.4
Ceriops tagal (Perr.) C.B. Rob.	−28.44	−66.9
Aims (9/81)		

Lysiana subfalcata ssp. *maritima*	−28.31	−67.6
Ceriops tagal	−29.13	−45.4
Aims (9/81)		

Africa and Madagascar

Bakerella sp.	−26.51	−71.5
Didierea trolli Cap. & Rauh (CAM)	−20.45	−47.1
Madagascar (Kluge leg., 88)		
Bakerella sp.	−32.40	−77.8
Weinmannia sp.	−26.63	−55.7
Madagascar (Kluge leg., 88)		
Odontella welwitschii (Engl.) Balle	−25.22	−57.7
Boscia tomentosa Tölken	−24.28	−67.5
Petr. Forest, Namibia (4/87)		
Septulina glauca (Thunb.) Tiegh.	−26.43	−72.3
Tamarix usneoides E. Mey. ex Bunge	−27.42	−69.5
Welwitschia Vlakte, Namibia (4/87)		
Tapinanthus discolor (Schinz) Dans.	−25.12	−54.6
Rhigozum trichotomum Burch.	−23.75	−67.9
Neuhof, Namibia (3/87)		
Tapinanthus discolor	−23.26	−53.5
Acacia reficiens Wawra	−24.35	−70.7
Uis, Namibia (3/87)		
Tapinanthus glaucocarpus (Peyr.) Dans.	−29.66	−63.7
Lannea discolor (Sonder) Engl.	−26.52	−62.1
Otavi Bergland, Namibia (4/87)		
Tapinanthus oleifolius (Wendl.) Dans.	−27.95	−66.6
Acacia hereroensis Engl.	−25.25	−80.3
Khomas Hochland, Namibia (4/87)		
Tapinanthus oleifolius	−26.06	−52.2
Acacia erubescens Welw. ex Oliver	−25.18	−85.0
Amaib, Namibia (4/87)		
Tapinanthus oleifolius	−29.20	−51.2
Euclea pseudebenus E. Mey. ex DC.	−25.42	−70.1
Oranje (3/87)		
Tapinanthus oleifolius	−25.65	−38.2
Parkinsonia africana Sonder	−23.91	−59.2
Fischfluß, Namibia (3/87)		
Tapinanthus oleifolius	−29.25	−49.0
Tamarix usneoides E. Mey ex Bunge	−26.13	−64.9
Welwitschia Vlakte, Namibia (3/86)		
Tapinanthus oleifolius	−24.50	−76.6
Euphorbia damarana Leach (CAM)	−11.95	−49.2
Namib (3/86)		

Table 19.3. *Continued*

Australia	$\delta^{13}C$	δD
Tapinanthus oleifolius	−17.60	−58.9
Aloe dichotoma Masson (CAM)	−13.79	−42.8
Neuhof, Namibia (3/87)		
Viscum biflorum	−28.70	−62.9
Acacia heterophylla Willd.	−28.65	−61.5
Maido, Reunion (4/89)		
Viscum capense L.f.	−26.53	−59.5
Boscia foetida Schinz	−23.52	−59.2
Kamelhof, Namibia (4/87)		
Viscum capense	−23.96	−46.5
Zygophyllum divaricatum	−24.74	−97.7
Bloomvelt (4/87)		
Viscum capense	−27.77	−40.3
Antizoma miersiana	−23.48	−73.3
Bloomvelt (4/87)		
Viscum capense	−25.90	−58.9
Euphorbia mauretanica L. (CAM)	−18.44	−63.4
Karroo Bot. Garden (3/86)		
Viscum crassulae Echl. & Zeyh.	−26.20	−6.2
Portulacaria afra Jacq. (CAM)	−22.54	−6.1
Karroo Bot. Garden (3/86)		
Viscum rotundifolium L.f.	−27.58	−52.9
Ehretia rigida (Thunb.) Druce	−23.94	−58.8
Khomas Hochland, Namibia (3/87)		
Viscum rotundifolium	−25.69	−28.6
Boscia foetida Schinz	−23.24	−59.2
Neuhof, Namibia (3/87)		
Viscum rotundifolium	−28.83	−38.5
Maerua schinzii Pax	−25.46	−56.5
Seeriem, Namibia (3/87)		
Viscum sp.	−30.79	−47.5
Macarisia sp.	−31.52	−69.1
Madagascar (Kluge leg., 88)		

The African C_3 host species are significantly ($p < 0.001$) richer in ^{13}C (less negative $\delta^{13}C$ values) than the ones from Australia. For the mistletoes, the difference is significant at the $p < 0.05$ level, but in the same direction.

When we consider the δD values of mistletoes and their hosts, it should be emphasized again that we did not analyze the tissue water of these partners, but the organic material.

The average δD of the Australian hosts (all C_3 plants) is $-72.8‰$ ($\pm$ 19.55), and of their mistletoes, $-51.1‰$ ($\pm$ 16.78). The difference (mistletoes richer in deuterium) is highly significant ($p < 0.001$). In the African host plants, the average δD values of the C_3 plants is -67.8 ($\pm$ 10.68) and -53.3 ($\pm$ 12.31) for their parasites. In the CAM hosts from Africa, the average δD value is -41.7 ($\pm$ 21.36), and $-54.4‰$ ($\pm$ 28.06) in their mistletoes. This means that the ratio mistletoe/host is reversed in the pairs with CAM hosts, compared with those with C_3 hosts.

There is no correlation between the $\delta^{13}C$ values and the δD values of the Australian ($K = 0.160$) or African hosts ($K = 0.382$) nor between the $\delta^{13}C$ values and the δD values of the Australian ($K = 0.259$) or African parasites ($K = 0.073$).

19.2.4 Isotope Contents of Parasitic or Saprophytic Fungi and Their Host Plants

The last combination we analyzed for isotope relationships was woody plants and associated wood parasites or saprophytes. In addition, a smut (order Ustilaginales), the common corn smut, *Ustilago maydis*, which infects maize plants and causes large boils or tumors, was included. In our experiment, the fungal spores were compared with the distorted leaves of the host plant.

When we consider the isotope contents of the host plants and in fungal fruiting bodies (basidiocarps) (Table 19.4), we come to the following conclusions:

The average $\delta^{13}C$ value of the eight hosts was $-25.3‰$ ($\pm$ 2.26), and $-24.3‰$ ($\pm$ 1.31) of the fungi. The difference is not statistically relevant.

The average of the host δD values is $-106.7‰$ ($\pm$ 16.16) and -50.9 ($\pm$ 15.95) for the parasites. This difference is statistically highly significant ($p < 0.001$).

We find a similar situation in the *Zea mays/Ustilago maydis* combination: while the $\delta^{13}C$ values are identical (C_4-typical), the relative deuterium content of the fungal spores is much higher (δD considerably less negative) than in the host tissue.

There is no correlation between the $\delta^{13}C$ and the δD values of the host (substrate) or between the $\delta^{13}C$ and δD values of the fungi.

19.3 What Are the Reasons for the Isotope Discriminations?

19.3.1 $\delta^{13}C$

The causes for differences in the $\delta^{13}C$ values between hosts and parasites (as far as they exist) are in general more easily explained than the differences in δD values.

Table 19.4. $\delta^{13}C$ (‰, PDB) and δD (‰, SMOW) values of fungi and their host plants (dry matter). All sample from surroundings of Munich

	$\delta^{13}C$	δD
Pseudohydnum gelatinosum (Scop. ex Fr.) Karst.	−24.04	−43.7
Pinus sylvestris L.	−27.68	−87.4
Thelephora terrestris Pers. ex Fr.	−24.38	−70.7
Picea abies (L.) Karst.	−25.81	−81.8
Piptoporus betulinus (Bull ex Fr.) Karst.	−22.81	−45.4
Betula pendula Roth	−26.12	−130.7
Trametes abietina (Dicks. ex Fr.) Pilát	−25.68	−48.5
Picea abies (L.) Karst.	−21.79	−104.8
Trametes hirsuta (Wulfen ex fr.) Pilát	−22.26	−36.9
Fagus sylvatica L.	−25.91	−116.0
Trametes abietina (Dicks. ex Fr.) Pilát	−25.68	−48.5
Picea abies	−21.79	−104.8
Schizophyllum commune Fr.	−24.34	−41.0
Fagus sylvatica L.	−27.23	−108.3
Hypoxylon fragiforme (Pers. ex Fr.) Kickx.	−25.59	−84.4
Fagus sylvatica L.	−25.88	−119.5
Ustilago maydis (DC) Tul. (distorted leaves of mays)	−11.23	−39.4
Zea mays L. (leaves)	−11.56	−62.2

In galls (Table 19.1) there is a slightly higher ^{13}C content than in the parent tissues. This could be due to a relative lower content of lipids (membranes) compared with cell wall material; lipids have a lower ^{13}C content (see Introduction).

Holoparasites (Table 19.2) have statistically identical $\delta^{13}C$ values with their hosts, whether the hosts have C_3-, C_4-, or CAM type photosynthesis. The same is true for parasitic fungi and their hosts (Table 19.4). This finding is not surprising. It indicates that hosts and parasites not only have the same carbon sources for their metabolism (the photosynthates of the hosts) but also handle them in a similar manner (without significant differences in isotope discrimination in the further anabolic or catabolic reactions).

The significant difference in the $\delta^{13}C$ content of mistletoes and their hosts is well known, and convincingly explained, and needs no further comment.

19.3.2 δD

Much more difficult to explain are the differences in the δD values of hosts and parasites. With the exception of mistletoes on CAM hosts, the parasites are always significantly richer in deuterium (in their dry mass) than their hosts.

The exception with the CAM hosts may be due to the fact that mistletoes obtain their tissue water (used in their own photosynthesis) from the transpiration stream of the host, which is not very much enriched in deuterium, while the CAM host also uses water from water-storage tissues or vacuoles, which can be considerably enriched in deuterium (cf. Ziegler 1988).

There is no explanation of the enrichment of deuterium in the biomass in galls, holoparasites, mistletoes, and fungi (here in contrast to the reduction of the ^{13}C content) on C_3 hosts, only hypotheses. The following may be mentioned:

- The tissue water in the galls or parasites could be richer in deuterium (less negative δD) than the tissue water in the host or substrate, leading to a deuterium-enriched organic matter. This could happen in the galls and mistletoes only, since the holoparasites and the fungi perform no photosynthesis.
- A deuterium enrichment of the tissue water of the galls and parasites could result in an increased deuterium content of some chemical components with exchangeable ($-OH$ or NH_2 bound) hydrogen groups (e.g., in cellulose or proteins (cf. Bonhoeffer 1934)).
- The galls and parasites could generally have much lower lipid contents than the source plants. This hypothesis seems quite improbable, since a lower lipid content would also mean a relative enrichment in ^{13}C and this is mostly not the case.
- The galls and parasites may have an especially intensive respiration and this process preferentially uses deuterium-poor substrates.

Since these hypotheses are experimentally accessible, we may soon have answers to our questions, even these may be only falsifications of the proposed hypotheses.

19.4 Conclusions

The differences, or lack of differences, in the $\delta^{13}C$ values between the hosts and the parasites can be explained satisfactorily.

Holoparasitic angiosperms and parasitic fungi (both lacking photosynthesis) have the same $\delta^{13}C$ values as their hosts (substrates) as expected. The small decrease in ^{13}C content in galls relative to the parent plants may be due to differences in chemical composition, but this remains to be shown. The higher ^{13}C content of mistletoes compared with their hosts was convincingly explained earlier.

There is no explanation for higher deuterium content of the organic material in tissues of parasites than in those of the hosts (exception: mistletoes on CAM hosts). However, since this pattern appears to be rather general, there may be the same process responsible for all cases.

Acknowledgments. For cooperation in mass spectrometric analyses I thank W. Stichler and P. Trimborn, Hydrologisches Institut der GSF-Forschungszentrum für Umwelt und Gesundheit. For help in collecting and identification of the plant material I am indebted to many colleagues (names in the publications on mistletoes), especially to O.L. Lange, E.-D. Schulze, J. Visser (+), and M. Kluge. The work was supported by the Deutsche Forschungsgemeinschaft and by the Fonds der Chemischen Industrie, FRG.

References

Bokhoven C, Theeuwen HHJ (1956) Deuterium content of some natural organic substances. Ned Akad Wet B 59: 78–83

Bonhoeffer KA (1934) Deuterium-Austausch in organischen Verbindungen. Z Elektrochem 40: 469–474

Craig H (1961) Isotopic variations in meteoric waters. Science 133: 1702–1703

Ehleringer JR, Schulze E-D, Ziegler H, Lange OL, Farquhar GD, Cowan IR (1985) Xylem-tapping mistletoes: water or nutrient parasites? Science 227: 1479–1481

Estep MF, Hoering TC (1981) Stable hydrogen isotope fractionation during autotrophic and mixotrophic growth of microalgae. Plant Physiol 67: 474–477

Kollmann R, Dörr I (1987) Parasitische Blütenpflanzen. Naturwissenschaften 74: 12–21

Marshall JD, Ehleringer JR (1990) Are xylem tapping mistletoes partially heterotrophic? Oecologia 84: 244–248

Osmond CB, Ziegler H, Stichler W, Trimborn H (1975) Carbon isotope discrimination in alpine succulent plants supposed to be capable of Crassulacean Acid Metabolism (CAM). Oecologia 28: 323–328

Schiegl W-G (1970) Natural deuterium in biogenic materials. Influence of environment and geophysical applications. PhD Thesis, University of South Africa, Pretoria

Schulze E-D, Lange OL, Ziegler H, Gebauer G (1991) Carbon and nitrogen isotope ratios of mistletoes growing on nitrogen and non-nitrogen fixing hosts and on CAM plants in the Namib desert confirm partial heterotrophy. Oecologia 88: 457–462

Smith BN, Epstein S (1970) Biogeochemistry of the stable isotopes of hydrogen and carbon in salt marsh biota. Plant Physiol 46: 738–742

Smith BN, Ziegler H (1990) Isotopic fractionation of hydrogen in plants. Bot Acta 103: 335–342

Ullmann I, Lange OL, Ziegler H, Ehleringer J, Schulze E-D, Cowan IR (1985) Diurnal courses of leaf conductance and transpiration of mistletoes and their hosts in Central Australia. Oecologia 67: 577–587

Urey HC (1932) The relative abundance of hydrogen isotopes in natural hydrogen. Phys Rev 40: 889

Washburn EW, Smith ER (1934) An examination of water from various natural sources for variation in isotope composition. J Res Nat Bur Stds 12: 305–311

Ziegler H (1955) Lathraea, ein Blutungssaftschmarotzer. Ber Dtsch Bot Ges 68: 311–318

Ziegler H (1986) Control of photosynthesis by variation of diffusion resistance in mistletoes and their hosts. In: Marcelle R, Clijstra H, Van Poucke M (eds) Biological control of photosynthesis. Nijhoff, Dordrecht, pp 171–185

Ziegler H (1988) Hydrogen isotope fractionation in plant tissues. In: Rundel PW, Ehleringer JR, Nagy KA (eds) Stable isotopes in ecological research. Ecological studies 68. Springer, Berlin Heidelberg New York, pp 105–123

20 Photosynthesis of Vascular Plants: Assessing Canopy Photosynthesis by Means of Simulation Models

W. Beyschlag, R.J. Ryel, and M.M. Caldwell

20.1 Introduction

Since the pioneering work of Monsi and Saeki (1953), Davidson and Philip (1958), and de Wit (1965), plant canopy models have been used to address numerous ecological questions (Monteith 1965; Norman 1978; Wang and Jarvis 1990). The level of detail of these models is largely dependent upon the objectives of the studies for which the models were designed although all contain routines to calculate net photosynthesis and light penetration into the canopy. In this chapter we will focus on a series of models which have been especially developed to address questions concerning competition for light in plant canopies.

We will start with a rather general description of this modeling approach and with the development of a family of canopy models for uniformly structured monotypic plant stands. The presentation will then move to more complex models for multispecies stands and canopies with nonhomogeneous structure. Several case studies in which the respective models were employed to address ecological questions pertaining to plant competition for light and associated issues will be illustrated.

20.2 General Structure of Canopy Photosynthesis Models

Canopy photosynthesis models generally consist of two components: a photosynthesis submodel and a microclimatic submodel. The photosynthesis submodel relates the response of leaf photosynthesis for a plant species to changes in microclimatic factors (e.g., light, air temperature, ambient CO_2 partial pressure, air humidity). The microclimatic submodel is used to calculate microclimatic conditions at any point within the canopy. Calculations of net photosynthesis are then made for representative portions of the canopy and the whole plant or canopy rates are calculated by summing these individual rates weighted by the distribution of foliage.

20.3   The Simple Case: Single-Species Homogeneous Canopies

The simple case is a canopy consisting of only one species which is relatively uniform in height. Foliage density may vary to a large extent vertically within the canopy but its horizontal distribution is relatively homogeneous. The canopy can be subdivided into several horizontal layers, each of which is assumed to be structurally similar and to have the same foliage orientation. This canopy is one-dimensional in design as differences in structure are only a function of height within the canopy. Several such models exist, but we will briefly describe one of Caldwell et al. (1986) used to determine the contribution of different foliage classes, such as shaded foliage or foliage at the top of the canopy, of a *Quercus coccifera* canopy in a Mediterranean macchia. Since all the models in this chapter use the same basic approach, this section begins with a short description of this model.

20.3.1   General Model Description

The canopy model of Caldwell et al. (1986) contains most of the important parts of a more complete canopy model (Norman 1978). The photosynthesis submodel calculates gross photosynthesis, dark respiration, and transpiration using as inputs incident light, leaf and air temperatures, relative humidity, ambient CO_2 concentration, and predawn soil moisture potential. It contains routines within the microclimatic submodel to calculate incident light on various classes of leaves, air temperature within the canopy, longwave radiation penetration into the canopy, and leaf temperature using as inputs solar

Table 20.1. Canopy and micrometeorological variables as input to the single species homogeneous canopy model. (Caldwell et al. 1986)

General characteristics

Latitude, Julian date, soil albedo, soil logwave emissivity, predawn water potential, coefficients for photosynthetic and stomatal characteristics in submodels, and barometric pressure

Canopy characteristics for each canopy layer

Leaf area, mean inclination and azimuth angles, and dispersion (degree of clustering); stem projected area, mean inclination angles, and dispersion; proportional reduction in maximum photosynthetic capacity, dark respiration capacity, and maximum leaf conductance relative to the top leaf layer; leaf absorptance, transmittance, and reflectance for 400–700 and 400–3000 nm; leaf longwave emissivity; leaf size

Micrometeorological variables for each time step

Total solar global shortwave radiation, soil surface temperature, air temperature, relative humidity, and wind speed above the canopy

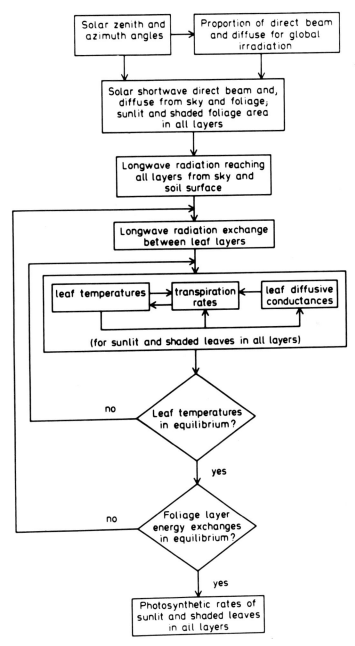

Fig. 20.1. Computation sequence of the single-species homogeneous canopy model. (Caldwell et al. 1986)

radiation, air temperature above the canopy, soil surface temperature, relative humidity, and foliage density and orientation. The distribution of foliage was then used to weight individual layer calculations to arrive at whole canopy rates. A list of input parameters for this model is given in Table 20.1 and the basic model computation flow is shown in Fig. 20.1. The structure of these models is modular, allowing for the replacement of specific routines by formulations. The time steps for the model are arbitrary but 1-h steps were used throughout the course of daylight hours in this work. Routines to calculate incident light on leaves, leaf temperature, and the gaseous exchange of CO_2 and H_2O, follow.

20.3.1.1 Incident Shortwave Radiation on Leaves

Incident shortwave radiation on leaves is calculated for direct beam, sky diffuse, and reflected and transmitted diffuse within the canopy. The canopy is assumed to be continuous without borders or edges that receive radiation. The penetration of direct beam solar radiation into the canopy is calculated following the classical gap probability functions (Monsi and Saeki 1953; Warren Wilson 1960; Duncan et al. 1967) and is a function of solar position in the sky and the surface area, inclinations, and dispersion patterns of foliage in each layer. While the intensity of direct beam shortwave radiation does not decrease through the canopy, the area of the beam is reduced by foliage interception. Thus, the fraction of sunlit leaves within each layer is calculated with the incident beam being calculated for both sides of the leaf for the sunlit portions with corrections made for the angle between the leaf surface and the direct beam (Burt and Luther 1979). Diffuse shortwave radiation from the sky reaching leaves in the different canopy layers is calculated in a manner analogous to that for direct beam; however, the intensity of sky diffuse decreases with foliage interception, not area. Different segments of the sky are taken as the source of radiation, with the hemisphere divided into nine annular rings of uniform brightness (Duncan et al. 1967). Diffuse radiation intensity through the canopy from the different sky segments is calculated separately and then integrated for all sky bands for foliage in each layer. All leaves within a layer are assumed to obtain the same incident sky diffuse with calculations made for both sides of a leaf (Duncan et al. 1967). Shortwave radiation is reflected from and transmitted through foliage as well as reflected from the soil surface. Contributions to the incident flux on foliage in each layer is calculated from these sources of diffuse shortwave radiation according to Duncan et al. (1967). The total incident shortwave radiation flux on leaves is then calculated as the sum of direct beam, sky diffuse, and reflected and transmitted diffuse of sunlit leaves and the same without direct beam fluxes for shaded leaves. Similar calculations can be made for stems if they have photosynthetic potential.

20.3.1.2   Energy Balance and Leaf Temperatures

The temperature of leaves within the canopy is a function of incident shortwave radiation, incident longwave radiation from other foliage, the soil surface and the sky, latent heat loss due to transpiration, and the emission of longwave radiation by the foliage (Norman 1978). Longwave radiation reaching leaves from the sky, the soil surface, and from surrounding foliage is computed using the same geometry as employed for diffuse shortwave radiation. Soil surface temperature is an input parameter and sky temperature is calculated from air temperature and vapor pressure (Brutsaert 1975). In dense canopies, such as considered in Caldwell et al. (1986), the longwave exchange is primarily between foliage elements and the process becomes an iterative in which leaf temperatures in each layer are affected by and have influence on temperatures of leaves in same and neighboring layers (Fig. 20.1). Additionally, because transpiration is a function of stomatal aperture (see next section), transpiration and the latent loss of heat is influenced by leaf temperature and such relationships must be added to the iterative calculation process. Probably the least developed aspects of the energy balance calculations involve the canopy boundary layer and turbulence within the stand since only a simply logarithmic attenuation of horizontal momentum measured above the canopy is assumed.

20.3.1.3   Leaf Conductance and Photosynthesis

Leaf stomatal conductance is calculated in response to incident shortwave radiation within 400–700 nm (PFD), predawn water potential, leaf temperature, and vapor pressure difference between leaf and air. The functional relationships between environmental parameters and stomatal conductance were first assembled from general concepts of stomatal function of plants (e.g., Schulze and Hall 1982) and then adapted to the species of concern by means of gas exchange cuvette measurements in different canopy layers. The rate of transpiration is then calculated from stomatal conductance, the boundary layer resistance (as affected by wind), and the vapor pressure difference between leaf and air.

The photosynthesis submodel is a modified version of earlier models developed by Tenhunen et al. (1976, 1980) and is similar to that described in Tenhunen et al. (1987). Input variables are total photon flux density (400–700 nm), leaf temperature, leaf conductance, and ambient CO_2 partial pressure. The CO_2-saturated photosynthetic rate is calculated as a function of leaf temperature and incident PFD (Tenhunen et al. 1976). The carboxylation efficiency, the initial slope of the relationship between photosynthesis, and leaf internal CO_2 partial pressure are computed as a function of maximum photosynthetic capacity, PFD, and leaf temperature at 21% O_2. The leaf photosynthetic response to internal CO_2 is nonlinear, as described by

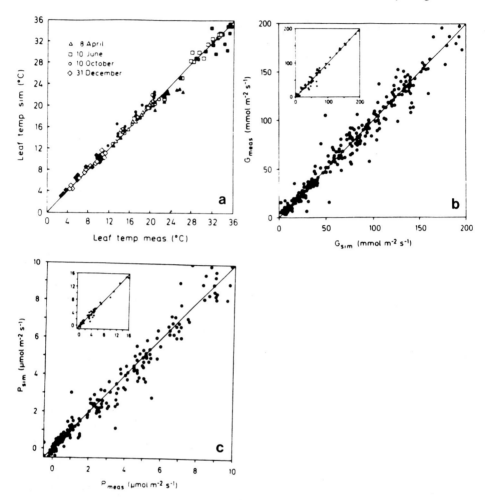

Fig. 20.2a–c. Validation of the single-species homogeneous canopy model for the Mediterranean evergreen shrub species *Quercus coccifera* (Caldwell et al. 1986). **a** Correspondence of measured and simulated leaf temperatures near the top (*solid symbols*) and bottom (*open symbols*) on clear days at four different times of year. The coefficient of determination is 0.90. **b** Correspondence between simulated and measured leaf conductance in different canopy layers for several dates in all four seasons of the year. The coefficient of determination based on 251 observations is 0.87. Most of the measured data were needed for parameter estimation in the conductance model. In the *inset* is a similar relationship between measured and simulated leaf conductance for a data set not used for parameter estimation. **c** Correspondence between simulated and measured net photosynthesis in different canopy layers for several dates in all four seasons of the year. The coefficient of determination based on 173 observations is 0.91. Most of the measured data were needed for parameter estimation in the conductance model. In the *inset* is a similar relationship between measured and simulated leaf conductance for a data set not used for parameter estimation

Smith (1937), with dark respiration represented by an Arrhenius equation and assumed to be light-independent. Iterative procedures then balance rates of net CO_2 fixation with rates of CO_2 entry into the leaf (Weber et al. 1985).

20.3.2 Model Validation

The value of a model beyond meeting its specified objectives is to generate time patterns of behavior that do not differ significantly from the real system (Forrester 1961). Such model "validation" provides a degree of confidence that the model can adequately perform to meet its objectives. Parameters for validation are selected based upon the model objectives and the ability to obtain reliable field measurements. Since these models calculate some outputs, such as total canopy gas exchange, that are difficult to measure in the field, a subset of model outputs is compared with more easily measured parameters to test their correspondence. Figure 20.2 shows the comparison between measured and simulated values of leaf temperature, leaf conductance, and net photosynthesis of *Q. coccifera* at four different times of the year using the model of Caldwell et al. (1986). Good correlation was found in all cases for these parameters, suggesting that the model can reliably calculate these parameter values under these and similar environmental conditions.

The approach of Caldwell et al. (1986) was to construct the model based on the knowledge of basic processes and to arrive at canopy function as a direct consequence of the interaction of the component parts. This approach is very instructive in the investigation of mechanistic interactions of model components as affected by the processes of light interception, energy balance, and diffusive conductance. One of the objectives of this model, to calculate parameters that cannot be measured, is illustrated in the following example.

20.3.3 Case Study: How Do Different Parts of the Canopy Contribute to Total Canopy Photosynthesis?

Figure 20.3 shows the contribution of different foliage classes to total canopy photosynthesis of *Q. coccifera* throughout the year as calculated with the model. In these situations, the net 24-h CO_2 gain contributed by all the shaded leaves in the canopy was nearly the same as that from leaves in the uppermost layer, which comprised only 13% of the total leaf area. The contribution of the lower 50% of the total canopy foliage is similar to that of all the shaded foliage during the summer months, but is much lower in the winter months.

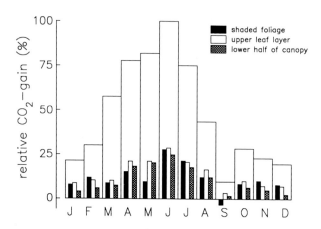

Fig. 20.3. The net CO_2 gain (per ground area surface) of a *Quercus coccifera* canopy for clear days in each month of the year for a canopy of 5.4 LAI (*large open bars*) and the contribution of different foliage classes (*thin bars*) under the same conditions: all shaded foliage (*solid*), the uppermost leaf layer (*open*) and the lower 50% of canopy foliage (*hatched*). Measured microclimate parameters for clear days in each months were used as input variables. (Caldwell et al. 1986)

20.4   Multispecies Homogeneous Canopies

The next step was to extend the canopy model approach towards assessments of photosynthesis for component species of mixed plant stands with homogeneous structure. This would then allow for the direct assessment of light competition between species, an important phenomenon which in some cases is directly responsible for the structure and composition of plant canopies (Wierman and Oliver 1979; Ford and Diggle 1981).

20.4.1   Description of the Model Extensions

Using the model structure of Caldwell et al. (1986), Ryel et al. (1990) developed a multispecies homogeneous canopy photosynthesis model which permits any number of specific classes of foliage for several species within each layer. These foliage classes can represent foliage of different species or, if desired, foliage of different orientation, morphology, or phenology. A detailed description as well as validation data of this model are given by Ryel et al. (1990). This model parallels the most complex structure described by Norman (1980) for single-species canopies. It also contains a significant increase in structural detail over multispecies models of Ross et al. (1972), McMurtrie and Wolf (1983), and Rimmington (1984). Sensitivity analyses presented by Beyschlag et al. (1990) for the photosynthetic parameters and by Barnes et al. (1990) for the structural input data clearly revealed (1) that single-leaf photosynthetic behavior is not necessarily representative of the integrated canopy behavior and (2) that in these dense grass stands, the structural canopy properties are much more important for determining the balance of light competition than the functional characteristics, such as photosynthetic capacity. The following two case studies will illustrate how effectively models of this type can be employed as a tool for analysis of aboveground competition.

20.4.2   Case Study: Symmetric Competition

Plants of similar stature can compete for light, and each plant exerts some shading effect on its neighbors, usually in proportion to its size. Such cases of symmetrical competition may involve mutual shading at several levels in the canopy rather than one plant simply overtopping another.

In field and glasshouse experiments, mixed canopies of wheat and wild oat were given supplemental UV-B radiation and the relative competitive status of wheat increased (Barnes et al. 1988). This was evaluated from aboveground biomass comparisons between monocultures and mixtures using relative crowding coefficients of de Wit (1960). Associated with the shift in the competitive balance under UV-B treatment were differential effects of UV-B on the growth form of the two species: leaf insertion heights and leaf blade lengths were reduced to a greater extent in wild oat than in wheat. In monocultures of these species, these growth form changes had no measurable bearing on the total production of the plant stands (Barnes et al. 1988). However, in mixtures, these changes in individual plant morphology resulted in shifts in foliage height distribution such that wheat tended to have its foliage somewhat higher in the canopy than did wild oat under UV-B enhancement. Assuming that carbon gain was related to biomass production, the question was whether these structural changes in the two species could quantitatively account for the changes in biomass of the two competitors.

Figures 20.4 and 20.5 show the results of model calculations using the multispecies version of the canopy photosynthesis model which had been parameterized either with the structural and photosynthetic data of control canopy mixtures without enhanced UV-B radiation or mixed canopies given the UV-B enhancement treatment. The results clearly demonstrate that a comparatively small but differential change in the structural properties of the two competing species can lead to changes in canopy net photosynthesis sufficient to explain the observed shift in the competitive balance. Barnes et al. (1990) further showed that the changes in leaf area were largely responsible for the effects, whereas changes in leaf angles had a negligible effect.

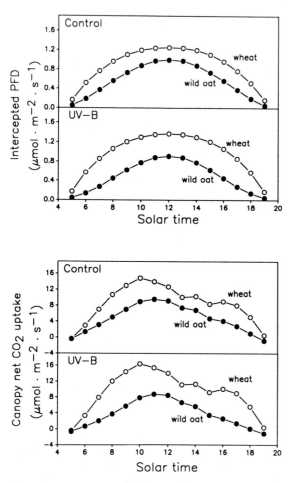

Fig. 20.4. Interception of PFD by foliage of wheat and wild oat (*top*) and canopy net CO_2 uptake for these two species (*bottom*) in 1:1 mixed canopies predicted for control and enhanced UV-B conditions during sunlit hours of the day in a glasshouse experiment. Over the entire day, wheat foliage intercepted 61% of the PFD in the control and 66% in the UV-B treatment canopies. Rates are expressed per ground area. (Ryel et al. 1990)

20.4.3   Case Study: Asymmetric Competition

When plants compete for light and one species clearly overtops another and is little influenced by the smaller competing species, asymmetric competition is considered to occur. Beyschlag et al. (1992) used the same model to analyze a case of severe asymmetric competition between the two roadside grass species *Elymus repens* and *Puccinellia distans*. While these two species are able to coexist under the stress conditions of road shoulders close to the pavement (Ullmann and Heindl 1989), they exhibited strongly asymmetric

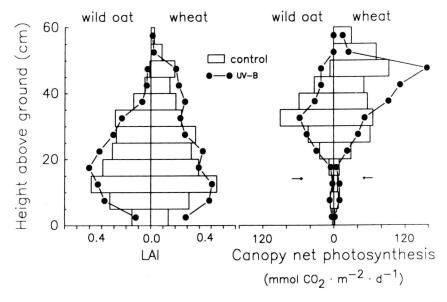

Fig. 20.5. LAI and daily net photosynthesis in different canopy layers of wheat and wild oat in 1:1 mixed canopies grown under control and enhanced UV-B conditions in a glasshouse experiment. *Below the arrows* canopy net photosynthesis is negative. Rates are expressed per ground area. (Ryel et al. 1990)

competition in a garden experiment where equiproportional mixtures of the two species were grown under optimum nutrient and water conditions (Beyschlag et al. 1992). A stable competitive balance was never reached in these experiments and *P. distans* was eliminated from the mixtures within the first vegetation period. Figure 20.6 shows the values for light interception and canopy photosynthesis calculated for three points of time during the vegetation period. In May (6 weeks after establishment of the plants) both species were growing vigorously, but as the summer progressed *E. repens* appeared to have an increasingly negative effect on the aboveground biomass production of *P. distans*. This reduction in biomass production corresponded with the continuous overtopping of *P. distans* by *E. repens*. The data in Fig. 20.6 suggest that the dieback of *P. distans* during the second half of the growing season may be largely explained by aboveground competition for light. Under the given experimental conditions, *E. repens* was obviously able to grow tall enough to severely shade the *P. distans* plants which resulted in a major reduction of net carbon gain of *P. distans* by the end of the season. The *P. distans* grew well in monocultures under the same conditions.

In order for *P. distans* to persist in these roadside communities, it would appear that periodic disturbance, such as mowing, would be necessary to keep *E. repens* from overtopping *P. distans*. The consequences of mowing on the light climate in mixed stands of the two species can be rather

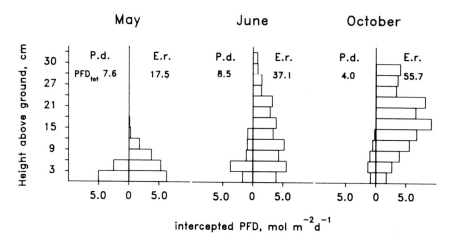

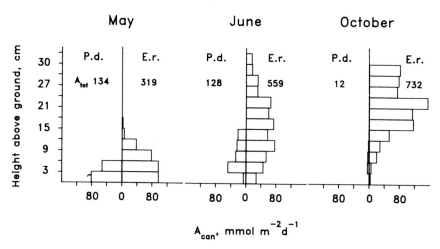

Fig. 20.6. Distribution of calculated daily intercepted PFD (*top*) and canopy net photosynthesis (*bottom*) in different canopy layers of 1:1 mixed canopies of *Puccinellia distans* (*P.d.*) and *Elymus repens* (*E.r.*) in mid-May, mid-June, and mid-October 1989. The *numbers* give the respective daily totals for each species. Rates are expressed per unit ground area. (Beyschlag et al. 1992)

explicitly simulated with the multispecies canopy model. Starting at the top of the canopy, leaf and stem area indices of both species were successively set to zero layer by layer in simulated mowing. In each case, canopy photosynthesis of the remaining canopy was calculated. The results (Fig. 20.7) clearly indicate that *P. distans* strongly benefits from the removal of the *E. repens* material from the top layers. A negative effect on *P. distans* is only apparent if the canopy is mowed below the height of the *P. distans* plants.

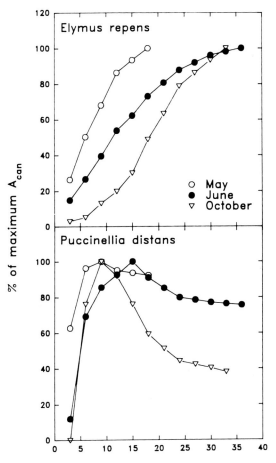

Fig. 20.7. Calculated relative changes in canopy net photosynthesis (A_{can}) of *Elymus repens* and *Puccinellia distans* in response to simulated mowing (see text) in mid-May, mid-June, and mid-October 1989. (Beyschlag et al. 1992)

Canopy height after simulated mowing, cm

20.5 Canopies with Nonhomogeneous Structure: Radiation Fluxes in Three Dimensions

Nonhomogeneous canopies occur where vegetation is substantially clumped and/or discontinuous in structure. Calculation of light interception and photosynthesis for the different elements of such canopies is necessarily much more complicated than for homogeneously structured canopies. The primary difficulty is that the light interception of a plant growing in such a canopy is affected by surrounding vegetation which may be present at various distances in different compass directions from the plant in question. In addition, light may reach the plants from the sides as well as from above and, thus, a three-dimensional modeling approach is needed.

20.5.1   Step 1: The Case of Single Plants

The simplest case of a nonhomogeneous canopy is a single plant growing on
a horizontal surface such as an isolated tussock of a desert bunchgrass. Ryel
et al. (1993) recently developed a photosynthesis model for this situation,
where the procedure used for calculating whole tussock-photosynthesis is
analogous to that for uniform canopies (see Fig. 20.1.) except in three
dimensions. The tussock is defined as cylindrical in shape with concentric
subcylinders defining regions of different foliage density and character (Fig.
20.8). These subcylinders are further subdivided into layers to allow additional subdivisions of foliage and physiological characteristics in the tussock.
Instead of using a single vertical array of points representing horizontal
canopy layers, light interception and photosynthesis are calculated for sunlit
and shaded foliage in a three-dimensional array of points in the cylinder
subsections. The points are uniformly arranged in equally spaced horizontal
planes. The array is set with the same number of points (n) in three
dimensions. Minimum values for n can be determined by sensitivity analysis.
Incident light and net photosynthetic rate are calculated separately for each
point. Mean net photosynthesis for each subsection is then calculated from
all points within the subsection, weighted by the size of the tussock subsection and fraction of sunlit and shaded foliage, and summed over all subsections to obtain whole-tussock photosynthesis.

   This three-dimensional canopy model and similar developments of
Norman and Welles (1983), Grace et al. (1987a), and Wang and Jarvis
(1990) are substantial departures from the homogeneous canopy models and
they represent a significant advantage for use with plants which are in
clumped or discontinuous distributions.

20.5.1.1   Case Study: Do Tussock Grasses Sacrifice Carbon Gain
from Having Steep Leaf Angles?

Compact, isolated tussock grasses often possess steeply inclined foliage, but
the implications of foliage orientation for this growth form have not been
thoroughly pursued. The tussock growth form is widely distributed in grass
steppe and desert regions, but the architectural advantages of these plants
are not always so obvious (Caldwell et al. 1983). Much of the daily
irradiation is intercepted by the lateral surfaces of the tussocks, and there
can still be considerable self-shading within the tussock (Caldwell et al.
1983). Steeply inclined foliage may have benefits for reduced potential for
photoinhibition and greater water-use efficiency, but there may also be
disadvantages in the effective harvesting of shortwave radiation at certain
times of the day (Werk and Ehleringer 1984).

   After parameterization and thorough validation, Ryel et al. (1993) used
the three-dimensional model to examine the effects of foliage orientation on

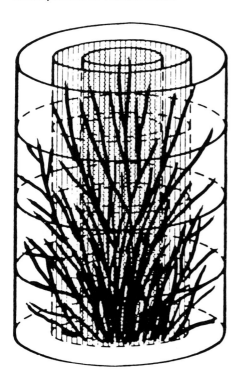

Fig. 20.8. Schematic depiction of a grass tussock placed in a cylindrical structure sub-divided into several layers and concentric subcylinders defining regions of different foliage density and character in the model. (Ryel et al. 1993)

the photosynthesis of two tussock grass species of the North American Intermountain West, the native grass, *Pseudoroegneria spicata*, and an exotic species of similar growth form in this region, *Agropyron desertorum*. Model simulations showed the effect of leaf orientation on whole tussock net photosynthesis ($A_{tu\,ss}$) to be minimal throughout the entire day (Fig. 20.9A,B) and it was concluded that *A. desertorum* and *P. spicata* do not appear to sacrifice carbon gain with steep foliage orientation. There were, however, pronounced differences in the diurnal courses of PFD intercepted by the tussocks (PFD_{tuss}) (Fig. 20.9C,D). Canopies with more horizontally oriented foliage had higher PFD_{tuss} during the midday hours than canopies with more vertically oriented foliage despite having less foliage exposed to direct sunlight (Fig. 20.9E,F). The reverse occurred during the early and late hours where canopies with more vertically oriented foliage intercepted more light but had less sunlit foliage. The distribution of incident PFD on green leaves (PFD_{int}) varied considerably with time of day and foliage orientation (Fig. 20.10). All canopies had little foliage with PFD_{int} above $1400\,\mu mol\,m^{-2}s^{-1}$ during early morning (0800 h), and late afternoon (1600 h) but the fraction of foliage with PFD_{int} above $1800\,\mu mol\,m^{-2}s^{-1}$ increased notably during the midday hours (1000 to 1200 h) for canopies with more horizontally oriented foliage (0° and 30°). Although photoinhibition has not yet been demonstrated in these tussock grass species, the high light environment coupled

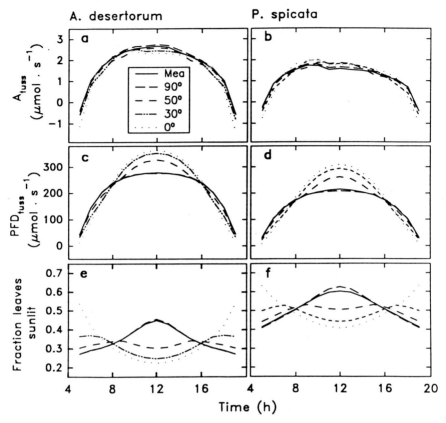

Fig. 20.9a–f. Simulated diurnal courses of whole-tussock photosynthesis (A_{tuss}), incident PFD (PFD_{tuss}) and fractions of leaves that are sunlit for single tussocks of *Agropyron desertorum* and *Pseudoroegneria spicata*. Simulations were conducted with leaf angles in the canopies as measured (*Mea*) and set at different leaf angles (0°, 30°, 50°, 90° from horizontal). (Ryel et al. 1993)

with high temperatures and drought makes these plant candidates for photo-inhibition (Powles 1984; Krause 1988). These tussock grasses have minimum amounts of foliage exposed to high incident PFD during periods of the day when temperatures are the highest (Fig. 20.10). Therefore, the potential for photoinhibition in these tussock grasses should be much lower than if the leaves were more horizontally oriented.

20.5.2   Step 2: Scaling up from Single Plants to Plant Neighborhoods

A single plant can be placed in an assemblage of other plants of the same species of in a mixed-species stand by defining other cylinders representing other plants. These cylinders can be arranged at any distance from each

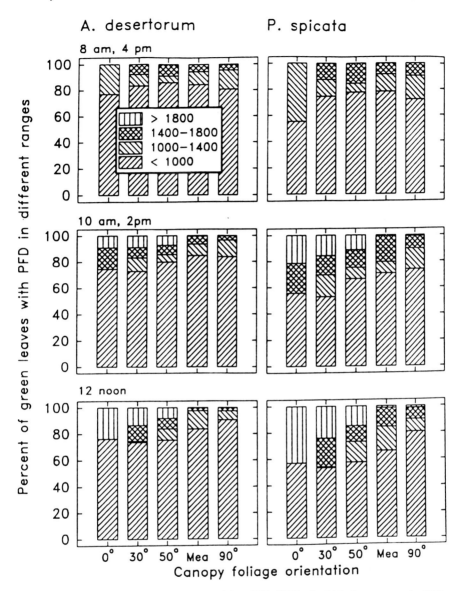

Fig. 20.10. Percent of green leaves with incident PFD (PFD_{int}) within four ranges (<1000, 1000–1400, 1400–1800, and >1800 μmol m^{-2} s^{-1}) for single tussocks of *Agropyron desertorum* and *Pseudoroegneria spicata* as calculated by model simulation for five time periods. Simulations were conducted with leaf angles in the canopies as measured (*Mea*) and set at different leaf angles (0°, 30°, 50°, 90° from horizontal). Light characteristics within the tussocks were identical at 0800 and 1600 h and at 1000 and 1400 h. (Ryel et al. 1993)

other, including situations where they overlap (Fig. 20.11). The structural
parameters needed for the other cylinders are the same as for the plant of
interest. Foliage of the neighboring plants can cause attenuation of direct
beam and sky diffuse shortwave radiation before this radiation reaches the
target plant (Fig. 20.11). This approach has been successfully used to determine that significant competition for light between adjacent tussocks does
not occur unless the plants are less than one diameter apart and that the
uniform placement of tillers (turf growth form) is much more efficient in
light interception and carbon gain than bunching of tillers unless the density
of bunches forms a nearly uniform canopy (Ryel et al., unpubl.). Recently,
a modified version of the model has been employed to evaluate the consequences of losing different age classes of needles for light interception and
canopy photosynthesis of dense declining stands of red spruce [*Picea abies*
(L.) Karst.]. Due to increased light levels within the tree crowns, a beneficial
effect on photosynthesis and water use efficiency of the remaining younger
needles results from the loss of older needle age classes. In terms of whole
tree photosynthesis, the reduction of photosynthetically active foliage area
can nearly be compensated by this phenomenon when there is substantial
tree overlap (Beyschlag et al., unpubl.). Similar analyses have been used in
assessing net photosynthesis in stands of *Picea sitchensis* (Norman and Jarvis
1975) and *Pinus radiata* (Grace et al. 1987b).

20.6 Conclusions

Although measurements of single-leaf photosynthesis can presently be performed in the field with a high degree of accuracy, the photosynthetic rate of
single leaves is often not representative of photosynthetic behavior for the
entire canopy. Simulation models which calculate light interception and net
photosynthesis provide an effective way to overcome this problem. As
discussed in this chapter, models are now available for both homogeneous
or nonhomogeneous, single-species or mixed-species canopies. They have
been developed with modular formats (Reynolds et al. 1987) which allows
replacement of model components by other routines which may be more
suitable in applications for other species or situations. While this family of
models has proved to be a useful tool to address many ecophysiological
questions, their outputs also provide the carbon gain inputs needed by
allocation and growth models (e.g., Johnson and Thornley 1985; Charles-
Edwards et al. 1986; Reynolds et al. 1987; Buwalda 1991; Webb 1991).

Precise calculations of canopy net photosynthesis may often be limited by
the quality of input parameter estimates. The time and effort involved in
obtaining model input parameters can be considerable, and there is potential
for cumulative errors to arise in these parameter estimates. However, these
limitations are less of a problem when the models are used to address

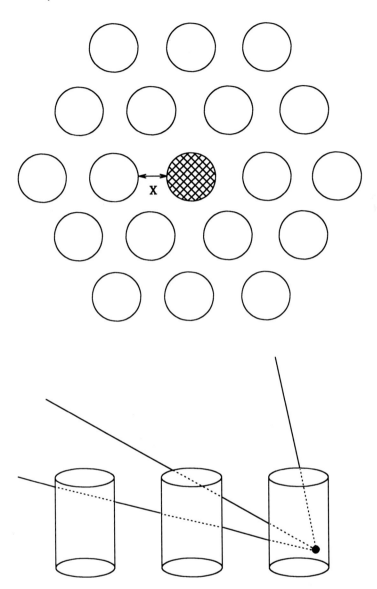

Fig. 20.11. *Top* Arrangement of cylinders representing plants either of the same species or of different species to simulate monotypic or mixed plant neighbourhoods. Cylinders can be arranged at any distance (*X*), and can overlap. Whole-plant photosynthesis is calculated for a target plant in the center. *Bottom* Radiation attenuation for any point within the foliage of the target plant on the right is affected by the foliage of the surrounding plants as well as the foliage of the target plant itself

questions of a comparative nature. For example, for the same canopy structure, how are photosynthesis and water loss influenced by changes in light conditions outside the canopy, air temperature, ambient-CO_2 partial pressures, or humidity? Or, how do changes in structural attributes (e.g., leaf orientation, foliage density distribution) influence canopy gas exchange or competition for light under given weather conditions? Such comparisons usually tend to reduce the effect of parameter errors since these inaccuracies are common to both canopy data sets being compared.

The confidence in these model outputs can be increased by performing sensitivity analyses of model outputs as a function of input parameter estimates. Although the possibilities for such sensitivity analyses over various ranges and interactions of parameter values, are seemingly infinite, procedures are available to limit such endeavors to appropriate bounds (e.g., see Forrester 1961; Innis 1978; Thornley and Johnson 1990). In addition, sensitivity analyses can help identify where parameters need to be more accurately defined.

Independent empirical measurements also provide a check of model performance. Whole-plant gas exchange measurements for individual plants provide one assessment in the case of the three-dimensional models (Ryel et al. 1993). For continuous canopies, eddy correlation techniques also provide a powerful approach to measuring whole-canopy fluxes if certain conditions are met (Verma 1990). Even though these flux measurements do not provide information on the contribution of different species or plant parts to canopy gas exchange, they do provide a means of assessing aspects of model performance.

Acknowledgments. We gratefully acknowledge funding of this work by the Deutsche Forschungsgemeinschaft (SFB 251, University of Würzburg) and the US National Science Foundation (BSR 87-05492).

References

Barnes PW, Jordan PW, Gold WG, Flint SD, Caldwell MM (1988) Competition, morphology and canopy structure in wheat (*Triticum aestivum* L.) and wild oat (*Avena fatua* L.) exposed to enhanced ultraviolet-B radiation. Funct Ecol 2: 319–330

Barnes PW, Beyschlag W, Ryel RJ, Flint SD, Caldwell MM (1990) Plant competition for light analyzed with a multispecies canopy model. III. Influence of canopy structure in mixtures and monocultures of wheat and wild oat. Oecologia 82: 560–566

Beyschlag W, Barnes PW, Ryel RJ, Caldwell MM, Flint SD (1990) Plant competition for light analyzed with a multispecies canopy model. II. Influence of photosynthetic characteristics on mixtures of wheat and wild oat. Oecologia 82: 374–380

Beyschlag W, Ryel RJ, Ullmann I (1992) Experimental and modelling studies of competition for light in roadside grasses. Bot Acta 105: 285–291

Brutsaert W (1975) On a derivable formula for long-wave adiation from clear skies. Water Resour Res 11: 742–744

Burt JE, Luther FM (1979) Effect of receiver orientation on erythema dose. Photochem Photobiol 29: 85–91

Buwalda JG (1991) A mathematical model of carbon acquisition and utilisation by kiwifruit vines. Ecol Model 57: 43–64

Caldwell MM, Dean TJ, Nowak RS, Dzurek RS, Richards JH (1983) Bunchgrass architecture, light interception, and water-use efficiency: assessment by fiber optic point quadrats and gas exchange. Oecologia 59: 178–184

Caldwell MM, Meister HP, Tenhunen JD, Lange OL (1986) Canopy structure, light microclimate and leaf gas exchange of *Quercus coccifera* L. in a Portuguese macchia: measurements in different canopy layers and simulations with a canopy model. Trees 1: 25–41

Charles-Edwards DA, Doley D, Rimmington GM (1986) Modelling plant growth and development. Academic Press, Sydney

Davidson JL, Philip JR (1958) Light and pasture growth. In: Climatology and microclimatology. UNESCO, Paris, pp 181–187

Duncan WC, Loomis RS, Williams WA, Hanau R (1967) A model for simulating photosynthesis in plant communities. Hilgardia 38: 181–205

Ford ED, Diggle PJ (1981) Competition for light in a plant monoculture modelled as a spatial stochastic process. Ann Bot 48: 481–500

Forrester JW (1961) Industrial dynamics. MIT Press, Cambridge, 464 pp

Grace JC, Jarvis PG, Norman JM (1987a) Modelling the interception of solar radiant energy in intensively managed stands. N Z J For Sci 17: 193–209

Grace JC, Rook DA, Lane PM (1987b) Modelling canopy photosynthesis in *Pinus radiata* stands. N Z J For Sci 17: 204–228

Innis GS (1978) Grassland simulation model, Ecological studies 26. Springer, Berlin Heidelberg New York, 298 pp

Johnson IR, Thornley JHM (1985) Dynamic model of the response of a vegetative grass crop to light, temperature and nitrogen. Plant Cell Environ 8: 485–499

Krause GH (1988) Photoinhibition of photosynthesis. An evaluation of damaging and protective mechanisms. Physiol Plant 74: 566–574

McMurtrie R, Wolf L (1983) A model of competition between trees and grass for radiation, water and nutrients. Ann Bot 52: 449–458

Monsi M, Saeki T (1953) Über den Lichtfaktor in den Pflanzengesellschaften und seine Bedeutung für die Stoffproduktion. Jpn J Bot 14: 22–52

Monteith JM (1965) Light distribution and photosynthesis in field crops. Ann Bot 29: 17–37

Norman JM (1978) Modelling the complete crop canopy. In: Barfield BJ, Gerber JF (eds) Modification of the aerial environment of plants. Am Soc Agric Eng, St. Joseph, pp 249–277

Norman JM (1980) Interfacing leaf and canopy light interception models. In: Hesketh JD, Jones JW (eds) Predicting photosynthesis for ecosystem models, vol II. CRC Press, Boca Raton, pp 49–67

Norman JM, Jarvis PG (1975) Photosynthesis in Sitka spruce (*Picea sitchensis* (Bong.) Carr.). V. Radiation penetration theory and a test case. J Appl Ecol 12: 839–878

Norman JM, Welles JM (1983) Radiative transfer in an array of canopies. Agron J 75: 481–488

Powles SB (1984) Photoinhibition of photosynthesis induced by visible light. Annu Rev Plant Physiol 35: 15–44

Reynolds JF, Acock B, Dougherty RL, Tenhunen JD (1987) A modular structure for plant growth simulation models. In: Pereira JS, Landsberg JJ (eds) Biomass production by fast growing trees, pp 123–134

Rimmington GM (1984) A model of the effect of interspecies competition for light on dry-matter production. Aust J Plant Physiol 11: 277–286

Ross PJ, Henzell EF, Ross DR (1972) Effects of nitrogen and light in grass-legume pastures, a systems analysis approach. J Appl Ecol 9: 535–556

Ryel RJ, Barnes PW, Beyschlag W, Caldwell MM, Flint SD (1990) Plant competition for light analyzed with a multispecies canopy model I. Model development and influence of enhanced UV-B conditions on photosynthesis in mixed wheat and wild oat canopies. Oecologia 82: 304–310

Ryel RJ, Beyschlag W, Caldwell MM (1993) Foliage orientation and carbon gain in two tussock grasses as assessed with a net canopy gas exchange model. Funct Ecol 7: 115–124

Schulze ED, Hall AE (1982) Stomatal responses, water loss and CO_2 assimilation rates of plants in contrasting environments. In: Lange OL, Nobel PS, Osmond CB, Ziegler H (eds) Encyclopedia of plant physiology, new series, vol 12B. Physiological plant ecology II. Springer, Berlin Heidelberg New York, pp 181–230

Smith E (1937) The influence of light and carbon dioxide on photosynthesis. Gen Physiol 20: 807–830

Tenhunen JD, Yocum CS, Gates DM (1976) Development of a photosynthesis model with an emphasis on ecological applications I. Theory. Oecologia 26: 89–100

Tenhunen JD, Hesketh JD, Gates DM (1980) Leaf photosynthesis models. In: Hesketh JD, Jones JW (eds) Predicting photosynthesis for ecosystem models, vol I. CRC Press, Boca Raton, pp 123–181

Tenhunen JD, Harley PC, Beyschlag W, Lange OL (1987) A model of net photosynthesis for leaves of the sclerophyll *Quercus coccifera*. In: Tenhunen JD, Catarino F, Lange OL, Oechel WC (eds) Plant response to stress – functional analysis in mediterranean ecosystems. Springer, Berlin Heidelberg New York, pp 339–354

Thornley JHM, Johnson IR (1990) Plant and crop modelling, a mathematical approach to plant and crop modelling. Clarendon Press, Oxford, 669 pp

Ullmann I, Heindl B (1989) Geographical and ecological differentiation of roadside vegetation in temperate Europe. Bot Acta 102: 261–269

Verma SB (1990) Micrometeorological methods for measuring surface fluxes of mass and energy. In: Goel NS, Norman JM (eds) Instrumentation for studying vegetation canopies for remote sensing in optical and thermal infrared regions. Remote sensing reviews, vol 5. Harwood Academic Publishers, Chur, pp 99–115

Wang YP, Jarvis PG (1990) Description and validation of an array model – MAESTRO. Agric For Meteorol 51: 257–280

Warren Wilson J (1960) Inclined point quadrats. New Phytol 58: 1–8

Webb WL (1991) Atmospheric CO_2, climate change, and tree growth: a process model I. Model structure. Ecol Model 56: 81–107

Weber JA, Tenhunen JD, Lange OL (1985) Effects of temperature at constant air dew point on leaf carboxylation efficiency and CO_2 compensation point of different leaf types. Planta 166: 81–88

Werk KS, Ehleringer JR (1984) Non-random leaf orientation in *Lactuca serriola* L. Plant Cell Environ 7: 81–87

Wierman CA, Oliver CD (1979) Crown stratification by species in even-aged mixed stands of Douglas-fir-western hemlock. Can J For Res 9: 1–9

Wit CT de (1960) On competition. Agric Res Rep 663, Versl Landbouwkd Oenderz, Wageningen, pp 1–57

Wit CT de (1965) Photosynthesis of leaf canopies. Agric Res Rep, Vers Landbouwkd Onderz, Wageningen

21 Effects of Phenology, Physiology, and Gradients in Community Composition, Structure, and Microclimate on Tundra Ecosystem CO_2 Exchange

J.D. Tenhunen, R.T.W. Siegwolf, and S.F. Oberbauer

21.1 "Phenomenological" or "Aggregate" Models of Ecosystem CO_2 Flux

Recently, attention among scientists has been focused on potential global climate change as well as on the deposition of pollutants and their impacts. These perspectives emphasize the role of ecosystems as exchange surfaces between atmosphere and vegetation and between vegetation and groundwater (Dickenson 1988; Bolin 1988; Ulrich 1987). Particularly with respect to northern taiga and tundra regions, it is important to determine whether climate change may have already altered or may in the future alter rates (positive or negative) of ecosystem carbon storage (Oberbauer et al. 1992; Oechel and Billings 1992). Furthermore, it is important to understand environmental controls on carbon fluxes and carbon storage, because the gradients in soil temperature, water availability, and available light energy in the Arctic are large and these will strongly affect the integrated values of net carbon dioxide (Tenhunen et al. 1992) and methane exchange (Whalen and Reeburgh 1988, 1990) in polar regions. Even when viewed simplistically and at the regional scale, temporal and spatial variation in ecosystem material exchange characteristics must be considered when estimating carbon balances (Miller et al. 1983). At smaller scales such as the watershed, temporal and spatial variation in ecosystem structure, species composition, physiology, and environmental conditions determine momentary net gas exchange rates, but also provide clues concerning the manner in which ecosystem properties may be shifted regionally in a future climate (Chapin et al. 1992).

Despite considerable effort having been devoted to the examination of carbon dioxide exchange in tundra regions (Kjelvik et al. 1975; Coyne and Kelley 1978; Billings et al. 1982, 1983; Eckardt 1982; Luken and Billings 1985; Hilbert et al. 1987; Grulke et al. 1990; Oberbauer et al. 1991, 1992), beginning with investigations near Barrow, Alaska, during the International Biological Program, we need to refine our understanding of the relationships between dynamic features of ecosystem structure and function and transfers at the atmosphere/vegetation interface. Limitations imposed by the small scale at which ecosystem-level experiments may usually be conducted and the nonacceptability of large-scale ecosystem manipulations force us to rely on "bottom-up" approaches when formulating models that will be useful under conditions of climate change, for example at elevated atmospheric CO_2 levels (Jarvis 1987). Essential are so-called "phenomenological" or "aggregate models" of tundra ecosystem CO_2 exchange that are simple, based on detailed process understanding, and include appropriate "responsiveness" (Rathstetter et al. 1992).

The detailed "bottom-up" model GAS-FLUX is described as developed for low stature tundra communities. The model integrates information on microclimate, species composition, spatial structure, and gas exchange. A similar model was developed by Miller et al. (1984) as a submodel driver for a simulator of tundra ecosystem processes (ARTUS). This earlier model had two main disadvantages that limited the potential for general application in other ecosystems as well as at different tundra sites. First, gas exchange physiology was described with empirical functions that are difficult to relate to biochemistry and available nutrients, thus posing great difficulties as soon as ecosystem-level feedbacks might be considered that modify nutrient availability. Secondly, vegetation structure was to a great extent ignored due to low leaf area index (LAI) in tundra regions, with all vascular plant material considered to be exposed to incident light and the total vascular plant material functioning as a neutral filter of changing density when determining incident light on the moss layer. The main disadvantage of this approach is that results become questionable as soon as LAI increases above 1.0, and the model is difficult to apply in other ecosystems of high LAI (e.g., shrublands or forests) without major modification.

An important advantage of the GAS-FLUX model is that estimates of gas exchange rates are based on mechanistic descriptions of photosynthetic processes (Harley and Tenhunen 1991) as well as on the strong correlation found between CO_2 assimilation rate and stomatal conductance. This modeling approach results in ecosystem "responsiveness" to atmospheric CO_2 concentration in terms of both CO_2 uptake and water loss (cf. Tenhunen et al. 1990). A new aspect of the model is that gas exchange from the moss component of the ecosystem is similarly treated, as moss is extremely important in tundra regions (Tenhunen et al. 1992). The low stature of tundra communities and limited depth to permafrost provide advantages for model verification by chamber measurements (Gillespie et al. 1993). The imperative need for verification of such simplified ecosystem-level models, for development of understanding of interactions at the vegetation/atmospheric interface, and for appropriate scaling up and adoption of common analytical philosophies among vegetation and atmospheric scientists strongly suggests that such formulations should be constructed in an interdisciplinary setting, e.g., in the case of gas flux models, development together with atmospheric scientists. Further verification, adjustment, and development of GAS-FLUX as a comparative analytical tool requires that structure, leaf-level gas exchange, large chamber experiments, xylem sap flow, and atmospheric estimates of gas exchange with the vegetation be studied simultaneously at selected sites in different climate zones. Illustrated below is the

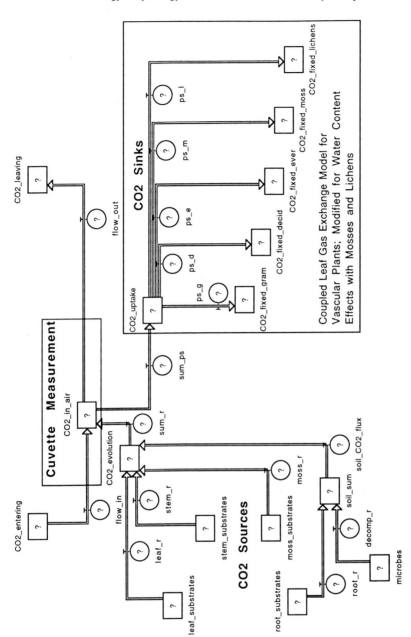

Fig. 21.1. Schematic diagram of the model GAS-FLUX. All CO_2 respiratory sources are indicated on the *left* of the figure, while all sinks for CO_2 due to photosynthesis are indicated on the *right*. *Boxes* indicate pools while *circles* indicate flux rates. Objective of the model is to quantify and sum all CO_2 fluxes. Net exchange or sum of the sinks and sources is apparent in the cuvette shown as the difference in CO_2 flowing through the measurement system, which provides for model verification. *Decomp* Decomposition; *ps* photosynthesis; *gram* graminoids; *decid* deciduous shrubs; *ever* evergreen shrubs

manner in which a simplified "phenomenological" or "aggregate" version of GAS-FLUX has been derived for CO_2 exchange of tundra communities under present climate conditions, and the manner in which it may be extended to consider spatial and temporal variation in ecosystem properties, including those which may result due to long-term climate change.

21.2 Concept and General Structure of the Stand Model GAS-FLUX

The concept of the GAS-FLUX model is extremely simple, as depicted in Fig. 21.1. The objective is to quantify all significant sources of respiratory CO_2 from an ecosystem, to quantify all significant CO_2 sinks as a result of carbon fixation in photosynthesis, and to obtain an estimate of ecosystem net CO_2 exchange by summing these fluxes. In practice, the calculation of each average flux rate shown is not simple, because these on a square meter ground surface basis are influenced by the momentary physiological status of many organisms as well as by microclimate conditions. Both physiological state and microclimate vary temporally, and many interactions occur, especially with respect to physical structure and vertical microclimate gradients. In the case of tundra ecosystems, separate considerations must be made for ecosystem components with differing physiological regulation of CO_2 flux, i.e., vascular plants (separating graminoid elements, deciduous shrubs, and evergreen shrubs), poikilohydric plants (mosses of different types and lichens), and soil microbes. As indicated in Fig. 21.1, verification of calculated net ecosystem CO_2 exchange is possible by comparison with measured net exchange rates obtained in small chambers (approximately 0.1 to $1.0\,m^3$ in size; Oechel et al. 1992) that isolate a three-dimensional block of tundra for short periods.

In the GAS-FLUX model, the vertical structure and microclimate of a definable area or community is considered to be relatively homogeneous. The structure of the model is described in Fig. 21.2. The vascular plant canopy is divided into a series of layers containing leaf, and possibly stem, material with a single type of physiological response (representative species – see discussion of physiological parameterization below). Consecutive layers may have material of different species or of the same species. Aspects of light interception, calculation of energy budget for leaves and estimation of leaf temperature, and calculation of photosynthetic CO_2 uptake as well as stomatal conductance and water loss for leaves in all canopy layers have been discussed previously (Caldwell et al. 1986; Reynolds et al. 1988; Tenhunen et al. 1990). Light interception by both stems and leaves is considered. With respect to gas exchange, stems are considered to remain at air temperature. Light incident on the understory moss is that passing through the lowest layer of the vascular plant canopy. Just as sunlit and

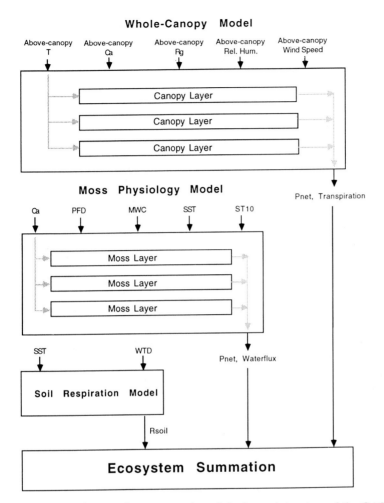

Fig. 21.2. Diagramatic representation of the layered structure of the GAS-FLUX model as applied to tundra plant communities. Processes in vascular plant canopy layers are driven by above-canopy air temperature (T), carbon dioxide concentration (Ca), short-wave radiation (Rg), air humidity, and wind speed and the predicted variables are net photosynthesis and transpiration rates. Processes in the moss layers are driven by Ca, photosynthetic photon flux passing through the vascular plant canopy (PFD), moss water content (MWC), and temperature within the moss layer estimated from soil surface temperature (SST) and temperature at 10 cm depth ($ST10$). Predicted variables are moss net photosynthesis and water loss rates. Soil CO_2 efflux is driven by soil surface temperature (SST) which was found to be best correlated with measured response and by water table (WTD) which influences the volume of respiring material. Summation of the flux rates provides estimates of ecosystem water and carbon balance

shaded leaf area is calculated in each canopy layer, sunlit and shaded portions of the ground surface are calculated. Considering the area of ground covered by *Sphagnum*, average sunlit and shaded gas exchange rates for each moss layer are determined in response to light intensity within the layer estimated from extinction coefficients for compact *Sphagnum* (Clymo and Hayward 1982; their Fig. 8.30 unshaded natural stand-extinction coefficient = −0.75). An average layer rate is obtained by weighting the two fluxes according to sunlit and shaded area. Soil respiratory flux is determined from empirical regression formulations established in field studies (Oberbauer et al. 1991, 1992). Temperature in the moss and upper soil layers is included in the factors influencing gas flux rates and is considered to decrease linearly with depth. Net ecosystem gas exchange is obtained by combining flux rates from all layers of the model. Input variables to the model for each diurnal simulation, for determining layered structure, and as hourly drivers are listed together with output variables of major importance in the Appendix.

The size of a "homogeneous" area to which the model may be applied will vary depending on particular applications and on the type of input data available to support model parameterization. In the application discussed here, structural homogeneity is assumed to occur over the entire area mapped for a particular tundra plant community within the Imnavait Creek watershed near Toolik Lake, Alaska (cf. vegetation map in Walker et al. 1989). Structural as well as functional homogeneity, which depends on momentary stand microclimate, is considered to occur for a ground surface area (pixel size) of at least 10×10 m. At such a scale, spatial patterning in variables, such as remotely sensed normalized difference vegetation index (Stow et al. 1989) and water table depth determined by hydrological models, may eventually be used to derive landscape patterns in tundra ecosystem CO_2 exchange (see below).

Simulations of diurnal courses of CO_2 exchange of three tundra communities in the Imnavait Creek watershed were conducted for clear and overcast days that might occur during the summer on June 15, July 1, July 15, August 1, and August 15. Three vegetation communities were considered, dry shrub tundra, tussock tundra, and wet sedge (riparian) meadow with a high presence of deciduous shrubs. These communities cover on the order of 85% of the surface within the watershed and exist as broad bands or zones along the topographic and water-availability gradients from the ridgetops to streamside within the basin. General structural and microclimatic gradients along this toposequence have been described (Walker et al. 1989; Tenhunen et al. 1992, see also below). Not considered here are water track communities (cf. Hastings et al. 1989) and a series of heath types occurring on the ridges (Hahn 1991).

Input weather conditions were based on extensive monitoring of stand microclimate carried out between 1985 and 1989 and included variation in above- and belowground temperature profiles within the natural range experienced. Humidity profile data were adjusted to be appropriate for the air temperature range imposed during a particular simulation run. Soil temperature profiles were adjusted such that they were compatible with soil moisture conditions in the simulations. Wind speed at the top of the canopy varied over the course of the day, as determined in field studies, and was assumed to decrease exponentially with canopy depth, thus decreasing leaf boundary layer conductance. The exponential decrease was adjusted to be compatible with measured wind profiles between 1 and 200 cm height above ground. Thirty-six sets of input weather data were utilized on each date, corresponding to three belowground moisture conditions (dry, moist, wet), three aboveground weather conditions (clear cold, clear warm, and overcast cold), and four levels of incident radiation. The 180 simulations for each of the three plant communities adequately revealed seasonal trends in potential canopy gas exchange. A 1-h time step was used, although other intervals are accommodated by the model. Iterative solutions were obtained for steady-state gas exchange and canopy microclimate at each hour, and these were integrated separately for sunlit and shaded leaves, for stem respiration, for aboveground vascular plant layers, for moss layers, for the soil compartment, and for the entire community (above- and belowground).

21.3   Structural Inputs to GAS-FLUX Along Water Gradients in Tundra

Biomass harvests were carried out in midsummer (1986 through 1989) after complete development of the vascular plant canopy in nine major vegetation types within the Imnavait Creek watershed (Hastings et al. 1989; Hahn 1991). General trends in the composition of vegetation along the water availability gradient from ridgetop to valley bottom were described by Tenhunen et al. (1992). In several cases, stratified harvests were conducted to determine the vertical distribution of stem and leaf biomass and to sample leaf angles. Based on these field studies, differences in canopy structure between the vegetation types simulated were summarized as shown in Table 21.1 for the mid-season or mature stage of phenological development. Vascular plant biomass is distributed among three functional groups or physiological types, deciduous shrubs, graminoids, and evergreen shrubs. Deciduous shrubs and graminoids occurred in significant quantities in all three vegetation zones, with lowest LAI in tussock tundra at mid-slope and highest LAI in riparian meadows of the lower slope. Stem area index (SAI) of the deciduous shrub component plays a role in light interception in all three communities, but decreases in riparian meadow "aboveground", apparently due to more rapid moss growth. Evergreen shrubs were equally present in upper and mid-slope locations but disappear in moist meadow sites.

Dry shrub and tussock tundra were considered to be very similar in structure with the main difference being that total LAI for deciduous shrubs

Table 21.1. Simplified canopy structural characteristics utilized in simulation of tundra ecosystem carbon dioxide exchange for three community types distributed along the topographic and water availability gradient occurring within the Imnavait Creek watershed. Structure indicated was assumed to occur during the mid-season or mature stage of phenological development from approximately July 15 to August 1

	Dry shrub tundra Slope crest		Tussock tundra Mid-slope		Riparian meadow Lower slope	
	LAI[a] (angle) width	SAI[b] (angle)	LAI (angle)	SAI (angle)	LAI (angle)	SAI (angle)
Deciduous shrub	0.24 (45) 1.0	0.174 (50–80)	0.12 (45)	0.174 (50–80)	0.35 (45)	0.151 (50–80)
Graminoid	0.24 (85) 0.3	–	0.12 (85)	–	0.6 (60–85)	–
Evergreen shrub	0.2 (30) 0.3	0.016 (30)	0.2 (30)	0.016 (30)	–	–
Total vascular Plant LAI	0.68		0.44		0.95	

[a] Leaf area and stem area indices given in $m^2 \, m^{-2}$; average angles from the horizontal in degrees; average leaf widths in cm considered the same in all three vegetation types; all leaves and stems were considered to be alive and nonclustered.
[b] All stems were less than or equal to 6 mm diameter and photosynthetically active.

and graminoid elements was reduced by half in tussock tundra. Harvests did not indicate any changes in either the amount of deciduous shrub stem material or biomass of evergreens between these communities. Evergreen shrubs were always considered to occur below deciduous shrub and graminoid biomass, spreading prostrate above the moss surface. Graminoid leaves were placed in a single layer above the evergreens. Deciduous leaves were distributed equally in eight layers above the graminoid and evergreen leaves. Leaf angle above the horizontal was maintained at 30° for evergreen leaves, at 45° for deciduous leaves, and 85° for graminoids (see also Miller et al. 1980; Berg et al. 1975). Deciduous shrub stem area index (SAI) decreased with height in subsequent canopy layers (eight in total) and elements became more vertically oriented (angle increasing from 50° to 80° with respect to the horizontal). Width of leaves, which influences boundary layer, and thus transpiration and leaf energy balance, was set at 1 cm for deciduous shrubs and 0.3 cm for graminoid and evergreen elements throughout.

Structure in the vascular plant canopy found at lower slope locations was described differently. Leaf and stem materials of deciduous shrubs were considered to be interspersed with graminoid leaf material in ten alternating layers. The uppermost canopy layer was occupied by relatively vertically oriented graminoid leaves (60°). The next lower layer was occupied by relatively vertically oriented twigs (50°) and deciduous shrub leaves (45°). In the subsequent sequence of alternating growth form layers, equal protions of leaves were maintained, but leaf angle of graminoid elements increased to 85° at the bottom of the canopy, stem angle of deciduous shrub material increased to 80° at the bottom, and SAI increased with canopy depth as determined in stratified harvests.

Based on phenological observations at the Imnavait Creek site, we assumed that canopy development was initiated only in mid-June. Therefore, on June 15, LAI of both deciduous shrubs and graminoids was zero in the simulations and carbon fixation was carried out only by the evergreen shrubs and mosses. Except for LAI of deciduous shrubs and graminoids, other structural components were held constant, as described above. For simulations on July 1 and August 15, LAI for deciduous shrubs and graminoids was increased to one half of that shown in Table 21.1. On the first date, this is the result of rapid leaf expansion occurring and on the latter date due to the onset of senescence as days become shorter. While variation in phenological pattern occurs depending on climate conditions in any particular year, the approach taken here adequately illustrates the critical manner in which phenological events influence the carbon balance of tundra ecosystems (Sect. 21.5.2; Fig. 21.8).

21.4 Ecophysiological Inputs to GAS-FLUX Along Water Gradients in Tundra

21.4.1 CO_2 Exchange of Vascular Plant Species of Differing Growth Forms

Photosynthetic response at the individual leaf level of vascular plants of the three growth forms discussed in Section 21.3 was described with equations proposed by Farquhar et al. (1980), based on ribulose-1,5-bisphosphate carboxylase-oxygenase (Rubisco) kinetics as mediated by (1) the concentrations of competing gaseous substrates, CO_2 and O_2, and (2) the ratio of ribulose-1,5-bisphosphate (RuBP) concentration to enzyme active sites. CO_2 assimilation is integrated with an empirical description of stomatal conductance (Ball et al. 1987). The specific equations as well as methods for application of the resulting leaf gas exchange model have been discussed in detail by Harley and Tenhunen (1991).

In order to use the model with tundra species, leaf gas exchange of representative deciduous shrub, graminoid, and evergreen shrub species was monitored with a Li-Cor portable gas exchange system (LI-6200) over 24-h periods and at a number of locations within the Imnavait Creek watershed (Gebauer et al., in prep.). Care was taken to obtain comparative measures of photosynthetic response to prevailing weather conditions by monitoring an

July   23,   1988

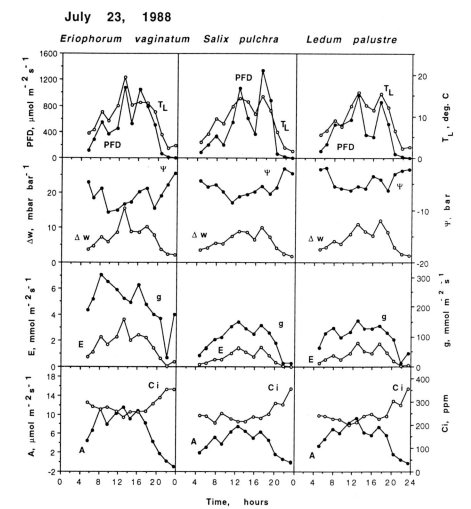

Fig. 21.3. Diurnal time course of gas exchange of *Eriophorum vaginatum*, *Salix pulchra*, and *Ledum palustre* leaves observed on July 23 1988. Shown are the naturally occuring time courses for photosynthetic photon flux density (*PFD*) incident on the leaves in the gas exchange cuvette, leaf temperature (T_L), water vapor mole-fraction difference between leaf air spaces and external chamber air (Δw), leaf xylem water potential (Ψ), total leaf conductance for water vapor (*g*), leaf transpiration rate (*E*), leaf net photosynthesis rate (*A*), and calculated internal leaf CO_2 concentration (C_i). (Gebauer et al. unpubl.)

equal number of leaves of the different growth form types during each observation period. Results from a typical experiment comparing *Eriophorum
vaginatum*, *Salix pulchra*, and *Ledum palustre* during July 1988 are shown in
Fig. 21.3. Transpiration rate under cuvette conditions and net photosynthesis
rate responded sensitively to changes in light and leaf temperature. Incident

light at the top of the plant canopy typically fluctuates between a photon flux density (PFD) of 1000 and 2000 μmol m^{-2} s^{-1} during the middle of clear days, while temperature varies from near zero °C to approximately 15 °C. Nevertheless, much higher temperatures of 25 to 30 °C may occur during warm weather periods. Transpiration rate and leaf conductance decrease to very low values between 2200 and 0400 h primarily in response to low incident PFD (below 10 μmol m^{-2} s^{-1}). Net photosynthesis becomes negative during these periods. Similar data have been reported by many researchers working in arctic regions (Stocker 1931; Kjelvik et al. 1975; Johansson and Linder 1975). The advantage provided here is that a truly comparative data base was constructed using plants of different growth forms growing adjacent to one another.

As seen in Fig. 21.3, maximum leaf conductance of the graminoid is approximately twice as large during midday as that observed for the deciduous and evergreen shrub species (300 versus 150 mmol m^{-2} s^{-1}). A difference in maximum net photosynthesis rate is also apparent, although this is not as large (10 versus 7 μmol m^{-2} s^{-1}). By pooling data from many experiments, verification data sets are obtained for the leaf gas exchange model that permit us to describe these individual species responses over a wide range of conditions.

Following the procedures outlined by Harley and Tenhunen (1991), parameter values for the gas exchange model were obtained for tundra species as indicated in Table 21.2 (see Harley and Tenhunen 1991 or Appendix of Tenhunen et al. 1990 for parameter definitions). Differences among species were easily described with only small changes in $C(P_{ml})$ and $C(V_{cmax})$, scaling parameters which describe the capacities for RuBP regeneration and enzymatic CO_2 fixation respectively. Parameters determining temperature response of photosynthesis and respiration were adjusted to obtain the best fits to verification data sets and to obtain Q_{10} values comparable to those generally found for arctic species (Limbach et al. 1982; Semikhatova et al. 1992). Differences in temperature response among species and growth forms included in the model at present are small, but this may easily be changed as appropriate generalizations as these ecophysiological responses are defined for particular sites.

The values shown (Table 21.2) for deciduous shrubs, graminoids, and evergreen shrubs are compatible with measured data for (1) *Betula nana*, (2) *Eriophorum vaginatum* and *Carex bigelowii*, and (3) *Ledum palustre* and *Vaccinium vitis-idaea*. However, response of the deciduous shrub *Salix pulchra* was found to be similar to the evergreen *Ledum palustre* with lower photosynthetic capacity than that of *Betula nana*. This suggests that generalizations based on functional groups must be made with caution. Due to the physiological differences established for *Betula* and *Salix*, canopy structure was described in such a manner that contributions of the two major deciduous shrubs were independently estimated. In other words, this is the reason for utilizing eight deciduous shrub layers above the graminoid

Table 21.2. Estimates of leaf gas exchange model parameters utilized in simulation of tundra ecosystem carbon dioxide exchange for three community types distributed along the topographic and water availability gradient occurring within the Imnavait Creek watershed. Parameters were derived by fitting the model to observed data obtained from the species described in the text. Parameters indicated were assumed to be valid during the mid-season or mature stage of leaves from approximately July 15 to July 31. For parameter definitions, see Harley and Tenhunen (1991) or Appendix of Tenhunen et al. (1990)

Parameter	Value				Units
	Deciduous shrub	Graminoid	Evergreen Shrub	Moss	
C (P_{ml})	10.9 (10.0)[a]	11.5 (11.0)	10.4 (9.8)[b]	f (H_2O)	–
ΔH_a (P_{ml})	33559	*	*	36206	J mol^{-1}
ΔH_d (P_{ml})	250000	*	*	100000	J mol^{-1}
ΔS (P_{ml})	805	*	*	333	J K^{-1} mol^{-1}
α	0.06	*	*	0.00012	mol CO_2/mol photons
f (R_d)	20.4	*	*	26.5	–
E_a (R_d)	51178	*	*	67145	J mol^{-1}
f (K_c)	31.95	*	*	*	–
E_a (K_c)	63500	*	*	*	J mol^{-1}
f (K_o)	19.61	*	*	*	–
E_a (K_o)	35000	*	*	*	J mol^{-1}
f (t)	-3.9489	*	*	*	–
E_a (t)	-28990	*	*	*	J mol^{-1}
C (V_{cmax})	34.3 (29.8)	35.0 (30.3)	32.3 (29.5)	f (H_2O)	–
ΔH_a (V_{cmax})	65000	*	*	*	J mol^{-1}
ΔH_d (V_{cmax})	250000	*	*	*	J mol^{-1}
ΔS (V_{cmax})	600	*	*	*	J K^{-1} mol^{-1}
Gfac	30	*	*	Not applicable	

* Indicates that parameter value is the same as indicated for deciduous shrubs.
[a] Value in parentheses is that applied to young and senescing leaves.
[b] Value for young leaves of *Salix*, not for evergreen shrubs – see explanation in text.

and evergreen elements in dry shrub and tussock tundra simulations. Average harvested leaf area for each species was distributed into four layers and these were then considered to be alternately interspersed. For simulations in wet meadow locations, the physiology established for *Salix* = *Ledum* was used together with that for graminoids, i.e., for *Eriophorum*.

Rapid phenological development during the Arctic summer influences not only the amount of leaf area exposed in the vascular plant canopy for carbon dioxide fixation (Sect. 21.3), but also photosynthetic capacity. Physiological effects are included in GAS-FLUX via temporal change in the parameters $C(P_{ml})$ and $C(V_{cmax})$. For the simulation on July 1 and August 15, these parameters were set at the values given in parentheses in Table 21.2 to describe the response of both young and senescent leaves. The values indicated are those that decrease photosynthetic capacity by approximately one half of that obtained during mid-season.

In the vegetation of Imnavait Creek watershed, deciduous shrubs are low in stature with stems of small diameter. Chlorophyll containing tissues in the bark of such shrubs may be important in providing for energy needs. Gas exchange of twigs and stems may be quantified using the approach described above for leaves. In the present version of GAS-FLUX, however, a simpler regression model for gas exchange of stems of the alpine shrub *Rhododendron ferrugineum* has been adopted from Siegwolf (1987). Stem gas exchange is a function of incident light on the stems in each layer, the air temperature of stems in the layer, and average stem diameter for the layer. Photosynthetic capacity of stems decreases with age. Stem photosynthesis in our simulations is large enough to approximately compensate for stem dark respiration.

21.4.2 CO_2 Exchange of Poikilohydric Plants

While GAS-FLUX at present includes only *Sphagnum* spp. as a poikilohydric plant element, it is constructed to simultaneously consider several poikilo-hydric functional groups distributed below the vascular plant canopy. Never-theless, *Sphagnum* alone is an extremely important tundra ecosystem structural component (Tenhunen et al. 1992), making up a major portion of the living biomass (Hastings et al. 1989), modifying soil characteristics and the accumulation of ground ice, strongly influencing habitat characteristics, and modifying biomass allocation and production in vascular plant species (Luken et al. 1985; Tenhunen et al. 1992). Modeling of CO_2 gas exchange of *Sphagnum*, as well as that of other poikilohydric plants, is made difficult by the dynamic response to multiple factors; light, temperature, CO_2, and water content. In addition, apparent long-term photoinhibitory effects can greatly modify photosynthetic capacity (Murray et al. 1993), such that the ecosystem role and vitality of *Sphagnum* is very dependent on structure within the plant community where it occurs (Murray et al. 1989b; Tenhunen et al. 1992).

Given the important role of *Sphagnum* in tundra ecosystems and recent research emphases on climate change effects, it seemed essential that the mechanistic approach of Farquhar et al. (1980) be utilized to describe gas exchange response of this ecosystem component. Derivation of the parameters in this model are based on analysis of the net photosynthesis de-pendency on carbon dioxide concentration. Experiments of the type shown in Fig. 21.4B were undertaken in which the carbon dioxide response curve of net photosynthesis was determined at a series of *Sphagnum* water con-tents. In such experiments, the *Sphagnum* shoots were vertically oriented but separated to ensure uniform lighting of the tissues. Air relative hunidity was maintained near saturation such that water content decreased slowly

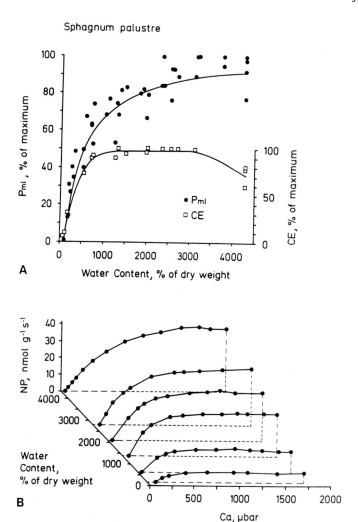

Fig. 21.4. A Estimates of the CO_2- and light-saturated rate of net photosynthesis (*Pml*) and the initial slope of the carbon dioxide response curve (*CE* carboxylation efficiency) at 15 °C as affected by tissue water content in *Sphagnum palustre*. Values are derived from curves shown in **B. B** Net photosynthesis (*NP*) response of *Sphagnum palustre* from foothills tundra north of the Brooks Range, Alaska, to air CO_2 level (*Ca*) and water content. Ca dependencies were measured in the laboratory on a sequence of days during drying (each response curve on a separate day). Incident PFD ($200\,\mu\text{mol}\,\text{m}^{-2}\text{s}^{-1}$) was close to saturation, but was kept low to avoid photoinhibition. Air temperature was 15 °C. (Tenhunen et al. 1992)

over several days. Thus, CO_2 could be varied over the range shown during a period when water content decreased very little. In *Sphagnum*, the maximum rate of potential carbon fixation at CO_2 saturation decreases as water content decreases (Fig. 21.4A). Apparent carboxylation efficiency (the initial

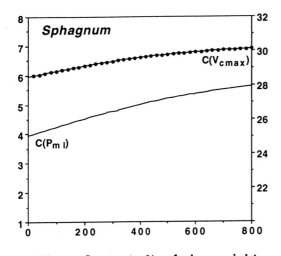

Water Content, % of dry weight

Fig. 21.5. Water content dependencies established as described in the text for the model parameters $C(P_{ml})$ (*left axis*) and $C(V_{cmax})$ (*right axis*) which scale electron transport capacity and carboxylation capacity for *Sphagnum*. Units given in Table 21.2

slope of the CO_2 response curve) stays constant between water contents of 700 and 3000% of dry weight and also decreases at lower water contents.

In order to simulate carbon dioxide fixation by *Sphagnum*, the water content dependency of RuBP regeneration rate or of its scaling constant $C(P_{ml})$ was determined (Fig. 21.5) such that it is compatible with water content effects on the CO_2-saturated fixation rate at 15 °C (Fig. 21.4A). The temperature dependency parameters for P_{ml} (ΔH_a, ΔH_d, and ΔS) were adjusted to fit a temperature response curve of CO_2-saturated fixation rate at high water content (2000 to 3000% of dry weight). Assuming that the enzymatic characteristics of carboxylase in *Sphagnum* are the same as in other Arctic and C_3 plant species, the water content dependency of carboxylation capacity [$C(V_{cmax})$] was derived (Fig. 21.5) from water content influences on carboxylation efficiency (Fig. 21.4A). Final parameter values given in Table 21.2 for *Sphagnum* were obtained with minor adjustment such that light and temperature and water content responses of net photosynthesis under ambient CO_2 conditions were reproduced (Harley et al. 1989; Murray et al. 1989a). Parameters for dark respiration were determined directly from measurements of carbon dioxide exchange in the dark (Harley et al. 1989). The low value of α occurs because the light utilization efficiency is a function of the geometrical orientation of the material. The value given is appropriate for tightly packed *Sphagnum* shoots in a layer approximately 1 cm thick.

As described above, the net carbon dioxide exchange rate ($nmol\,g^{-1}\,s^{-1}$) for an approximately $1\,cm^3$ sample of *Sphagnum* is obtained at a particular value of light, temperature, $[CO_2]$, and water content. Community flux rate for the moss component was obtained by multiplying by the bulk density and ground area covered by *Sphagnum* (40, 50, and 60% cover in dry shrub, tussock, and riparian tundra respectively). This was done for layers 1 cm in thickness, determining the average light and temperature values for each layer as described in Section 21.2 and calculating rates for sunlit and shaded *Sphagnum* separately. Water content of the moss was varied from 750, to 500, and to 250% of dry weight in dry shrub tundra for wet, normal, and dry conditions based on field observations. The corresponding values for tussock tundra and riparian meadow were 1100, 750, and 500% of dry weight. Gradients in $[CO_2]$ within moss cushions were not considered, rather $[CO_2]$ was maintained in all simulations at ambient concentration (340 ppm).

21.4.3 CO_2 Exchange of the Soil

The integrated response of soil microbial and root respiration is approximately equal in magnitude to total canopy carbon fixation. Since one objective of the GAS-FLUX model in applications for tundra ecosystems is to understand controls on overall ecosystem carbon balance, relatively extensive monitoring studies were undertaken to examine carbon dioxide flux rates from soils of different communities and to correlate these flux rates with soil environmental factors (Oberbauer et al. 1991, 1992). The results of these studies indicated that CO_2 efflux from tundra communities with deep organic soils, such as those for which simulations have been conducted, varies primarily in response to changes in the depth to water table and/or soil moisture and changes in soil temperature. Presumably the mechanism controlling CO_2 loss in response to depth to water table is related to oxygen diffusion limitation. Correlations between soil moisture and CO_2 efflux were very high, but explained less of the site variability in CO_2 effluxes than depth to water table. Soil temperature is a factor which modifies the respiration rates permitted by prevailing water table or thaw depth. Since seasonal increases in mean soil temperature are small, the primary effect of temperature is seen in diurnal changes in CO_2 efflux.

   The CO_2 efflux data were fit to an empirical model incorporating the Arrhenius function for temperature and an asymptotic function for water table depth or thaw depth (Oberbauer et al. 1992). An asymptotic model was chosen based on preliminary examination of the data, which suggested that at depths greater than 10 cm, depth to water table had little effect. This simple mathematical description for CO_2 efflux in response to soil environmental factors was included as a subroutine of GAS-FLUX. In simulations, soil water content to 10 cm depth was assumed to be the same as that of *Sphagnum* (See Sect. 21.4.2 – only very small differences were found in the field). Permafrost at shallow depth that might reduce respiring soil volume occurs only very early in June, and thus was not included here as a factor influencing tundra carbon balance. High water tables that result in decreased soil respiration were observed only in conjunction with the wettest conditions in tussock tundra and riparian meadow. In the results shown below, high water table situations decreased soil respiration by approximately 50% when moss and soil water content was equal to 1100% of dry weight.

21.5 Simulations of Ecosystem CO_2 Exchange

21.5.1 Diurnal Course of Gas Exchange of Major Tundra Structural Components

Examples of the diurnal course for carbon dioxide exchange of vascular plant leaves, *Sphagnum* moss, vascular plant twigs and stems, and of the soil are presented additively in Fig. 21.6, with net carbon fixation by leaves and moss shown as positive fluxrs and respiration of moss, twigs, and soil shown as negative fluxes. The left panel illustrates the simulated results for a clear, warm day at midseason with moist soil (water content 750% of dry weight), the right panels for an overcast day. The leaves of vascular plants fix

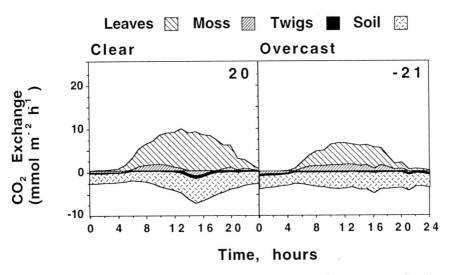

Fig. 21.6. Simulated CO_2 gas exchange of tussock tundra vegetation on a warm clear day at midseason. The fluxes for leaves, *Sphagnum* moss, twigs, and soil are presented additively, with carbon fixation by leaves and moss shown as positive fluxes and respiration of moss, twigs, and soil shown as negative. The *left panel* illustrates simulations for a clear day at midseason, the *right panel* for an overcast day. Soil was considered to be moist and well aerated. *Sphagnum* water content was set at 750% of dry weight. The integrated total daily CO_2 flux ($\text{mmol m}^{-2}\text{d}^{-1}$) is shown for each situation in the *upper right corner* of each graph

considerably more carbon than mosses because they have a relatively high photosynthetic capacity and are exposed to high light. Lower light on overcast days reduces carbon fixation by leaves. Respiratory losses from twigs are unimportant, for most twigs of tundra species are small in diameter and fix enough carbon to compensate for their own respiration. *Sphagnum* carbon dioxide exchange is strongly influenced by variation in light and temperature during the day (at lower temperatures, moss maintained positive photosynthesis despite lower light on the overcast day), but most strongly by long-term changes in soil water content (not shown). Although the model predicts ecosystem net carbon dioxide exchange (diurnal sum of all components shown in upper right of each panel) to be positive on clear days, the carbon balance is sensitive to daily changes in environmental conditions, with carbon fixation by both leaves and *Sphagnum* reduced enough on overcast days so that the daily integrated carbon dioxide flux may be negative (cf. Grulke et al. 1990).

"Coarse scale data", i.e., measured diurnal course data on net ecosystem carbon dioxide exchange of the tussock tundra plant community within the Imnavait Creek watershed assessed with small chambers as indicated in Fig. 21.7, were gathered to examine potential "aggregation error" in the model (cf. Rathstetter et al. 1992) during summer 1990 (Gillespie et al. 1993). Chambers were placed in the field to observe both individual tussocks and intertussock areas separately, and community gas exchange was observed in both light and dark chambers. Subtraction of the measured dark CO_2 efflux from net exchange rates provided estimates of total carbon fixation or "CO_2 uptake". Diurnal changes in net ecosystem CO_2 uptake were strongly correlated with diurnal changes in PFD. A scatter plot of measured CO_2 uptake values versus incident PFD at the time of measurement indicates that uptake is sensitive to PFD below $500 \, \mu mol \, m^{-2} s^{-1}$ but tends toward saturation at higher flux levels (Fig. 21.7). In agreement with differences in LAI, CO_2 uptake in intertussock areas was in general lower than that of tussocks. We compared the estimated uptake rates with the net aboveground flux rate predicted by GAS-FLUX. As seen in Fig. 21.7, the GAS-FLUX model similarly predicts large influences of PFD below $550 \, \mu mol \, m^{-2} s^{-1}$ and uptake fluxes on the same order of magnitude as those observed when soil respiration is "removed". The differences in CO_2 uptake observed between tussock and intertussock areas suggest that while the model is reasonably calibrated (sensu Rathstetter et al. 1992) for the Imnavait Creek watershed, these independent calibration measurements are an important consideration in each individual application of the model. Tussock density and development are factors that may deserve further consideration. While "calibration measurements" were not undertaken in dry shrub tundra and riparian meadow vegetation, they are highly desirable.

Billings et al. (1982) found daily net carbon dioxide exchange of coastal tundra microcosms to vary between -2 and $+10 \, g \, CO_2 \, m^{-2} \, d^{-1}$, the highest values being obtained with a water table near the ground surface, i.e.,

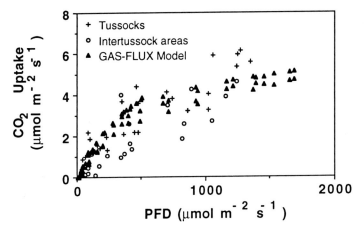

Fig. 21.7. Scatter plot of estimated CO_2 uptake of *Eriophorum vaginatum* tussocks and of intertussock areas as a function of PFD at the time of measurement. Comparison with the mean tussock tundra ecosystem uptake estimated for GAS-FLUX under a variety of weather conditions. (Gillespie et al., unpubl.)

with little soil CO_2 efflux. The apparent major difference between tussock tundra and coastal wet tundra is that carbon dioxide efflux in the dark is considerably greater in tussock tundra areas. Nevertheless, Oberbauer et al. (1992) found ecosystem CO_2 efflux in the dark in riparian meadow areas of the Imnavait Creek watershed that was equal in magnitude to that found in tussock tundra when water tables fell to low levels. The estimated carbon dioxide uptake rates as well as the range of daily net carbon dioxide exchange values predicted by GAS-FLUX agree well with both studies (considering riparian meadow as comparable with coastal tundra). The diurnal time courses of CO_2 efflux shown in Fig. 21.6 are similar to those measured by Oberbauer et al. (1992). The general conclusion from studies at the Imnavait Creek watershed (Oberbauer et al. 1991, 1992; Gillespie et al. 1993) is that respiratory capacity is determined in tundra organic soils by the depth of aeration, that water content usually remains available to support soil microorganisms, and that instantaneous rates over the diurnal course are a function of the prevailing soil temperature profile. Such behavior is simulated quite well by the soil respiration subroutine of GAS-FLUX.

21.5.2 Environmental Effects on Diurnal CO_2 Exchange and Aggregate Formulations

While many applications of an aggregate version of GAS-FLUX may be visualized, one goal was to provide a driving function for plant growth simulations within the spatial confines of the Imnavait Creek watershed. An appropriate time step in growth simulations is daily, since both ecosystem feedback effects, i.e., changes in nutrient availability, and structural change, i.e., phenology, that influence carbon dioxide uptake and growth occur on such a time scale (Reynolds and Leadley 1992). It is essential in the aggregate model to separate carbon fixation from CO_2 efflux, since the former is directly related to the growth process. At present the entire community net aboveground carbon dioxide flux is treated as a lumped variable (comparable to the data in Fig. 21.6). Including carbon dioxide efflux from the soil is important, since there is much interest in the potentials for carbon storage in tundra soils, especially as related to climate change. An aggregate "semi-mechanistic" model such as the aggregate GAS-FLUX may be very useful in examining climate effects on tundra carbon balance. Finally, the separation of above- and belowground carbon dioxide flux rates is essential in the aggregate GAS-FLUX, since they are sensitive to different microclimatic driving variables.

The response of integrated diurnal aboveground net carbon dioxide exchange (labeled as uptake, since adequate light and photosynthetic capacity are available on all days to permit aboveground net uptake) to integrated diurnal PFD input is shown for the three vegetation communities in Fig. 21.8. Simulations conducted for differing stages of phenological development are indicated with different symbols. Thirty-six symbols of each of five types (five different dates) are plotted in each panel of Fig. 21.8. The 36 diurnal simulations extensively sample the range of environmental conditions occurring in the Imnavait Creek watershed. From Fig. 21.8, the major factors influencing aboveground CO_2 uptake are easily recognized. At any point in time, diurnal CO_2 uptake (U) in response to integrated PFD may be described with a relatively simple hyperbolic function of the type described by Smith (1937):

$$U = \frac{sI}{\left(1 + \frac{s^2I^2}{U_{max}^2}\right)^{1/2}},\tag{1}$$

where I is integrated incident photosynthetic photon flux density, s is the initial slope of the curve relating CO_2 uptake to irradiance, and U_{max} is the maximum potential uptake rate of the community for a particular stage in phenological development (point in time). U_{max} may not actually be attainable, since radiation input is limited by daylength and sun angle.

Along with radiation input, phenological development of the vegetation strongly affects CO_2 uptake. Three light response curves are in general evident in each panel of Fig. 21.8 which are the result of both structural and physiological aspects of phenology included in the model (Sects. 21.3 and 21.4). Lowest carbon fixation potential occurs before leaves expand and the moss cushions grow. Carbon fixation potential on July 1 and August 15 are approximately equal due in the first case to changes in the developing canopy and later to the onset of senescence. During a short period at mid-season, carbon fixation potential remains high. Scatter of the symbols of each type at a constant light value is the result of aboveground temperature

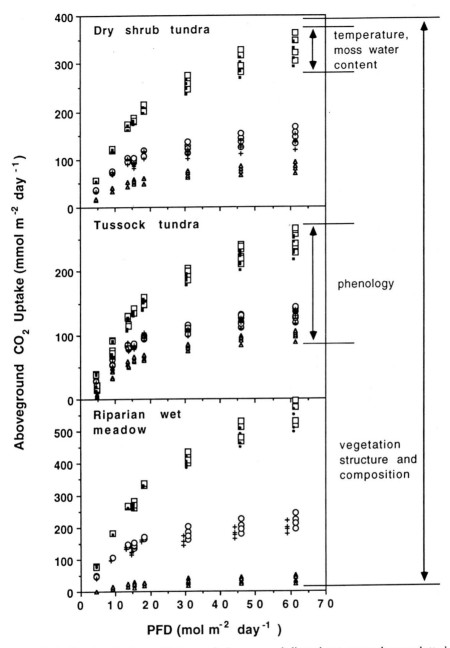

Fig. 21.8. Simulated values of integrated aboveground diurnal net gas exchange plotted versus total integrated PFD incident on the plant canopy. Results are shown for three tundra communities and at five times during the summer growing season (differing symbols) as discussed in the text. *Triangles* are for young leaves on June 15, *open circles* for young leaves on July 1, *open squares* for mature leaves on July 15, *small closed squares* for mature leaves on August 1, and + symbols for senescing leaves on August 15. Thirty-six symbols of each of five types are plotted in each panel. The 36 diurnal simulations extensively sample the range of environmental conditions occurring in the Imnavait Creek watershed with respect to variation in temperature, humidity and moss water content

Table 21.3. Parameter values for the initial slope of the curve (s) relating diurnal CO_2 uptake to integrated irradiance and for maximum potential uptake rate of the community (U_{max}) for three community types distributed along the topographic and water availability gradient occurring within the Imnavait Creek watershed. Parameters are given for the average response at five times during the season, thus eliminating variation due to aboveground temperature and moss water content

	Dry shrub tundra Slope crest		Tussock tundra Mid-slope		Riparian meadow Lower slope	
	s	U_{max}	s	U_{max}	s	U_{max}
June 15	6.3	106	4.5	104	1.4	41
July 1	14	182	7.0	136	12.7	218
July 15	18.9	472	9.8	270	20.1	603
August 1	19.0	445	9.6	258	19.9	573
August 15	13.7	160	6.7	134	11.4	195

and moss water content effects on photosynthesis. Both factors appear relatively unimportant over the range of conditions observed during 1985 through 1990, at least in comparison to radiation input and seasonal phenology. Parameter values for Eq. (1) are given for each simulation date and for each community in Table 21.3.

"Apparent photosynthesis" and total ecosystem respiration rates measured in a wet meadow at Hardangervidda, Norway (Kjelvik et al. 1975) compare reasonably with the rates shown in Fig. 21.8, as do gas exchange rates of heavily grazed *Salix glauca* stands at Upernaviarsuk, Greenland (Eckardt et al. 1982). Estimated daily canopy net carbon exchange (aboveground balance separate from soil respiration) of *Salix glauca* stands at Upernaviarsuk ranged up to $350\,mmol\,CO_2\,m^{-2}\,d^{-1}$ with LAI of approximately 1.0 and increased to greater than $450\,mmol\,CO_2\,m^{-2}\,d^{-1}$ with LAI of the mature stand equal to approximately 2.0. While direct comparison with published studies is difficult due to the complexity of factors influencing response, the somewhat greater efficiency of carbon fixation in GAS-FLUX might be (1) the result of over-estimation of available PFD due to homogeneity assumptions above a flat surface, (2) due to flaws in the radiation interception and radiation balance models, (3) due to underestimates of aboveground respiration fluxes, or (4) due to other site differences affecting physiological response of the vascular plants. The need for further validation studies, especially with large data sets providing information on variation in integrated diurnal fluxes, is apparent.

The driving variables for predicting diurnal soil CO_2 efflux in longer-term (seasonal) applications of the aggregate model are daily values for water table depth (constant over the day) and mean temperature of the soil at 1 cm depth (cf. Oberbauer et al. 1992). Results from the simulations indicate that depending on water table depth, 24-h soil CO_2 efflux may vary from 0 to $200\,mmol\,m^{-2}\,d^{-1}$. With a fully aerated profile in tussock tundra, diurnal

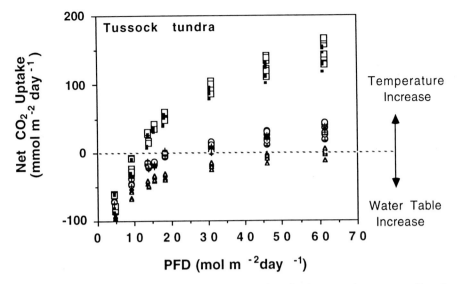

Fig. 21.9. Diagramatic presentation of simulated values for integrated ecosystem diurnal net gas exchange plotted versus total integrated PFD incident on the plant canopy. Results are shown for tussock tundra at five times during the summer growing season and for a relatively high water table. The shifts in ecosystem carbon balance (evidenced as shift in the location of the zero point on the y-axis) which would result due to temperature or water table increase are indicated by *arrows*. Symbols as in Fig. 21.8

soil CO_2 efflux was approximately $170 \, mmol \, m^{-2} \, d^{-1}$ with a mean soil temperature of $4 \, °C$ and increased to $210 \, mmol \, m^{-2} \, d^{-1}$ at $8.5 \, °C$. The effect of soil CO_2 efflux on diurnal tundra carbon balance in the aggregate version of GAS-FLUX may be visualized as shown in Fig. 21.9. In any particular community, the effect of soil respiration is to shift the zero point of the y-axis in Fig. 21.8, depending on daily belowground microenvironment. Increasing temperature raises the zero or break even point and reduces net CO_2 uptake, while increasing water table has the opposite effect. Thus, overall diurnal carbon balance is negative or positive depending on the interplay of incident light, soil temperature, water table, and aboveground phenological development.

21.6 Conclusions: Future Directions of GAS-FLUX Development

While considerable progress has been made in synthesizing results of vegetation and ecophysiological studies within the Imnavait Creek watershed, the model GAS-FLUX could be significantly improved with acquisition of further verification data sets. Such data sets include organ level observations of stem respiration, the monitoring of gas exchange of whole moss cushions, observations of whole-canopy and moss-cushion water loss, and further measurements of community carbon dioxide exchange in the light and dark. Extremely detailed studies of both lichen community structure, microclimate, and CO_2 gas exchange behavior in four ridgetop heath sites (Hahn 1991) provide the opportunity to extend the model to such vegetation units of the watershed. To our knowledge, studies of ecosystem carbon dioxide exchange at the large plot scale and examinations of temporal changes simultaneously in aboveground and belowground gas exchange processes have not been carried out in heath vegetation. These studies are made difficult by the rapid changes in water content of poikilohydric plants that occur in the open vegetation and correlated change in physiological response. Recent improvements in measuring techniques allow us to develop a realistic picture of the dynamic changes occurring and the resulting data help establish criteria for improvement of the model.

The driving variables for predicting ecosystem net carbon dioxide exchange, incident light, soil temperature, water table, and aboveground phenological development, are all variables for which spatial patterns in the landscape may be defined as a function of time. Radiation input is predictable from physical models. Our field studies provide reasonable information on soil temperature shifts along topographic and vegetation gradients within the Imnavait Creek watershed and on the relationship of soil temperature to water table depth (which is related to watershed hydration and water movement). Water table depth is a critical predictor of carbon balance which apparently may be described spatially by further development of a terrain-based hydrological model (Ostendorf and Reynolds 1993). Patterns in phenological development have been observed but may also be definable from time-dependent changes in vegetation index sensed remotely. Thus, it may be possible to link the aggregate version of GAS-FLUX to routines providing the driving variables and mapping carbon balance results for the entire watershed in a dynamic fashion on a daily time step. This would result in an interesting opportunity to view carbon exchange properties both with fine scale resolution ($10 \times 10\,m$ pixels) and at the integrated landscape level. The approach to an aggregate model description described above is a pixel-based approach. Whether such an approach is adequate to link simple descriptions of carbon uptake to simulations via other models of growth and storage in plant biomass will require further study.

While there are many pitfalls in both scaling up and in extrapolation of results, tundra ecosystems, due to their unique structure, offer us the possibility to verify models at several scales. Tundra is one of the few natural ecosystems, perhaps the most important one to date, for which ecosystem response to long-term elevated CO_2 exposure has been examined (Oechel and Billings 1992). The GAS-FLUX model is a simulator based on mechanistic descriptions, such that essential and basic characteristics of the response to elevated CO_2 are included. While ecosystem feedback effects in response to elevated CO_2 that potentially lead to redistribution of nutrients and modification of structure are beyond the present scope of this model, the equations used to describe CO_2 exchange are expected to be generally applicable. Furthermore, as demonstrated, the aggregate GAS-FLUX model is compatible with including aspects of vegetation change and can be developed to describe the response of a variety of community types under elevated CO_2 environmental conditions, including the effects of long-term temperature increase and precipitation effects on water table.

Although tundra ecosystems are unique in certain characteristics, such as limited rooting volume and extremely cold soil temperatures due to permafrost, most aspects of the GAS-FLUX model are of a general nature, making it a useful tool for analysis in other ecosystems. Similar types of data are required to establish parameter values, to drive the model, and to validate output. Present research efforts are directed toward model applications in temperate and mediterranean forests and shrublands. In these situations, stress factors other than those found in tundra modify physiological response, i.e., stomatal closure in response to decreased water availability or due to cold temperature in conifers. As a result of these species-specific behaviors, it is possible that aggregate versions of the model may become more complex and that other approaches are required. In forest situations, small chambers provide important information on understory fluxes, while xylem sap flow methods and eddy correlation techniques potentially provide validation for the overstory and at the stand level. Potential to evaluate the impact of pollutants on mosses and to quantify moss physiological response is an important new component of the model, since mosses in terms of pollutant deposition provide the last buffer between atmosphere and microbial and root responses in the soil. Models such as GAS-FLUX will help us to describe physiological controls on exchange rates, to understand the partitioning of total flux to the many sinks in the ecosystem (both for carbon dioxide and gaseous pollutants), and to assess potential impacts or response at different locations within the ecosystem structure. GAS-FLUX will help us demonstrate a number of ways in which ecophysiological information may be used to better understand relevant responses of ecosystems important to us on a daily basis.

Acknowledgments. This research was supported by the US Department of Energy R4D Program, the Bayreuth Institute for Terrestrial Ecosystem Research, and the German Federal Ministry for Research and Technology (grant no. BEO-0339476A).

References

Ball JT, Woodrow IE, Berry JA (1987) A model predicting stomatal conductance and its contribution to the control of photosynthesis under different environmental conditions. In: Binggins I (ed) Progress in photosynthesis research, vol IV.5. Proc VII Int Photosynthesis Congr. Nijhoff, Dorrdrecht, pp 221–224

Berg A, Kjelvik S, Wielgolaski FE (1975) Measurement of leaf areas and leaf angles of plants at Hardangervidda, Norway. In: Wielgolaski FE (ed) Fennoscandian tundra ecosystems. Part 1. Plants and microorganisms. Ecological Studies 16. Springer, Berlin Heidelberg New York, pp 279–286

Billings WD, Luken JO, Mortenson DA, Peterson KM (1982) Arctic tundra: a source or sink for atmospheric carbon dioxide in a changing environment? Oecologia 53: 7–11

Billings WD, Luken JO, Mortenson DA, Peterson KM (1983) Increasing atmospheric carbon dioxide: possible effects on arctic tundra. Oecologia 58: 286–289

Bolin B (1988) Linking terrestrial ecosystem process models to climate models. In: Rosswall T, Woodmansee RG, Risser PG (eds) Scales and global change. John Wiley and Sons, New York, pp 109–124

Caldwell MM, Meister HP, Tenhunen JD, Lange OL (1986) Canopy structure, light microclimate and leaf gas exchange of *Quercus coccifera* L. in a Portuguese macchia: measurements in different canopy layers and simulations with a canopy model. Trees 1: 25–41

Chapin FS, Jefferies RL, Reynolds JF, Shaver GR, Svoboda J, Chu EW (eds) (1992) Arctic ecosystems in a changing climate. Academic Press, New York, 469 pp

Clymo RS, Hayward PM (1982) The ecology of *Sphagnum*. In: Smith AJE (ed) Bryophyte ecology. Chapman and Hall, London, pp 229–289

Coyne PI, Kelley JJ (1978) Meteorological assessment of CO_2 exchange over an Alaskan arctic tundra. In: Tieszen LL (ed) Vegetation and production ecology of an Alaskan arctic tundra. Ecological Studies 19. Springer, Berlin Heidelberg New York, pp 299–319

Dickenson (1988) Atmospheric systems and global change. In: Rosswall T, Woodmansee RG, Risser PG (eds), Scales and global change. Wiley, New York, pp 57–80

Eckardt FE, Heerfordt L, Jørgenson HM, Vaag P (1982) Photosynthetic production in Greenland as related to climate, plant cover, and grazing pressure. Photosynthetica 16: 71–100

Farquhar GD, von Caemmerer S, Berry JA (1980) A biochemical model of photosynthetic CO_2 assimilation in leaves of C_3 species. Planta 149: 78–90

Gillespie CT, Oberbauer SF, Gebauer R, Sala A, Tenhunen JD (1993) Climate effects on ecosystem carbon balance of tussock tundra in the Philip Smith Mountains, Alaska (in prep.)

Grulke NE, Riechers GH, Oechel WC, Hjelm U, Jaeger C (1990) Carbon balance in tussock tundra under ambient and elevated atmospheric CO_2. Oecologia 83: 485–494

Hahn S (1991) Photosynthese und Wasserhaushalt von Flechten in der Tundra Alaskas: Gaswechselmessungen unter natürlichen Bedingungen und experimentelle Faktorenanalyse. PhD Thesis, University of Würzburg, 127 pp

Harley PC, Tenhunen JD (1991) Modeling the photosynthetic response of C_3 leaves to environmental factors. In: Boote KJ, Loomis RS (eds) Modeling crop photosynthesis – from biochemistry to canopy. ASA, Madison, Wisconsin, pp 17–39

Harley PC, Tenhunen JD, Murray KJ, Beyers J (1989) Irradiance and temperature effects on photosynthesis of tussock tundra *Sphagnum* mosses from the foothills of the Philip Smith Mountains, Alaska. Oecologia 79: 251–259

Hastings SJ, Luchessa SA, Oechel WC, Tenhunen JD (1989) Standing biomass and production in water track drainages of the foothills of the Philip Smith Mountains, Alaska. Holarct Ecol 12: 304–311

Hilbert DW, Prudhomme TI, Oechel WC (1987) Response of tussock tundra to elevated carbon dioxide regimes: analysis of ecosystem CO_2 flux through nonlinear modeling. Oecologia 72: 466–472

Jarvis PG (1987) Water and carbon fluxes in ecosystems. In: Schulze E-D, Zwölfer H (eds) Potentials and limitations of ecosystem analysis. Ecological studies 61. Springer, Berlin Heidelberg New York, pp 50–67

Johansson L-G, Linder S (1975) The seasonal pattern of photosynthesis of some vascular plants on a subarctic mire. In: Wielgolaski FE (ed) Fennoscandian tundra ecosystems, part 1. Plants and microorganisms. Ecological studies 16. Springer, Berlin Heidelberg New York, pp 194–200

Kjelvik S, Wielgolaski FE, Jahren A (1975) Photosynthesis and respiration of plants studied by field technique at Hardangervidda, Norway. In: Wielgolaski FE (ed) Fennoscandian tundra ecosystems, part 1. Plants and microorganisms. Ecological studies 16. Springer, Berlin Heidelberg New York, pp 184–193

Limbach WE, Oechel WC, Lowell W (1982) Photosynthetic and respiratory responses to temperature and light of three Alaskan tundra growth forms. Holarct Ecol 5: 150–157

Luken JO, Billings WD (1985) The influence of microtopographic heterogeneity on carbon dioxide efflux from a subarctic bog. Holarct Ecol 8: 306–312

Luken JO, Billings WD, Peterson KM (1985) Succession and biomass allocation as controlled by *Sphagnum* in an Alaskan peatland. Can J Bot 63: 1500–1507

Miller PC, Webber PJ, Oechel WC, Tieszen LL (1980) Biophysical processes and primary production. In: Brown J, Miller PC, Tieszen LL, Bunnell FL (eds) An arctic tundra ecosystem, the coastal tundra at Barrow, Alaska. Dowden, Hutchinson, and Ross, Stroudsburg, Pennsylvania, pp 66–101

Miller PC, Kendall R, Oechel WC (1983) Simulating carbon accumulation in northern ecosystems. Simulation 40: 119–141

Miller PC, Miller PM, Blake-Johnson M, Chapin.III FS, Everett KR, Hilbert DW, Kummerow J, Linkins AE, Marion GM, Oechel WC, Roberts SW, Stuart L (1984) Plant-soil processes in *Eriophorum vaginatum* tussock tundra in Alaska: a systems modeling approach. Ecol Monogr 54: 361–405

Murray KJ, Harley PC, Beyers J, Walz H, Tenhunen JD (1989a) Water content effects on photosynthetic response of *Sphagnum* mosses from the foothills of the Philip Smith Mountains, Alaska. Oecologia 79: 224–250

Murray KJ, Tenhunen JD, Kummerow J (1989b) Limitations on moss growth and net primary production in tussock tundra areas of the foothills of the Philip Smith Mountains, Alaska. Oecologia 80: 256–262

Murray KJ, Tenhunen JD, Nowak RS (1993) Photoinhibition as a control on photosynthesis and production of *Sphagnum* mosses, Oecologia (in press)

Oberbauer SF, Tenhunen JD, Reynolds JF (1991) Environmental effects on CO_2 efflux from water track and tussock tundra in arctic, Alaska, USA Arct Alp Res 23: 162–169

Oberbauer SF, Gillespie CT, Cheng W, Gebauer R, Sala Serra A, Tenhunen JD (1992) Environmental effects on CO_2 efflux from riparian tundra in the northern foothills of the Brooks Range, Alaska, USA. Oecologia 92: 568–577

Oechel WC, Billings WD (1992) Effects of global change on the carbon balance of arctic plants and ecosystems. In: Chapin FS, Jefferies RL, Reynolds JF, Shaver GR, Svoboda J, Chu EW (eds) Arctic ecosystems in a changing climate. Academic Press, New York, pp 139–168

Oechel WC, Riechers G, Lawrence WT, Prudhomme TI, Grulke N, Hastings SJ (1992) CO_2 LT' an automated, null-balance system for studying the effects of elevated CO_2 and global climate change on unmanaged ecosystems. Funct Ecol 6: 86–100

Ostendorf B, Reynolds JF (1993) Relationships between a terrain-based hydrologic model and patch-scale vegetation pattern in an Arctic tundra landscape. Landscape Ecology (in press)

Rathstetter EB, King AW, Cosby BJ, Hornberger GM, O'Neill RV, Hobbie JE (1992) Aggregating fine-scale ecological knowledge to model coarser-scale attributes of ecosystems. Ecol Appl 2: 55–70

Reynolds JF, Leadley PW (1992) Modeling the response of arctic plants to changing climate. In: Chapin FS, Jefferies RL, Reynolds JF, Shaver GR, Svoboda J, Chu EW (eds) Arctic ecosystems in a changing climate. Academic Press, New York, pp 413–438

Reynolds JF, Dougherty RL, Tenhunen JD, Harley PC (1988) PRECO: plant response to elevated CO_2 Simulation Model, parts I–III. Report #042, Response of Vegetation to carbon dioxide series, Carbon Dioxide Research Division. US Department of Energy, Washington, DC, 102 pp

Semikhatova OA, Gerasimenko TV, Ivanova TI (1992) Photosynthesis, respiration, and growth of plants in the Soviet arctic. In: Chapin FS, Jefferies RL, Reynolds JF, Shaver GR, Svoboda J, Chu EW (eds) Arctic ecosystems in a changing climate. Academic Press, New York, pp 169–192

Siegwolf R (1987) CO_2-Gaswechsel von *Rhododendron ferrugineum* L. im Jahresgang an
  der alpinen Waldgrenze. PhD Thesis, University of Innsbruck, 283 pp
Smith E (1937) The influence of light and carbon dioxide on photosynthesis. Gen Physiol
  20: 807–830
Stocker O (1931) Transpiration und Wasserhaushalt in verschiedenen Klimazonen. I.
  Untersuchungen an der arktischen Baumgrenze in Schwedisch-Lappland. Jahrb Wiss
  Bot 75: 494–549
Stow D, Burns B, Hope AS (1989) Mapping arctic tundra vegetation using digital
  SPOT/HRV-XS data: a preliminary assesment. Int J Remote Sens 10: 1451–1457
Tenhunen JD, Sala Serra A, Harley PC, Dougherty RL, Reynolds JF (1990) Factors
  influencing carbon fixation and water use by mediterranean sclerophyll shrubs during
  summer drought. Oecologia 82: 381–393
Tenhunen JD, Lange OL, Hahn S, Siegwolf R, Oberbauer SF (1992) The ecosystem role
  of poikilohydric tundra plants. In: Chapin FS, Jefferies RL, Reynolds JF, Shaver GR,
  Svoboda J, Chu EW (eds) Arctic ecosystems in a changing climate. Academic Press,
  New York, pp 213–237
Ulrich B (1987) Stability, elasticity, and resilience of terrestrial ecosystems with respect to
  matter balance. In: Schulze E–D, Zwölfer H (eds) Potentials and limitations of eco-
  system analysis. Ecological sudies 61. Springer, Berlin Heidelberg New York, pp 11–49
Walker DA, Binnian E, Evans BM, Lederer ND, Nordstrand E, Webber PJ (1989)
  Terrain, vegetation, and landscape evolution of the R4D research site, Brooks Range
  foothills, Alaska. Holarct Ecol 12: 238–261
Whalen SC, Reeburgh WS (1988) A methane flux time series for tundra environments.
  Global Biogeochem Cycles 2: 399–409
Whalen SC, Reeburgh WS (1990) A methane flux transect along the trans-Alaska pipeline
  haul road. Tellus 42B: 237–249

Appendix: Gas-Flux Model Variables

General Input Variables Read Once for Each Diurnal Simulation

Barometric pressure [mbar]
Albedo of ground
Ambient CO_2 [ppm]
Julian day
Latitude of site [degrees]
Indicator of clear or overcast sky
Soil water content at 0–10 cm depth [% of dry weight]
Depth to permafrost [cm]
Depth of water table [cm]
Extinction coefficient of the moss
Number of moss layers
Water content of the moss [% of dry weight]
Moss bulk density [$g\,cm^{-3}$]
Percentage of moss cover on the ground [%]
Initial simulation time [hour]
Final simulation time [hour]
Time step [hour]
Time that solar noon occurs in the input data [hour]
Number of canopy layers
Number of species in the simulation
Number of stem diameter classes
Species identification number for each layer

Input Variables Specific for Each Canopy Layer

Leaf absorptance for photosynthetic photon flux density [PPFD]
Leaf reflectance for PPFD
Leaf transmittance for PPFD
Leaf absorptance for shortwave radiation
Leaf reflectance for shortwave radiation
Leaf transmittance for shortwave radiation
Leaf absorptance for infra-red radiation
Projected leaf area [m^2/m^2]
Mean leaf angle [degrees]
Leaf clustering factor
Portion of leaves that are live
Width of average leaf [cm]
Convective heat coefficient
Stem area index [m^2/m^2]
Stem inclination angle [degrees]
Stem clustering factor

Input Variables Read Hourly

Time [hour]
Incident shortwave radiation on canopy [W/m^2]
Air temperature above canopy or air temperature profile [deg. C]
Wind speed [cm/s]
Relative humidity above canopy or humidity profile [%]
Temperature of top cm of soil [deg. C]
Temperature of 10 cm soil depth [deg. C]
Soil water content [% of dry weight]
Moss water content [% of dry weight]

Output Variables for Each Layer

Average net CO_2 exchange of leaves on an area basis [μmol m^{-2} s^{-1}]
Net CO_2 exchange of leaves on ground area basis [μmol m^{-2} s^{-1}]
Average transpiration on leaf area basis [mol m^{-2} s^{-1}]
Transpiration on a ground area basis for layer [mol m^{-2} s^{-1}]
Temperature of air in layer [deg. C]
Average PPFD on sunlit and shaded leaves [μmol m^{-2} s^{-1}]
Average temperature of sunlit and shaded leaves [deg. C]
Average leaf conductance calculated from transpiration [mol m^{-2} s^{-1}]
Relative humidity of air in layer [%]
Average leaf internal CO_2 [μbar]
Average net stem CO_2 exchange on stem area basis [μmol m^{-2} s^{-1}]
Net stem CO_2 exchange on ground area basis [μmol m^{-2} s^{-1}]
Average PPFD on stems [μmol m^{-2} s^{-1}]
Average temperature of sunlit and shaded stems [deg. C]
Leaf area index for layer [m^2 leaves/m^2 ground]
Net moss CO_2 exchange for sunlit and shaded areas [μmol g^{-1} dry weight s^{-1}]
Total net moss CO_2 exchange per hour for sun lit and shaded areas [mmol m^{-2} ground surface h^{-1}]
PPFD in the sunlit and shaded areas of mosslayer [μmol m^{-2} s^{-1}]
Temperature in the sunlit and shaded areas of each mosslayer [deg. C]
Moss water content [% of dry weight]
Sunlight and shaded fraction of moss area
Sum of net gas exchange per hour for sunlit plus shaded area [mmol m^{-2} ground surface h^{-1}]
Net CO_2 exchange of soil on a ground surface area basis [μmol m^{-2} s^{-1}]

Daily Output Over Entire Canopy for All Species and Compartments

Daily sum of the whole canopy gas exchange [mmol m^{-2} d^{-1}]
Daily sum of moss gas exchange [mmol m^{-2} d^{-1}]
Daily sum of soilrespiration [mmol m^{-2} d^{-1}]
Daily sum of all primary producers [mmol m^{-2} d^{-1}]
Daily total net ecosystem production [mmol m^{-2} d^{-1}]

**Part D
Global Aspects
of Photosynthesis**

22 Leaf Diffusive Conductances in the Major Vegetation Types of the Globe

Ch. Körner

22.1 The Significance of Leaf Conductances in Vegetation Modeling

Water availability is a key determinant of growth and distribution of individual plants as well as vegetation. In turn, plant coverage of landscapes influences terrestrial water storage, and – by controlling evapotranspiration – feeds back on the climate and soil moisture regimes. Because of this interrelation of plant functioning with climate and soils, plant control of vapor loss is an important component of biosphere as well as global circulation models.

By far the largest fraction of terrestrial evaporative water loss passes through the microscopic stomata pores of plant leaves. It is estimated that this pathway accounts for 70% of the evaporative vapor loss in Europe (Maniak 1988), and similar values may apply to other regions. The degree of stomatal opening, expressed as the diffusive conductance of leaves for water vapor (g), is controlled by physiological processes, which themselves respond to environmental conditions such as photon flux density, soil, or atmospheric moisture. The maximum rate of vapor loss of vegetation for any given meteorological situation is determined by the maximum diffusivity of leaves (i.e., the number and geometry of stomatal pores), and the total amount of leaf area per unit of land area (the leaf area index, LAI). At any given leaf conductance the most important climatic determinants of transpiration are the moisture deficit of the atmosphere and the aerodynamic conditions in and above the plant canopy. The latter conditions are strongly determined by the geometry and density of the plant stand. Thus, leaf conductance by itself is insufficient to predict transpiration, but it is, in addition to LAI, the major component that is under biological control.

22.2 Constraints of Utilizing Leaf Conductances in Vegetation Modeling

In this contribution, key components of leaf diffusive conductance will be summarized for a number of globally important biomes. However, it is clear from the above that the applicability of these numbers for leaf conductance is only given in connection with a number of other biotic and climatic variables. Canopy structure and canopy climate may, in some instances, become stronger biotic determinants of vapor flux than leaf conductance itself (e.g., McNaughton and Jarvis 1991). For example, Roberts et al. (1990) provide field data about the relative significance of stomata and aerodynamic boundary layer conductance within a tropical rainforest canopy. On an average, boundary layer conductance of leaves in the understory was similar to g_{max}, but was six times larger than g_{max} (i.e., of little significance) at the top of the canopy. Kelliher et al. (1993) illustrated that the aerodynamic conductance for water vapor of grass swards is about eight times smaller than that of forest canopies. As a consequence, they found the proportional sensitivity of canopy transpiration to changes in g to be >0.7 in conifers and as low as 0.3 in grassland. This indicates that the significance of g for predicting fluxes will vary with canopy structure, in particular with leaf area density. The often unknown relative abundance of species within a certain type of vegetation also provides a limitation to the utilization of species based data for the estimation of vegetation means.

Furthermore, plant canopies are not only composed of various species, but bear leaves of various age and leaves which are exposed to contrasting "microclimates" that influence their morphology and physiology. Thus, the main problem associated with this attempt at characterizing leaf conductance is the overwhelming diversity of plant species and functional leaf types (Körner 1993) that cover the continents, contrasted by an otherwise small number of adequate field studies. Therefore, besides providing a rather provisional catalog of conductance parameters for certain vegetation types, the second aim of this chapter is to highlight the major gaps on the global vegetation map that need to receive top priority in future field measurement campaigns.

22.3 How Was the Data Set Compiled?

In order to characterize stomatal control of vapor diffusion on a leaf basis, three types of information are required:

- the maximum diffusive conductance of leaves,
- the minimum diffusive conductance of leaves, and
- the functions that control conductance between these extremes.

Coarse ranges of maximum and minimum conductances for certain groups of species can be distilled from the literature on relatively safe grounds (Körner et al. 1979; Körner 1993). However, on the scales required in biosphere modeling, the third requirement, the functions for stomatal dynamics, is difficult to fulfill. Despite the fact that some very detailed studies on stomatal responses to climatic variations were conducted in leaves of crop species and some temperate zone woody species, the information available for the majority of other vegetation types is, at best, anecdotal. Short-term stomatal responses to light and moisture supply are the simpler parts of relationships. More complicated are predictions of seasonal changes and phenological changes. On a vegetation basis, further complications arise from mutual leaf shading and aerodynamic uncoupling from the climate measured outside the plant canopy by meteorological stations. The present survey does not account for these additional determinants of evapotranspiration. Attempts at modeling these interactions at the canopy level were published, for instance, by Federer (1979), Hall (1982), Kaufmann (1983), Jarvis et al. (1985), Caldwell et al. (1986), Landsberg (1986), Baldocchi et al. (1987) and Baldocchi (1989).

The present chapter is divided in two major parts. The first part provides a reassessment of the knowledge on maximum leaf conductance incorporating the information that accumulated in addition to that compiled in the 1979 review (Körner et al. 1979). This will be supplemented with a brief account on representative values for minimum conductance. The second part contains a collection of stomatal response characteristics observed in plants from major vegetation types. This latter part may assist in designing approximative stochastic functions of stomatal responses.

Five types of leaf conductances (g) are distinguished throughout this chapter:

g_{max} The absolute maximum g achieved in the field following the criteria listed in the next paragraph.

g_{pot} The potential daily maximum of g reached under a given situation. g_{pot} varies throughout the year depending on phenology, soil moisture, and low temperature stress. Only under optimum, peak-season conditions will g_{pot} equal g_{max}.

g_{min} The lowest conductance a leaf can reach when stomata are completely closed as a result of desiccation stress. Other environmental influences (e.g., high CO_2, low light) do not normally lead to the same low g_{min}.

g_{low} The lowest conductance normally found in the field. In almost all cases g_{low} will be higher than g_{min}.

g_{act} The actual leaf conductance occurring under a given climatic constellation during a certain day. The maximum value of g_{act} during a specific day equals g_{pot}.

All values for conductance are presented as molar flux values (mmol $m^{-2} s^{-1}$). For the sake of comparison, the interconversion of conductances from velocity values (cm s^{-1}) was set to 20 °C and a barometric pressure of 1000 hPa. Under these conditions the numerical value of g in molar units is 410 times greater than in cm s^{-1}. For other conditions the equation g (mmol $m^{-2} s^{-1}$) = g (cm s^{-1}) · k · (To/T) · (P/Po) needs to be applied. k is a constant derived from the mole volume of air and has the value of 1/22.73 for

T = To and P = Po, To and T are the absolute temperature at 0°C and the real temperature in K, P and Po are the local pressure and the standard pressure of 1000 both in hPa (=mbar).

22.3.1 Definition of Maximum Leaf Conductance for Water Vapor

Since diffusion porometers became easily available by the end of the 1970s, thousands of plant studies on leaf conductance were undertaken all over the world. In fact, about 90% of the database currently available dates from after 1979, when our first review was completed (Körner et al. 1979). However, a closer look at this novel flood of porometer studies showed that only a small fraction suits the current modeling needs. In order to explain this dilemma, I need to define the criteria under which data were accepted for the present purpose.

Maximum diffusive conductance is defined here as the seasonal maximum conductance found in fully developed but not senescent leaves of adult plants, growing in their natural environment. Unlike the first review (Körner et al. 1979), all data were omitted from plants grown in nurseries, plantations (with few exceptions), greenhouses, growth chambers, all data from seedlings, potted plants, fertilized or watered plots, all data taken from shaded plant parts (e.g., understory) or exposed to any sort of manipulation. This strict treatment of the literature narrows the database to less than one-quarter of the published work. In some cases, as for instance in prairie or steppe grasses, only three papers out of several dozens heading under these key words passed the "sieve". However, it is imperative to adopt such a discriminating approach, since it is well documented that leaves of young plants or plants grown under protected conditions, as well as leaves in the lower part of plants, exhibit stomatal (and photosynthetic) responses rather different from those that control the major part of vapor flux in the upper part of natural vegetation (e.g., Franich et al. 1977; Hinckley et al. 1978; Körner et al. 1979; Leverenz et al. 1983; Beadle et al. 1985; Lichtenthaler 1985). With the above criteria, we concluded with 380 g_{max} values, some species with multiple references, and with the abundance of data biased toward a few biomes.

A closer inspection of this data set showed that numerous maximum leaf conductances were totally unrealistic. For instance, the relatively large data set for evergreen conifers includes g_{max} values between 12 and 584 mmol m^{-2} s^{-1}, while the majority of values range between 100 and 300. If, for instance, one would believe all published numbers for *Pinus banksiana*, g_{max} of this species varied between 29 and 404 during peak season (with no drought stress), leaving the reviewer with the choice of ranging *P. banksiana* among the conifers with the smallest or highest conductances ever recorded. Similar curiosities have been published for all vegetation types and reflect either a still widespread ignorance of porometer calibration or calculation errors. Since there is an obvious bias of such erroneous values towards rather small conductances, their inclusion would have resulted in unrealistically low average maximum leaf conductances, even when using the median instead of the arithmetic mean.

Although I have tried to omit only the most extreme values, the data set may still include a few unsecure extreme values. Therefore arithmetic means have been supplemented with median values where sample size was sufficient. Where multiple reports for a single species were available, the mean from all these reports was used to calculate the means for a vegetation type. A full documentation of all individual original data is not possible here; however, I have selected a few references as examples for each biome. Nevertheless, the means I present include more information than is presented in these references.

A key problem in defining g in a comparative survey is the choice between projected leaf area or total leaf surface area as the reference for leaf transpiration. Papers with unclear information in this respect had to be disregarded. All values presented here are based on a projected leaf area basis. Although this may affect the comparative value in an ecological context (Körner et al. 1979) this reference was chosen, because the leaf area index (LAI) is defined as projected leaf area per unit of land area, and LAI is the second important vegetation variable that controls total canopy conductance (Schulze 1982). In plant species with thick leaves such as needles, ericoid or festucoid leaves, g may become three to five times as large as if based on a total surface area basis. The dilemma is particularly serious in conifers, for which most researchers have adopted total surface area, while most people working with flat leaves use projected leaf areas.

22.3.2 Definition of Minimum Leaf Conductance for Water Vapor

Minimum conductance (complete stomatal closure) is difficult to measure. For most species, porometers or gas exchange systems are not accurate enough to determine the low vapor fluxes that occur after stomatal closure. Furthermore, neither darkness (night values) nor high CO_2 levels or low ambient humidity can mimic the effect of desiccation stress in nature, where minimum conductance becomes ecologically meaningful. Therefore repeated weighing of detached, and thus desiccating, leaves (and thereby forcing stomata to complete closure) has often been adopted as a useful approximation. On the other hand, such severe and fast desiccation again is rare in the real world, and thus is of more theoretical value. In the course of severe drought, most species reduce their total leaf area and control vapor loss by reduced leaf area index. Hence, there will be a difference between the lowest value of leaf conductance observed in the field, g_{low} (with a large measurement error), and the much lower "lowest" leaf conductance g_{min} that can be determined (technically quite accurately) in detached leaves in the laboratory. For modeling purposes g_{low} is more useful, while text books usually refer to g_{min}. Wherever possible, I will include both values in order to avoid confusion between these two "minimum" conductances. The data base for g_{min} is much smaller than for g_{max} and is only approximate.

22.3.3 Definition of Stomatal Response Functions

The prediction of leaf conductances smaller than g_{max} in the field requires a hierarchy of response functions. First is the seasonal variation of the diurnal maximum value of g, g_{pot}, which is a function of the overall predisposition of a plant – a combined effect of phenological control (including photo-periodism effects) and environmental prehistory. In deciduous vegetation such as drought or temperate deciduous forests and woodlands, as well as grasslands, the annual course of the leaf area index is by far the most important variable, periodically setting g_{max} to zero, simply because leaves are absent. When leaves are present, soil moisture is the prime determinant of g_{pot} in these areas. In zonal grasslands phenological changes in plants usually parallel the course of seasonal soil moisture, but, in the absence of drought, reproductive development may set limits to leaf function, simply by leaf dieback in the course of diaspore ripening.

In evergreen vegetation the major determinants of seasonal variations in g_{pot} are again soil moisture and, in the temperate, boreal, and subarctic zone, low temperatures. The prediction of soil drought stress for plants requires a large set of subroutines, including those for plant water consumption. There is great uncertainty about how the moisture-related signal ultimately controls g_{pot} and how this may vary among species (see Sect. 22.8).

Predictions of g_{pot} on the basis of minimum temperatures are easier, since they do not depend on a capacitor such as the soil water reserve, and response thresholds are narrower, simply because soil water freezes near $0\,°C$. Some examples for low temperature control of g_{pot} will be provided in Section 22.8.

Once the highest value of g that may be reached on a given day ($= g_{pot}$) is set, the diurnal climatic and soil moisture conditions become the main determinants of the actual g, g_{act}. The major problem then is that all these environmental variables are effective only over certain ranges and inter-active influences may differ from single-factor influences. Besides assigning response functions for the "responsive range", thresholds and saturation levels need to be defined. Again, a few examples of response characteristics for species from important types of vegetation will be presented.

Table 22.1. Approximative areal extent of vegetation categories mentioned in this chapter. (Whittaker and Likens 1975)[a]

	mio km^2
Tundra (including shrub-, graminoid-, bog-tundra; wet as well as dry tundra and narrow high altitude vegetation)	8
Coniferous forests of the temperate and boreal zone (mainly boreal)	12
Deciduous and mixed forests of the temperate zone (mainly deciduous)	7
Temperate grasslands (steppe, prairie, semi-natural grasslands)	9
Mediterranean shrublands	1
Warm temperate evergreen broad-leaved forests (including eucalyptus forests, laurophyllous forests of SE Asia a.o.)	5
Seasonal subtropical/tropical forests (including monsoonal forests)	7
Humid tropical/subtropical forests	17
Semi-arid subtropical/tropical shrub- and woodlands (including semi-deserts)	25
Tropical savannas	15
Cultivated land	14
Deserts (hot and cold)	24
Undefined (coastal, freshwater, urban, high mountain)	5
Total land area	149

[a] There is no consensus in the recent literature about the separation of growth forms (plant types) and climatic zones. Therefore various (partly curious) types of shrub- and woodlands or grasslands and various types of semi-arid vegetation are listed in different categories by different authors. Arid land data have therefore been pooled here into only two categories, namely savannas and all the rest (mainly shrub- and woodlands), although even this separation seems to be doubtful. These numbers should serve only as a coarse background information that permits ranking of the conductance data presented here with respect to areal extent of the relevant vegetation.

22.4 Selection of Vegetation Types

Various catalogs of the vegetation types of the earth have been published over the years. Faced with the problem of assigning g values to certain vegetation types, it soon became apparent, that most of these systems do not match with the categories used by authors who publish conductance values. Therefore, I will use the very simple and general categories familiar to most people (Table 22.1) and leave it to the user to decide which of these categories may also apply to more specific typologies or which types mentioned here may perhaps apply to other types not presently covered by data. For instance, a separation of tropical nonforested wetlands and tropical nonforested floodplains typically goes beyond the resolution that makes sense in the present context. Most readers might be confused with names like "wet savannas", "temperate savannas", distinctions such as "xeric woodlands" versus "xeric shrublands", or the mention of "mediterranean shrublands", which, except soon after fire, in most parts of the world are dominated by sclerophyll trees of small stature. The terms "wet" and "moist" often used in finer-scale distinctions of vegetation types also do not find adequate reflection in published g-data, further restricting the possibility of compiling g_{max} lists coherent with common vegetation typologies such as the one by Holdridge (1947).

22.5 Maximum Leaf Diffusive Conductances in Important Vegetation Types

Most readers familiar with leaf gas exchange parameters have the vision of a clear separation of typical conductance values for the various types of plants that dominate different vegetation (Körner et al. 1979; Schulze and Hall 1982; Körner 1993). The grouping of conductance data for important vegetation types rather than by plant morphotypes yields a surprise (Table 22.2).

Since morphotypes such as succulents, herbaceous wild plants, and various types of cultivated plants that form the "outlayers" in morphotype-based arrays of g are relatively unimportant with respect to global land coverage, the "nice" cascades of g_{max} on a morphotype basis (Körner 1993)

Table 22.2. Maximum leaf conductance ($mmol\,m^{-2}\,s^{-1}$, projected leaf area) in different types of vegetation. n is the number of species for which reliable field data were available

Type of vegetation	Number of species	Mean ± SD	Median[a]
Woody vegetation			
1a Tundra, deciduous shrubs	8	270 ± 91	–
1b Tundra, evergreen shrubs	6	235 ± 127	–
2 Coniferous forests (mainly boreal)	26	234 ± 99	252
3 Temperate deciduous forests	22	190 ± 71	183
4a Mediterranean deciduous shrubs/trees	6	235 ± 87	–
4b Mediterranean evergreen shrubs/trees	35	203 ± 108	–
5 Eucalyptus forests	6	218 ± 124	–
6 Monsoonal forest (only one reference, possibly shade plants)	5	$(138 ± 19)^b$	–
7a Hot desert shrubs, drought deciduous	4	202 ± 83	–
7b Hot desert shrubs, evergreen	3	222 ± 86	–
7c Cold desert shrubs (Great Basin data only)	2	$(177 ± 18)^b$	–
8 Semi-arid subtropical/tropical shrub and woodland vegetation	16	198 ± 58	195
9 Seasonal tropical forest	4	$(211 ± 144)^b$	–
10a Humid tropical forests	17	249 ± 133	–
10b Humid tropical forests including extremes such as Tectona grandis (g_{max} 1043) and Tabebuia rosea (g_{max} 49)	19	280 ± 228	193
11 Mangrove	4	$(168 ± 49)^b$	–

Nonwoody vegetation

12	Graminoid tundra	7	273 ± 130	254
13	Temperate dry continental grassland (prairie, steppe)	5	326 ± 163	–
14	Desert annuals	17	302 ± 107	–
15	Desert succulent vegetation	4	$(115 \pm 27)^c$	–
16	Graminoid tropical swamp vegetation	2	$(250 \pm 20)^b$	–
17	Semi-arid tropical grasslands	(No adequate field data)		

Anthropogenic vegetation (Körner et al. 1979)

Humid temperate grassland species	ca. 400 (200–1000)
Cereals	ca. 450
Broadleaved herbaceous crops	ca. 500

Index to Table 22.2

Reference examples: (1) Miller et al. 1978, 1980; Oberbauer and Oechel 1989. (2) Running 1976; Hellkvist et al. 1980; Beadle et al. 1985; Goldstein et al. 1985. (3) Federer and Gee 1976; Elias 1979; Jurik 1986; Reich and Hinckley 1989. (4) Mooney 1982; Miller 1983; Davis and Mooney 1985; Blake-Jacobson 1987; Rhizopoulou and Mitrakos 1990. (5) Connor et al. 1977; Sinclair 1980; Körner and Cochrane 1985. (6) De Lillis and Sun 1990. (7) Schulze et al. 1973; Szarek u. Woodhouse 1976; Forseth et al. 1984. (8) Meinzer et al. 1983; Ullmann 1985; Ullmann et al. 1985; Goldstein et al. 1986; Wright and Howe 1987. (9) Fetcher 1979. (10) Grace et al. 1982; Aylett 1985; Roberts et al. 1990; Dolman et al. 1991. (11) Miller et al. 1975; Attiwill and Clough 1980. (12) Miller et al. 1978. (13) Kuhn 1983; Knapp 1985; Monson et al. 1986; Nowak and Caldwell 1986. (14) Tieszen et al. 1979; Goldstein et al. 1985. (15) Nobel 1977a,b. (16) Jones 1988.

Plant species examples, g_{max} in mmol m^{-2} s^{-1}: (1) *Empetrum* sp. 127, *Vaccinium vitis idea* 155, *Andromeda polifolia* 385, *Loiseleuria procumbens* 409; (2) *Picea glauca* 207, *Pinus ponderosa* 247, *Pinus contorta* 277, *Pinus sylvestris* 281; (3) *Acer rubrum* 99, *Quercus rubra* 152, *Fagus sylvatica* 180, *Quercus alba* 250; (4) *Heteromeles arbutifolia* 176, *Calliguaya odorifera* 198, *Ceanotus velutinus* 286, *Quercus ilex* 286, *Quercus coccifera* 389; (5) *Eucalyptus microcarpa* 140, *Eucalyptus regnans* 176, *Eucalyptus obliqua* 225; (6) *Castanopsis, fissa* 135, *Linderia chunii* 160; (7) *Prosopis juliflorae* 217, *Larrea divaricata* 268, *Hamada scoparia* 274; (8) mean for six African *Acaia* sp. 223, mean for seven Australian *Acacia* sp. 143, *Curatella americana* 256; (9) *Faramea occidentalis* 123, *Trichilia cipo* 390; (10) *Bocoa viridiflora* 170, *Coccoloba liebmannii* 236, extremes for *Gmelina arborea* (means of two studies) 675 and *Tectona grandis* 1043; (11) *Avicennia germinans* 207, *Rhizophora mangle* 203; (12) *Carex aquatilis* 199, *Dupontia fisheri* 307; (13) *Bouteloua gracilis* 200, *Bromus erectus* 409, *Agropyron smithii* 507; (14) *Eriogonum inflatum* 163, *Camissonia boothii* 369, *Malvastrum* sp. 491; (15) *Opuntia compressa* 82, *Ferrocactus acanthodes* 135, *Agave desertii* 139; (16) *Cyperus papyrus* 230, *Cyperus latifolius* 270.

[a] Median is listed only for groups with sufficient sample size.
[b] Insufficient amount of information for a representative mean.
[c] Values not directly comparable with the rest of the data set, because the reference area is different; original surface related data from authors were multiplied by 2.5 to match the projected reference area used here, but this is only a coarse approximation.

become insignificant on a vegetation-type based grouping. In short, there is little, if any, difference in g_{max} for the major biomes of the world. Conifers, often regarded as "low conductance" plants, fit this picture only if numbers are expressed per total needle surface area, but not if converted to projected area (as here), where they reach higher mean g_{max} than some deciduous trees of the temperate zone. Several re-assessments of the available data did not change this picture.

Means are well supported by individual data for tundra, boreal coniferous forests, temperate deciduous forests and mediterranean vegetation. Thanks to some very intensive studies in S. America, Africa, and Australia, a good data set for trees and shrubs from semi-arid shrub- and woodlands is available. A fair number of values is available for humid tropical evergreen forests, but given the diversity and geographic extent of this type of vegetation the mean presented here still seems to be weakly founded. Evidently "THE" humid tropical rainforest does not exist, and many readers may miss a finer distinction, but neither were such finer categories supported by sufficient data, nor did the ones that were available indicate any difference, e.g., between mountain and lowland tropical forests. Completely insufficient information is available for dry natural grasslands, both in the continental temperate zone and the subtropics and tropics. Research with grasses from these regions, with very few exemptions, was done in nurseries or greenhouses irrelevant for the present purpose. Many types of vegetation have not been studied at all, but their areal extent may be relatively small and categories presented here may serve as first approximation values for these missing categories.

In conclusion, the numbers for g_{max} available at this time do not provide a justification to apply different g_{max} for the major biomes of the globe that are dominated by woody plants. The global mean for the most important groups of woody vegetation (Table 22.3) is $218 \pm 24 \, mmol \, m^{-2} \, s^{-1}$, i.e., the variation between the mean values for each vegetation type is $\pm 11\%$. As a consequence, canopy conductance will mainly be determined by variations in LAI and by the seasonal (g_{pot}) and diurnal (g_{act}) variations of g. Major deviations from the $218 \, mmol \, m^{-2} \, s^{-1}$ mean are found in grasslands and herbfields, agricultural crop plants and succulents.

Table 22.3. Summary of maximum leaf conductance for woody vegetation ($mmol \, m^{-2} \, s^{-1}$, from Table 22.2)

	g_{max}	Number of species
Tundra shrub vegetation	253	14
Coniferous forests	234[a]	26
Temperate deciduous forests	190	22
Mediterranean shrub vegetation	219	41
Eucalyptus forests	218	6
Hot and cold desert shrublands	200	9
Semi-arid, subtropical and tropical shrub and woodlands	198	16
Humid tropical rainforests	249	17
Mean for all eight groups (n = 8)	218[b] ± 24	151

[a] Note that all data presented here relate to the projected leaf area which, in conifers, is ca. 2.6 times smaller than the overall surface area of needles.
[b] Analysis of variance showed that there is no significant difference between these eight groups of plant/vegetation types ($p = 0.726$).

22.6 Maximum Leaf Diffusive Conductance and Maximum Rate of Leaf Photosynthesis

A number of publications checked for g_{max} also contained photosynthesis data. Table 22.4 summarizes the results of a paired comparison of g_{max} and A_{max} for this subsample data set. A_{max} represents the maximum rate of photosynthesis per unit of leaf area observed under nonlimiting light and moisture conditions in the field. Since the species investigated for g_{max} usually represent dominant elements of the respective flora, A_{max} should be fairly representative as well. Regression analysis with all 73 individual data pairs as well as regression with group means yield the known positive correlation between g_{max} and A_{max} (Körner et al. 1979; Wong et al. 1979; Schulze and Hall 1982). The slope of the correlations differs between herbaceous and woody species, the latter achieving 1/3 lower rates of photosynthesis at any given leaf conductance (Table 22.4, bottom). Within the

Table 22.4. Correlation of g_{max} (mmol H_2O m^{-2}s^{-1}) with A_{max} (μmol CO_2 m^{-2}s^{-1}). This table is based on those publications for g_{max} only that were also providing photosynthesis data (hence g_{max} deviates from the more representative means in Tables 22.2 and 22.3)

	g_{max}	A_{max}	n
Woody vegetation			
Tundra, evergreen shrubs	207 ± 120	7.5 ± 1.3	4
Tundra, deciduous shrubs	296 ± 106	11.0 ± 2.1	7
Boreal conifer forest trees	233 ± 50	11.2 ± 3.5	10
Temperate deciduous forest trees	167 ± 40	8.7 ± 3.8	11
Mediterranean evergreen shrubs	178 ± 90	9.7 ± 2.9	11
Semi-arid shrub- and woodlands	158 ± 95	8.4 ± 7.5	3
Tropical rainforest trees	185 ± 106	7.1 ± 3.4	9
Herbaceous/graminoid vegetation			
Tundra, graminoid plants	292 ± 149	11.8 ± 1.8	5
Desert annuals	365 ± 116	33.6 ± 12.1	6
Savanna grasses	85 ± 32	8.0 ± 2.9	4
Tropical swamp graminoids	333 ± 146	22.1 ± 4.5	3
Total number of data pairs			73

Linear regressions forced through zero; all correlations are highly significant ($p < 0.001$):

			r^2	A_{max} for $g_{max} = 200$
Woody species alone	(n = 55)	$A_{max} = 0.042\,g_{max}$	0.48	8.4
Herbaceous/graminoid species alone	(n = 18)	$A_{max} = 0.064\,g_{max}$	0.85	12.8
All species	(n = 73)	$A_{max} = 0.050\,g_{max}$	0.64	10.0
Mean A_{max} for different woody vegetation types	(n = 7)	9.1 ± 1.6		

seven woody vegetation types considered, mean values for A_{max} vary insignificantly (p = 0.127) and average at $9.1 \pm 1.6 \, \mu mol \, m^{-2} s^{-1}$. Although this subsample contains only field data for 55 woody species, it appears that these types of vegetation do not differ significantly in leaf-based rate of maximum photosynthesis, similar to what applies to g_{max}. Only by including species or morphotypes with extreme high or extreme low g_{max} and A_{max} such as succulents or the various types of herbaceous plants (as was done in our review, Körner et al. 1979), a significant variation in g_{max} and A_{max} among plant groups is achieved. It seems that the inclusions of those other life forms has overshadowed the rather small differences that exist among the major types of woody vegetation. In their review on canopy photosynthesis Ceulemans and Saugier (1991) present data about meteorologically measured net CO_2 fluxes above various forest types which – although at a rather different scale – lead to a similar conclusion: various types of deciduous forest, spruce and pine forest, and the tropical rain forest average at $20 \pm 3 \, \mu mol \, m^{-2} s^{-1}$ land area (extremes for a ponderosa pine forest of 12 and a young rubber plantation of 40 not included). Taken as a maximum flux rate (not specified by Ceulemans and Saugier), these canopy CO_2 fluxes are about twice as high as the mean A_{max} obtained here.

The surprisingly high value for photosynthesis in conifers again reflects the fact that the projected leaf area is about 2.6 times larger than the overall needle surface area on which photosynthesis data are traditionally based. If one assumes a mean CO_2 mixing ratio in the atmosphere of $340 \, \mu mol$ per mol during the period 1975–1985 in which most of these data were collected, the "all species regression" of A_{max} versus g_{max} indicates a mean c_i/c_a ratio of about 0.77 ($c_i = 262 \, \mu mol \, mol^{-1}$). In conclusion, it can be assumed in accordance with earlier attempts that conductance and assimilation data can be approximately interconverted using the functions listed in Table 22.4.

22.7 Minimum Leaf Diffusive Conductances

The minimum leaf conductance of plants varies between 1/20 in herbaceous plants and almost 1/100 of maximum conductance in evergreen conifers (Körner 1993; Table 22.5). Most broadleaved shrubs and trees can reduce their leaf conductance to 1/40 to 1/60 of the maximum value. Succulents form an exception with their ability to reduce the conductance of their surfaces by a factor of 300 and more.

However, except for CAM plants, such reductions rarely operate in the field. Minimum daytime leaf conductances in the field (g_{low}) in the most important woody vegetation types of more humid regions rarely drop below 5% of the maximum and often stay above 25% of the maximum (Table 22.5). Diurnal courses of leaf conductances in plants from arid regions

Table 22.5. Minimum leaf conductance in the field (g_{low}) and under experimental desiccation stress (g_{min}, from Körner 1993) for various vegetation types (very rough estimates, n for number of species checked, mmol m^{-2} s^{-1}, projected leaf area)

Type of vegetation	g_{low}	n	g_{min}	n
Woody vegetation				
1 Tundra, deciduous shrubs	ca. 80[a]		ca. 10	
Tundra, evergreen shrubs	ca. 50		ca. 5	
2 Boreal forest conifers	38 ± 16	3	2.9 ± 0.8	9
3 Temperate deciduous forests	50 ± 40	6	5.8 ± 0.8	16
4 Mediterranean vegetation	77 ± 60	10	2.9 ± 0.8	7
5 Eucapytus forests	ca. 10		ca. 2–5	
6 Monsoonal forests	ca. 30		ca. 5	
7 Desert shrubs (evergreen)	ca. 5		5.2 ± 0.9	7
8 Semi-arid subtropical/tropical shrub/tree vegetation	67 ± 29	4	ca. 5	
10 Humid tropical forests	30 ± 21	6	5.0 ± 0.8	5[b]
Nonwoody vegetation				
12 Temperate grassland	ca. 100		23 ± 4	3
13 Desert annuals	136 ± 68	12	ca. 10	
14 Desert succulents	ca. 1		0.5 ± 0.4	2
Herbaceous crops (cereals plus dicots) (Körner 1993)	>50		13 ± 4	8

[a] "ca." values estimated from various sources and own experience.
[b] But surprisingly high values between 15 and 30 mmol m^{-2} s^{-1} have been found in humid tropical sclerophyll shrubs and small trees in montane New Guinea (Körner et al. 1983), indicating that this value may represent a lower limit rather than a representative mean.

indicate that g_{low} approaches g_{min} more closely, but in many cases values between 5 and 10% of g_{max} are retained. Under persistent, severe drought g_{min} will determine the onset of leaf dieback or leaf shedding, which forms the ultimate control of water loss. Table 22.5 provides some approximate values for g_{low}, that may assist in model parametrization. It should be recalled that g_{low} tends to be underestimated since most gas exchange systems become increasingly inaccurate when g approaches values as low as 10 mmol m^{-2} s^{-1}.

22.8 Stomatal Responses in the Field

22.8.1 Long-Term Trends and Seasonal Changes

Soil moisture and low temperatures are the main preconditioning environmental factors that reduce g_{max} to g_{pot} for a specific day. Predawn leaf water potential or accumulative values of predawn water potentials over a certain number of days have often been found to correlate with the highest g reached during a day (e.g., Hinckley et al. 1978; Schulze et al. 1980; Graham and Running 1984; Ullmann 1985; Davis and Mooney 1986; Grieu et al. 1988). Rhizosphere moisture content may also affect g_{pot} either directly by a hydraulic signal or via a hormonal signal (e.g., Turner et al. 1985; Gollan et al. 1986; Schulze 1986; Wartinger et al. 1990; Bennett 1990). However, response functions are available for only few plant species, not permitting a parametrization on a global scale. Several researchers have studied the control of g_{pot} during the cold season (e.g., Fahey 1979; Smith et al. 1984; Graham and Running 1984; DeLucia and Smith 1987; Carter et al. 1988; Körner and Perterer 1988; Day et al. 1989; Haellgren et al. 1990), and a close relationship with photosynthetic capacity is well established (e.g., Fig. 22.1a); but information on long-term responses (weeks, months) of g in the field is scarce. The following examples provide some trends for low temperature responses in evergreen conifers (temperature, seasonality) and on soil moisture responses for boreal, temperate, and mediterranean plants.

In evergreen conifers low night temperatures have been shown to linearly affect g on the following day (e.g., Fahey 1979; Graham and Running 1984; Smith et al. 1984). Correlations improve if the mean for the last 3 nights or rhizosphere temperatures are utilized instead of last night minima. Between daily minimum air temperatures of $+5$ and $-6\,°C$ the following functions

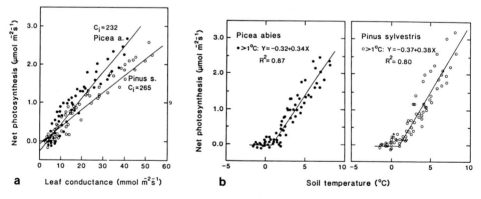

Fig. 22.1. a Daily maxima of net photosynthesis correlate with g_{pot}. Each *point* was taken from ca. 50 diurnal courses of shoot gas exchange collected between November and April during four consecutive winter/spring seasons (1985–1989, J. Perterer, A. Meusburger, M. Zöschg and Ch. Körner, unpubl.). Linear regressions are y = 0.069 × −0.25 (r^2 = 0.89) for *Picea abies* and y = 0.045 × −0.1 (r^2 = 0.94) for *Pinus sylvestris*. **b** Correlation of soil temperature (10 cm depth) measured beneath the closed forest canopy, and maximum rate of photosynthesis. Since maximum net photosynthesis correlates linearly with g_{pot}, these regressions also hold for maximum leaf conductance. The same source of data presented in **a**, but including additional days where g (in contrast to A) could not be measured because it was too cold or branches were covered by frost

have been found for *Pinus contorta* (projected needle area, g in mmol m^{-2} s^{-1}, T_{min} in °C):

$$g_{pot} = 14.0\ T_{min} + 85 \quad \text{Fahey (1979)}$$

$$g_{pot} = 17.3\ T_{min} + 139 \quad \text{Graham and Running (1984)}.$$

On an average, mean minimum air temperatures for the preceding night of less than −6 °C cause g_{pot} to approach g_{min} independently of daytime temperature. However, as mentioned above, soil temperature is a much more precise predictor of g_{pot} which approaches g_{min} in *Pinus sylvestris* and possibly other conifers whenever soil temperatures decrease to +1 °C (Fig. 22.1b).

Predawn water potentials also correlate with g_{pot} in conifers. Graham and Running (1984) report a linear decline in *Pinus contorta* following the equation

$$g_{pot} = 171\Psi + 320.$$

Thus, g_{pot} is approaching g_{min} at predawn potentials of less than −1.6 MPa. DeLucia et al. (1988) report similar slopes for three pine species of which two (*P. monophylla*, *P. ponderosa*) reached g_{min} at predawn water potentials below −2.0 MPa. Figure 22.2 shows the seasonal course of g_{pot} for two boreal pine species.

For deciduous forest trees of the temperate zone, temperature-induced reductions of g_{pot} are of less importance and/or are mediated mainly via seasonal leaf appearance and leaf senescence. According to Jurik (1986), within-season variations of g_{pot} are rather small. Similar observations are reported by Turner and Heichel (1977). However, there is a strong seasonal dependence of g_{pot} in predawn water potential, as indicated by the work of Reich and Hinkley (1989). These authors found a linear reduction of g_{pot}

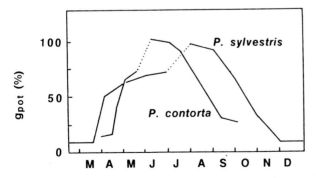

Fig. 22.2. Examples of seasonal courses of the daily maximum leaf conductance (g_{pot}) in evergreen conifers *Pinus contorta* (Montana, US, Graham and Running 1984) and *P. sylvestris* (Tyrol, Austria, J. Perterer and Ch. Körner, unpubl.). The *dotted line* indicates the transition from old to new foliage

between predawn water potentials of -0.2 and $-2.0\,$MPa. At the latter value g_{pot} reached 25% of g_{max} (still substantially higher than g_{min}).

Seasonal courses of g_{pot} in mediterranean ecosystems are well documented (e.g., Miller and Poole 1979; Roberts et al. 1981; Tenhunen et al. 1987a) and mainly follow water availability. According to Tenhunen et al. (1987a), g_{pot} declines fast between predawn water potentials of -0.3 and $-0.7\,$MPa and thereafter remains at about 25% of g_{max} until very low values are reached ($<3.0\,$MPa).

In woody plants from semi-arid or arid regions, g_{pot} is predominantly controlled by water availability, but predawn water potential does not seem to be a good indicator for g_{pot}. Schulze et al. (1980) found accumulative predawn water potentials to be of better predictive value. It should be re-emphasized that most plants in dry regions (except succulents) reduce their leaf area in response to water stress. Leaf sheding and leaf withdrawal in graminoids are thus by far the most important means of controlling plant transpiration on a land area basis in these types of land coverage. As long as leaves are active in drought-deciduous species, g_{pot} does not vary a lot (yet diurnal courses may change). Also for continental dry grassland (steppe, prairie) presence or absence of green leaf area is the main means of control of water loss, and Monson et al. (1986) found very little variation in g_{pot} during the active growing season. Relatively stable predawn water potentials in the course of the season in a desert environment have been reported by Nilsen et al. (1983). Accordingly, they found little seasonal variation in g_{pot} in active leaves. In the humid tropics, seasonal variation in g_{pot} may not be synchronized among species and may largely reflect flushing rhythms of species or even individuals.

22.8.2 Short-Term and Diurnal Changes

A large number of diurnal courses of leaf diffusive conductance in the field have been published from which response functions were distilled. Some examples (rather than an exhaustive review) will be presented below. These should serve as a source of information on typical response phenomena likely to occur in the field.

22.8.2.1 Responses to Photon Flux Density (PFD)

Responses of g_{act} to PFD generally follow saturation functions comparable to the hyperbolic functions known for net photosynthesis. The two key determinants of such functions are the dark minimum of g (g_{low}) and the PFD required to achieve g_{pot}. Leaf conductance in the dark varies considerably, and values may range from 5 to 30% of g_{pot}, with values between

5 and 10% found most frequently. Generally, stomata are not completely closed in the dark, and some species – not only CAM species – keep them almost fully open (e.g., Stalfelt 1963; Gross and Pham-Nguyen 1978). In CAM species, parasitic species, and subarctic species, "night" time stomatal opening is the rule (e.g., Nobel and Hartsock 1984; Press et al. 1988; Gauslaa and Odasz 1990). PFD required for attaining g_{pot} under otherwise unrestricted environmental conditions is difficult to define, since PFD response curves are usually published only for photosynthesis. g tends to peak sharply at somewhat lower PFDs than photosynthesis, which increases more assymptotically. As a rule, most late successional woody species reach 50% of g_{pot} around 3–5% of maximum daily PFD (ca. 50–100 µmol photons $m^{-2} s^{-1}$) and reach >90% of g_{pot} between 10 and 20% of maximum PFD (e.g., Benecke et al. 1981; Whitehead et al. 1981; Goldstein et al. 1985; Ullmann et al. 1985). In graminoid leaves growing at steep angles, light saturation of g occurs at somewhat higher levels (ca. 40% of maximum PFD, e.g., Monson et al. 1986). It should be noted that these light responses are dependent on leaf angle and competitive shading within shoots or tussocks. A clear distinction needs to be made between data for (1) leaves or shoots "bent" into a horizontal position during porometry or cuvette enclosure, (2) leaves or shoots illuminated in their natural position when measured, and (3) whole canopy light responses (Kimes et al. 1980; Carter and Smith 1985; Körner and Cochrane 1985; Oker-Blom and Smolander 1988). The above estimates of PFD responses of g are valid for unshaded leaves measured in an approximately natural position. In dense conifer shoots this will allways include some self-shading between adjacent needles. In addition, stomata were found to respond very sensitively to dawn light and may not perfectly track PFD during the day. Instead, they may remain in an open "waiting" position dispite periodic shading, which frequently occurs in dense plant canopies (e.g., Young and Smith 1979; Weber et al. 1985; Knapp and Smith 1987; Hollinger 1987; Pearcy and Seemann 1990).

22.8.2.2 Moisture Responses

As mentioned in Section 22.8.1, the most important long-term moisture-related control of vegetation water loss operates via LAI and/or species replacement. Only within a certain range of moderate water stress and over short periods of time, stomatal control, plays an important role. Although a number of further details of moisture control of stomata opening still need to be clarified (Schulze et al. 1987; Passioura 1988; Bennett 1990), at least three mechanisms seem to be employed (with decreasing frequency of employment): (1) direct responses to the speed of vapor loss and/or ambient vapor pressure deficit (sensu Lange et al. 1971; Hall and Kaufmann 1975; Lösch and Tenhunen 1981; Turner et al. 1984; a possible mechanism is

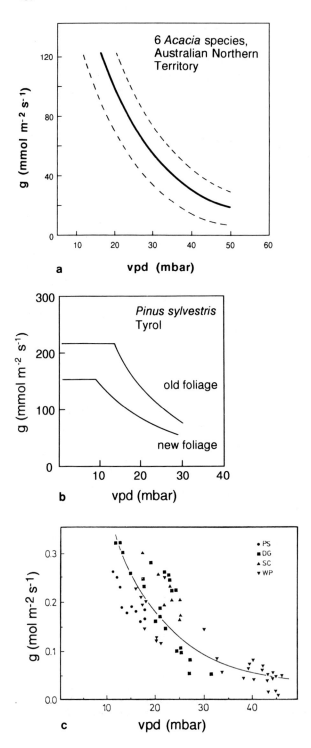

a

b

c

described by Nonami et al. 1990), (2) responses to rhizosphere moisture depletion mediated by a root signal (Gollan et al. 1985, 1986), and (3) leaf water potential. The mechanism mentioned last seems to be rarely employed and, within a given day, water potential tends to reach its lowest value when g reaches its maximum (e.g., Schulze et al. 1972, 1975; Schulze and Hall 1982; Nilsen et al. 1983). There seems to be a surprisingly wide range of water potentials over which no feedback control of g occurs (e.g., for *Pinus sylvestris* and *Picea abies*, values down to −1.8 and −2.0 MPa, respectively, Körner and Perterer, unpubl.). Tenhunen et al. (1987b) conclude "under natural conditions, diurnal changes in Ψ seem not to exert a major influence on stomatal aperture". Since rhizosphere moisture is difficult to define in a diurnal cycle (the questions of soil layer and microcompartments of soil around roots), its usefulness for modeling g lies more in the long-term determination of g_{pot} (Sect. 22.8.1). Therefore, ambient humidity, expressed as vapour pressure deficit (vpd) of the air (or more exactly between air and the substomatal airspace in leaves) becomes the most significant moisture-related predictor of g during a day. Figure 22.3 shows some examples. Most studies showed nearly linear negative responses once vpd exceeded a threshold of 5 to 12 hPa (mostly around 10, Körner 1985; Körner and Cochrane 1985; Roberts et al. 1990).

22.8.2.3 Time-Dependent Processes

In addition to light and moisture effects, almost all plants investigated show hysteresis phenomena in their diurnal course of leaf conductance, generally not taken into account in modeling. The same set of climatic conditions could "produce" a higher leaf conductance during the morning than in the afternoon, (e.g., Jarvis 1976; Lange et al. 1982; Körner and Cochrane 1985; Küppers et al. 1986; Pereira et al. 1987). Such unexplained time-dependent responses are often not found under laboratory conditions. Evidently it will not be possible to recommend a unique set of response characteristics for each type of vegetation. This afternoon reduction of g compared to morning values under equal climatic conditions is in the order of 20% (see above references).

Fig. 22.3a–c. Examples of humidity (vpd) responses of g in woody plant species in the field. **a** Range of responses for six *Acacia* species in semi-arid shrubland (Ullmann et al. 1985). **b** Response of old and new foliage in *Pinus sylvestris* at low altitude in the Alps (derived by boundary analysis from several hundred data points collected over two years, J. Perterer, unpubl.). **c** Response of *Eucalyptus pauciflora* from four different altitudes in the Australian Snowy Mountains. Data from phase II (Fig. 22.6) only. (Körner and Cochrane 1985)

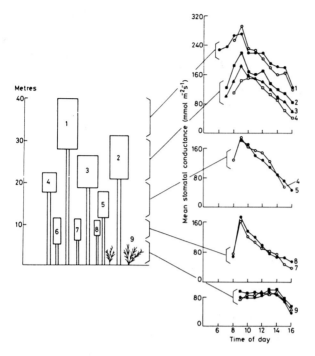

Fig. 22.4. Daily variation of stomatal conductance for Amazonian rainforest measured at several levels within the canopy. (Dolman et al. 1991)

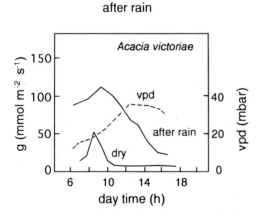

Fig. 22.5. Examples of diurnal courses of leaf diffusive conductance in semi-arid shrubland of Northern Australia before and after rain, but otherwise similar diurnal variation of vpd. (Ullmann et al. 1985)

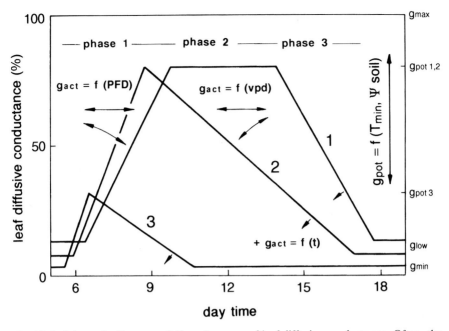

Fig. 22.6. Schematic diagrams of diurnal courses of leaf diffusive conductance. Often, the dominating determinants of g_{act} change from PFD during the morning (*phase 1*) to vpd (*phase 2*) and to time-dependent (unexplained) late afternoon responses (*phase 3*). In many instances, the highest g during a specific day (g_{pot}) is lower than g_{max} and is determined mainly by soil moisture or preceding low temperatures. Both PFD and vpd act gradually (*slope*) as well as threshold factors (saturation or critical value). *Curves 1* and *2* are for high soil moisture but different ambient vpd, *curve 3* represents a situation of severe soil drought

22.8.2.4 Diurnal Courses

While g_{pot} largely responds to the thermal prehistory and soil moisture conditions, g_{act} will deviate from g_{pot} in the diurnal course mainly in response to PFD and vpd. Figures 22.4 and 22.5 depict examples out of hundreds of diurnal courses of g found in the literature. The example from the Amazonian forest was selected because it illustrates the vertical variation in leaf conductance within the canopy. Figure 22.6 illustrates the principles that govern the diurnal course of g_{act} in an unshaded leaf in the field. In the simplest case, the day can be divided in three phases where g is controlled by a different predominant factor in each phase (Körner and Cochrane 1985). During the early morning (phase 1) stomatal opening is, in most cases, largely controlled by PFD until g_{pot} is reached. Thereafter (phase 2), vpd may come into play, reducing g, and thus constraining further increases of transpiration. Leaf water potential often plateaus during this phase. In late afternoon, before PFD limitation comes into play again, unexplained "time-dependent" reductions in g relative to morning values frequently occur at equal PFD and vpd (phase 3, see previous paragraph). The resultant diurnal curve for g most frequently found in the field approximates a triangular shape with a steep short ascent and a variable descent, which, in the case of humid weather (with vpd never falling below a critical level) may even form a trapezoid. Symmetrical bell-shaped diurnal courses (midday peaks) which were reported in the earlier literature are actually very rare in the field. The same is true for the so-called midday depression of g. In some regions, diurnal variations of g_{act} in response to environmental influences are quite regular, so that even a simple statistical function with daytime as the only independent variable can be utilized. This was shown to work in the Amazonian forest by Dolman et al. (1991).

22.9   Conclusions and Recommendations for Further Research

The purpose of this reconsideration of the variation of leaf conductance measured for the major types of global vegetation in the field was to provide a simplified and reliable summary of the appropriate data to be used in large-scale models. It became apparent that to achieve this goal, it was not necessary to differentiate among types of woody vegetation in terms of maximum leaf diffusive conductance. The major gaps of knowledge with respect to leaf conductance are in the tropics and the subtropics, especially in tropical grasslands and tropical seasonal forests. Humid tropical forests are also under-represented in the literature, considering their great significance for global carbon storage, energy balance, and their great susceptibility to changing moisture supply.

Since g_{max} rarely occurs in the field, the seasonal variation of g_{pot} and the diurnal reduction of g induced by vpd and/or transpirational flux per se become major determinants of the water loss from vegetation. However, further refinements of our knowledge of leaf conductance will be of limited value unless we achieve a better data base of leaf area index and its seasonal variation. Future campaigns for improving the current database of leaf conductance should thus include protocols for seasonal leaf area dynamics. Studies also need to account for the relative abundance of species within a certain vegetation type. Priority should be given to screening programs which include large numbers of species and individuals rather than very detailed studies on only a few species and sites (Körner 1991).

Acknowledgments. I am grateful to Susanna Pelaez-Riedl for her assistance in literature search, to Bernd Schäppi for compiling new conductance data, to J. Arnone and the editors for their helpful comments, and to L. Strasser for her unwavering effort in typing succeeding versions of this manuscript. Figure 22.4 is reproduced with permission from Elsevier Science publishers B.V., Amsterdam.

References

Attiwill PM, Clough BF (1980) Carbon dioxide and water vapour exchange in the white mangrove. Photosynthetica 14: 40–47

Aylett GP (1985) Irradiance interception, leaf conductance and photosynthesis in Jamaican upper montane rain forest trees. Photosynthetica 19: 323–337

Baldocchi D (1989) Canopy-atmosphere water vapour exchange: can we scale from a leaf to a canopy? Estimation of areal evapotranspiration. Proc workshop, Vancouver/Canada, August 1987. IAHS Publ 177: 21–41

Baldocchi DD, Hicks BB, Camara P (1987) A canopy stomatal resistance model for gaseous deposition to vegetated surfaces. Atmos Environ 21: 91–101

Beadle CL, Neilson RE, Talbot H, Jarvis PG (1985) Stomatal conductance and photosynthesis in a mature scots pine forest. I. Diurnal, seasonal and spatial variation in shoots. J Appl Ecol 22: 557–571

Benecke U, Schulze ED, Matyssek R, Havranek WM (1981) Environmental control of CO_2-assimilation and leaf conductance in *Larix decidua* Mill. I. A comparison of contrasting natural environments. Oecologia 50: 54–61

Bennett JM (1990) Problems associated with measuring plant water status. HortScience 25: 1551–1554

Blake-Jacobson ME (1987) Stomatal conductance and water relations of shrubs growing at the chaparral-desert ecotone in California and Arizona. In: Tenhunen JD, Catarino FM, Lange OL, Oechel WC (eds) Plant response to stress. Functional analysis in mediterranean ecosystems. NATO ASI Series, Springer, Berlin Heidelberg New York, pp 223–245

Caldwell MM, Meister HP, Tenhunen JD, Lange OL (1986) Canopy structure, light microclimate and leaf gas exchange of *Quercus coccifera* L. in a Portuguese macchia: measurements in different canopy layers and simulations with a canopy model. Trees 1: 25–41

Carter GA, Smith WK (1985) Influence of shoot structure on light interception and photosynthesis in conifers. Plant Physiol 79: 1038–1043

Carter GA, Smith WK, Hadley JL·(1988) Stomatal conductance in three conifer species at different elevations during summer in Wyoming. Can J For Res 18: 242–246

Ceulemans RJ, Saugier B (1991) Photosynthesis. In: Raghavendra AS (ed) Physiology of trees. Wiley, London, pp 21–50

Connor DJ, Legge NJ, Turner NC (1977) Water relations of mountain ash (*Eucalyptus regnans* F. Muell.) forests. Aust J Plant Physiol 4: 753–762

Davis SD, Mooney HA (1985) Comparative water relations of adjacent Californian shrub and grassland communities. Oecologia 66: 522–529

Davis SD, Mooney HA (1986) Water use patterns of four co-occurring chaparral shrubs. Oecologia 70: 172–177

Day TA, DeLucia EH, Smith WK (1989) Influence of cold soil and snowcover on photosynthesis and leaf conductance in two Rocky Mountain conifers. Oecologia 80: 546–552

De Lillis M, Sun GC (1990) Stomatal response patterns and photosynthesis of dominant trees in a tropical monsoon forest. Acta Oecol 11: 545–555

DeLucia EH, Smith WK (1987) Air and soil temperature limitations on photosynthesis in Engelmann spruce during summer. Can J For Res 17: 527–533

DeLucia EH, Schlesinger WH, Billings WD (1988) Water relations and the maintenance of Sierran conifers on hydrothermally altered rock. Ecology 69: 303–311

Dolman AJ, Gash JHC, Roberts J, Shuttleworth WJ (1991) Stomatal and surface conductance of tropical rainforest. Agric For Meteorol 54: 303–318

Elias P (1979) Stomatal activity within the crowns of tall deciduous trees under forest conditions. Biol Plant (Prague) 21: 266–274

Fahey TJ (1979) The effect of night frost on the transpiration of *Pinus contorta* ssp. *latifolia*. Oecol Plant 14: 483–490

Federer CA (1979) A soil-plant-atmosphere model for transpiration and availability of soil water. Water Resour Res 15: 555–562

Federer CA, Gee GW (1976) Diffusion resistance and xylem potential in stressed and unstressed northern hardwood trees. Ecology 57: 975–984

Fetcher N (1979) Water relations of five tropical tree species on Barro Colorado Island, Panama. Oecologia 40: 229–233

Forseth IN, Ehleringer JR, Werk KS, Cook CS (1984) Field water relations of Sonoran Desert annuals. Ecology 65: 1436–1444

Franich RA, Wells LG, Barnett JR (1977) Variation with tree age of needle cuticle topography and stomatal structure in *Pinus radiata* D. Don. Ann Bot 41: 621–626

Gauslaa Y, Odasz AM (1990) Water relations, temperatures, and mineral nutrients in *Pedicularis dasyantha* (Scrophulariaceae) from Svalbard, Norway. Holarct Ecol 13: 112–121

Goldstein GH, Brubaker LB, Hinckley TM (1985) Water relations of white spruce [*Picea glauca* (Moench) Voss] at tree line in north central Alaska. Can J For Res 15: 1080–1087

Goldstein G, Sarmiento G, Meinzer F (1986) Patrones diarios y estacionales en las relaciones hidricas de arboles siempreverdes de la sabana tropical. Acta Oecol Oecol Plant 7: 107–119

Gollan T, Turner NC, Schulze ED (1985) The responses of stomata and leaf gas exchange to vapour pressure deficits and soil water content. III. In the sclerophyllous woody species *Nerium oleander*. Oecologia 65: 356–362

Gollan T, Passioura JB, Munns R (1986) Soil water status affects the stomatal conductance of fully turgid wheat and sunflower leaves. Aust J Plant Physiol 13: 459–464

Grace J, Okali DUU, Fasehun FE (1982) Stomatal conductance of two tropical trees during the wet season in Nigeria. J Appl Ecol 19: 659–670

Graham JS, Running SW (1984) Relative control of air temperature and water status on seasonal transpiration of *Pinus contorta*. Can J For Res 14: 833–838

Grieu P, Guehl JM, Aussenac G (1988) The effects of soil and atmospheric drought on photosynthesis and stomatal control of gas exchange in three coniferous species. Physiol Plant 73: 97–104

Gross K, Pham-Nguyen T (1978) Verlauf des Xylem-Wasserpotentials und des Öffnungszustandes der Stomata von Nadeln junger Fichten (*Picea abies* (L.) Karst.) am Tage und in der Nacht bei unterschiedlicher Wasserversorgung. Forstwiss Centralbl 97: 32–334

Haellgren JE, Lundmark T, Strand M (1990) Photosynthesis of Scots pine in the field after night frosts during summer. Plant Physiol Biochem 28: 437–445

Hall AE (1982) Mathematical models of plant water loss and plant water relations. In: Lange OL, Nobel PS, Osmond CB, Ziegler H (eds) Physiological plant ecology. II. Encyclopedia of plant physiology. Springer, Berlin Heidelberg New York, pp 231–261

Hall AE, Kaufmann MR (1975) Regulation of water transport in the soil-plant-atmosphere continuum, In: Gates DM, Schmerl RB (eds) Perspection of biophysical ecology. Ecological Studies 17. Springer, Berlin Heidelberg New York, pp 187–202

Hellkvist J, Hillerdal-Hagstromer K, Mattson-Djos E (1980) Field studies of water relations and photosynthesis in Scots pine using manual techniques. In: Persson T (ed) Structure and function of northern coniferous forests – an ecosystem study. Ecol Bull 32: 183–204

Hinckley TM, Lassoie JP, Running SW (1978) Temporal and spatial variations in the water status of forest trees. For Sci 24, Monogr 20

Holdridge LR (1947) Determination of world plant formations from simple climatic data. Science 105: 367–368

Hollinger DY (1987) Photosynthesis and stomatal conductance patterns of two fern species from different forest understoreys. J Ecol 75: 925–935

Jarvis PG (1976) The interpretation of the variations in leaf water potential and stomatal conductance found in canopies in the field. Philos Trans R Soc Lond B 273: 593–610

Jarvis PG, Miranda HS, Muetzelfeldt RI (1985) Modelling canopy exchanges of water vapor and carbon dioxide in coniferous forest plantations. In: Hutchinson BA, Hicks BB (eds) The forest-atmosphere interaction. Reidel, Dordricht pp 521–542

Jones (1988) Photosynthetic responses of C_3 and C_4 wetland species in a tropical swamp. J Ecol 76: 253–262

Jurik TW (1986) Seasonal patterns of leaf photosynthetic capacity in successional northern hardwood tree species. Am J Bot 73: 131–138

Kaufmann MR (1983) A canopy model (RM-CWU) for determining transpiration of subalpine forests. I. Model development. Can J For Res 14: 218–226

Kelliher FM, Leuning R, Schulze ED (1993) Evaporation and canopy characteristics of coniferous forest and grassland. Oecologia 95: 153–163

Kimes DS, Ranson KJ, Smith JA (1980) A Monte Carlo calculation of the effects of canopy geometry on PhAR absorption. Photosynthetica 14: 55–64

Knapp AK (1985) Effect of fire and drought on the ecophysiology of *Andropogon gerardii* and *Panicum virgatum* in a tallgrass prairie. Ecology 66: 1309–1320

Knapp AK, Smith WK (1987) Stomatal and photosynthetic responses during sun/shade transitions in subalpine plants: influence on water use efficiency. Oecologia 74: 62–67

Körner Ch (1985) Humidity responses in forest trees: precautions in thermal scanning surveys. Arch Meteorol Geophys BioKlimatol, Ser B 36: 83–98

Körner Ch (1991) Some often overlooked plant characteristics as determinants of plant growth: a reconsideration. Funct Ecol 5: 162–173

Körner Ch (1993) Scaling from species to vegetation: The usefulness of functional groups. In: Schulze ED, Mooney H (eds) Biodiversity and ecosystem function ecological. Studies Springer, Berlin Heidelberg New York, pp 117–142

Körner Ch, Scheel JA, Bauer H (1979) Maximum leaf diffusive conductance in vascular plants. Photosynthetica 13: 45–82

Körner Ch, Cochrane PM (1985) Stomatal responses and water relations of *Eucalyptus pauciflora* in summer along an elevational gradient. Oecologia 66: 443–455

Körner Ch, Perterer J (1988) Nehmen immergrüne Waldbäume im Winter Schadgase auf? Ber Ges Strahlen- Umweltforschung (GSF) Neuherberg/München, ARGE-ALP Symp Garmisch, pp 1–15

Körner Ch, Allison A, Hilscher H (1983) Altitudinal variation of leaf diffusive conductance and leaf anatomy in heliophytes of montane New Guinea and their interrelation with microclimate. Flora 174: 91–135

Kuhn U (1983) Koexistenz durch verschiedene Strategien des Wasserhaushaltes. Eine Untersuchung an sechs Halbtrockenrasenarten. Verh Ges Ökol 10: 201–209

Küppers M, Matyssek R, Schulze ED (1986) Diurnal variations of light-saturated CO_2 assimilation and intercellular carbon dioxide concentration are not related to leaf water potential. Oecologia 69: 477–480

Landsberg JJ (1986) Physiological ecology of forest production. 4. The carbon balance of leaves. Academic Press, London, pp 69–164

Lange OL, Lösch R, Schulze ED, Kappen L (1971) Responses of stomata to changes in humidity. Planta 100: 76–86

Lange OL, Tenhunen JD, Braun M (1982) Midday stomatal closure in mediterranean type sclerophylls under simulated habitat conditions in an environmental chamber. I. Comparison of the behaviour of various European mediterranean species. Flora 172: 563–579

Leverenz J, Deans JD, Ford ED, Jarvis PG, Milne R, Whitehead D (1983) Systematic spatial variation of stomatal conductance in a sitka spruce plantation. J Appl Ecol 19: 835–851

Lichtenthaler HK (1985) Differences in morphology and chemical composition of leaves grown at different light intensities and qualities. In: Baker NR, Davies WJ, Ong CK (eds) Control of leaf growth. Cambridge University Press, SEB Seminar Ser 27: 201–221

Lösch R, Tenhunen JD (1981) Stomatal responses to humidity – phenomenon and mechanism. In: Jarvis PG, Mansfield TA (eds) Stomatal physiology. CUP, London, pp 137–161

Maniak U (1988) Hydrologie und Wasserwirtschaft – Einführung für Ingenieure. Springer, Berlin Heidelberg New York, pp 24–63

McNaughton KG, Jarvis PG (1991) Effects of spatial scale on stomatal control of transpiration. Agric For Meteorol 54: 279–301

Meinzer F, Seymour V, Goldstein G (1983) Water balance in developing leaves of four tropical savanna woody species. Oecologia 60: 237–243

Miller PC (1983) Canopy structure of mediterranean-type shrubs in relation to heat and moisture. In: Kruger FJ, Mitchell DT, Jarvis JUM (eds) Mediterranean-type ecosystems. Ecological Studies 43. Springer, Berlin Heidelberg New York, pp 133–166

Miller PC, Poole DK (1979) Patterns of water use by shrubs in Southern California. For Sci 25: 84–98

Miller PC, Hom J, Poole DK (1975) Water relations of three mangrove species in south Florida. Oecol Plant 10: 355–367

Miller PC, Stoner WA, Ehleringer JR (1978) Some aspects of water relations of arctic and alpine regions. In: Tieszen LL (ed) Vegetation and production ecology of an Alaskan arctic tundra. Ecological studies 29. Springer, Berlin Heidelberg New York, pp 343–357

Miller PC, Webber PJ, Oechel WC, Tieszen LL (1980) Biophysical processes and primary production. In: Brown J, Miller PC, Tieszen LL, Bunnell FL (eds) An arctic ecosystem: the coastal tundra at Barrow, Alaska. Dowden, Hutchinson & Ross, Stroudsburg, pp 66–107

Monson RK, Sackschewsky MR, Williams GJ III (1986) Field measurements of photosynthesis, water-use efficiency, and growth in *Agropyron smithii* (C3) and *Bouteloua gracilis* (C4) in the Colorado shortgrass steppe. Oecologia 68: 400–409

Mooney HA (1982) Habitat, plant form, and plant water relations in mediterranean-climate regions. Ecol Mediterr 8: 481–488

Nilsen ET, Sharifi MR, Rundel PW, Jarrell WM, Virginia RA (1983) Diurnal and seasonal water relations of the desert phreatophyte *Prosopis glandulosa* (honey mesquite) in the Sonoran Desert of California. Ecology 64: 1381–1393

Nobel PS (1977a) Water relations of flowering of *Agave deserti*. Bot Gaz 138: 1–6

Nobel PS (1977b) Water relations and photosynthesis of a barrel cactus, *Ferocactus acanthodes*, in the Colorado Desert. Oecologia 27: 117–133

Nobel PS, Hartsock TL (1984) Physiological responses of *Opuntia ficus-indica* to growth temperature. Physiol Plant 60: 98–105

Nonami H, Schulze ED, Ziegler H (1990) Mechanism of stomatal movement in response to air humidity, irradiance and xylem water potential. Planta 183: 57–64

Nowak RS, Caldwell MM (1986) Photosynthetic characteristics of crested wheatgrass and bluebunch wheatgrass. Range Manage 39: 443–450

Oberbauer SF, Oechel WC (1989) Maximum CO_2-assimilation rates of vascular plants on an Alaskan arctic tundra slope. Holarct Ecol 12: 312–316

Oker-Blom P, Smolander H (1988) The ratio of shoot silhouette area to total needle area in Scots pine. For Sci 34: 894–906

Passioura JB (1988) Response to Dr. P.J. Kramer's article, "Changing concepts regarding plant water relations", 11 (7): 565–568. Plant Cell Environ 11: 569–571

Pearcy RW, Seemann JR (1990) Photosynthetic induction state of leaves in a soybeans canopy in relation to light regulation of ribulose-1-5-biphosphate carboxylase and stomatal conductance. Plant Physiol 94: 628–633

Pereira JS, Tenhunen JD, Lange OL (1987) Stomatal control of photosynthesis of *Eucalyptus globulus* Labill. trees under field conditions in Portugal. J Exp Bot 38: 1678–1688

Press MC, Graves JD, Stewart GR (1988) Transpiration and carbon acquisition in root hemiparasitic angiosperms. J Exp Bot 39: 1009–1014

Reich PB, Hinckley TM (1989) Influence of pre-dawn water potential and soil-to-leaf hydraulic conductance on maximum daily leaf diffusive conductance in two oak species. Funct Ecol 3: 719–726

Rhizopoulou S, Mitrakos K (1990) Water relations of evergreen sclerophylls. I. Seasonal changes in the water relations of 11 species from the same environment. Ann Bot 65: 171–178

Roberts SW, Miller PC, Valamanesh A (1981) Comparative field water relations of four co-occurring chaparral shrub species. Oecologia 48: 360–363

Roberts J, Cabral OMR, Ferreira de Aguiar L (1990) Stomatal and boundary-layer conductances in an Amazonian terra firme rain forest. J Appl Ecol 27: 336–353

Running SW (1976) Environmental control of leaf water conductance in conifers. Can J For Res 6: 104–112

Schulze ED (1982) Plant life forms and their carbon, water and nutrient relations. In: Lange OL, Nobel PS, Osmond CB, Ziegler H (eds) Encyclopedia of plant physiology, New Ser 12B. Springer, Berlin Heidelberg New York, pp 616–676

Schulze ED (1986) Carbon dioxide and water vapor exchange in response to drought in the atmosphere and in the soil. Annu Rev Plant Physiol 37: 247–274

Schulze ED, Hall AE (1982) Stomatal responses, water loss and CO_2 assimilation rates of plants in contrasting environments. In: Lange OL, Nobel PS, Osmond CB, Ziegler H (eds) Encyclopedia of plant physiology, New Ser 12B. Springer, Berlin Heidelberg New York, pp 181–230

Schulze ED, Lange OL, Buschbom U, Kappen L, Evenari M (1972) Stomatal responses to changes in humidity in plants growing in the desert. Planta 108: 259–270

Schulze ED, Lange OL, Kappen L, Buschbom U, Evenari M (1973) Stomatal responses to changes in temperature at increasing water stress. Planta 110: 29–42

Schulze ED, Lange OL, Evenari M, Kappen L, Buschbom U (1980) Long-term effects of drought on wild and cultivated plants in the Negev Desert. II. Diurnal patterns of net photosynthesis and daily carbon gain. Oecologia 45: 19–25

Schulze ED, Robinchaux RH, Grace J, Rundel PW, Ehleringer JR (1987) Plant water balance. BioScience 37: 30–37

Sinclair R (1980) Water potential and stomatal conductance of three Eucalyptus species in the Mount Lofty Ranges, South Australia: responses to summer drought. Aust J Bot 28: 499–510

Smith WK, Young DR, Carter GA, Hadley JL, McNaughton GM (1984) Autumn stomatal closure in six conifer species of the central Rocky Mountains. Oecologia 63: 237–242

Stalfelt MG (1963) Diurnal dark reactions in the stomatal movements. Physiol Plant 16: 756–766

Szarek SR, Woodhouse RM (1976) Ecophysiological studies of Sonoran Desert plants. I. Diurnal photosynthesis patterns of Ambrosia deltoidea and Olneya tesota. Oecologia 26: 225–234

Tenhunen JD, Beyschlag W, Lange OL, Harley PC (1987a) Changes during summer drought in leaf CO_2 uptake rates of macchia shrubs growing in Portugal: limitations due to photosynthetic capacity, carboxylation efficiency, and stomatal conductance. In: Tenhunen JD, Catarino FM, Lange OL, Oechel WC (eds) Plant responses to stress. NATO ASI Series G15, Springer, Berlin Heidelberg New York, pp 305–327

Tenhunen JD, Pearcy RW, Lange OL (1987b) Diurnal variations in leaf conductance and gas exchange in natural environments. In: Zeiger E, Farquhar GD, Cowan IR (eds) Stomatal function. Standford University Press, Stanford, pp 321–351

Tieszen LL, Hein D, Qvortrip SA, Troughton JH, Imbamba SK (1979) Use of $\delta^{13}C$ values to determine vegetation selectivity in East African herbivores. Oecologia 37: 351–359

Turner NC, Heichel GH (1977) Stomatal development and seasonal changes in diffusive resistance of primary and regrowth foliage of red oak (Quercus rubra L.) and red maple (Acer rubrum L.). New Phytol 78: 71–81

Turner NC, Schulze ED, Gollan T (1984) The responses of stomata and leaf gas exchange to vapour pressure deficits and soil water content. I. Species comparisons at high soil water contents. Oecologia 63: 338–342

Turner NC, Schulze ED, Gollan T (1985) The responses of stomata and leaf gas exchange to vapour pressure deficits and soil water content. II. In the mesophytic herbaceous species *Helianthus annuus*. Oecologia 65: 348–355

Ullmann I (1985) Tagesgänge von Transpiration und stomatärer Leitfähigkeit Sahelischer und Saharischer Akazien in der Trockenzeit. Flora 176: 383–409

Ullmann I, Lange OL, Ziegler H, Ehleringer J, Schulze ED, Cowan IR (1985) Diurnal courses of leaf conductance and transpiration of misteltoes and their hosts in central Australia. Oecologia 67: 577–587

Wartinger A, Heilmeier H, Hartung W, Schulze ED (1990) Daily and seasonal courses of leaf conductance and abscisic acid in the xylem sap of almond trees [*Prunus dulcis* (Miller) D.A. Webb] under desert conditions. New Phytol 116: 581–587

Weber JA, Jurik TW, Tenhunen JD, Gates DM (1985) Analysis of gas exchange in seedlings of *Acer saccharum*: integration of field and laboratory studies. Oecologia 65: 338–347

Whitehead D, Okali DUU, Fasehun FE (1981) Stomatal response to environmental variables in two tropical forest species during the dry season in Nigeria. J Appl Ecol 18: 571–587

Whittaker RH, Likens GE (1975) The biosphere and man. In: Lieth H, Whittaker RH (eds) Primary productivity of the biosphere. Ecological Studies 14. Springer, Berlin Heidelberg New York, pp 305–328

Wong SC, Cowan IR, Farquhar GD (1979) Stomatal conductance correlates with photosynthetic capacity. Nature 282: 424–426

Wright SJ, Howe HF (1987) Pattern and mortality in Colorado desert plants. Oecologia 73: 543–552

Young DR, Smith WK (1979) Influence of sunflecks on the temperature and water relations of two subalpine understory congeners. Oecologia 43: 195–205

23   Predictions and Measurements
of the Maximum Photosynthetic Rate, A_{max},
at the Global Scale

F.I. Woodward and T.M. Smith

23.1   Introduction

The study of plant photosynthesis has an exceptionally long pedigree, because photosynthesis is the primary source of chemical energy for plants,
animals, and for integrated ecosystem functioning. The central importance
of photosynthesis in ecology was clearly stated by Lange et al. (1987) "Thus
photosynthetic primary production is not only a requirement for the single
plant but the essential energy-harvesting process for the total biosphere. If
we want to understand how ecosystems function, we must analyze the
photosynthetic performance of the relevant plants".

This view is generally held and, as a consequence, strong research schools
and approaches have developed, based on photosynthesis as the central
process. The measurement of photosynthesis in the field and in relation to
plant distribution has been dominated in Europe by Lange and his co-
workers (e.g., Lange 1965; Schulze and Hall 1982; Lange et al. 1984), with
significant studies also by North Americans (Mooney and West 1964; Mooney
et al. 1966). These pioneering studies have been followed by studies and
approaches which aimed to compare and contrast photosynthetic properties
in very different ecosystems but with the same mechanistic base. The concentration of leaf nitrogen has been the dominant base, as it is an easy
measure which is correlated with the concentration of the photosynthetic
apparatus and enzymes (Field 1983; Field and Mooney 1986; Evans 1989).

The search for a general mechanistic base to photosynthesis was further
extended by the mathematical modeling of the rate-limiting processes
(Farquhar and von Caemmerer 1982; Farquhar 1988), an approach which
through its mechanistic rather than empirical base has greatly improved the
predictive capacity of photosynthetic models (Leuning 1990).

Given the extensive interest in predicting the effects of global climate
change on the ecosystems of the World (Melillo et al. 1990) and the central
importance of photosynthesis to ecosystem processes (Lange et al. 1987), it
is important to take stock of the current range of photosynthetic capacities
at a global scale so that the responses to environmental change can be
predicted from a sound process and data base. This chapter aims to develop
a general model for predicting the maximum rate of terrestrial photo-

synthesis (A_{max}) at a global scale and to follow this by validating the model against global scale observations of A_{max} in the field.

23.2 Philosophy

Field and Mooney (1986) have argued that the correlation between the concentration of leaf nitrogen and A_{max} is a general case which applies to plants across a wide variety of plant communities. Indeed they state that: "From a purely predictive viewpoint, the correlation should prove a useful tool in the preliminary characterization of photosynthetic capacities", although they move on to indicate that the mechanistic basis to the correlation must be understood in order to establish a firm, functional foundation to ecological understanding.

The ranges of the correlations between leaf N and A_{max} are shown in Fig. 23.1 for the mean case (Field and Mooney 1986) and for the extremes of observations (from Evans 1989). There is clearly a large range of possible predictions of A_{max} from a single value of leaf N. This range is too large for satisfactory prediction. Field and Mooney (1986) and Evans (1989) interpret the large range of A_{max} predictions as variations in different patterns of N partitioning between species and within plants, and as inter- and intraspecific differences in plant responses to environmental constraints. As Field and Mooney (1986) succinctly conclude: "A_{max} is not an ecologically meaningless physiological parameter, but is an index of integrated natural constraints on photosynthesis". However, species differ in their capacities to accumulate and store N, particularly nitrate, (Millard 1988), in addition to the influences of rates of leaf expansion on the concentration of leaf N (Evans 1989). Therefore it is the complexity of function of leaf N which may account for much of the scatter of the relationship between leaf N and A_{max} and it is this aspect which will be investigated further.

The particular aim of the work presented here is to provide a general model which can predict the effects of changes in soil N on A_{max}, with a simplicity which cannot be currently achieved from the traditional approach relating the concentration of total N to A_{max}, such as shown in Fig. 23.1. In this chapter, A_{max} is taken to be the highest rate of photosynthesis observed in a vegetation, or biome type, and which is characteristic of the major dominant species of the biome. Therefore A_{max} excludes the effects of reduced irradiance, drought, vapor pressure deficit, and nonoptimal temperatures. These effects can be incorporated as environmental constraints on A_{max}, always acting to reduce the realized value of A_{max} in the field, but this aspect will not be presented in this chapter.

The overall philosophy is that as leaf N is mechanistically related to A_{max} (Evans 1989) and that leaf N originates from the soil, then the first step in predicting A_{max} should be based on the capacity of the soil to provide N to

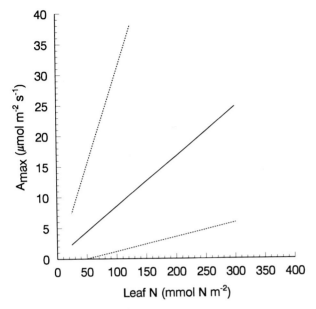

Fig. 23.1. The relationship between the maximum rate of photosynthesis (A_{max}) and the concentration of nitrogen in the leaf. (----), extremes of the relationship from Evans (1989); (—), average relationship from Field and Mooney (1986)

the root and the capacity of the root to supply the whole of the plant (Hunt et al. 1988). The rate of this availability should set an upper bound to A_{max} (Lawlor et al. 1988). This upper bound can then be reduced by environmental constraints, such as temperature, vapor pressure deficit, and irradiance. These constraints can be added via the conventional Farquhar model (e.g., Farquhar and von Caemmerer 1982; Farquhar 1988; Leuning 1990). However, these features will not be presented here.

At a global scale, and in natural and semi-natural vegetation the uptake of N into the plant depends on the concentration and type of N in the soil and on the type of mycorrhiza associated with the plant (Read 1990). At this large scale a model needs a simple but robust guide to the potential rates of N uptake, from as small a global data set as possible. The next section (23.3) describes the experimental basis for the general model, while the extension and testing of this approach, at the global scale, is presented in Section 23.4.

23.3 Experimental Evidence for the Soil N Supply Constraint on A_{max}

23.3.1 Introduction

Read (1990) described how the rate of N supply to the plant from the soil decreases through an ecological succession from an old field system, through a forested early succession to a late successional forest. In addition, if ericoid species dominate, for example in late succession or on poor soil, then N supply rate may be even lower. Through the succession, the rate of phosphorus supply to the plant will be influenced by plant associations with mycorrhizal mutualists, in addition to the pH of the soil. It has been recently discovered that ectomycorrhizal and ericoid mycorrhizas but probably not vesicular-arbuscular mycorrhizas are capable of taking up N from organic N in the soil (Stribley and Read 1980; Abuzinadah and Read 1986). These organic sources are not available to the host plant, particularly at low pH (Abuzinadah and Read 1989).

An important feature of mycorrhizas, particularly ectomycorrhizas, is their low host specificity, therefore mycorrhizas may connect between the roots of different species in the same habitat, ensuring a similar capacity to forage for N (Read 1990). From a modeling viewpoint this is an excellent simplifying property, as the properties of the mycorrhizal associations of different species in the same habitat can be considered the same.

The utility of the host-mycorrhizal association for predicting A_{max} will be tested by an experimental examination of the effects of mycorrhizal type on N supply to the plant and its associated A_{max}. It is hypothesized that the rate of N supply to the plant will decrease in proportion to the increased reliance of the plant on N supply from organic sources in the soil. Modeling of nitrogen supply by mycorrhizas in grasslands (Hunt et al. 1991) supports this premise. Therefore, through the series from soils with high nutrients, where plants are predominantly nonmycorrhizal to moderate nutrients (vesicular-arbuscular type dominates) to soils with refractory nutrient supply (ectomycorrhizal and ericoid mycorrhizas) it is predicted that the rate of N supply to the plant and A_{max} will both decrease.

23.3.2 Experimental Detail

Fourteen species of herbs and shrubs were grown from seeds on their natural soils (four sites of acid peat, beech forest soil, limestone grassland, and old field, in all cases after the removal of surface litter) in the glasshouse and controlled environmental facility of the Department of Botany, University of Cambridge, over the period from 1984 to 1989. Species were grown for either 1 year (annual species) or 2 years (perennial species) and on all combinations of soil type. In each case plants were harvested for growth analysis and analyzed for gas exchange in the period from May to July and in the second year for the perennial species. Natural sunlight was used for the glasshouse experiments, with mean temperatures over the growth period varying from 18 to 26 °C and relative humidities maintained at or above 80%. Maximum photosynthetic rates and stomatal conductances were measured in July (ADC LCA 2, portable infrared gas analyzer, or Leybold-Heraus Binos 2 portable analyzer).

The measurements of photosynthesis were made over a range of temperatures at saturating irradiance. This provided the temperature response of photosynthesis about its optimum. The effects of vpd on photosynthesis were removed by calculating the rate of photosynthesis at the maximum observed stomatal conductance (G_{max}), when vpd was less than 0.75 kPa (Friend and Woodward 1990).

Plants were harvested in July and analyzed for total dry weight, leaf, and whole-plant N and soil N (Kjeldahl total N analysis, Wild 1988). All soils in the experiments have more than 800 g m^{-2} of N. The occurrence and type of mycorrhizal infection was determined by staining for all species (Harley and Smith 1983). The bulk density of the soil was measured and converted to an estimate of soil C (Wild 1988). In brief, the soil types may be described as an acid peat soil (under *Calluna vulgaris* and *Vaccinium myrtillus* heath), a deep gray rendzina (under *Fagus sylvatica* woodland), a shallow rendzina (under *Anthoxanthum adoratum* and *Festuca rubra* grassland) and a fertile loam (previously under wheat but following the end of production, now under *Arrhenatherum elatius* and *Lolium perenne*).

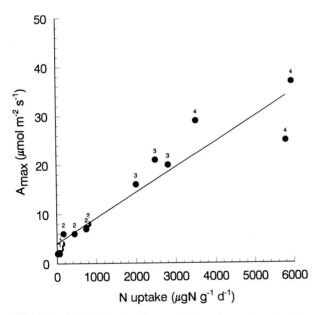

Fig. 23.2. The influence of rate of N uptake on A_{max} in 14 species. Regression line A_{max} = 3.996 + 0.00521 N uptake, r = 0.949, n = 14. *Numbers on symbols* indicate mycorrhizal status of species; *1* ericoid; *2* ectomycorrhizal; *3* vesicular-arbuscular; *4* non-mycorrhizal. Species studied in order of increasing A_{max}: *Vaccinium vitis-idaea (1)*, *Calluna vulgaris (1)*, *V. myrtillus (1)*, *Rhododendron ponticum (1)*, *Betula pendula (2)*, *Fagus sylvatica (2)*, *Quercus robur (2)*, *Acer pseudoplatanus (2)*, *Eupatorium cannabinum (3)*, *Potentilla reptans (3)*, *Verbena officinalis (3)*, *Senecio vulgaris (4)*, *Lolium perenne (4)*, *Triticum aestivum (4)*

23.3.3   Results

The relationship between N uptake by the whole plant and A_{max} is well described by a linear regression (Fig. 23.2, r = 0.949, n = 14). In general, when plants were grown in soils from which they did not originate, then mycorrhizal development was poorer than on their native soils and, in addition, the growth of the ericoid and ectomycorrhizal species was poor. Therefore, the data from these plants have not been included here.

The one regression covers four major soil and mycorrhizal types from high nutrients and nonmycorrhizal, through vesicular-arbuscular mycorrhizal to low quality and high refractory N soils with ectomycorrhizal and ericoid mycorrhizal associations. The relationship supports the hypothesis that A_{max} increases with the rate of N uptake and with a decreased reliance on the mycorrhizal association for N supply. The relationship between leaf N and A_{max} (Fig. 23.3) has much greater scatter than between N uptake and A_{max}. Leaf N would clearly be a poor predictor of A_{max}. Nevertheless, the relationship between leaf N and A_{max} observed by Field and Mooney (1986) is similar to that observed in the experiment described here, indicating the generality of the relationship between leaf N and A_{max}.

A clear relationship between A_{max} and soil carbon also emerged (Fig. 23.4). It is not supposed that the soil C is a direct determinant of A_{max}, it is

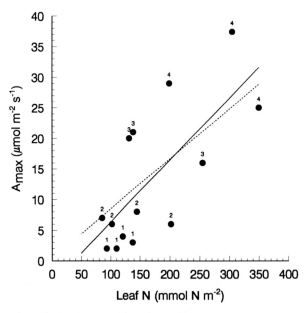

Fig. 23.3. The relationship between leaf N concentration and A_{max} for the species listed under Fig. 23.2. Regression line (—) A_{max} = −3.775 + 0.101 · N conc., r = 0.723, n = 14. Regression line (----) from Field and Mooney (1986), A_{max} = 0.342 + 0.0814 · N conc

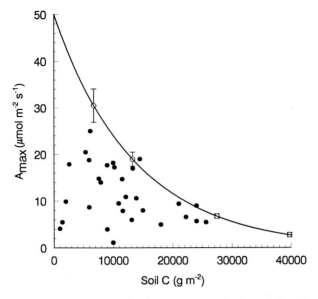

Fig. 23.4. A_{max} for species (mean plus standard error) listed in Fig. 23.2 in relation to soil carbon ($g\,C\,m^{-2}$ soil surface). Soil C increases in the sequence nonmycorrhizal, vesicular-arbuscular, ectomycorrhizal, and ericoid mycorrhizal. Equation of the regression line, $A_{max} = 50 \cdot 0.999927^{SoilC}$. (●) Field observations from the biomes listed in Table 23.1

used to measure the accumulation of organic sources of N and to indicate the reliance of the plant host on the mycorrhizal uptake of N. The accumulation of soil organic C indicates the capacity of the soil flora and fauna to decompose plant litter and for the roots and mycorrhizas to take up nutrients (Read 1990; Schlesinger 1991). The highest soil C corresponds to the acid peat soil where ericoid mycorrhizas are necessary for N uptake by the plant and the lowest soil C corresponds with the old-field (nonmycorrhizal plants).

23.3.4 Discussion

An important feature of the relationship between A_{max} and soil C (Fig. 23.4) is that soil C data are available globally by biome (Post et al. 1982, 1985; Zinke et al. 1984). Therefore the relationship between soil C and A_{max} (Fig. 23.4) can also be applied globally to make predictions of A_{max}. In addition the mechanistic relationship between soil C, mycorrhizal type and nutrient supply rate (Figs. 23.2 to 23.4 and Schlesinger 1991) provides a testable hypothesis which can be logically investigated in the case of failure.

23.4   Modeling A_{max} at the Global Scale

23.4.1   Introduction

Post et al. (1982, 1985) and Zinke et al. (1984) related estimates of soil carbon and nitrogen to global patterns of climate and vegetation by classifying each soil profile by the Holdridge Life-Zone (1947, 1967) in which it was found. The Holdridge Life-Zone Classification (Holdridge 1967) is a bioclimatic classification scheme relating the distribution of major ecosystem complexes or biomes to the climatic variables of biotemperature, mean annual precipitation, and the ratio of potential evapotranspiration (PET) to precipitation. The scheme defines a total of 36 biomes (or climatically defined life zones) at a global scale. The classifications are geographically independent, assuming that the same biome will occur on differing continents given the same climatic conditions.

  In the Holdridge scheme, each particular life zone (biomes/vegetation types) is uniquely occurring in a zone of climate, defined by average annual biotemperature and average annual precipitation. The average biotemperature is the average temperature over a year above a threshold of 0 °C. The average biotemperature and average annual precipitation (both logarithmic scales) form two sides of an equilateral triangle. A logarithmic axis for the potential evapotranspiration (PET) ratio (effective humidity) forms the third side of the triangle. By marking equal intervals along the three axes, hexagons are defined within which a particular life-zone or biome occurs.

  One additional division in the scheme is based on the occurrence of potentially killing frosts. This division is along a critical temperature line that divides hexagons between 12 and 24 °C into warm temperate and subtropical zones. Smith et al. (1992) provide a complete description and evaluation of the Holdridge scheme.

  Although errors, e.g., no consideration of seasonality, correlative not mechanistic base, are involved in this simple classification (Woodward 1987), it provides a broadly accurate method of representing the correlation between biome distribution and climate at a global scale (Smith et al. 1992). In addition, it is a useful approach for extrapolating point data (e.g., soil C and N) from various biomes to climate, in order to provide global coverage from small data sets for representative biomes. The areal extent and global classification of biome types are presented in Table 23.1.

23.4.2   Method of Predicting A_{max} from Soil C

The experimentally established relationship between soil C and A_{max} (Fig. 23.4) will be used to predict A_{max} without environmental constraints. The procedure for providing a global coverage of A_{max} predictions uses the global climate database of Leemens and Cramer (1990). The data base

Table 23.1. Biome type, global area (km^2 × 10^4) and A$_{max}$ observed and predicted (μmol m^{-2}s^{-1})

No.	Biome Description	Area	A$_{max}$ (obs.)	A$_{max}$ (pred.)
1	Polar dry tundra	69.42	1.2	6.7
2	Polar moist tundra	251.91	9.5	22.6
3	Polar wet tundra	465.81	6.6	11.0
4	Polar rain tundra	171.97		3.5
5	Boreal desert	41.47	4.0	19.9
6	Boreal dry bush	188.73		23.7
7	Boreal moist forest	970.78	7.9	16.1
8	Boreal wet forest	441.08	6.0	8.0
9	Boreal rainforest	90.79	5.5	4.8
10	Cool temperate desert	149.91		20.4
11	Cool temperate desert bush	251.51	18.2	24.1
12	Cool temperate steppe	739.15	17.0	18.9
13	Cool temperate moist forest	821.35	10.9	20.7
14	Cool temperate wet forest	148.04	10.6	13.9
15	Cool temperate rainforest	24.56	5.7	11.3
16	Warm temperate desert	67.91	5.5	7.9
17	Warm temperate desert bush	181.85	8.7	16.0
18	Warm temperate thorn steppe	228.00	14.8	25.6
19	Warm temperate dry forest	329.80	14.0	27.3
20	Warm temperate moist forest	292.56	18.8	25.4
21	Warm temperate wet forest	20.45	8.0	7.1
22	Warm temperate rainforest	4.23	5.0	6.9
23	Subtropical desert	742.59	4.1	7.5
24	Subtropical desert bush	540.10	9.9	13.3
25	Subtropical thorn steppe	438.58	20.5	21.2
26	Subtropical dry forest	822.87	17.9	21.6
27	Subtropical moist forest	1512.81	17.7	25.5
28	Subtropical wet forest	284.70	19.0	25.2
29	Subtropical rainforest	12.58	9.0	16.7
30	Tropical desert	394.96		3.9
31	Tropical desert bush	157.96		3.9
32	Tropical thorn steppe	190.12		18.1
33	Tropical very dry forest	327.26	25.0	29.9
34	Tropical dry forest	663.06	17.3	23.7
35	Tropical moist forest	526.29	14.7	21.8
36	Tropical wet forest	22.14	9.4	17.3

provides estimates of mean monthly temperature, rainfall, humidity and irradiance based on a network of over 6000 meteorological stations interpolated to provide global coverage at a 0.5° × 0.5° (latitude and longitude) resolution. Values of annual precipitation, biotemperature, and PET ratio estimated from mean monthly temperature and precipitation are used to classify each 0.5° × 0.5° land cell using the Holdridge Scheme (Smith et al. 1992). Estimates of soil C and N are then assigned using the values reported by Post et al. (1982, 1985) for each Holdridge category. The equation describing the experimentally-established relationship between soil C and A$_{max}$ (Fig. 23.4) is then applied to the soil carbon value to predict A$_{max}$.

23.4.3   Validating A_{max} Predictions

The experimental points presented for 14 species (Fig. 23.4) were the source of the relationship between soil C and A_{max}. The scheme described in Section 4.2 provides the capacity to A_{max} from soil C, at a global scale. This approach can be tested against field observations of A_{max} (A). Unfortunately, reliable measurements of A are not available for all biomes. The Appendix shows the sources of field measurements of A which have been used to test the model predictions. Data have been found for 29 of the 36 biome types classified by the Holdridge scheme. In order to provide a global coverage, it must be assumed that measurements of A in a particular biome on one continent are equally applicable to the same biome on another continent.

The points presented on Fig. 23.4 indicate field observations of A in 29 biomes (data listed in Table 23.1). It may be seen that the experimentally derived line describes the upper boundary of the relationship between A and soil C. However, at low values of soil C (usually semi-arid and arid areas), observations of A are much lower than expected from soil C alone. It was considered that the total pool of available N in these soils may also be a limitation to A. Therefore this limitation was also devised to be included in the model.

23.4.4   Predicting A_{max} from Soil C and Soil N

The soils used in the experiments all possessed soil N levels greater than $800 \, g \, m^{-2}$. However in arid regions total soil N may be considerably less (Post et al. 1985), imposing a limitation on A_{max}. Therefore the experimentally derived relationship between soil C and A_{max} needs to account for the total pool of soil N. In this case, theoretical estimates of the required N supply for growth (Moll et al. 1982; Evans 1989) and field observations of soil N limitations of plant growth (Wild 1988; Rendig and Taylor 1989; Schlesinger 1991), have been used to establish a simple relationship between soil N and the multiplier of A_{max} (Fig. 23.5). These theoretical and experimental approaches indicate a simple linear relationship between increasing A_{max} and soil N, up to a total soil N of $600 \, g \, m^{-2}$. Beyond that concentration, there is little evidence for any further responses to total soil N. Therefore the simple model for A_{max} is modified so that the estimate of A_{max} from soil C (Fig. 23.4) is reduced, if soil N is less than $600 \, g \, m^{-2}$, by a soil N-dependent multiplier (Fig. 23.5).

The ratio between field observations of A and A_{max} predicted from soil C and N indicates that the theoretical relationship between the multiplier and soil N defines the upper boundary of the relationship, indicating the suitability of the approach for at least defining the maximum potential photosynthetic rate based on soil C (i.e., mycorrhizal supply rate) and soil N (total N pool).

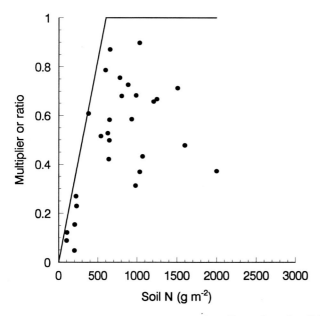

Fig. 23.5. Multiplier to incorporate the effect of total soil N on A_{max}. If soil N > 600 g m^{-2} then multiplier (m) = 1. If soil N < 600 g m^{-2} then m = (1.66 E −0.3 · soil N). (●) Ratio of field observed A to A_{max} predicted from soil C and N

Post et al. (1985) also classify soil N globally and by biome type. The limitations due to soil N on predicted A_{max} is applied as a multiplier to the A_{max} established from soil C.

Soil phosphorus may also influence A_{max} and it would be possible to define an A_{max} multiplier in the same manner as for N (Rendig and Taylor 1989: Schlesinger 1991). In addition, all mycorrhizal types have the capacity to take up phosphorus (Read 1990). Therefore the treatment of soil C and N, based on mycorrhizal activity would be equally applicable to phosphorus. However there is a major limitation over the measurements of soil P. Different techniques often produce different estimates (Wild 1988) and as a consequence the global coverage for soil P is inadequate.

23.5   Global Predictions and Tests of Soil-Based A_{max}

The relationship between the maximum values of the field observations of A and A_{max} based on soil C and N are shown in Fig. 23.6, for the 29 biome types for which field measurements of A have been found in the literature. Sixty eight percent of the variance in A_{max} can be accounted for by the linear

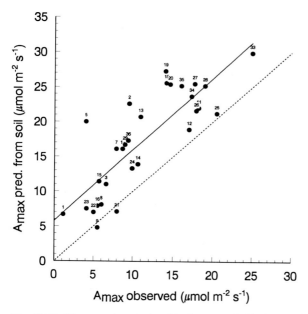

Fig. 23.6. Biome-scale relationship between field observations of A_{max} and predictions based only on soil carbon and nitrogen. Numerical key to biomes as follows: *1* polar dry tundra; *2* polar moist tundra; *3* polar wet tundra; *5* boreal desert; *7* boreal moist forest; *8* boreal wet forest; *9* boreal rainforest; *11* cool temperate desert bush; *12* cool temperate steppe; *13* cool temperate moist forest; *14* cool temperate wet forest; *15* cool temperate rainforest; *16* warm temperate desert; *17* warm temperate desert bush; *18* warm temperate thorn steppe; *19* warm temperate dry forest; *20* warm temperate moist forest; *21* warm temperate wet forest; *22* warm temperate rainforest; *23* subtropical desert; *24* subtropical desert bush; *25* subtropical thorn steppe; *26* subtropical dry forest; *27* subtropical moist forest; *28* subtropical wet forest; *29* subtropical rainforest; *33* tropical very dry forest; *34* tropical dry forest; *35* tropical moist forest; *36* tropical wet forest. Equation for regression line (—) A_{max} predicted $= 5.762 + 1.024 \cdot A_{max}$ observed, $r = 0.822$, $n = 30$. (----) slope of 1:1 line

regression. The slope is not significantly different from unity, indicating that A_{max} based on soil C and N closely parallels changes of A observed in different biomes. The intercept of the linear regression is significantly greater than zero. This is to be expected because no environmental constraints have been imposed on the predictions of A_{max} but they are incorporated in the field measurements.

In addition, it is also to be expected that the rate of mycorrhizal and soil N supply is temperature-dependent, a feature which may explain the wide discrepancy between predicted A_{max} and observed A in the cooler climates of the tundra and boreal biomes. It would be interesting and useful, therefore, to consider modifying the model, which is one essentially of predicting the rate of N supply, to take into account the direct effects of temperature on supply rate.

23.6   Conclusions: Global Scale Maps of Observed and Predicted A_{max}

The final approach of this chapter is to present a global map of predicted A_{max} and observed values of A, by biome type. As for the estimates of soil C and N, it has been assumed that the same biome type has the same value of A_{max} or A on different continents, and even if field measurements are only available from one continent or region.

The overall impression from the two maps (Figs. 23.7 and 23.8) is of a general agreement between observations (Fig. 23.7) and predictions (Fig. 23.8). A_{max} tends to increase in an equatorial direction, for both observations and predictions. Both sets of measurements are in agreement over the major subtropical desert regions of the world, such as the northern Sahara. No field observations of A in the tropical desert regions have been found, while the predictions are for low rates in soils which are virtually skeletal with little C or N (Post et al. 1982, 1985).

The agreement between the high values of A_{max} in the tropical very dry forests of the Sahel, India, and Australia may be fortuitous as grasses with C_4 photosynthesis form a very significant part of this savanna vegetation (Ellis et al. 1980; Livingstone and Clayton 1980). C_4 grasses have high values of A_{max}, compared with species with the C_3 pathway of photosynthesis (Pearch and Ehleringer 1984) and have not been modeled in the predictions. However, a source of N is also required for species with C_4 photosynthesis and, in spite of higher nitrogen use efficiency (Pearcy and Ehleringer 1984), it is to be expected that the rate of N supply from the soil is still the major determinant of A_{max} (Wallace et al. 1982). Therefore A_{max} could be very similar for species with C_3 and C_4 photosynthesis co-occurring, a feature which has been observed (Caldwell et al. 1977).

The values of A_{max} (Figs. 23.7 and 23.8) have been observed and predicted for conditions in which water supply by precipitation is least limiting. If the effects of precipitation are included to provide an annual time course, then the maps will appear very different from month to month and for the annual average (Woodward 1987). This will be the next and future development of the model, in parallel to predictions of leaf area index (Woodward 1987), because it is the product of leaf area index and net photosynthesis which will predict net primary productivity, a property which has also been measured and which is a vital ecosystem property. Lange et al. (1987) have the final words for this endeavor: "However, photosynthetic carbon gain at the level of the single leaf does not necessarily reflect plant characteristics at higher levels of organization. At the whole plant or even at the community level additional regulation occurs which may be of even greater significance. Caution is warranted in generalizing about the ecological meaning of leaf physiological processes for production or even plant success in nature".

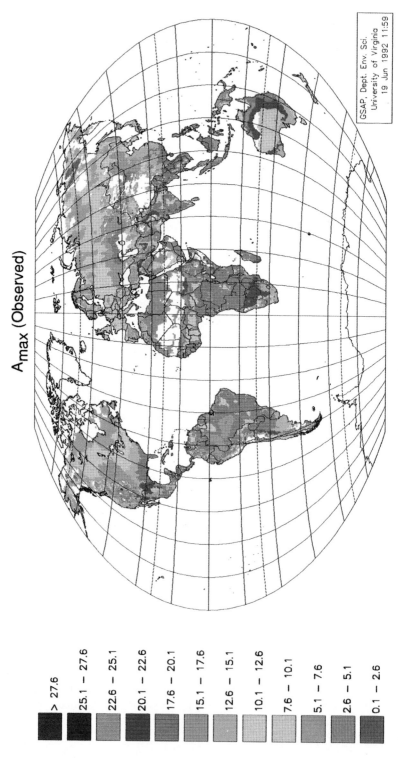

A$_{max}$ (Observed)

> 27.6
25.1 – 27.6
22.6 – 25.1
20.1 – 22.6
17.6 – 20.1
15.1 – 17.6
12.6 – 15.1
10.1 – 12.6
7.6 – 10.1
5.1 – 7.6
2.6 – 5.1
0.1 – 2.6

GSAP, Dept. Env. Sci.
University of Virginia
19 Jun 1992 11:59

Fig. 23.7. Field observations of A. Color code shows scale of A (μmol m^{-2} s^{-1})

A_max (Based on Soil C and N)

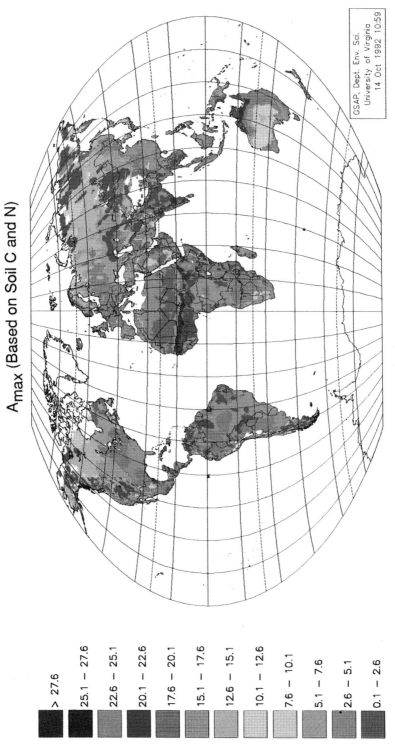

GSAP, Dept. Env. Sci.
University of Virginia
14 Oct 1992 10:59

> 27.6

25.1 – 27.6

22.6 – 25.1

20.1 – 22.6

17.6 – 20.1

15.1 – 17.6

12.6 – 15.1

10.1 – 12.6

7.6 – 10.1

5.1 – 7.6

2.6 – 5.1

0.1 – 2.6

Fig. 23.8. A_{max} predictions from soil C and N. Color code scale (μmol m^{-2} s^{-1}).

Acknowledgments. We are grateful for ideas, comments, and feedback about this work from Professors D.J. Read, D.H. Lewis, E.D. Schulze, and H.H. Shugart.

Appendix: References for Global Measurements of A_{max}

The references presented are in addition to those used in the main text and in some cases also refer to other work. In addition data from, as yet, unpublished work of D.S. Schimel and J.C. Menaut are gratefully acknowledged. Unpublished field measurements from FIW are also included.

Atkinson CJ et al. (1988) Oecologia 75: 386–393
Aylett GP (1985) Photosynthetica 19: 323–337
Bazzaz FA (1979) Annu Rev Ecol Syst 10: 351–371
Bazzaz FA, Pickett STA (1980) Ann Rev Ecol Syst 11: 287–310
Benecke U, Havranek WM (1980) NZ For Serv Pap 70: 195–212
Beyschlag W et al. (1987) Flora 179: 399–420
Carter GA, Smith WK (1988) Can J Bot 66: 963–969
Chambers JL et al. (1985) For Sci 31: 437–450
Chazdon RL, Field CB (1987) Oecologia 73: 222–230
Comstock J, Ehleringer JR (1986) Oecologia 68: 271–278
Davies WJ, Kozlowski TT (1974) Can J Bot 52: 1525–1534
Day TA et al. (1989) Oecologia 80: 546–552
Day TA et al. (1990) Oecologia 84: 474–481
DeLucia EH et al. (1989) Oecologia 78: 184–190
Doley D et al. (1987) Oecologia 74: 441–449
Ehleringer JR (1988) Oecologia 76: 553–561
Jurik TW et al. (1990) Oecologia 82: 180–186
Grace J et al. (1982) J Appl Ecol 19: 659–670
Guehl JM, Aussenac G (1987) Plant Physiol 83: 316–322
Harley PC et al. (1987) Oecologia 74: 380–388
Heckathorn SA, DeLucia EH (1991) Bot Gaz 152: 263–268
Hicks RA et al. (1990) Photosynthetica 24: 63–74
Hoogesteger J, Karlsson PS (1992) Funct Ecol 6: 317–323
Jane GT, Green TGA (1985) Bot Gaz 146: 413–420
Kappen L et al. (1990) Oecologia 82: 311–316
Knapp AK (1986) Oecologia 71: 69–74
Knapp AK, Smith WK (1991) Bot Gaz 152: 269–274
Koike T (1987) Photosynthetica 21: 503–508
Koike T (1988) Plant Spec Biol 3: 77–87
Koike T (1990) Tree Physiol 7: 21–32
Koike T, Sakagami Y (1985) Can J For Res 15: 631–635

Körner C et al. (1979) Photosynthetica 13: 45–82
Küppers M (1984) Oecologia 64: 332–343
Küppers M et al. (1986) Oecologia 70: 273–282
Laijtha K, Whitford WG (1989) Oecologia 80: 341–348
Lange OL, Redon J (1983) Flora 174: 245–284
Langenheim JH et al. (1984) Oecologia 63: 215–224
Leverenz JW (1981) Can J Bot 59: 349–356
Leverenz JW (1981) Can J Bot 59: 2568–2576
Marek M (1988) Photosynthetica 22: 179–183
Martin CE et al. (1991) Bot Gaz 152: 257–262
Matthes-Sears U, Larson DW (1991) Bot Gaz 152: 500–508
Medina E (1987) In: Walker BH (ed) Determinants of savannas, pp 39–65
Meinzer FC et al. (1988) Oecologia 77: 480–486
Midgley GF et al. (1992) Funct Ecol 6: 334–345
Monson RK et al. (1986) Oecologia 68: 400–409
Natarajan K et al. (1990) Photosynthetica 24: 459–467
Nilson ET et al. (1989) Oecologia 79: 193–197
Nilson ET et al. (1990) Oecologia 82: 299–303
Nowak RS et al. (1988) Oecologia 77: 289–295
Oberbauer SF et al. (1987) Oecologia 71: 369–374
Osmond CB et al. (1987) Oecologia 72: 542–549
Pathre UV et al. (1990) Photosynthetica 24: 151–154
Pezeshki SR, Hinckley TM (1982) Can J For Res 12: 761–771
Reich PB et al. (1990) Ecology 71: 2179–2190
Reich PB et al. (1991) Plant Cell Environ 14: 251–259
Riddoch I et al. (1991) J Ecol 79: 491–503
Roberts J et al. (1990) J Appl Ecol 27: 336–353
Roessler PG, Monson RK (1985) Oecologia 67: 380–387
Senock RS et al. (1991) Bot Gaz 152: 275–281
Slatyer RO, Morrow PA (1977) Aust J Bot 25: 1–20
Smith SD et al. (1983) Oecologia 60: 10–17
Sternberg LSL et al. (1987) Oecologia 72: 457–460
Tenhunen JD et al. (1985) Oecologia 67: 23–30
Tenhunen JD et al. (1987) In: Zeiger E et al. (eds) Stomatal function, pp
  323–351
Tenhunen JD et al. (1990) Oecologia 82: 381–393
Thomas CM, Davis SD (1989) Oecologia 80: 309–320
Tissue DT, Oechel WC (1987) Ecology 68: 401–410
Trlica MJ, Biondini ME (1990) Plant Soil 126: 187–201
Tuohy JM et al. (1991) Oecologia 88: 378–382
Ullmann I et al. (1985) Oecologia 67: 577–587
Walters MB, Field CB (1987) Oecologia 72: 449–456
Waring RH, Franklin JF (1979) Science 204: 1380–1386
Williams WE (1983) Plant Cell Environ 6: 145–151
Wong SC, Dunin FX (1987) Aust J Plant Physiol 14: 619–632

References

Abuzinadah RA, Read DJ (1986) The role of proteins in the nitrogen nutrition of ectomycorrhizal plants. I. Utilization of peptides and proteins by ectomycorrhizal fungi. New Phytol 103: 481–493

Abuzinadah RA, Read DJ (1989) The role of proteins in the nitrogen nutrition of ectomycorrhizal plants. V. Nitrogen transfer in birch (*Betula pendula*) grown in association with mycorrhizal and non-mycorrhizal fungi. New Phytol 112: 61–68

Caldwell MM, White RS, Moore RT (1977) Carbon balance, productivity, and water use of cold-winter shrub communities dominated by C_3 and C_4 species. Oecologia 29: 275–300

Ellis RP, Vogel JC, Fuls A (1980) Photosynthetic pathways and the geographical distribution of grasses in south west Africa/Namibia. S Afr J Sci 76: 307–314

Evans JR (1989) Photosynthesis and nitrogen relationships in leaves of C_3 plants. Oecologia 78: 9–19

Farquhar GD (1988) Models relating subcellular effects of temperature to whole plant responses. In: Long SP, Woodward FI (eds) Plants and temperature. Soc Exp Biol Symp 42: 395–409

Farquhar GD, Caemmerer S von (1982) Modelling of photosynthetic response to environmental conditions. In: Lange OL, Nobel PS, Osmond CB, Ziegler H (eds) Encyclopedia of plant physiology, vol 12B. Physiological plant ecology, vol II. Springer, Berlin Heidelberg New York, pp 550–587

Field C (1983) Allocating leaf nitrogen for the maximisation of carbon gain: leaf age as a control on the allocation program. Oecologia 56: 341–347

Field C, Mooney HA (1986) The photosynthesis-nitrogen relationship in wild plants. In: Givnish TJ (ed) On the economy of form and function. Cambridge University Press, Cambridge, pp 25–55

Friend AD, Woodward FI (1990) Evolutionary and ecophysiological responses of mountain plants to the growing season environment. Adv Ecol Res 20: 59–124

Harley JL, Smith SE (1983) Mycorrhizal symbiosis. Academic Press, London

Holdridge LR (1947) Determination of world plant formations from simple climatic data. Science 105: 367–368

Holdridge LR (1967) Life zone ecology. Tropical Science Centre, Costa Rica

Hunt HW, Ingham ER, Coleman DC, Elliott ET, Reid CPP (1988) Nitrogen limitation of production and decomposition in prairie, mountain meadow and pine forest. Ecology 69: 1009–1016

Hunt HW, Trlica MJ, Redente EF, Moore JC, Detling JK, Kittel TGF, Walter DE, Fowler MC, Klein DA, Elliott ET (1991) Simulation model for the effects of climate change on temperate grassland ecosystems. Ecol Model 53: 205–246

Lange OL (1965) Der CO_2-Gaswechsel von Flechten bei tiefen Temperaturen. Planta 64: 1–19

Lange OL, Kilian E, Meyer A, Tenhunen JD (1984) Measurement of lichen photosynthesis in the field with a portable steady-state CO_2-porometer. Lichenologist 16: 1–9

Lange OL, Beyschlag W, Tenhunen JD (1987) Control of leaf carbon assimilation – input of chemical energy into ecosystems. In: Schulze E-D, Zwölfer H (eds) Potentials and limitations of ecosystem analysis. Ecological studies 61. Springer, Berlin, Heidelberg New York, pp 149–163

Lawlor DW, Boyle FA, Keys AJ, Kendall AC, Young AT (1988) Nitrate nutrition and temperature effects on wheat: a synthesis of plant growth and nitrogen uptake in relation to metabolic and physiological processes. J Exp Bot 39: 329–343

Leemens R, Cramer W (1990) The IIASA climate database for land area on a grid of $0.5°$ resolution. WP-41, International Institute for Applied Systems Analysis, Laxenburg, Austria

Leuning R (1990) Modeling stomatal behaviour and photosynthesis of *Eucalyptus grandis*. Aust J Plant Physiol 17: 159–175

Livingstone DA, Clayton WD (1980) An altitudinal cline in tropical African grass floras and its paleoecological significance. Quat Res 13: 392–402

Melillo JM, Callaghan TV, Woodward FI, Salati E, Sinha SK (1990) Effects on ecosystems. In: Houghton JT, Jenkins GJ, Ephraums JJ (eds) Climate change. The IPCC scientific assessment. Cambridge University Press, Cambridge, pp 283–310

Millard P (1988) The accumulation and storage of nitrogen by herbaceous plants. Plant Cell Environ 11: 1–8

Moll RH, Kamprath EJ, Jackson WA (1982) Analysis and interpretation of factors which contribute to efficiency of nitrogen utilization. Agron J 74: 562–564

Mooney HA, West M (1964) Photosynthetic acclimation of plants of diverse origin. Am J Bot 51: 825–827

Mooney HA, Strain BR, West M (1966) Photosynthetic efficiency at reduced carbon dioxide tensions. Ecology 47: 490–491

Pearcy RW, Ehleringer J (1984) Comparative ecophysiology of C_3 and C_4 plants. Plant Cell Environ 7: 1–13

Post WM, Emanuel WR, Zinke PJ, Stangenberger AG (1982) Soil carbon pools and world life zones. Nature 298: 156–159

Post WM, Pastor J, Zinke PJ, Stangenberger AG (1985) Global patterns of soil nitrogen storage. Nature 317: 613–616

Read DJ (1990) Mycorrhizas in ecosystems. Experientia 47: 376–391

Rendig VV, Taylor HM (1989) Principles of soil-plant interrelationships. McGraw-Hill, New York

Schlesinger WH (1991) Biogeochemistry an analysis of global change. Academic Press, San Diego, USA

Schulze E-D, Hall AE (1982) Stomatal responses, water loss and CO_2 assimilation of plants in contrasting environments. In: Lange OL, Nobel PS, Osmond CB, Ziegler H (eds) Encyclopedia of plant physiology, vol 12B. Physiological plant ecology, vol II. Springer, Berlin Heidelberg New York, pp 181–230

Smith TM, Shugart HH, Bonan GB, Smith JB (1992) Modeling the potential response of vegetation to global climate change. Adv Ecol Res 22: 93–116

Stribley DP, Read DJ (1980) The biology of mycorrhiza in the Ericacea. VII. The relationship between mycorrhizal infection and the capacity to utilize simple and complex organic nitrogen sources. New Phytol 86: 365–371

Wallace LL, McNaughton SJ, Coughenour MB (1982) The effects of clipping and fertilization on nitrogen nutrition and allocation by mycorrhizal and nonmycorrhizal *Panicum coloratum* L., a C_4 grass. Oecologia 54: 68–71

Wild A (ed) (1988) Russell's soil conditions and plant growth. Longman Scientific and Technical, London

Woodward FI (1987) Climate and plant distribution. Cambridge University Press, Cambridge

Zinke PJ, Stangenberger AG, Post WM, Emanuel WR, Olson JS (1984) Worldwide organic soil carbon and nitrogen data. ORNL/TM-8857. Oak Ridge National Laboratory, Tennessee

24 Remote Sensing of Terrestrial Photosynthesis[1]

C.B. Field, J.A. Gamon, and J. Peñuelas

24.1 Remote Sensing, from the Leaf to the Globe

Most photosynthesis measurements involve remote determinations. In gas-exchange systems, including those based on chambers, atmospheric gradients, and eddy correlation, photosynthesis determinations are remote in the sense that the measurements are based on effects of leaves, plants, or canopies on the gaseous environment. In radiation-based remote sensing, the subject of this chapter, photosynthesis determinations are based on interactions between leaves, plants, or canopies and the radiation environment. The wavelength bands potentially useful for measurements related to photosynthesis range from the visible through the thermal and microwave regions. While remote sensing generally connotes large-scale satellite measurements, radiation-based remote sensing can be effectively utilized to address questions in photosynthesis research ranging in spatial scale from the chloroplast to the globe.

For this brief chapter, we review some major approaches to assessing photosynthesis using radiation-based remote sensing. We discuss measurements at a range of spatial scales and develop the theme that understanding the concepts, constraints, and challenges of making accurate photosynthesis measurements at leaf and plant scales is an important foundation for regional and global estimates.

The motivation for developing remote sensing approaches to photosynthesis measurements comes from at least four areas. First, remote sensing-based assessments of photosynthesis may supplement alternative approaches with additional information or offer less invasive and/or expensive methodologies. Second, remote sensing of photosynthesis provides opportunities to develop and test our understanding of the upward scaling of physiological phenomena. Just as the controls on photosynthesis may be different at the leaf than at the chloroplast level (Woodrow et al. 1990), they may also differ between the leaf and higher levels of organization. Third, a clearer picture of broad-scale patterns of photosynthesis can suggest new hypotheses that

---

[1] This is CIWDPB Publication number 1151

lead to improved understanding of the ecological and physiological regulation of CO_2 exchange at the level of the leaf and the individual plant.

Fourth, increasing evidence that human activity is changing the earth places a high priority on an extensive approach to ecophysiology. Photosynthesis determinations of global extent are critical both for assessing a key quantity in the global carbon budget and for regional monitoring for change detection. The role of CO_2 as a climate-forcing factor (Lashof and Ahuja 1990) and the current uncertainty concerning the direction and quantity of carbon flux from the terrestrial biosphere (Tans et al. 1990; Enting and Mansbridge 1991; Quay et al. 1992) combine to make terrestrial photosynthesis and the sensitivity of terrestrial photosynthesis to future change central issues in the global change agenda.

24.1.1   A Range of Approaches

Remote sensing products potentially useful for assessing photosynthesis fall into three categories. One class of remote sensing products is related to photosynthetic capacity or the potential for CO_2 uptake. Examples of these products include leaf area index (LAI), total canopy nitrogen, and the fraction of incoming solar radiation absorbed. A remote sensing-based assignment of ecosystem type also falls into this class. A second class of products is related to instantaneous CO_2 exchange. Methods that probe fluorescence, the status of the xanthophyll cycle pigments, or that use canopy temperature in a calculation of stomatal conductance are examples of this class. A third class of products provides environmental parameters needed in photosynthesis models. Remote sensing products can lead to estimates of surface temperature, incoming solar radiation, and soil moisture.

Products from these three classes can serve as inputs to different types of photosynthesis models operating at a range of temporal scales. Indices related to photosynthetic capacity need to track phenomena for which the dominant dynamics tend to be seasonal. In contrast, instantaneous CO_2 exchange has important dynamics on time scales from seasonal to brief shading by passing clouds. The temporal resolution required for environmental parameters in photosynthesis models ranges from minutes for biochemically based models to a month or more for ecosystem production models.

24.2   Models: from Radiance to CO_2 Exchange

Every remote sensing-based estimate of photosynthesis involves a model calculation – remote sensing, as we use the term here, measures radiances, which are never sufficient in themselves for determining CO_2 flux. Models for calculating photosynthesis from remote sensing products can be as simple as a description of a linear relationship between absorbed solar radiation and net primary production (e.g., Monteith 1977) or can include more or less detailed treatments of plant physiology (e.g., Sellers et al. 1992a) and radiation (e.g., Gerstl and Borel 1992) and turbulent transfer (e.g., Dickenson et al. 1986).

The central role of models highlights the three major challenges in the remote sensing of photosynthesis. First, ecological and physiological models of photosynthesis were not generally developed to utilize remote sensing data, resulting in a mismatch between the inputs required by some of the existing models and the outputs available from satellite observations. The process of reformulating physiological and ecological models in terms of parameters easier to relate to radiances needs to be balanced with the development of new sensors and algorithms for deriving relevant parameters (Ustin et al. 1993). Second, the quality of the remote sensing data must be uniformly high. Effects of calibration changes, atmospheric transmission, solar elevation, and canopy architecture still present major challenges. Third, many remote sensing techniques, especially satellite techniques, provide data on spatial scales where even the smallest spatial unit can be very heterogeneous, consisting of many plants or even contrasting ecosystems.

24.3   Remote Sensing of Photosynthetic Capacity

Ecosystem photosynthetic capacity, potential photosynthetic CO_2 fixation under optimal conditions, is controlled by LAI, the photosynthetic capacity of individual leaves, and the spatial organization of leaves. Each of these is potentially accessible through remote sensing. In addition, a number of physical principles and ecological regularities often appear to simplify the situation, such that a single index may capture the combined effects of all three variables. The mechanistic basis for these simplifications appears to involve a combination of physical, biochemical, and ecological factors, discussed below. To understand these simplifications and some of their implications, it is useful to first consider the way the measurements are made.

24.3.1   Absorbed Radiation

Leaf area index is one of a number of parameters that can be derived from a comparative assessment of reflectance in the red, where green leaves are strong absorbers, and the near infrared, where green leaves are highly reflective (Fig. 24.1). In addition to LAI, methods based on the red-near infrared contrast can yield estimates of crop yield (Fig. 24.2; Monteith 1977), green biomass (Fig. 24.2), annual net primary production in natural

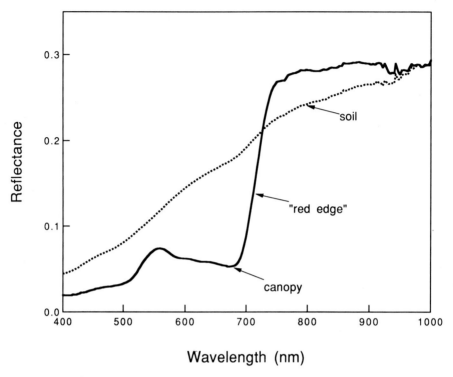

Fig. 24.1. Typical spectral reflectances of a dense vegetation canopy (sunflower, in this case) and bare soil in the visible and near infrared

ecosystems (Goward et al. 1985; Running and Nemani 1988), and canopy photosynthetic capacity (Fig. 24.2; Sellers 1985; Choudhury 1987; Sellers et al. 1992a). In at least some cases, indices based on the red near-infrared contrast are highly correlated with short-term canopy photosynthesis (Fig. 24.2). In others, especially in ecosystems with evergreens exposed to un-favorable periods, these indices do not track seasonal changes in photo-synthesis (Fig. 24.2).

The identity of the physical parameter or parameters measured by the red-near infrared contrast is not completely clear, but the dominant driver of the contrast is the fraction of photosynthetically active radiation absorbed by green leaves (FPAR). Because the contrast in reflectance between the near-infrared and red is large in leaves but small in soils and most other materials, the amount of contrast increases with increasing FPAR.

A number of vegetation indices (Jackson 1983) capture the red-near infrared contrast or FPAR in different ways. The most widely used indices are the simple ratio (SR)

$$SR = \frac{R_{IR}}{R_R} \qquad (1)$$

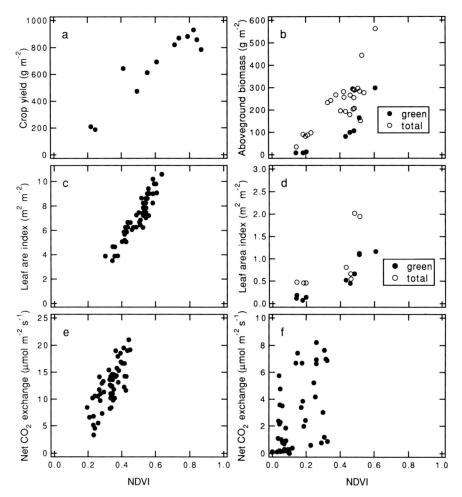

Fig. 24.2a–f. Relationships between the Normalized Difference Vegetation Index (*NDVI*) and several concurrently measured or modeled indices of canopy structure and function. They are: **a** Wheat yield, in plots grown at a range of fertilizer inputs (NDVI from an instrument at 2 m; after Ashcroft et al. 1990), **b** Aboveground green and total biomass in grassland (NDVI from an instrument at 1 m; after Gamon et al. 1993), **c** Leaf area index estimated from an ecosystem simulation (NDVI from satellite data; after Nemani and Running 1989b), **d** Green and total leaf area index in grassland (NDVI from an instrument at 1 m; after Gamon et al. 1993), **e** Grassland canopy photosynthesis, measured with an eddy correlation aircraft (NDVI from a spectroradiometer in the aircraft; redrawn from Desjardins et al. 1992), **f** Conifer canopy photosynthesis from a cold dry site, estimated with an ecosystem simulation model. (NDVI from satellite data; based on data in Running and Nemani 1988)

and the normalized difference vegetation index (NDVI)

$$\text{NDVI} = \frac{R_{IR} - R_R}{R_{IR} + R_R}, \tag{2}$$

which are algebraically interconvertible. Here, R_{IR} is the reflectance in the near infrared and R_R is the reflectance in the red. One reason that these indices are attractive is that they work well with broad-band radiances (Sellers 1987). The current primary source of global vegetation indices is the Advanced Very High Resolution Radiometer (AVHRR) on the NOAA series of weather satellites, in which the red band (AVHRR 1) is sensitive ($>\frac{1}{2}$ peak sensitivity) over the 550–670 nm range and the near infrared band (AVHRR 2) is sensitive from 710 to 980 nm. On the basis of model results, Sellers (1987) concluded that AVHRR bands 1 and 2 are nearly optimal for global inventories of photosynthetic capacity and maximum canopy conductance.

Narrow band spectra, with bands from a few to a few tens of nm in width, provide other options for quantifying the red near-infrared contrast in leaf reflectance. Yoder's (1992) experimental studies indicate that SR and NDVI calculated from narrow bands in the red are most highly correlated with FPAR and LAI. Indices calculated from broad bands in the red or green are, however, better predictors of canopy photosynthetic capacity. In these broad bands, the slower saturation of absorption with increasing total chlorophyll results in sensitivity to a combination of LAI and chlorophyll concentration per unit of leaf area. The location of the "red edge", the wavelength at which the first derivative of reflectance as a function of wavelength reaches a maximum, varies with total canopy chlorophyll in some studies (Curran et al. 1990), but with the chlorophyll concentration of individual leaves (Horler et al. 1983) in others. Sellers (1985) and Choudhury (1987) observed that the relationship between FPAR and NDVI or SR should be quite sensitive to the color of the soil; this sensitivity is greatly reduced for a vegetation index calculated as the second derivative of reflectance versus wavelength at 690 or 704 nm (the bottom and top, respectively, of the red-near infrared contrast) (Hall et al. 1990).

Other useful indices of FPAR do not use the red-near infrared contrast as the basis for the determination. Passive microwave emissions from soils are highly polarized, while plant canopies are strong depolarizers. A satellite index based on this effect, which is due mainly to canopy water, can be an accurate predictor of biomass and yield, especially in ecosystems with low standing biomass (Choudhury 1988).

Another promising approach to determining FPAR involves deconvolving a reflectance spectrum into a number of components. This spectral mixture analysis is a powerful way to compress the information in many spectral bands into a combination of a small number of component spectra (Adams et al. 1986). At the landscape scale, spectral mixture analysis using component spectra representing green leaves, soil, wood, and litter can capture many ecosystem features (Ustin et al. 1993). The green vegetation fraction is highly correlated with NDVI (Gamon et al. 1993), and may provide estimates of FPAR, along with additional information.

The question of which vegetation indices are related to which vegetation parameters and why has several facets. Some of these are biophysical, some are physiological, and some are ecological. Choudhury (1987), Kumar and Monteith (1981), Sellers (1985), and Sellers (1987) have explored the biophysical basis of the relationships and more or less agree on the conclusion that FPAR should be more nearly linear with SR than with NDVI. FPAR gradually saturates as LAI increases above 1, implying that the ability of the vegetation indices to predict LAI should decrease at high LAI. Because canopy photosynthesis also saturates with increasing LAI, the two curvilinear functions tend to cancel each other, making photosynthesis a linear or nearly linear function of FPAR. This is the conceptual basis for the numerical results of Choudhury (1987), Sellers (1985), and Sellers (1987), who used models that assumed constant photosynthetic capacity per unit of leaf area, from the top to the bottom of the canopy. With a gradient in photosynthetic capacity tuned to parallel the gradient in average illumination, the linear SR/photosynthesis relationship is preserved, and the slope of the relationship becomes much less sensitive to the spatial organization of the canopy within the view area (Sellers et al. 1992a,b).

From the ecological perspective, parallel saturation in FPAR and canopy photosynthesis with increasing LAI are not entirely sufficient to account for a strong relationship between a vegetation index and yield or net primary production. This kind of relationship also implies that photosynthesis and growth vary in parallel with light absorption. The explanation for this phenomenon has trivial and deep components, with relative importances that remain to be quantified. The trivial component is related to the fact that growth drives light absorption. Light absorption, therefore, varies in parallel with size in anything that grows, including animals and inanimate objects. The deep component is related to the observation that the efficiency of light utilization in plant growth (the ratio of annual plant yield to absorbed radiation) is surprisingly constant across a range of ecosystems (Russell et al. 1989) and over a range of environmental stresses (Fig. 24.3). This observation implies that plants generally limit investments in light harvesting (canopy development) to levels consistent with the availability of the factor that limits yield or net primary production, independent of whether that factor is a resource (e.g., water or nutrients) or an inherent constraint of growth potential (Field 1991).

The balancing of canopy development to the availability of the resource or resources that limit growth is likely to be regulated on different time scales and with different fidelities in different ecosystems. In herbaceous ecosystems, FPAR, growth, and photosynthetic capacity potentially track growth-limiting resources on a time scale of days to weeks (Figs. 24.2 and 24.3). In evergreen ecosystems, persistence of a canopy through periods

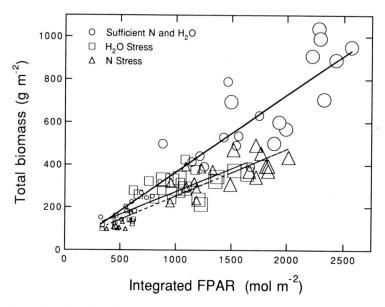

Fig. 24.3. Relationship between total (above- and belowground) biomass and total intercepted photosynthetically active radiation (FPAR) integrated through the growing season, for sunflower canopies grown under contrasting resource availabilities. *Larger points* indicate harvests later in the growing season. (G. Joel, unpublished data)

unfavorable for photosynthesis or growth requires regulation on an annual or multi-year time scale. The hypothesis of long-term regulation of light harvesting in evergreens is consistent with observations indicating strong correlations between NDVI and net primary production on an annual but not on shorter time scales (Running and Nemani 1988).

24.3.2 Photosynthetic Pigments

Photosynthetic pigments give green leaves their characteristic reflectance spectrum and serve as the energy conduit from incident light to the biochemical reactions of photosynthesis, making them obvious targets for remote sensing. The chlorophyll content of a leaf or canopy clearly influences FPAR, but the generally poor correlation between leaf chlorophyll content and photosynthetic capacity (Björkman 1981), as well as the difficulty of quantifying chlorophyll deep in a canopy, limits the utility of remote sensing of chlorophyll for the purpose of estimating terrestrial photosynthesis. For the purpose of estimating canopy photosynthesis, a more useful chlorophyll index might be total chlorophyll weighted by PAR or photosynthetic capacity at each level in the canopy, an index very similar to FPAR. In marine ecosystems, remote sensing of chlorophyll and covarying pigments provides the primary probe for photosynthesis, though estimating total column pigment from measures indicating primarily surface conditions presents comparable problems for marine and terrestrial environments (Parslow and Harris 1990).

While chlorophyll estimates are not widely used for remote sensing-based estimates of photosynthetic capacity in terrestrial ecosystems, they are incorporated in several kinds of estimates of environmental stress. For example, a blue shift in the position of the "red edge" can indicate a decrease in leaf chlorophyll characteristic of some instances of forest decline (Rock et al. 1988). A second example involves the ratio of carotenoids to chlorophyll. Because carotenoids increase in plants under stress and persist longer than chlorophyll in senescing leaves (Peñuelas 1984), estimates of the ratio of carotenoids to chlorophyll provide an additional window on plant function. Peñuelas et al. (1993) found that, for a number of aquatic angiosperms, another normalized difference ($R_{680} - R_{430}/R_{680} + R_{430}$), was highly correlated with the ratios of total carotenoids to chlorophyll a and neoxanthin to chlorophyll a.

24.3.3 Other Compounds

In laboratory settings, near infrared reflectance spectroscopy is the basis for accurate, highly repeatable assays of many biological materials, including digestibility, nitrogen, energy content, moisture, ash, crude fats, total reducing sugars, alkaloids, and a number of other compounds and classes of compounds in plant matter (Clark 1989). Remote sensing indices of some of these, for example total nitrogen, protein, or Rubisco, could be extremely valuable as probes for photosynthetic capacity. While large-scale remote sensing indices for these compounds are not yet available, some of the results to date are encouraging.

Wessman et al. (1988) used the IR reflectance measured with NASA's Airborne Imaging Spectrometer (AIS), an instrument with moderately narrow wavebands, to develop an index for canopy lignin in Wisconsin forest sites. The index accurately predicted not only lignin but also annual net nitrogen mineralization. In ecosystems where production is limited by nitrogen availability, remote sensing of lignin, often a major control on decomposition rate (Melillo et al. 1982), may provide an indirect approach to estimating canopy nitrogen, as well as growth and production.

Successful application of this technology in the laboratory and in the field suggests that the near IR (700–2400 nm) may provide a fertile band in the search for useful remote sensing indices of ecosystem function. Most of the promising techniques use a large number of narrow spectral bands, beyond the capabilities of current satellites but within those of instruments scheduled for launch in the next decade (Wickland 1991).

24.3.4   Vegetation Type

Large differences in the productive capacities of different ecosystem or biome types (Whittaker and Likens 1975) makes an accurate map of vegetation distributions an important prerequisite for large-scale assessments of photosynthesis and production. Most of the global-scale vegetation maps in use today are based on surface observations (Matthews 1983) or climate data (Holdridge 1947), but remote sensing is increasingly important. The standard techniques for vegetation classification from satellite data require knowledgeable individuals to identify training sites with known vegetation types and then use statistical techniques, especially maximum likelihood, to assign unknown areas (Richards and Kelly 1984). Mixture analysis, in which the spectrum or time series of, for example, vegetation indices from each pixel is deconvolved into a small set of endmember components, provides another approach to land-surface classification. It offers several advantages, of which one of the most promising for ecological studies is that pixels can be treated as mixtures of components that may or may not be traditional ecosystem types (Ustin et al. 1993).

24.4   Remote Sensing of Physiological Status

24.4.1   Fluorescence

Chlorophyll fluorescence provides an extremely useful window on photosynthetic mechanisms, especially with recent improvements in the techniques for pulse-modulated measuring systems and analysis of fluorescence lifetimes (Krause and Weis 1991). Almost all fluorescence measurements are remote sensing, according to the definition we use here, but the diversity of laboratory and field-oriented leaf-level techniques is beyond the scope of this chapter.

In general, extending fluorescence studies from the leaf level to larger scales has been difficult, because most techniques require controlled excitation. Extensions to larger scales are, however, beginning to look increasingly tractable. These extensions involve both passive techniques that assess chlorophyll fluorescence under ambient conditions and active techniques that provide controlled excitation. In marine ecosystems, ambient chlorophyll fluorescence creates increased upwelling in the red that can be used to estimate chlorophyll concentration (Parslow and Harris 1990). Fluorescence detection in the Fraunhofer lines, where atmospheric oxygen absorbs essentially all of the downwelling sunlight, provides a potentially useful approach to eliminating interference from reflected light (Carter et al. 1990). Techniques for measuring laser-induced fluorescence (Chappelle and Williams 1987) are currently under development for canopy-scale and aircraft measurements (Lichtenthaler 1988).

24.4.2   Xanthophyll Pigments

The carotenoid pigments that comprise the xanthophyll cycle are clearly implicated in the dissipation of absorbed PAR not used in photosynthesis (Björkman and Demmig-Adams, Chap. 2, this Vol.). The total pool of xanthophyll pigments increases with growth PAR. The quantity of pigment in the de-epoxidized (energy-dissipating) form increases strongly as non-photochemical quenching increases (Björkman and Demmig-Adams, this Vol.) and as the realized photon efficiency (O_2 evolution or CO_2 uptake per unit of absorbed PAR) decreases (Demmig-Adams and Adams 1992; Thayer and Björkman 1990). Remote sensing of the status of the xanthophyll pigments could lead to remote estimates of in situ photosynthetic efficiency, which, in combination with estimates of downwelling PAR, could lead to a new approach for estimating photosynthetic CO_2 uptake.

Bilger et al. (1989) first reported that the status of the xanthophyll cycle pigments influences leaf spectral absorptance. Following sudden illumination, changes in leaf spectral absorptance in the green region represent the combined effect of redistribution of the xanthophyll pigments, plus changes in light scattering resulting from chloroplast movements and changes in thylakoid conformation (Brugnoli and Björkman 1992).

Using sunflower canopies, Gamon et al. (1990) observed two regions of unstable spectral reflectance following sudden illumination. One region (from about 670 to 800 nm) had the spectral characteristics of chlorophyll fluorescence, while the other (from about 500 to 560 nm) had a wavelength range expected for changes due to the interconversion of the xanthophyll pigments and chloroplast movements. A strong correlation between the reflectance at 531 nm and the epoxidation state of the xanthophyll pigments in upper canopy leaves, as well as the elimination of the spectral instability in the green in leaves fed dithiothreitol (an inhibitor of the deepoxidation reaction responsible for the production of the energy dissipating xanthophyll pigment), confirmed the involvement of the xanthophyll cycle pigments in the spectral instability.

Gamon et al. (1992) considered approaches for quantifying the status of the xanthophyll pigments without canopy-scale PAR manipulations. A physiological reflectance index or PRI,

$$PRI = \frac{R_{REF} - R_{531}}{R_{REF} + R_{531}}, \qquad (3)$$

where R_{REF} is a reference wavelength (550 nm for sunflower canopies), closely tracked diurnal changes in the realized photon efficiency (CO_2 uptake/absorbed PAR) in the upper leaves of sunflowers grown under unstressed conditions or with nutrient stress. For sunflowers grown under water stress, PRI varied over a narrow range and was poorly correlated with realized photon efficiency (Fig. 24.4). The different response in the water-stressed plants was probably due to a combination of stomatal closure,

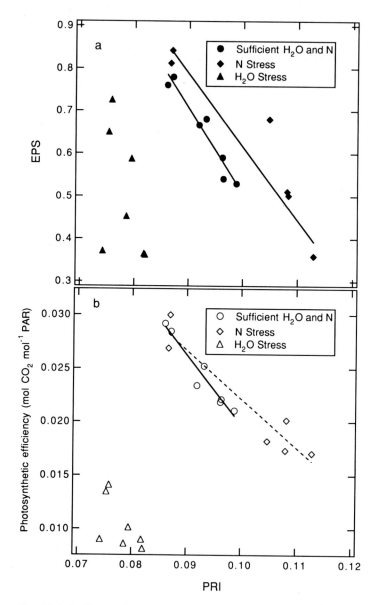

Fig. 24.4a,b. For sunflower canopies grown under contrasting resource availabilities, the relationship between a Physiological Vegetation Index (*PRI*) and **a** the epoxidation state (*EPS*) of the xanthophyll cycle, and **b** the efficiency of light utilization in net CO_2 uptake. The epoxidation state of the xanthophyll is defined as $(V + \frac{1}{2}A) / (V + A + Z)$, where V is violaxanthin (2 epoxide groups), A is antheraxanthin (1 epoxide group), and Z is zeaxanthin (0 epoxide groups). (After Gamon et al. 1992)

which can disrupt the relationship between electron transport and CO_2 uptake, and midday wilting.

Improved spectral indices may allow more general and robust estimates of photosynthesis based on remote sensing of the xanthophyll cycle. The relatively small magnitude of the spectral signal, the requirement for narrow band spectra, and complications from factors like water stress, however, make it likely that practical applications of this approach will be limited to scales ranging from the leaf to the landscape for the next several years.

24.4.3   Canopy Temperature

The difference between canopy and air temperature increases as canopy conductance decreases. This effect creates the potential for monitoring canopy conductance with a measure of canopy temperature. Infrared temperature measurements are increasingly used to schedule crop irrigation, and Peñuelas et al. (1992) showed that even mild water stress can lead to significant elevations in canopy temperature. At larger scales, Nemani and Running (1989a) demonstrated that the ratio of canopy temperature to NDVI increases with decreasing canopy conductance.

24.5   Remote Sensing of Environmental Factors

Models of photosynthesis and production typically require several kinds of environmental data, including incoming PAR, temperature, humidity, soil moisture, and CO_2. For large-scale assessments, remote sensing offers increasingly attractive options for obtaining some of these data.

Downwelling PAR can be calculated from satellite-based estimates of cloud cover. Bishop and Rossow (1991) developed a method that determines surface solar radiation by combining data from the International Satellite Cloud Climatology Project (ISCCP, which uses data from a number of geostationary and polar orbiting satellites) with calculations from a general circulation model. The method of Dye (1992) uses data from an ultraviolet channel on the Total Ozone Mapping Spectrometer (TOMS) to calculate cloud optical thickness.

Several sources yield temperature information. The microwave sounding units on the NOAA weather satellites yield middle troposphere temperatures with a monthly precision approaching 0.01 °C (Spencer and Christy 1990). Several sensors yield surface temperature, but the interpretation of radiometric surface temperature is difficult enough that few ecological models use it as a source of temperature data.

The polarization of passively emitted microwaves is sensitive to soil moisture (Schmugge et al. 1986), but the strong effect of vegetation makes soil moisture difficult to quantify, except in sites with very little vegetation (Choudhury et al. 1987). Microwave reflectance in active, synthetic aperture radar (SAR) systems is also sensitive to soil moisture, but only the longer wavelengths penetrate to the soil. Detection of soil moisture with experimental SAR systems is a rapidly developing, promising area (Ustin et al. 1991). Passive microwave emissions can also be used for the remote sensing of precipitation (Barrett 1991), offering an alternative approach to calculating soil moisture.

24.6 Conclusions

Remote sensing of photosynthesis represents a variety of techniques applied over a variety of scales. From the leaf to the global scale, however, the methods provide information that is difficult or impossible to obtain with other approaches, but that also tends to be difficult to interpret. The difficulty of interpretation has been and must continue to be addressed on several fronts. A clear physical understanding of the rediative transfer is important, but it is also critical to recognize that the factors controlling a remote sensing signature can be physical, biochemical, or ecological in nature. The development of new remote sensing instruments and techniques will provide improved access to some of these factors, but it is also important to continually rethink physiological and ecolgical models in the context of the nature and quality of the remote sensing data. Extending small-scale techniques to large spatial scales involves a number of challenges that can only be addressed with theoretical and empirical studies at a range of scales.

Remote sensing has led to important advances in the understanding of photosynthesis, and it offers exciting prospects for the future. The advances of the past have come from studies at a range of spatial scales and especially from studies in which the analysis of remote sensing data is combined with analysis of data from other sources. This paradigm provides a blueprint for future progress.

Acknowledgments. This chapter is dedicated to Professor Otto Lange, whose pioneering work in applying photosynthesis measurements to ecological questions laid many of the foundations for the work described here. Thanks to NASA (EOS/IDS award) for stimulating our studies in remote sensing and to awards from Fundación Conde de Barcelona and CICYT (Spain) for allowing JP to participate in them. Comments from J. Randerson were very useful.

References

Adams JB, Smith MO, Johnson PE (1986) Spectral mixture modeling: a new analysis of rock and soil types at the Viking Lander I Site. J Geophys Res 91: 8098–8112

Ashcroft PM, Catt JA, Curran PJ, Munden J, Webster R (1990) The relation between reflected radiation and yield on the Broadbalk winter wheat experiment. Int J Remote Sens 11: 1821–1836

Barrett EC (1991) Diagnostic, historic and predictive analyses of rainfall data using passive microwave image data. Paleogeogr Paleoclimatol Paleoecol 90: 99–106

Bilger W, Björkman O, Thayer SS (1989) Light-induced spectral absorbance changes in relation to photosynthesis and the epoxidation state of xanthophyll cycle components in cotton leaves. Plant Physiol 91: 542–551

Bishop JKB, Rossow WB (1991) Spatial and temporal variability of global surface solar irradiance. J Geophys Res 96(16): 839–858

Björkman O (1981) Responses to different quantum flux densities. In: Lange OL, Nobel PS, Osmond CB, Ziegler H (eds) Encyclopedia of plant physiology, vol 12A. Plant physiological ecology I. Springer, Berlin Heidelberg New York, pp 57–107

Brugnoli E, Björkman O (1992) Chloroplast movements in leaves: influence on chlorophyll fluorescence and measurements of light-induced absorbance changes related to ΔpH and zeaxanthin formation. Photosynth Res 32: 23–35

Carter GA, Theisen AF, Mitchell RJ (1990) Chlorophyll fluorescence measured using the Fraunhofer line-depth principle and relationship to photosynthetic rate in the field. Plant Cell Environ 13: 79–83

Chappelle EW (1987) Laser induced fluorescence (LIF) from plant foliage. IEEE Trans Geosci Remote Sens GE-25: 726–736

Chappelle EW, Williams DL (1987) Laser induced fluorescence (LIF) from plant follage. IEEE Trans Geosol Remote Sens GE-25: 726–736

Choudhury BJ (1987) Relationships between vegetation indices, radiation absorption, and net photosynthesis evaluated by a sensitivity analysis. Remote Sens Environ 22: 209–233

Choudhury BJ (1988) Relating Nimbus-7 37 GHz data to global land-surface evaporation, primary productivity and the atmospheric CO_2 concentration. Int J Remote Sens 9: 169–176

Choudhury BJ, Tucker CJ, Golus RE, Newcomb WW (1987) Monitoring vegetation using Nimbus-7 multichannel microwave radiometer's data. Int J Remote Sens 8: 533–538

Clark DH (1989) History of NIRS analysis of agricultural products. In: Marten GC, Shenk JS, Barton FE II (eds) Near infrared reflectance spectroscopy (NIRS): analysis of forage quality. US Department of Agriculture (Agriculture Handbook 643), Washington, DC, pp 7–11

Curran PJ, Dungan JL, Gholz HL (1990) Exploring the relationship between reflectance red edge and chlorophyll content in slash pine. Tree Physiol 7: 33–48

Demmig-Adams B, Adams WW III (1992) Carotenoid composition in sun and shade leaves of plants with different life forms. Plant Cell Environ 15: 411–419

Desjardins RL, Schuepp PH, MacPherson JI, Buckley DJ (1992) Spatial and temporal variations of the fluxes of carbon dioxide and sensible and latent heat over the FIFE site. J Geophys Res 97: 18467–18475

Dickenson RE, Henderson-Sellers A, Kennedy PJ, Wilson MF (1986) Biosphere/atmosphere transfer scheme (BATS) for the NCAR community climate model. NCAR Technical Note TN275. National Center For Atmospheric Research, Boulder

Dye DG (1992) Satellite estimation of global distribution and variability of incident photosynthetically active radiation. PhD Thesis, University of Maryland at College Park

Enting IG, Mansbridge JV (1991) Latitudinal distribution of sources and sinks of CO_2: results of an inversion study. Tellus 43B: 156–170

Field CB (1991) Ecological scaling of carbon gain to stress and resource availability. In: Mooney HA, Winner WE, Pell EJ (eds) Integrated responses of plants to stress. Academic Press, San Diego, pp 35–65

Gamon JA, Field CB, Bilger W, Björkman A, Fredeen AL, Peñuelas J (1990) Remote sensing of the xanthophyll cycle and chlorophyll fluorescence in sunflower leaves and canopies. Oecologia 85: 1–7

Gamon JA, Peñuelas J, Field CB (1992) A narrow-waveband spectral index that tracks diurnal changes in photosynthetic efficiency. Remote Sens Environ 41: 35–44

Gamon JA, Field CB, Roberts DA, Ustin SL, Valentini R (1993) Functional patterns in an annual grassland during an AVIRIS overflight. Remote Sens Environ 44: 239–253

Gerstl SAW, Borel CC (1992) Principles of the radiosity method versus radiative transfer for canopy reflectance modeling. IEEE Trans Geosci Remote Sens 30: 271–274

Goward SN, Tucker CJ, Dye DG (1985) North American vegetation patterns observed with the NOAA-7 advanced very high resolution radiometer. Vegetatio 64: 3–14

Hall FG, Huemmrich KF, Goward SN (1990) Use of narrow-band spectra to estimate the fraction of absorbed photosynthetically active radiation. Remote Sens Environ 34: 273–288

Holdridge LR (1947) Determination of world plant formations from simple climate data. Science 105: 367–368

Horler DNH, Dockray M, Barber J (1983) The red edge of plant leaf reflectance. Int J Remote Sens 4: 273–288

Jackson RD (1983) Spectral indices in n-space. Remote Sens Environ 14: 409–421

Krause GG, Weis E (1991) Chlorophyll fluorescence and photosynthesis: the basics. Annu Rev Plant Physiol Plant Mol Biol 42: 313–349

Kumar M, Monteith JL (1981) Remote sensing of crop growth. In: Smith H (eds) Plants and the daylight spectrum. Academic Press, London, pp 133–144

Lashof DA, Ahuja DR (1990) Relative contributions of greenhouse gas emissions to global warming. Nature 344: 529–531

Lichtenthaler HK (1988) Remote sensing of chlorophyll fluorescence in oceanography and in terrestrial vegetation: an introduction. In: Lichtenthaler JK (eds) Applications of chlorophyll fluorescence. Kluwer, Dordrecht, pp 287–297

Matthews E (1983) Global vegetation and land use: new high-resolution data bases for climate studies. J Climatol Appl Meteorol 22: 474–487

Melillo JM, Aber JD, Muratore JF (1982) Nitrogen and lignin control of hardwood leaf litter decomposition dynamics. Ecology 63: 621–626

Monteith JL (1977) Climate and the efficiency of crop production in Britain. Philos Trans R Soc Lond B 281: 277–294

Nemani R, Running SW (1989a) Estimating regional surface resistance to evapotranspiration from NDVI and thermal-IR AVHRR data. J Appl Meteorol 28: 276–284

Nemani R, Running SW (1989b) Testing a theoretical climate-soil-leaf area hydrologic equilibrium of forests using satellite data and ecosystem simulation. Agric For Meteorol 44: 245–260

Parslow JS, Harris GP (1990) Remote sensing of marine photosynthesis. In: Hobbs RJ, Mooney HA (eds) Remote sensing of biosphere function. Springer, Berlin Heidelberg New York, pp 269–290

Peñuelas J (1984) Pigment and morphological response to emersion and immersion of some aquatic and terrestrial mosses in N.E. Spain. J Bryol 13: 115–128

Peñuelas J, Savé R, Marfá O, Serrano L (1992) Remotely measured canopy temperature of greenhouse strawberries as indicator of water status and yield under mild and very mild water stress conditions. Agric For Meteorol 58: 63–77

Peñuelas J, Gamon JA, Griffin KL, Field CB (1993) Assessing community type, plant biomass, pigment composition, and photosynthetic efficiency of aquatic vegetation from spectral reflectance. Remote Sens Environ (in press)

Quay PD, Tilbrook B, Wong CS (1992) Oceanic uptake of fossil fuel CO_2: carbon-13 evidence. Science 256: 74–79

Richards JA, Kelly DJ (1984) On the concept of the spectral class. Int J Remote Sens 5: 987–991

Rock BN, Hoshizaki T, Miller JR (1988) Comparison of in situ and airborne spectral measurements of the blue shift associated with forest decline. Remote Sens Environ 24: 109–127

Running SW, Nemani RR (1988) Relating seasonal patterns of the AVHRR vegetation index to simulated photosynthesis and transpiration of forests in different climates. Remote Sens Environ 24: 347–367

Russell G, Jarvis PG, Monteith JL (1989) Absorption of radiation by canopies and stand growth. In: Russell G, Marshall B, Jarvis PG (eds) Plant canopies: their growth, form and function. Cambridge University Press, Cambridge, pp 21–39

Schmugge T, O'Neill PE, Wang JR (1986) Passive microwave soil moisture research. IEEE Trans Geosci Remote Sens 24: 12–22

Sellers PJ (1985) Canopy reflectance, photosynthesis and transpiration. Int J Remote Sens 6: 1335–1372

Sellers PJ (1987) Canopy reflectance, photosynthesis, and transpiration. II. The role of biophysics in the linearity of their interdependence. Remote Sens Environ 21: 143–183

Sellers PJ, Berry JA, Collatz GJ, Field CB, Hall FG (1992a) Canopy reflectance, photosynthesis and transpiration, III. A reanalysis using enzyme kinetics-electron transport models of leaf physiology. Remote Sens Environ 42: 187–216

Sellers PJ, Heiser MD, Hall FG (1992b) Relations between surface conductance and spectral vegetation indices at intermediate ($100\,m^2$ to $15\,km^2$) length scales. J Geophys Res 97: 19033–19059

Spencer RW, Christy JR (1990) Precise monitoring of global temperature trends from satellites. Science 247: 1558–1562

Tans PP, Fung IY, Takahashi T (1990) Observational constraints on the global CO_2 budget. Science 247: 1431–1438

Thayer SS, Björkman O (1990) Leaf xanthophyll content and composition in sun and shade determined by HPLC. Photosynth Res 23: 331–343

Ustin SL, Wessman CA, Curtiss B, Kasischke E, Way J, Vanderbilt VC (1991) Opportunities for using the EOS imaging spectrometers and synthetic aperture radar in ecological models. Ecology 72: 1934–1945

Ustin SL, Smith MO, Adams JB (1993) Remote sensing of ecological processes: a strategy for developing and testing ecological models using spectral misture analysis. In: Ehleringer JR, Field CB (eds) Scaling physiological processes: leaf to globe. Academic Press, San Diego, pp 339–357

Wessman CA, Aber JD, Melillo JM (1988) Remote sensing of canopy chemistry and nitrogen cycling in temperate forest ecosystems. Nature 335: 154–156

Whittaker RH, Likens GE (1975) Primary production: The biosphere and man. In: Lieth H, Whittaker RH (eds) Primary productivity of the biosphere. Springer, Berlin Heidelberg New York, pp 305–328

Wickland DE (1991) Mission to planet earth: the ecological perspective. Ecology 72: 1923–1933

Woodrow IE, Ball JT, Berry JA (1990) Control of photosynthetic carbon dioxide fixation by the boundary layer, stomata and ribulose 1,5-bisphosphate carboxylase/oxygenase. Plant Cell Environ 13: 339–347

Yoder BJ (1992) Photosynthesis of conifers: influential factors and potentials for remote sensing. PhD Thesis, Oregon State University

25 Are C_4 Pathway Plants Threatened by Global Climatic Change?

S. Henderson, P. Hattersley, S. von Caemmerer, and C.B. Osmond

25.1 Introduction

During his decades of leadership in plant photosynthetic ecophysiology Otto Lange has seen many major developments in the field, and participated in most of them. Perhaps none of these fields has been more exciting than the recognition of the ecophysiological implications of the C_4 pathway of photosynthetic carbon metabolism. Perhaps none will be so profound as the response of vegetation to the now well established, inexorable increase in atmospheric CO_2 concentration, the major greenhouse gas and the principal driver of impending global climatic change. Man, the dominant mammal, has effectively accelerated global respiration about 10^6 fold, by the combustion of several billion years' worth of accumulated photosynthate and other organic carbon in the course of a few hundred years. Here we ask if this activity which is leading to a rapid increase in atmospheric CO_2 concentration will directly affect the relative fitness of C_4-pathway plants, which may have evolved in response to low atmospheric CO_2 concentration.

The potential threat posed to C_4 pathway plants by these changes has been raised frequently. The concern arises because the advantages in productivity, water-use efficiency, and temperature tolerance that the CO_2 concentrating mechanism confers on C_4 plants over C_3 plants, may be diminished at elevated CO_2. In this chapter we will retrace, briefly, our understanding of photosynthetic ecophysiology that leads us to pose the title question and explore some of its dimensions. We will conclude that plants which evidently evolved an internal CO_2-concentrating mechanism to insulate themselves against low external CO_2 concentrations may be well equipped to handle rising atmospheric CO_2 concentrations. Indeed, the magnitude of the threat is not likely to be determined by effects on fitness of C_4 plants at all, but upon whether C_3 plants can take advantage of elevated CO_2 and thereby improve their fitness and competitive abilities.

25.2   Low Atmospheric CO_2 Concentrations and Evolution
of C_4 Pathway Photosynthesis

Low atmospheric CO_2 concentrations in the geologically recent past are commonly regarded as a major selective pressure leading to the evolution of the C_4 pathway of photosynthesis (Ehleringer et al. 1991). Values of 1500 to $3000 \, \mu l \, l^{-1}$ CO_2 are thought to have prevailed throughout much of angiosperm evolution during the Cretaceous (Budyko et al. 1987). Minima approaching present day concentrations are thought to have occurred in the Paleocene (7×10^7 years ago) and the Miocene (2×10^7 years ago), and these are thought to be the periods in which C_4 plants evolved. However, we need to bear in mind that stomatal closure in plants of xeric habitats may have exaggerated the selective pressure of low internal CO_2 concentrations over a longer period of time.

These speculations are confirmed by a few palaeontogical data. Fossil grass leaves from the Miocene in western North America are anatomically similar to extant C_4-pathway grasses (Thomasson 1987; Tidwell and Nambudiri 1989). Isotopic evidence from paleosols (carbonate deposits in soils formed about plant roots) suggest a dramatic change from predominantly C_3 to predominantly C_4 vegetation in the Padawar Plateau of Pakistan, again in the Miocene, some 10^7 years ago (Quade et al. 1989). This change is believed to mark a change in seasonality of rainfall, and the development of the Asiatic monsoon as a major feature of global climate, adding credence to the notion that seasonal water stress was an important factor associated with evolution of C_4-pathway photosynthesis.

Reliable data for atmospheric CO_2 concentration extends back only 160000 years. Figure 25.1 shows the concentration of CO_2 in ice of the Vostock and Siple cores, and that measured recently in the atmosphere itself (at Manna Loa, Hawaii, Boden et al. 1990). Although atmospheric CO_2 concentration has increased more or less logarithmically for about 10^5 years, the present concern arises because the rate of change in atmospheric CO_2 concentration is now possibly the most rapid in the history of angiosperm evolution. The Vostock core shows that CO_2 concentrations as low as $200 \, \mu l \, l^{-1}$ prevailed at several times in the last 160000 years and that an increase from 200 to $300 \, \mu l \, l^{-1}$ has occurred twice, with a half-time of about 10000 years. At the present time, with an atmospheric CO_2 concentration midpoint between 300 to $400 \, \mu l \, l^{-1}$, the rate of change is 100 to 1000 times more rapid. The increase in global atmospheric CO_2 concentration is possibly greater and more rapid than anything experienced by land plants since the end of the Cretaceous.

The notion that low atmospheric CO_2 concentrations should play such a dominant role in evolution of C_4 plants is based on the extensive biochemical, physiological, and anatomical evidence for the central importance of a CO_2 concentrating mechanism in C_4-pathway photosynthesis. First

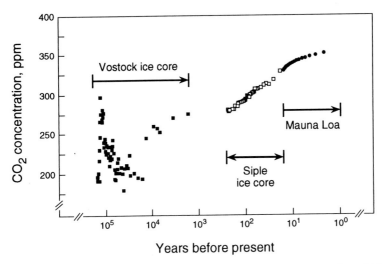

Fig. 25.1. Changes in atmospheric CO$_2$ concentrations in recent time. (After Boden et al. 1990)

proposed specifically by Björkman (1971) and Hatch (1971), a generalized model of the CO$_2$-concentrating mechanism, based on enzyme compartmentation and metabolite transport, is now widely accepted (Leegood and Osmond 1991). The overall processes of "cooperative photosynthesis" (Karpilov 1970) serve to elevate the CO$_2$ concentration in bundle-sheath cells of C$_4$ plants, thereby enabling ribulose-1,5-bisphosphate carboxylase/oxygenase (Rubisco) in these cells to function nearer to CO$_2$ saturation. The concentrations achieved (>2000 μl l^{-1}; Furbank and Hatch 1987) are about 10–100 times greater than those experienced by Rubisco in C$_3$ plants. In effect, evolution of C$_4$ photosynthesis has restored to the photosynthetic carbon reduction cycle in bundle-sheath cells an internal CO$_2$ environment remarkably similar to the external atmospheres of the Cretaceous.

The extent to which C$_4$ plants are threatened by climatic change in the magnitude confronting the biosphere in the next 100 years depends upon physiological flexibility in the face of changed climates, as well as on genetic diversity, population size and area, dispersal, and rate of evolution. Although we will be able to specify several dimensions of physiological flexibility so far as C$_4$ plants are concerned, it is important to stress that we know very little about the other factors. Recent studies of the evolution of C$_4$ plants (Osmond et al. 1980; Sanderson et al. 1990) and of C$_3$–C$_4$ intermediate species indicate hitherto unsuspected natural genetic diversity with respect to photosynthesis (Edwards and Ku 1987; Monson and Moore 1989). Furthermore, artificial hybrids between C$_3$ and C$_4$ species of *Atriplex* L. (Björkman et al. 1971) show immense physiological diversity, comparable to that for artificial hybrids obtained with *Flaveria* A.L. Juss (Huber et al.

1989). Although we know something of the population size and area for many C_4 species, little is known of contemporary evolution of C_4 plants. However, in the grass *Alloteropsis* C. Presl., there is evidence from recent collections that evolution of C_4 photosynthesis is in process (P.W. Hattersley and C. Long 1992, unpubl.).

25.3   Physiological Flexibility in C_4 Plants Under High CO_2 Concentrations

It is often assumed that the advantages which the C_4 photosynthetic pathway confers on plants over the C_3 photosynthetic pathway will be somewhat diminished under elevated global concentrations of CO_2. This conclusion is based on observations that short-term CO_2 assimilation by leaves of C_4 species is almost completely saturated at current CO_2 concentrations. In leaves of C_3 species it continues to increase with increasing external CO_2 concentration as Rubisco becomes increasingly saturated by CO_2. Increased rates of carboxylation and lower rates of photorespiration result in larger net CO_2 assimilation rates and improved quantum efficiencies at high temperatures for C_3 plants. However, growth and ecological advantage are not simply related to photosynthetic assimilation. For example, the advantages which elevated CO_2 concentrations may confer on rate of C_3 photosynthesis can be negated by their limited ability to use or store the extra carbohydrate produced. In this section we shall argue that although there are no obvious benefits of elevated CO_2 concentrations to plants that possess the C_4 photosynthetic pathway, the physiology of these plants may be well suited to these environments.

25.3.1   Coordination of Metabolism

The dramatic specialization of photosynthetic functions between mesophyll and bundle-sheath cells in C_4 plants means that CO_2 fixation depends upon regulation of metabolite transport both within and between cells. The C_4 cycle can be considered simply as a multifaceted, ATP-driven CO_2 pump which concentrates CO_2 in the bundle sheath. Yet its very complexity seems to confer a large degree of flexibility. For this system to operate efficiently, CO_2 supply must match consumption by the C_3 cycle with the avoidance of excessive leakage of CO_2 from the bundle sheath. In NADP malic enzyme species, such as maize and sorghum, the C_4 cycle is obligatorily coupled to the C_3 cycle because NADPH, generated in the reactions catalyzed by NADP malic enzyme, is re-oxidized in the bundle-sheath chloroplast by the reduction of glycerate-3-phosphate (PGA). However, in all the C_4 plants, a PGA/triose-P shuttle operates between the mesophyll and bundle sheath.

Thus, electron transport in the mesophyll chloroplasts not only powers regeneration of phosphoenolpyruvate (PEP) in the C$_4$ cycle but also drives regeneration of PGA in the mesophyll cells so that interdependence between the C$_3$ and C$_4$ cycle exists, irrespective of the type of decarboxylation pathway (Hatch and Osmond 1976).

Further metabolic communication between the two cycles occurs through the interconversion of PGA and PEP catalyzed by phosphoglycerate mutase and enolase (Furbank and Leegood 1984). The relationship between the PGA and PEP pools is shown in Fig. 25.2. This figure demonstrates the strong positive correlation between the two metabolite pools in leaves of *Zea mays* L. and *Amaranthus edulis* L. under different flux conditions generated by short-term variations in irradiance, CO$_2$ partial pressures and leaf temperatures (Leegood and von Caemmerer 1988, 1989; Labate et al. 1990). In both, the size of the PGA pool was found to be closely related to the rate of PGA production. Since PEP carboxylase is a highly regulated enzyme, with a high affinity for bicarbonate and a relatively low affinity for PEP (O'Leary 1982), the amount of PEP is likely to be of crucial importance

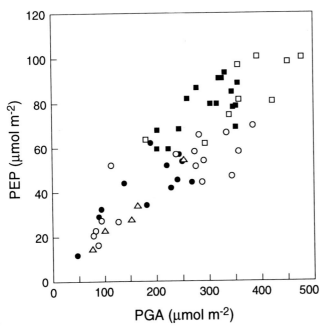

Fig. 25.2. The relationship between the steady-state contents of PGA and PEP in leaves of *Zea mays* and *Amaranthus edulis* with varying intercellular partial pressure of CO$_2$ (p$_i$), irradiance and leaf temperature. The data are from Leegood and von Caemmerer (1988, 1989) for *Z. mays* L. [p$_i$ varied ($\bigcirc$), irradiance varied ($\square$)] and *A. edulis* [p$_i$ varied ($\bullet$), irradiance varied ($\blacksquare$)], and from Labate et al. (1990) for *Z. mays* [leaf temperature varied ($\triangle$)]. Data from Labate et al. (1990) are converted from nmol mg Chl^{-1} to µmol m^{-2} assuming 363 mg Chl m^{-2} for *Zea mays*. (Leegood and von Caemmerer 1988)

in controlling the rate of CO_2 assimilation (Leegood and Osmond 1991). Although one can argue that such short-term perturbations (Fig. 25.2) have a more severe effect on photosynthetic metabolism than when plants are allowed to acclimate to changes in growth conditions, the fact that large changes in metabolite pool sizes did not occur when leaves were exposed to CO_2 concentrations above ambient suggests that extensive metabolic regulation occurs during C_4 photosynthesis which may provide mechanisms that allow it to adjust readily to elevated CO_2 concentrations. As yet no studies have addressed metabolite regulation during C_4 photosynthesis in plants grown in elevated CO_2 concentrations.

25.3.2 Leakage of CO_2 from the Bundle Sheath

The efficiency of the CO_2 pump in the C_4 cycle is influenced by the extent of passive leakage of CO_2 from the bundle sheath. This leakage of CO_2 represents an energy cost to the leaf because ATP is required for the regeneration of PEP. The rate of leakage of CO_2 from the bundle sheath cannot be measured directly but estimates have been made in various ways. Hatch and Osmond (1976) estimated that, on diffusional grounds, the back flux of CO_2 could be about 10% of the C_4 acid flux, a value confirmed in recent biochemical studies (Furbank and Hatch 1987; Jenkins et al. 1989).

Theoretical treatments of carbon isotope discrimination that occurs during C_4 photosynthesis have shown that the extent of discrimination is linked to the rate of leakage of CO_2 from the bundle sheath (O'Leary 1981; Peisker 1982; Farquhar 1983; Deleens et al. 1983; Peisker and Henderson 1992). Henderson et al. (1992) used direct measurements of carbon isotope discrimination made in conjunction with gas-exchange to examine leakiness (the ratio of leakage of CO_2 from the bundle sheath to the rate of PEP

Table 25.1. Estimates of bundle-sheath leakiness at low and high CO_2 concentrations

Species	Pa (μbar)	Leakiness[a]
Amaranthus edulis	340	0.21
	570	0.21
Sorghum bicolor	342	0.21
	566	0.23

[a] Leakiness, the fraction of CO_2 derived from C_4 acid decarboxylation which is lost by leakage from the bundle sheath, was estimated from simultaneous measurements of gas-exchange and carbon isotope discrimination. Measurements were made at a leaf temperature of 29 °C, irradiance of 1600 μmol quanta $m^{-2}s^{-1}$ and a leaf-to-air vapor pressure difference of approximately 14 mbar and two CO_2 concentrations, p_a. For further details, see Henderson et al. (1992).

carboxylase) under a variety of short-term environmental perturbations of CO$_2$ partial pressure, irradiance, and leaf temperature. The near constancy of 21% leakiness over the wide variety of conditions imposed is greater than that estimated previously by Hatch and Osmond (1976). Evidently many factors contribute to the regulation that occurs between the mesophyll and the bundle-sheath cells. Table 25.1 also shows that leakiness did not vary when measured during short-term variation in CO$_2$ concentration with *Sorghum bicolor* (L.) Moench and *Amaranthus edulis*. However, there was a small increase in leakiness at irradiances of less than 250 µmol quanta

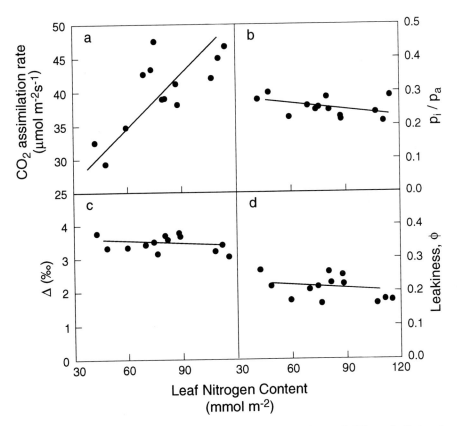

Fig. 25.3a–d. The effect of leaf nitrogen content on **a** Rate of CO$_2$ assimilation in *Sorghum bicolor* (L.) Moench (cv. T×610SR). **b** The ratio of intercellular to ambient partial pressure of CO$_2$, p_i/p_a. **c** Carbon isotope discrimination measured during CO$_2$ exchange, Δ. **d** Bundle-sheath leakiness, ϕ. Leakiness was calculated from the measurements of Δ in **c**, and measurements of p_i/p_a in **b** using the equation $\Delta = 4.4 + (28.2\phi - 9.6)\, p_i/p_a$. The *lines* are drawn by eye. Measurements were made at an irradiance of 1600 µmol quanta m^{-2}s^{-1}, an ambient CO$_2$ partial pressure of 330 µbar and a leaf-to-air vapor pressure difference of 20 mbar and each symbols is the average of three to five measurements per leaf. Leaf nitrogen was determined on an auto-analyzer after a Kjehldahl digestion

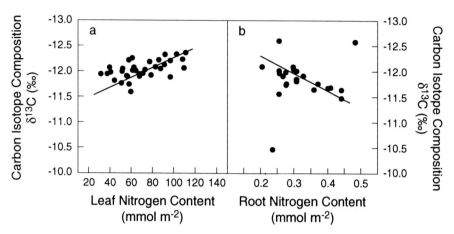

Fig. 25.4a,b. The effect of nitrogen nutrition on **a** Carbon isotope composition, $\delta^{13}C$, of whole plant leaf dry matter of *Sorghum bicolor* (cv. T×610SR) plants plotted as a function of their leaf nitrogen content. **b** Carbon isotope composition of root dry matter plotted as a function of root nitrogen content. Root and leaf nitrogen were determined on an auto-analyzer after a Kjehldahl digestion

$m^{-2}s^{-1}$, which is consistent with the observations that C_4 species in general do not grow well in deep shade.

Bowman et al. (1989) found that high salinity and water stress increased leakiness in *Andropogon glomeratus* Walter, and Wong and Osmond (1991) reported what appeared to be increased leakiness in *Echinochloa frumentacea* Link at low nitrogen nutrition. Leakiness is the product of the physical conductance of the bundle sheath to CO_2 leakage and the relative biochemical capacities of the C_3 and C_4 cycles. Indeed, once bundle-sheath conductance is sufficiently high, leakiness may be largely determined by the relative activities of Rubisco and PEP carboxylase. Interestingly, the ratio of these two enzymes changes when C_4 plants are grown with different levels of nitrogen nutrition (Sugiyama et al. 1984; Sage et al. 1987). However, Henderson (1992) found little variation in leakiness when *Sorghum bicolor* was grown with different levels of nitrogen nutrition, despite an almost twofold variation in CO_2 assimilation (Fig. 25.3). Instead, carbon isotope composition of shoot and root dry matter responded differently to changes in their respective nitrogen contents (Fig. 25.4). Evidently, we cannot infer leakiness in the shoot from the isotopic composition of roots, but we do not understand the basis of this tissue-specific effect.

We suggest that the ability to maintain constant leakiness over such a range of environmental perturbations demonstrates that the C_4 photosynthetic metabolism already possesses the regulatory mechanisms to cope with increasing atmospheric CO_2 concentration or increasing temperature. Indeed, measurements of leakiness may provide an indication of the performance of C_4 plants when grown at high CO_2 concentrations.

25.3.3 Translocation of Carbohydrate

In C$_4$ plants the C$_3$ cycle in bundle-sheath cells is exposed to high CO$_2$ concentrations which support greater rates of sucrose synthesis than in C$_3$ species (Kalt-Torres et al. 1987; Furbank and Foyer 1988), but the extra carbohydrate does not accumulate because the rate of assimilate export is greater than in C$_3$ leaves (Stephenson et al. 1976; Wardlaw 1982; Kalt-Torres et al. 1987). The higher capacity for sucrose synthesis in maize may be supported by the tenfold lower sensitivity of enzymes of sucrose synthesis to their regulators (Stitt and Heldt 1985). The higher capacity for sucrose translocation in C$_4$ plants may be supported by the closer proximity of mesophyll cells to the vascular system. Indeed, several authors have used this or related anatomical properties for taxonomic and predictive purposes (Crookston and Moss 1974; Hattersley and Watson 1975).

Thus we also believe that C$_4$ plants will cope with elevated CO$_2$ because they have mechanisms to deal with high rates of carbohydrate synthesis from a C$_3$ cycle exposed to high concentrations of CO$_2$. Indeed, short-term exposures (h) of C$_4$ leaves to $\times 2$ current CO$_2$ concentrations had little effect on the amount of carbohydrate accumulated (Wardlaw 1982; Blechschmidt-Schneider et al. 1989). On the other hand, the ability of C$_3$ plants to adapt and utilize high CO$_2$ concentrations is often limited by their capability to store, translocate, or use the extra carbohydrate produced under high CO$_2$ concentrations (Stitt 1992).

25.3.4 Water Use Efficiency

Another advantage which the C$_4$ photosynthetic pathway confers under present CO$_2$ concentrations is a higher water use efficiency compared to C$_3$ species, and it is often proposed that this advantage may diminish under elevated CO$_2$. However, available data suggest that C$_4$ plants may retain most of this advantage (Table 25.2). The instantaneous water use efficiency (ratio of CO$_2$ assimilation rate to transpiration rate for single leaves) or whole plant transpiration efficiency (unit dry matter produced per unit water used) increases when both C$_3$ plants and C$_4$ plants are grown at $\times 2$ current CO$_2$ concentration. In C$_3$ plants this is generally due to increases in the CO$_2$ assimilation rate, whereas in C$_4$ plants it is due to a decrease in stomatal conductance at elevated CO$_2$.

25.3.5 Nitrogen Use Efficiency

The nitrogen use efficiency of C$_4$ plants is generally held to be superior to C$_3$ plants because Rubisco functions under CO$_2$ saturation in the bundle sheath (Brown 1978) and less Rubisco protein is required to sustain high rates of

Table 25.2. Comparison of leaf, W^a, and whole plant water use efficiencies, W^b, in C_3 and C_4 species grown at current and $\times 2$ ambient CO_2 concentrations. The table shows the ratios of W at $\times 2$ and ambient CO_2 concentrations

Measurement	C_3 (High/ambient)	C_4 (High/ambient)
Instantaneous W^a (range)	2.24 (1.44–3.75)	1.58 (1.43–1.85)
Whole plant W^b (range)	1.58 (1.29–2.01)	1.38 (1.22–1.53)

[a] Leaf water use efficiency was calculated from leaf gas-exchange measurements of CO_2 assimilation and transpiration (mmol C mol^{-1} H_2O) made at the CO_2 concentration which the plants were subjected to during growth. Data were taken from Wong (1979), Rogers et al. (1983), Smith et al. (1987), and Ziska et al. (1991).
[b] Morison and Gifford (1984) measured whole plant water use efficiencies (g dry matter kg^{-1} H_2O).

photosynthesis. This advantage of C_4 plants may be lost under elevated CO_2 because Rubisco sparing is often observed in C_3 plants grown under high CO_2 (Stitt 1992). There have been few studies on the effect of elevated CO_2 on nitrogen use efficiency of C_4 plants (Wong 1979), and counter-intuitive responses abound when one attempts to assess interactions between high CO_2 and nitrogen use efficiency. In one set of experiments, elevated CO_2 stimulated growth of C_4 plants in mixed cultures at high irradiance/high nitrogen, and in all low nitrogen treatments, but there was a marked decline in growth of C_4 plants with at an increasing proportion of C_3 plants in the mixture under the low nitrogen treatments (Wong and Osmond 1991).

25.4 Growth and Competition Between C_4 and C_3 Plants Under Elevated CO_2

A recent comprehensive literature survey of 156 species grown singly under $\times 2$ current CO_2 reported a significant average increase in dry matter accumulation for C_4 plants of 22%, compared to 41% for C_3 plants (Poorter 1993). There is a popular perception that these growth enhancements under elevated CO_2 are not entirely expected on the basis of leaf level responses of photosynthesis in C_4 plants. However, the photosynthetic responses to CO_2 in C_4 plants are not necessarily completely saturated at current CO_2 concentrations. Poorter, who explored a range of physiological interactions which could account for the growth responses, concluded that an increase throughout the growth period of only 2% in CO_2 assimilation is enough to explain the increased growth of C_4 plants. Differences in growth response to elevated CO_2 between C_3 and C_4 plants are much smaller than frequently supposed, and should be taken into account when speculating on competitive interactions.

Studies of C$_3$ and C$_4$ plants in competition under elevated CO$_2$ show variable outcomes. Patterson et al. (1984) found that elevated CO$_2$ improved the competitive ability of soybean [*Glycine max* (L.) Merrill, C$_3$] against its major weed Johnson grass [*Sorghum halepense* (L.) Pers., C$_4$]. Wray and Strain (1987) found that the "old field" competitor *Aster pilotus* Willd. (C$_3$) became more competitive with *Andropogon virginicus* L., (C$_4$) at elevated CO$_2$. Other studies of annual species grown together have demonstrated that the advantages conferred under elevated CO$_2$ on C$_3$ plants are not always so marked, and that it is difficult to predict competitive outcomes. For example, Bazzaz and Carlson (1984) found that the competitive advantage of *Amaranthus retroflexus* L. (C$_4$) over *Polygonum pensylvanicum* (C$_3$) in the current atmosphere and low soil moisture was only partially offset by elevated CO$_2$. In experiments with "mini-communities" Zangerl and Bazzaz (1984) found that elevated CO$_2$ concentrations increased relative dominance of C$_3$ species, with differing, species-dependent effects on vegetative biomass and resource allocation to seed; however, C$_4$ species also increased their biomass and seed production at elevated CO$_2$. Growth analysis of competition experiments demonstrated that responses to elevated CO$_2$ were greatest during early growth (Bazzaz et al. 1989). Surprisingly, these effects were greater in *Amaranthus retroflexus* (C$_4$) than *Abutilon theophrasti* medikus (C$_3$), and this determined the competitive outcome.

In many cases, the above experiments are difficult to interpret because of pot size effects (Arp 1991). Competition belowground is likely to be crucial, and few experiments to date have addressed this problem. Such studies are greatly facilitated by the measurements of $\delta^{13}C$ value of total root biomass, a technique already well proven in the field (Tieszen and Archer 1990). Wong and Osmond (1991) followed interactions between the improbable competitors *Echinochloa frumentacea* (C$_4$) and *Triticum aestivum* L. (C$_3$) which had shoot/root ratios of 2 to 12, so shoot responses dominated. Shoot growth of the C$_3$ plants was routinely stimulated by elevated CO$_2$, but under low nitrogen the C$_4$ plants also responded. Surprisingly, C$_4$ plants gained the advantage belowground only under high nitrogen treatments, suggesting that root access to nitrogen supply, rather than shoot nitrogen use efficiency, determined competitive outcome. The conclusion of Bazzaz et al. (1989) sums up our present position in that: "The precise competitive relationships among and between the two groups will not be predictable from individual responses as these will be modified by the other environmental factors that impinge upon the plant community."

25.5 Present Distributions and Diversity of C$_4$ Plants

The grass family is the most comprehensively studied plant family with respect to the biology of C$_4$ photosynthesis, even though it is one of only two monocotyledonous plant families to contain C$_4$ species (compared with more than 15 dicotyledonous families). C_4 species comprise about half of the approximately 10000 grass species known worldwide (Hattersley 1987; Hattersley and Watson 1992). The latter authors concluded that the diversification of grasses at subfamily level may have been mainly associated with divergence in photosynthesis in response to declining CO_2 concentrations and associated changes in temperature and rainfall. It is striking that the climatic variables that seem likely to have been important in the evolutionary diversification of grasses, are the same as those which are now being altered anthropogenically; CO_2, temperature and precipitation. Temperature and to a lesser extent the seasonality of rainfall seem to have been most important for the divergence of C_3 and C_4 plants in response to lowering of atmospheric CO_2 partial pressure, while amount of rainfall (in tropical and subtropical climates) may have been most significant for diversification of the C_4 pathway itself (Hattersley and Watson 1992). By no means all tropical plants are C_4, but where C_4 photosynthesis evolved in a group (such as the grasses), the outcome of intense competition between species, otherwise similarly adapted, seems to have favored the C_4 pathway.

Comparative present-day distributions of C_3 and C_4 grasses, for example for North America (Teeri and Stowe 1976), Australia (Hattersley 1983), Japan (Takeda et al. 1985) and Argentina (Cavagnaro 1988), suggest that temperature is the key climatic variable which determines the presence or absence of species with the two photosynthetic pathways. The correlation is often best with minimum summer temperature as shown for Australian grasses (Fig. 25.5a). Biochemical diversity within C_4 grasses, on the other hand, is best correlated with rainfall parameters. Thus percent C_4 acid decarboxylation type in Australian C_4 grass floras correlates most closely with median annual rainfall or median summer (December + January + February) rainfall (Fig. 25.5b). It is fascinating to reflect upon the water relations parameters which might underscore Fig. 25.5b.

There may well be no simple explanation for Fig. 25.5a. Although it is commonly accepted that C_4 plants may outcompete C_3 plants at high temperatures because the CO_2 concentrating mechanism renders them insensitive to increased photorespiration, other factors could be just as, or even more important. Thus, minimum temperature during the growth season might limit migration of C_4 species of tropical origin because minimum (night) temperatures limit development of the photosynthetic apparatus (Slack et al. 1974) or translocation and/or utilization of assimilates (Potvin et al. 1985). This would provide an alternative to the explanation that limitations on the spread of C_4 species is due to the increasing competitiveness of C_3 plant growth rates at lower temperatures which prevail at higher latitudes and altitudes.

In either case, reproductive success presumably underlies the presence or absence of species in a particular habitat, and almost nothing is known of, for example, relationships between Summer minimum temperature and reproductive success in the following Autumn. Although floral emergence in

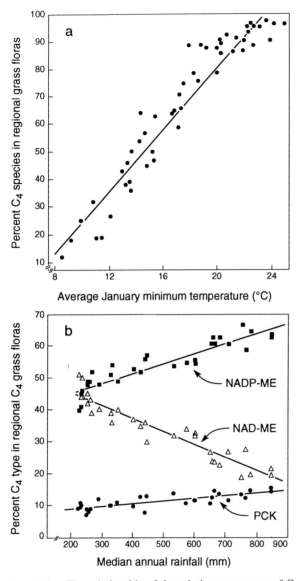

Fig. 25.5a. The relationship of the relative occurrence of C_3 and C_4 native grass species in regional grass floras in Australia and average January minimum temperature (°C) (subdivisional area-weighted mean) taken from Hattersley (1983). Linear regression y = $-32.46 + 5.71 \times$ ($r^2 = 0.92$). The regional grass floras correspond to subdivisions on a State by State basis, as used by Hattersley (1983). **b** The relationships of the relative occurrence of native C_4 grass species of the NADP-ME, NAD-ME and PCK C_4 types in Australia, on a regional C_4 grass flora basis, and median annual rainfall (mm, as lowest value per subdivision) taken from Hattersley (1992). C_4 type for species was assigned on a biochemical basis (that is, by assay of the activity of the relevant C_4 acid decarboxylating enzyme: NADP-malic enzyme (NADP-ME), NAD-malic enzyme (NAD-ME) or phosphoenolpyruvate carboxykinase (PCK). Linear regressions NADP-ME, y = 39.5 + 0.031x, NAD-ME y = 53.3 − 0.040x, PCK y = 7.06 + 0.009x

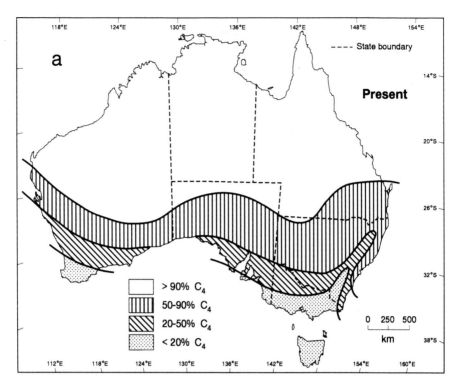

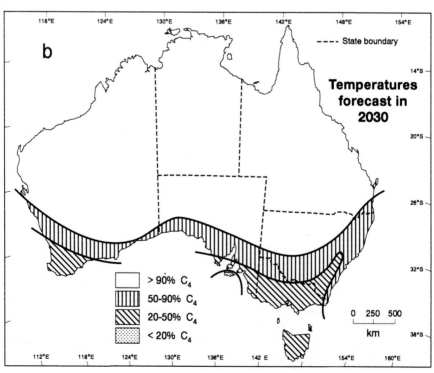

the weed *Echinochloa crus-galli* (L.) P. Beauv. in North America was best predicted by August maximum temperature, plants from the coolest treatments produced a small number of heavy seeds of superior germinabilty (Potvin 1991). Precisely which trait, or which combination of traits, is most relevant to the relative fitness of C_3 and C_4 species remains baffling.

25.6 Future Distributions of C_4 Plants

In spite of our limited insights into processes which underlie the relationships between temperature parameters and the representation of C_4 plants in regional floras, these relationships are remarkably robust. Although there is no compelling evidence that C_4 plants will be physiologically debilitated by elevated CO_2 itself, and although competitive interactions remain poorly understood, it seems reasonable, nevertheless, to use present day algorithms for C_4 grass distribution vs. temperature to predict changes in composition which might follow increases in temperature attributable to greenhouse gases.

Global warming is the least disputed outcome from increases in greenhouse gases and there is a high degree of concurrence among different global circulation models so far as future temperature in Australia is concerned. According to the *Climate Change Scenario for the Australian Continent* for 2030, issued by the CSIRO Climate Impact Group in November 1991, mean annual temperature is estimated to increase by 1 to 2°C in northern coastal areas, by 1 to 3°C in southern coastal areas, and by 2 to 4°C inland, with the larger increases being observed in dry regions. These projections were used to adjust the average January minimum temperatures of each region considered by Hattersley (1983), and then changes in percent C_4 grasses of regional grass floras were calculated using the linear regression equation of Fig. 25.5a.

The present relative distribution of C_3 and C_4 native grasses in Australia is shown in Fig. 25.6a, redrawn from Hattersley (1983). The isolines for projected relative C_3/C_4 grass species' occurrence for the temperatures

Fig. 25.6a,b. Maps of Australia showing 90, 50, and 20% C_4 isolines. **a** At present temperatures. **b** At temperatures forecast for the year 2030. Percent C_4 refers to the percentage of native species in a regional grass flora that is C_4. **a** is based on Fig. 2 in Hattersley (1983), with isolines interpolated between the centers of regional areas for which percent C_4 values are available. Isolines are not based on point data and their positions are therefore only approximate (refer to Hattersley 1983 for source data). **b** is a first approximation of the change in the position of the isolines using continental temperatures forecast by the CSIRO's Climate Impact Group (1991) for the year 2030, adjusting the January minimum temperature by this amount, and using the linear regression equation from Fig. 25.5a to calculate percent C_4 values

forecast in 2030 are shown in Fig. 25.6b. This projection suggests that the 90% C_4 isoline would extend southwards by 200–400 km, and that the grass flora of more than 80% of the continental area could be almost totally dominated by C_4 species. The magnitude of the potential shift is un-expected, until one realises that an increase in temperature of 4 °C is 25% of the present continental range in average January minimum temperature (Fig. 25.5a). Whether this changed grass distribution will be realized, and when, depends on many factors, especially species migration rate, barriers to migration and present land use. We cannot predict the extent to which elevated CO_2 itself may modulate the temperature response of C_3 and C_4 species, but some territorial advance of C_4 grasses still seems likely.

Factors other than the direct effect of CO_2 and temperature and phy-sical migration will also alter relative distributions of C_3 and C_4 plants. These include intensity and seasonality of rainfall, the response of other components of the vegetation to climatic change, and changes in land man-agement practices as weather changes. Rainfall predictions, even for a physiographically simple landscape such as the Australian continent, are less certain than temperature projections in climate change. Current regional climatic change projections for Australia (using several global climate models) suggest that the summer rainfall region will become more extensive over the continent (A.B. Pittock 1992, pers. comm.). This would again favor C_4 grasses. In fact, the southerly movement of summer rainfall dominance is likely to complement to the temperature-driven southerly migration of C_4 plants. However, other potential changes, for example in the Darwin-Gulf region of the Northern Territory in Australia could exclude C_4 grasses. If rainfall was more uniform over the year in a tropical or subtropical area that had previously been a "wet-dry tropics" area, then closed forest might replace open woodland or savanna woodlands, and C_4 grasses (which are high light-requiring plants) might be excluded.

It is likely that representation of the three C_4 types in grass floras will also change in response to changed rainfall regimes. Based on regional generalizations for New South Wales (Pittock 1991) and the whole Australian continent (Pittock et al. 1991), we have estimated the percentage increase or decrease in summer and annual rainfall for some selected regions. The equations derived by Hattersley (1992) (Fig. 25.5b) were used to predict changes in percent C_4 type in grass floras. Preliminary analyses of the effects of changed precipitation suggest little effect on percent C_4 type in any subdivision of the continent.

25.7 Conclusion

This essay is a crude, preliminary attempt to consider some of the dimen-sions of what, in effect, will be a global experiment in photosynthetic ecophysiology over which we have almost no control, but from which we may learn much. In the preparation of the essay we have come to appreciate the evolutionary significance of the biochemical and physiological cooperation which allows C$_4$ plants to sequester the photosynthetic carbon reduction cycle in the high CO$_2$, Cretaceous-like, atmosphere of bundle-sheath cells. We have become less concerned about coordination and fine tuning, and more appreciative of the potential for compensation and compromise that arises as a consequence of metabolic complexity. We conclude that, so far as making the best use of elevated atmospheric CO$_2$ concentrations is concerned, C$_4$ plants have "been there, done that". Beyond the biochemical flexibility of CO$_2$ assimilation, we also suspect that the physiological and anatomical requirements for handling high rates of assimilation, especially at high temperatures, have been adequately selected during evolution of C$_4$ plants. However, we have to concede that we know almost nothing about the photosynthetic ecophysiology of C$_4$ plants growing under $\times 2$ current CO$_2$ concentrations. Much experimental research needs to be done to explore an array of counter intuitive physiological responses, such as the significance of nitrogen use efficiency (Wong and Osmond 1991). Most of all, we need to explore the effects of elevated CO$_2$ concentration on assimilate partitioning and its reproductive outcome through viable seed and vegetative development (Potvin 1986, 1991).

Although some of the relative advantage that the CO$_2$-concentrating mechanism may confer on the photosynthetic rate of C$_4$ plants in present atmospheres may diminish in the face of global climatic change, it is by no means clear that C$_3$ plants will uniformly exploit accelerated CO$_2$ fixation in the long term. In fact C$_3$ plants face an array of limitations on translocation and utilisation of assimilates, of respiration at elevated temperatures and of unpredictable reproductive outcomes (Smith et al. 1989; Potvin 1991) that may constitute larger threats to fitness than they do to C$_4$ plants. We are tempted to conclude that the outcomes of the contest between plants with the C$_3$ and C$_4$ photosynthetic pathway under elevated CO$_2$ will depend more on the extent of advantages gained by the former than on those surrendered by the latter.

Indeed, when considering the increase in temperature occurring with increasing CO$_2$, assuming that C$_4$ grasses will not be significantly debilitated by elevated CO$_2$, and that response to elevated temperature can be projected on the basis of current correlations, we conclude that there may well be a significant increase in the representation of C$_4$ grasses in the Australian flora. The southerly and easterly expansion in the range of C$_4$ grasses in Australia may match the northerly expansion of the range of North American grasses such as *Echinochloa*. It is evident that well-designed monitoring programs to follow abundance and frequency of C$_3$ and C$_4$ plants in key locations could provide valuable information for photosynthetic ecophysiology in the decades ahead and serve as useful indicators of the impact of global climatic change. Our results suggest that it may be more appropriate to ask not whether C_4 grasses will be endangered by global climatic change, but whether C_3 grasses will be endangered, especially in Australia.

References

Arp WJ (1991) Effects of source-sink relations on photosynthetic acclimation to elevated CO_2. Plant Cell Environ 14: 869–875

Bazzaz FA, Carlson RW (1984) The response of plants to elevated CO_2. I Competition among an assemblage of annuals at two levels of soil moisture. Oecologia 62: 196–198

Bazzaz FA, Garbutt K, Reekie EG, Williams WE (1989) Using growth analysis to interpret competition between a C_3 and C_4 annual under ambient and elevated CO_2. Oecologia 62: 196–208

Björkman O (1971) Comparative photosynthetic CO_2 exchange in higher plants. In: Hatch MD, Osmond CB, Slatyer RO (eds) Photosynthesis and photorespiration, Wiley, New York, pp 18–32

Björkman O, Nobs M, Pearcy R, Boynton J, Berry J (1971) Characteristics of hybrids between C_3 and C_4 species of *Atriplex*. In: Hatch MD, Osmond CB, Slatyer RO (eds) Photosynthesis and photorespiration. Wiley, New York, pp 105–119

Blechschmidt-Schneider S, Ferrar P, Osmond CB (1989) Control of photosynthesis by the carbohydrate level in leaves of the C_4 plant *Amaranthus edulis* L. Planta 177: 515–525

Boden TA, Kanciruk P, Farrell MP (1990) Trends '90. A compendium of data on global change. ORNL/CDLAC 36 Oak Ridge National Laboratory, Oak Ridge

Bowmann WD, Hubick KT, von Caemmerer S, Farquhar GD (1989) Short-term changes in leaf carbon isotope discrimination in salt- and water-stressed C_4 grasses. Plant Physiol 90: 162–166

Brown RH (1978) A difference in nitrogen use efficiency in C_3 and C_4 plants and its implications in adaption and evolution. Crop Sci 18: 93–98

Budyko MI, Ronov AB, Yanshin YL (1987) History of the earth's atmosphere. Springer, Berlin Heidelberg New York

Cavagnaro JB (1988) Distribution of C_3 and C_4 grasses at different altitudes in a temperate arid region of Argentina. Oecologia 76: 273–277

Climate Impact Group (1991) Current Climate Change Scenario for the Australian Region-2030. Climate Impact Group, CSIRO Division of Atmospheric Research, Mordiallic Victoria, 4 pp

Crookston RK, Moss DN (1974) Interveinal distance for carbohydrate transport in leaves of C_3 and C_4 grasses. Crop Sci 14: 123–125

Deleens E, Ferhi A, Queiroz O (1983) Carbon isotope fractionation by plants using the C_4 pathway. Physiol Veg 21: 897–905

Edwards GE, Ku MSB (1987) Biochemistry of C_3–C_4 intermediates. In: Hatch MD, Boardman NK (eds) The biochemistry of plants: a comprehensive treatise, photosynthesis, vol 14. Academic Press, New York, pp 275–325

Ehleringer JR, Sage RF, Flanagan LB, Pearcy RW (1991) Climate change and the evolution of C_4 photosynthesis. Trends Ecol Evol 6: 95–99

Farquhar GD (1983) On the nature of carbon isotope discrimination in C_4 species. Aust J Plant Physiol 10: 205–226

Furbank RT, Foyer C (1988) C_4 plants as valuable model experimental systems for the study of photosynthesis. New Phytol 109: 265–277

Furbank RT, Hatch MD (1987) Mechanism of C_4 photosynthesis. The size and composition of the inorganic carbon pool in bundle-sheath cells. Plant Physiol 85: 958–964

Furbank RT, Leegood RC (1984) Carbon metabolism and gas-exchange in leaves of *Zea mays* L. Interaction between the C$_3$ and C$_4$ pathways during photosynthetic induction. Planta 162: 475–462

Hatch MD (1971) Mechanism and function of the C$_4$-pathway of photosynthesis. In: Hatch MD, Osmond CB, Slatyer RO (eds) Photosynthesis and photorespiration. Wiley, New York, pp 139–152

Hatch MD, Osmond CB (1976) Compartmentation and transport in C$_4$ photosynthesis. In: Stocking CR, Heber U (eds) Transport in plants III. Encyclopedia of plant physiology, vol 3. Springer, Berlin Heidelberg New York, pp 144–184

Hattersley PW (1983) The distibution of C$_3$ and C$_4$ grasses in Australia in relation to climate. Oecologia 57: 113–128

Hattersley PW (1987) Variations in photosynthetic pathway. In: Soderstrom TR, Hilu KW, Campbell CS, Barkworth ME (eds) Grass systematics and evolution. Smithsonian Institution Press, Washington DC, pp 49–64

Hattersley PW (1992) C$_4$ photosynthetic pathway variation in grasses (Poaceae): its significance for arid and semi-arid lands. In: Chapman GP (ed) Desertified grasslands: Their biology and management. Linnean Society Symposium Series 13, Academic Press, London, pp 181–212

Hattersley PW, Watson L (1975) Anatomical parameters for predicting photosynthetic pathways of grass leaves: the "maximum lateral cell count" and the "maximum cells distant count." Phytomorphology 25: 325–333

Hattersley PW, Watson L (1992) Diversification of photosynthesis. In: Chapman GP Grass evolution and domestication. Cambridge University Press, Cambridge, pp 38–116

Henderson SA (1992) Carbon isotope discrimination in a number of C$_4$ species. PhD Thesis, Australian National University, Canberra, Australia

Henderson SA, Caemmerer S von, Farquhar GD (1992) Short-term measurements of carbon isotope discrimination in several C$_4$ species. Aust J Plant Physiol 19: 263–285

Huber SC, Brown RH, Bouton JH, Sternberg LO'R (1989) CO$_2$ exchange, cytogenetics and leaf anatomy of hybrids between photosynthetically distinct *Flaveria* species. Plant Physiol 89: 839–844

Jenkins CLD, Furbank RT, Hatch MD (1989) Mechanism of C$_4$ photosynthesis: A model describing the inorganic carbon pool in bundle-sheath cells. Plant Physiol 91: 1372–1381

Kalt-Torres W, Kerr PS, Usuda H, Huber SC (1987) Diurnal changes in maize leaf photosynthesis. I. Carbon exchange rate, assimilate export rate and enzyme activities. Plant Physiol 83: 283–288

Karpilov YS (1970) Cooperative photosynthesis in xerophytes. Proc Mold Inst Irrigation Vegetable Res 11: 3–66

Labate CA, Adcock MD, Leegood RC (1990) Effects of temperature on the regulation of photosynthetic carbon assimilation in leaves of maize and barley. Planta 181: 547–554

Leegood RC, Caemmerer S von (1988) The relationship between contents of photosynthetic metabolites and the rate of photosynthetic carbon metabolism in leaves of *Amaranthus edulis* L. Planta 174: 253–262

Leegood RC, Caemmerer S von (1989) Some relationships between contents of photosynthetic metabolites and the rate of photosynthetic carbon assimilation in leaves of *Zea mays* L. Planta 178: 258–266

Leegood RC, Osmond CB (1991) The flux of metabolites in C$_4$ and CAM plants. In: Dennis DT, Turpin DH (eds) Plant physiology, biochemistry and molecular biology VI. Chloroplast and cytosol interactions. Longman, London, pp 274–298

Monson RK, Moore BD (1989) On the significance of C$_3$–C$_4$ intermediate photosynthesis to the evolution of C$_4$ photosynthesis. Plant Cell Environ 12: 689–699

Morison JIL, Gifford RM (1984) Plant growth and water use with limited water supply in high CO$_2$ concentrations. II. Plant dry weight, partitioning and water use efficiency. Aust J Plant Physiol 11: 375–384

O'Leary MH (1981) Carbon isotope fractionation in plants. Phytochemistry 20: 553–567

O'Leary MH (1982) Phosphoenolpyruvate carboxylase: an enzymologist's view. Annu Rev Plant Physiol 33: 297–315

Osmond CB, Bjorkman O, Anderson DJ (1980) Physiological processes in plant ecology. Toward a synthesis with *Atriplex*. Springer, Berlin Heidelberg New York

Patterson DT, Flint EP, Beyers JL (1984) Effects of CO_2 enrichment on competition between a C_4 weed and a C_3 crop. Weed Sci 32: 101–105

Peisker M (1982) The effect of CO_2 leakage from the bundle-sheath cells on carbon isotope composition. Photosynthetica 16: 533–541

Peisker M, Henderson SA (1992) Carbon: terrestrial C_4 plants: a review. Plant Cell Environ 15: 987–1004

Pittock AB (1991) Developing regional climate change senarios: their reliability and seriousness. In: Osborne PL, Burgin S (eds) Climate change: Implications for natural resources conservation. Hawkesbury Centenary Conf, Univ Western Sydney, Richmond, NSW Australia, pp 3–4

Pittock AB, Fowler AM, Whetton PH (1991) Probable changes in rainfall regimes due to enhanced grenhouse effect. In: Challenges to sustainable development. Int Hydrology and Water Resources Symp, Institution of Engineers, Perth, Australia, vol 1, pp 182–186

Poorter H (1993) Interspecific variation in the growth response to an elevated ambient CO_2 concentration. Vegetatio 104/105: 77–97

Potvin C (1986) Biomass allocation and phenological differences among southern and northern populations of the C_4 grass *Echinochloa crus-galli*. J Ecol 74: 915–923

Potvin C (1991) Temperature-induced variation in reproductive success: field and control experiments with the C_4 grass *Echinchloa crus-galli*. Can J Bot 69: 1577–1582

Potvin C, Strain BR, Goeschel JD (1985) Low night temperatures' effect on photosynthetic translocation of two C_4 grasses. Oecologia 67: 305–309

Quade J, Cerling TE, Bowman JR (1989) Development of Asian monsoon revealed by marked ecological shift during the latest Miocene in northern Pakistan. Nature 342: 163–166

Rogers HH, Bingham GE, Cure JD, Smith JM, Surano KA (1983) Responses of selected plant species to elevated carbon dioxide in the field. J Environ Qual 12: 569–574

Sage RF, Pearcy RW, Seeman JR (1987) The nitrogen use efficiency of C_3 and C_4 plants. III. Leaf nitrogen effects on the activity of carboxylating enzymes in *Chenopodium album* L. and *Amaranthus retroflexus* L. Plant Physiol 85: 355–359

Sanderson SC, Stutz HC, McArthur ED (1990) Geographic differentiation in *Atriplex confertifolia*. Am J Bot 77: 490–498

Slack CR, Roughan RG, Bassett HCM (1974) Selective inhibition of mesophyll chloroplast development in some C_4 pathway species by low night temperatures. Planta 118: 67–73

Smith SD, Strain BR, Sharkey TD (1987) Effects of CO_2 enrichment on four Great Basin grasses. Funct Ecol 1: 139–143

Stephenson RA, Brown RH, Ashley DA (1976) Translocation of ^{14}C assimilate and photosynthesis in C_3 and C_4 species. Crop Sci 16: 285–288

Stitt M (1992) Rising CO_2 levels and their potential significance for carbon flow in photosynthetic cells. Plant Cell Environ 14: 741–762

Stitt M, Heldt HW (1985) Control of photosynthetic sucrose synthesis by fructose-2,6-bisphosphate. Intercellular metabolite distribution and properties of the cytoplasmic fructose bisphosphatase in leaves of *Zea mays* L. Planta 164: 179–188

Sugiyama T, Mizuno M, Hayashi M (1984) Partitioning of nitrogen among ribulose-1,5-bisphosphate carboxylase/oxygenase, phosphoenolpyruvate carboxylase and pyruvate orthophosphate dikinase as related to biomass productivity in maize seedlings. Plant Physiol 75: 665–669

Takeda T, Tanikawa T, Agata W, Hakoyama S (1985) Studies on the ecology and geographical distribution of C_3 and C_4 grasses. I. Taxonomic and geographical distribution of C_3 and C_4 grasses in Japan with special reference to climatic conditions. Jpn J Crop Sci 54: 54–64 (in Japanese)

Teeri JA, Stowe LG (1976) Climatic patterns and the distribution of C$_4$ grasses in North America. Oecologia 23: 1–12

Thomasson JR (1987) Fossil grasses: 1820–1986 and beyond. In: Soderstrom TR, Hilu KW, Campbell CS, Barkworth ME (eds) Grass systemics and evolution. Smithsonian Institute Press, Washington DC, pp 159–167

Tidwell WD, Nambudiri EMV (1989) *Thomlinsonia thomassonii, gen. et sp. nov.*, a premineralized grass from the upper Miocene Ricardo formation, California. Rev Palaeobot Palynol 60: 165–177

Tieszen LL, Archer S (1990) Isotopic assessment of vegetation changes in grassland and woodland systems. In: Osmond CB, Pitelka LF, Hidy GM (eds) Plant Biology of the basin and range. Springer, Berlin Heidelberg New York, pp 293–322

Wardlaw IF (1982) Assimilate movement in *Lolium* and *Sorghum* leaves. III. Carbon dioxide concentration effects on the metabolism and translocation of photosynthate. Aust J Plant Physiol 9: 705–714

Wong SC (1979) Elevated atmospheric partial pressure of CO$_2$ and plant growth I. Interactions of nitrogen nutrition and photosynthetic capacity in C$_3$ and C$_4$ plants. Oecologia 44: 68–74

Wong SC, Osmond CB (1991) Elevated atmospheric partial pressures of CO$_2$ and plant growth III. Interactions between *Triticum aestivum* (C$_3$) and *Echinochloa frumentacea* (C$_4$) during growth in mixed culture under different CO$_2$, N nutrition and irradiance with emphasis on below ground responses estimated using the $\delta^{13}C$ of root biomass. Aust J Plant Physiol 13: 37–152

Wray SM, Strain BR (1987) Competition in old field perennials under CO$_2$ enrichment. Ecology 68: 1116–1120

Zangerl AR, Bazzaz FA (1984) The response of plants to elevated CO$_2$. II. Competitive interactions between annual plants under varying light and nutrients. Oecologia 62: 412–417

Ziska H, Hogan KP, Smith AP, Drake BG (1991) Growth and photosynthetic response of nine tropical species with long-term exposure to elevated carbon dioxide. Oecologia 86: 383–389

Part E
Perspectives in Ecophysiological Research
of Photosynthesis

26 Overview: Perspectives in Ecophysiological Research of Photosynthesis

E.-D. Schulze and M.M. Caldwell

26.1 Introduction: A Historic Perspective

The beginning of the ecophysiology of photosynthesis is generally associated with Lundegårdh (1929), Boysen-Jensen and Müller (1929), Harder et al. (1931), Stocker (1929, 1935), and others who used the weighing method for transpiration and CO_2 absorption by KOH for assessing photosynthesis in the Sahara desert, the Scandinavian tundra, and in a Javan tropical forest. However, it was the invention of the infrared-gas analyzer (Egle and Ernst 1949) that propelled this field of study into an active area of research. The on-line measurements of photosynthesis by Tranquillini (1957) on *Pinus cembra*, by Mooney et al. (1964) on *Pinus aristata*, and by Lange et al. (1968) on lichens were pioneer pieces of research. Much of the initial work was devoted to technical problems of gas exchange chamber environments, etc. (Sesták et al. 1971). The methodological workshop of the International Biological Program held in Trebon, Czechoslovakia, centered around questions such as whether physiological or meteorological approaches were more appropriate (Málek 1970).

Following this "methodological age" in the ecophysiology of photosynthesis, measurements of gas exchange became an almost routine undertaking in the 1970s. Nearly all major plant types in most of the world's biomes have been enclosed in chambers or porometers at one time or another to test their gas exchange characteristics. Much of this was summarized in the Encyclopedia of Plant Physiology 12 (Lange et al. 1982). Screening for response curves and daily courses of gas exchange lost much of its momentum in the 1980s since ecophysiology was taking different, and very divergent, courses pursuing: (1) molecular bases of photosynthetic response to the environment such as accommodation of excessive light, (2) whole-plant phenomena including controls exerted by and of root/shoot hormonal signals, carbon allocation, storage, and interactions with mineral nutrition, (3) characterization of photosynthetic behavior elucidating major plant groups, such as aquatic plants, poikilohydric plants, and C_3 or C_4 species. In this endeavor, use of stable isotopes has become a powerful tool in extending our understanding of relationships between CO_2 and water vapor exchange to more expansive scales of time and geographical distribution, and (4) scaling information on photosynthetic behaviour of leaves to canopies, landscapes, and even attempts to a global level.

Global concerns of climate change and man's land-use patterns, high-lighted in the forthcoming International Geosphere-Biosphere Programme, are requiring answers from the ecophysiology community on questions such as vegetation response to elevated levels of CO_2. How plants respond to greater CO_2 in the field would seem to be rather straightforward. However, to our surprise, we find that we are not capable of making predictions for the scales of time and space needed to address the questions presented.

This book attempts to summarize the research that can serve as a basis to the larger questions of the future.

26.2 Methodology

Following the flourish of technological development during the "gas-exchange cuvette era" of photosynthetic ecophysiology, several new method-ological achievements have been made:

- Automated porometers have made gas exchange measurements globally accessible (Pearcy et al. 1989). Large mobile field laboratories seem to be relics of the past.
- Florescence techniques are giving insight into the biochemical processes of photosynthesis even under field conditions (see Schreiber et al., Chap. 3, this Vol.).[1]
- Molecular biology and the production of "antisense" mutants are provid-ing a powerful tool to separate coincidence and correlation from mech-anistic function (Stitt and Schulze 1993; see also Trebst, Chap. 1; Schulze and Schulze, Chap. 6). For example, studies with mutants have shown that previous models of photosynthesis were of a correlative nature and, thus, not suitable to make predictions in the future if the baseline con-ditions such as atmospheric CO_2 change (Fichtner et al., Chap. 7).
- The use of stable isotopes has allowed investigation of plant photo-synthetic behaviour on a broad geographic scale, which would never have been possible by use of gas exchange chambers (Ehleringer, Chap. 18).
- Simulation modeling has reached a level such that highly interactive processes can be examined at canopy and landscape scales (Beyschlag et al., Chap. 20; Tenhunen et al., Chap. 21).
- Remote sensing is being increasingly used to monitor global surface processes; attempts are even being made to measure transpiration, photosynthesis, and protective pigment conversions from space (Field, Chap. 24; see also Fig. 26.1).

---

[1] All chapters referred to here are in the present volume.

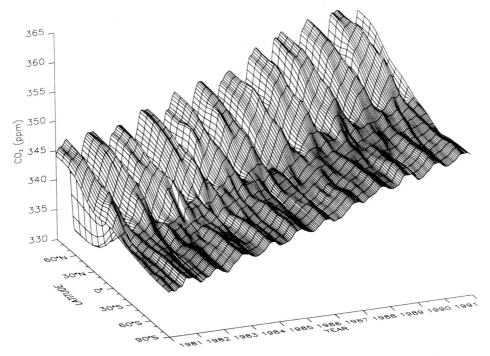

Fig. 26.1. Seasonal fluctuations of atmospheric CO_2 concentrations during the years 1981 to 1984. Smaller fluctuations occur in the southern hemisphere, reaching peak concentrations during the northern minima. The fluctuations are more pronounced in the north because of the size of the continents. Fluctuations level off in the tropical climates of the equator. (After Conway et al. 1988)

Although these new approaches are receiving considerable attention, and rightly so, one should not forget the value of long-term monitoring using the tried and proven methodologies. For example, the long-term observations on Mauna Loa of atmospheric carbon dioxide concentrations using classical infrared gas analysis has produced the most important global data set showing the trend in atmospheric CO_2 in recent history (Fig. 26.1). Only with long-term observations can one separate local or occasional trends from long-term changes, especially with regard to phenomena which are based on land-use change and air pollution (Schulze 1989; Last and Watling 1991; Teller et al. 1992).

26.3 The Molecular and Biochemical Venue of Photosynthetic Ecophysiology

Chlorophyll fluorescence has permitted insight into in vivo photosynthesis ‘nondestructively which had not been previously possible (Schreiber et al.,

Chap. 3, this Vol.). A remarkable array of mechanisms regulate the interception and dissipation of energy, both short- and long-term. What had not been clear from gas exchange measurements is becoming very obvious, namely that the overexcitation of the light-harvesting system must be avoided, or damage will result (Björkman and Demmig-Adams, Chap. 2, this Vol.). Thus, plants have evolved structural mechanisms, such as leaf movements or changes in reflectance to reduce absorption of light, and physiochemical mechanisms to dissipate excess absorbed energy in the form of heat or fluorescence. However, despite these mechanisms, photochemical damage is in fact a very common event (Trebst, Chap. 1, this Vol.) and primarily located in a single protein of the light-harvesting complex, the D1 protein. This protein has an unexpectedly high turnover rate because of the direct damage by light. The sensitivity of this protein is highly dependent on environmental conditions. Any stress, such as air pollution and mineral deficiency, will enhance D1 protein turnover. It is also very dependent on the nitrogen and carbohydrate status of the plant (Schäfer 1993). If the light-dependent repair mechanism fails, photosystem II may be completely inactivated, and in a chain reaction this can lead to the sudden and complete death of the assimilatory organ. Trebst (Chap. 1) indicates that needle loss of conifers, which is a major symptom of forest decline in Europe (see also Heber et al., Chap. 14), could be related to this molecular event, a fact which could never have been discovered by gas exchange or fluorescence measurements alone.

It appears that plant growth in full sunlight is associated with the ability to avoid photoinhibition and yet to make use of the photon flux density for carbon gain (Björkman 1981). Shade plants encounter a similar problem, which can be even more serious as they experience highly variable light conditions between shade light and sunflecks (Pearcy and Pfitsch, Chap. 17). These plants must make use of the photon flux during light flecks in order to avoid starvation, yet they are adapted to deep shade as their average light climate and, thus, they are more susceptible to photoinhibition than plants occurring in full sunlight. Pearcy and Pfitsch (Chap. 18) explain the balance between these risks of starvation by lack of photosynthates and photoinhibition during periods of high illumination by minimizing the induction period of photosynthesis.

For obvious reasons gas exchange research has concentrated on photosynthesis and much less on respiration. We presently lack important information on the regulation of respiratory processes (Amthor, Chap. 4). Respiration, photorespiration, and photosynthesis interact since these processes share intermediates such as ATP, reductants, and carbon skeletons, and products of the individual processes are substrates or cofactors for the others. Thus, all three processes are well coordinated and regulated at the whole-plant level. But it is still unclear how this is achieved. It appears that accumulation of starch "protects" carbohydrates from being respired (Schulze and Schulze, Chap. 6), and this may promote growth in a day-night cycle. It appears that this coordination is keyed to the carbohydrate status, and that intermediates of carbohydrate metabolism such as sucrose and fructose-1, 6-bisphosphate have special regulatory capacity (Stitt 1993a).

The problem of regulating respiration and sugar metabolism leads immediately to the question of how growth is linked to photosynthesis. The correlation between both processes is poor (Peirera, Chap. 8). Growth varies much more than photosynthesis. Only when nitrogen is limiting does a correlation between growth and photosynthesis emerge (Fichtner et al., Chap. 7). Obviously, the partitioning of carbon and nitrogen between root and shoot has a long-lasting effect on whole-plant carbon relations and on photosynthesis, which overrides in feedback the actual effect of carbon gain. Storage of carbohydrates, such as starch, and of nitrate are intermediate accumulatory processes, which essentially exclude each other. Maximum growth rate occurs when both these accumulatory processes reach a minimum (Stitt and Schulze 1993; Fichtner et al., Chap. 7).

Thus far, ecological models fail to explain partitioning on a mechanistic basis. Molecular tools involving use of "antisense" mutants of carboxylase, invertase, etc. are beginning to lift "the lid of this black box" (Stitt 1993a,b). It appears that sucrose has a regulatory function at the genome level and determines, along with the fructoses, partitioning between root and shoot (Stitt 1993b). When nitrate and carbohydrate mutants were investigated at different nutrition levels, a linear relationship of shoot and root growth emerges when plotted against free sucrose (W. Schulze et al. 1993).

Monitoring photosynthetic integrity, such as by leaf fluorescence characteristics, provides early indications and assessment of plant stress at extreme temperatures (Larcher, Chap. 13). Membranes are sensitive sites of temperature injury and photosynthetic function is easily impaired when membranes associated with the photosynthetic apparatus are damaged. Thus, much can be learned about the progress of temperature acclimation, avoidance or tolerance of temperature extremes and critical threshold temperatures by measuring leaf fluorescence.

26.4   Balancing Photosynthesis and Transpiration

Since the early days of ecophysiology, it has been recognized that leaves must make a compromise in regulating stomatal aperture to allow carbon gain without undue water loss. Both fluxes are directly coupled by the diffusive path through the stomatal pore. Therefore, much effort has gone into understanding stomatal function (Lösch and Schulze, Chap. 9). Direct observations of stomatal movement (Kappen et al., Chap. 11) and theoretical considerations (Cowan, Chap. 10) suggest that feedback responses through plant internal water status are more important than previously thought. Even the apparent direct response to humidity appears to be regulated by local epidermal water stress which is created by water loss from the inner side of the guard.cell complex. This is an apparent contradiction of the observation that large variations in leaf water potential can occur without stomatal effects (Schulze 1993). This leads to questions of internal plant controls, which include mineral nutrition and regulation by plant hormones.

One key factor in this complex appears to be the regulation of cellular and xylem pH (Pfanz, Chap. 5). The pH gradient causes changes in the distribution and protonation of abscisic acid, which is probably the most potent stress hormone and which directly affects stomatal function. The pH regulation is dependent on many factors such as nutrition and carbon and acid metabolism. Lösch and Schulze (Chap. 9) show that water stress at the root level will cause roots to produce ABA which moves through the xylem and reaches the leaf, where (1) it may be recirculated into the phloem, or (2) it may enter into the mesophyll cells depending on the pH gradient, where it is stored or metabolized in the light, or (3) it reaches the stomata and interacts with the ion channels of the guard cell plasmalemma. Plant nutrition is involved because the balance in uptake of anions and cations will determine the xylem pH and thus determine the protonation of ABA and movement into mesophyll cells. Apart from its effects on stomatal function, ABA emerges as a very potent phytohormone which affects extension growth (Schulze 1993).

Although it was largely the ecophysiology of photosynthesis that initiated this line of research, it has revealed an aspect of whole-plant physiology that formerly was not recognized, namely root-shoot signals that coordinate shoot and root function in addition to carbohydrates. It is just such research that demonstrates that photosynthesis can no longer be studied in isolation when in the context of the whole plant. Photosynthesis is a highly interactive process which involves not only water but also nutrient relations and hormonal signals.

26.5 Photosynthetic Performance of Different Plant Groups

The extensive data set of plant photosynthetic behaviour under a broad range of conditions has enabled us to recognize plant patterns and specific features of different plant groups.

Aquatic plants are much more diverse than terrestrial plants in terms of light-harvesting mechanisms, which has been apparent from their diverse pigment systems. Aquatic plants have evolved a CO_2-concentrating mechanism which is different from that observed in terrestrial plants (Raven, Chap. 15). This mechanism is not known in terrestrial plants, except in lichens (Green and Lange, Chap. 16). Thus, it is with aquatic plants where basically new observations on carbon fixation mechanisms were made in recent years (Raven, Chap. 15).

Just as exciting is the world of poikilohydric plants, which received general scientific attention following the work of Otto Lange (1953, 1965). Presently, molecular biologists are searching for protein features in poikilohydric plants in order to achieve improved desiccation tolerance in crop plants. However, even in comparing two groups of plants which are ecologically closely related, such as mosses and lichens, their whole-plant gas exchange processes turn out to be quite complicated. Both plant types inhabit similar niches; the lichens probably extending more to the dry side and the mosses to the wet side of this spectrum. Mosses exhibit a low gas-phase resistance and a high cellular resistance for carbon dioxide movement, while lichens exhibit a high gas phase and a low cellular resistance for the carbon-concentrating mechanism.

The distribution and competition of C_3 and C_4 plants is a problem similar to that of the moss/lichen comparison. It has fascinated ecologists and physiologists since the recognition of the C_4 pathway. Study of competition and distribution of C_3 and C_4 plants was greatly facilitated by use of carbon stable isotope natural abundance levels in plant tissues. These allow not only the identification of photosynthetic pathway types, but also other aspects of photosynthesis, stomatal function, and water use. For example, it has been possible to estimate season-long comparative water-use efficiencies of different plants occurring in the same environments, or water-use behavior of stems and leaves of the same plants. Assessment of deuterium natural abundance levels in plant tissues makes it possible to make inferences about the source water for plants in different habitat positions. Some surprises have emerged; for example, streamside trees do not always obtain their water from the stream (Ehleringer, Chap. 18). The combined information on carbon, water, and nitrogen relations makes the use of stable isotopes an attractive approach. However, many problems remain unsolved, such as the heavy water abundance in parasitic plants relative to their hosts (Ziegler, Chap. 19).

26.6 Photosynthesis and Global Climate Change: Making Global Predictions

The question being put to ecophysiologists most commonly at present relates to the expected changes of photosynthesis and plant growth with global climate change, which includes elevated CO_2 and temperatures. One of the central issues is whether long-term exposure of plants to elevated CO_2 results in a down-regulation of photosynthesis, i.e., that photosynthetic rates at elevated CO_2 are lower than would be expected based on short-term assessments of photosynthetic rates as a function of CO_2. If down-regulation occurs, plants might constitute much less of a sink for CO_2 than previously thought. These and other questions are the subject of much controversy and certainly the case is not decided. Other changes are also expected. For example, there are indications that, depending on the mineral nutrition of plants, the photosynthetic protein content will decrease (Stitt 1991, 1993b). This would be consistent with a down-regulation. In this book, the biochemical issues are of less concern than the whole-plant behavior of photosynthesis using mangroves and C_4 plants are case examples.

The mangroves are taken as representing plants in a stressful environment. It appears (Ball and Passioura, Chap. 12) that mangroves probably will save water rather than increase their photosynthetic rates. However, this will not change the distribution of mangroves in the saline portions of their distribution but would likely alter species composition in the less saline habitats. This is an example of how difficult it is to make predictions on the effects of global change. One must begin with species-specific predictions and pursue broader generalizations with considerable caution and circumspection.

An equally difficult question relates to the future of C_4 plants (Henderson et al., Chap. 25). One might expect that C_3 plants will gain dominance because they profit in their photosynthetic rates with increasing CO_2 while C_4 plants do not. However, the picture is again more complicated because in addition to CO_2, temperatures are also expected to rise. This may lessen the increase of photosynthetic rates of C_3 plants while having a negligible effect on C_4 plants. At present, there are many examples of C_4 species spreading at the expense of C_3 species.

The chapters by Ball and Passioura (Chap. 12) and by Henderson et al. (Chap. 22) venture into the future with predictions of whole- plant performance. However, in order to do this, we need not only to extrapolate processes but also to scale up from leaves to canopies (Beyschlag et al., Chap. 20) and landscapes (Tenhunen, Chap. 21). The problem of scaling has led to a whole new area of global modeling in the ecophysiology of photosynthesis. It is quite clear that the integration of our knowledge and contribution to global climate predictions will only be achieved through modeling. However, there is the risk of transmitting, and sometimes amplifying, errors when scaling up. Thus, scaling problems remain a major challenge at present. In the future, however, it may in fact be possible to directly measure global processes. Conway et al. (1988) determined atmospheric CO_2 concentrations in remote areas, and thereby made the seasonal pulses of photosynthesis and respiration of the globe visible (Fig. 26.1). Wintertime increases in CO_2 concentration alternate with CO_2 depletion during summers of the northern hemisphere. These fluctuations in the southern hemisphere are smaller and opposite in phase in the northern hemisphere. The differences between maximum and minimum CO_2 concentrations in the winter climates of higher latitudes would be a measure of photosynthesis and respiration (and other CO_2-emitting processes). In this case, the whole globe may be viewed as a "closed cuvetts", and one can visualize gas exchange at a global scale, which has never before been possible. At this scale species may not be important in describing the present status of the globe. On the other hand, if one is to make realistic predictions of species' migrations during global climate change and their effect on assimilation, then issues of biodiversity must be addressed because species will not move as communities but as solitary species, but questions of biodiversity and ecosystem function in a changing world remain elusive at this point (Schulze and Mooney 1993). Nevertheless, some strides in this general direction have been made. For example, global maps of photosynthetic capacity have been drawn (Woodward, Chap. 23) which necessarily incorporates the general distribution of major vegetation types with representative species. Surprisingly, the average rates of gas exchange (not the photosynthetic capacity) are remarkably constant. This has been demonstrated for leaf conductance and net CO_2 uptake (Körner, Chap. 22) as well as for canopy conductance (Kelliher et al. 1993) and even for canopy photosynthesis (Mooney and Field 1989). If this is true, then global modeling would be greatly eased. Yet, it still would not answer questions of how biodiversity interacts with ecosystem function. Schulze et al. (1993) concluded, that the instability at lower scales (species diversity and its density-related interactions) are the driving mechanisms that lead to the observed stability at the higher scales of ecosystems. Thus, we may be able to use canopy models ignoring species composition, but it is doubtful that these models would be truly predictive because of secondary and tertiary effects of species changes.

26.7 Where Will Ecophysiology of Photosynthesis Venture in the Coming Decade? We Offer Some Thoughts

- The link between molecular biology and ecology will be strengthened and, by using molecular tools, much progress can be made beyond simply correlative evidence. This is especially needed with respect to whole-plant carbon allocation and growth, because this feeds directly back into photosynthesis at the canopy scale.
- Research of recent years has shown that studying photosynthesis in isolation is no longer acceptable. Interactions with several other major plant processes must be understood. For example, nutrition has emerged as one prominent parameter and not just its influence on photosynthetic capacity. Water relations, which has always been a central theme in photosynthetic research, is losing some importance because it is well

recognized that most of the production of photosynthates occurs only when water is available. Thus, responses of plants to water deficits are more important with respect to survival rather than with respect to more production. Water-use efficiency is important and stomatal behavior receives attention; however, this may slacken. Kelliher et al. (1993) showed that functional rooting depth may be more important than canopy conductance in avoiding water stress. In turn, functional rooting depth is related to carbon partitioning and nutrition.

– Species-specific interactions in ecosystem function will remain a major challenge. The competitive interactions among plants usually result in a rather complete use of resources. However, predicting changes in species composition in future environmental scenarios is only in the early stages of development.

– Despite all the knowledge collected concerning photosynthetic processes, there is still considerable uncertainty in making predictions on whole-plant performance at elevated CO_2 due to feedbacks within the plant. Even the question of some degree of down-regulation is far from settled. If less protein is needed in order to achieve a certain carbon gain, then down-regulation might be the appropriate response, as it would release plant internal resources for further growth and carbon investment in other components of the plant. This, in turn, may affect water relations and mineral nutrition at the whole-plant level. We cannot model this interaction without experimental evidence on plant response at high CO_2 under a range of field conditions.

– Although the ecophysiology of photosynthesis has come a long way, it is still far from being predictive at a global scale. The challenges of scaling up from plants to canopies, landscapes, and even a global level will drive many efforts in modeling and innovative approaches to validation of these models.

References

Björkman O (1981) Responses to different quantum flux densities. In: Lange OL, Nobel PS, Osmond CB, Ziegler H (eds) Encyclopedia of plant physiology vol 12A. Springer, Berlin Heidelberg New York, pp 57–108

Boysen-Jensen P, Müller D (1929) Die maximale Ausbeute und der tägliche Verlauf der Kohlensäureassimilation. Jahrb Wiss Bot 70: 493–502

Conway TJ, Tans P, Waterman LS, Thoning KW, Masarie KA, Gammon RH (1988) Atmospheric carbon dioxide measurements in the remote global troposphere 1981–1984. Tellus 40B: 81–115

Egle K, Ernst A (1949) Die Verwendung des Ultrarotabsorptionsschreibers für die vollautomatische und fortlaufende CO_2-Analyse bei Assimilations- und Atmungsmessungen an Pflanzen. Z Naturforsch 4b: 351–360

Harder R, Filzer P, Lorenz A (1931) Über Versuche zur Bestimmung der Kohlensäureassimilation immergrüner Wüstenpflanzen während der Trockenzeit in Beni Unif (algerische Sahara). Jahrb Wiss Bot 75: 45–194

Kelliher FM, Leuning R, Schulze E-D (1993) Evaporation and canopy characteristics of coniferous forest and grassland. Oecologia 95: 153–163

Lange OL (1953) Hitze- und Trockenresistenz der Flechten in Beziehung zu ihrer Verbreitung. Flora 140: 39–97

Lange OL (1965) Der CO_2-Gaswechsel von Flechten bei tiefen Temperaturen. Planta 64: 1–19

Lange OL, Schulze E-D, Koch W (1968) Photosynthese von Wüstenflechten am natürlichen Standort nach Wasserdampfaufnahme aus dem Luftraum. Naturwissenschaften 12: 658–659

Lange OL, Nobel PS, Osmond CB, Ziegler H (1982) Physiological plant ecology II: Water relations and carbon assimilation. In: Lange OL, Nobel PS, Osmond CB, Ziegler H (eds) Encyclopedia of plant physiology N.S., vol 12B. Springer, Berlin Heidelberg New York, 747 pp

Last FT, Watling R (1991) Acid deposition: its nature and impacts. Royal Society of Edinburgh, Edinburgh, 343 pp

Lundegårdh H (1929) Ecological studies in the assimilation of certain forest plants and shore plants. Sven Bot Tidskr 15: 46–85

Málek I (1970) Prediction and measurement of photosynthetic productivity. Proceedings of the IBP/PP Technical Meeting, Trebon, 14.–21. September 1969. PUDOK, Wageningen, 432 pp

Mooney HA, Field CB (1989) Photosynthesis and plant productivity – calling to the biosphere. In: Briggs WR (ed) Photosynthesis. Alan R Liss, New York, 19–44

Mooney HA, Wright RD, Strain BR (1964) The gas exchange capacity of plants in relation to vegetation zonation in the White Mountains of California. Am Midl Nat 72: 281–297

Pearcy RW, Ehleringer J, Mooney HA, Rundel PW (1989) Plant physiological ecology; field methods and instrumentation. Chapman and Hall, London, 457 pp

Schäfer C (1993) Controlling the effects of excessive light energy fluxes: Dissipative mechanisms, repair processes and long-term acclimation. In: Schulze E-D (ed) Flux control in biological systems: from the enzyme to the population and ecosystem level. Academic Press (in press)

Schulze E-D (1989) Air pollution and forest decline in a spruce (Picea abies) forest. Science 244: 776–783

Schulze E-D (1993) The regulation of plant transpiration: Interactions of feedforward, feedback and futile cycles. In: Schulze E-D (ed) Flux control in biological systems: from the enzyme to the population and ecosystem level. Academic Press (in press)

Schulze E-D, Mooney HA (eds) (1993) Biodiversity and ecosystem function. Ecological studies 99. Springer, Berlin Heidelberg New York, 525 pp

Schulze W, Schulze E-D, Stadler J, Heilmeier H, Mooney HA (1994) Growth and reproduction of Arabidopsis thaliana in relation to storage of starch and nitrate in wild types, starch-deficient and nitrate-uptake-deficient mutants. Plant cell Environ (in press)

Sesták Z, Catsky J, Jarvis PG (1971) Plant photosynthetic production, manual of methods. Junk, The Hague, 818 pp

Stitt M (1991) Rising CO_2 levels and their potential significance for carbon flow in photosynthetic cells. Plant Cell Environ 14: 741–762

Stitt M (1993a) Flux contrl at the level of the pathway: studies with mutants and transgenic plants having a decreased activity of enzymes involved in photosynthesis ddjpartitioning. In: Schulze E-D (ed) Flux control in biological systems: from the enzyme to the population and ecosystem level. Academic Press (in press)

Stitt M (1993b) Enhanced CO_2, photosynthesis and growth; what should we measure to gain a better understanding of the plant's response? In: Schulze ED, Coldwell ED (eds) Design and execution of experiments at elevated CO_2. European Community. Brussels (in press)

Stitt M, Schulze E-D (1993) Plant growth, storage and resource allocation – from flux control in a metabolic chain to the whole plant level. In: Schulze E-D (ed) Flux control in biological systems: from the enzyme to the population and ecosystem level. Academic Press (in press)

Stocker O (1929) Eine Feldmethode zur Bestimmung der momentanen Transpirations- und Evaporationsgröße. Ber Dtsch Bot Ges 47: 126–136

Stocker O (1935) Assimilation und Atmung westjavanischer Tropenbäume. Planta 24: 402–445

Teller A, Mathy P, Jeffers JNR (1992) Responses of forest ecosystems to environmental changes. Elsevier, London, 1009 pp

Tranquillini W (1957) Standortklima, Wasserbilanz und CO_2-Gaswechsel junger Zirben an der alpinen Waldgrenze. Planta 49: 612–661

Subject Index

Species Index

(Reference only to Latin names in the text)

Printing: Mercedesdruck, Berlin
Binding: Buchbinderei Lüderitz & Bauer, Berlin